International Conference Wood Science and Engineering in the Third Millennium - ICWSE 2023

International Conference Wood Science and Engineering in the Third Millennium - ICWSE 2023

Editors

Lidia Gurau
Mihaela Câmpean
Emilia-Adela Salca

MDPI

Basel • Beijing • Wuhan • Barcelona • Belgrade • Novi Sad • Cluj • Manchester

Editors
Lidia Gurau
Transilvania University
of Brasov
Brasov
Romania

Mihaela Câmpean
Transilvania University
of Brasov
Brasov
Romania

Emilia-Adela Salca
Transilvania University
of Brasov
Brasov
Romania

Editorial Office
MDPI AG
Grosspeteranlage 5
4052 Basel, Switzerland

This is a reprint of articles from the Special Issue published online in the open access journal *Applied Sciences* (ISSN 2076-3417) (available at: https://www.mdpi.com/journal/applsci/special_issues/548H40ZI59).

For citation purposes, cite each article independently as indicated on the article page online and as indicated below:

Lastname, A.A.; Lastname, B.B. Article Title. *Journal Name* **Year**, *Volume Number*, Page Range.

ISBN 978-3-7258-1783-2 (Hbk)
ISBN 978-3-7258-1784-9 (PDF)
doi.org/10.3390/books978-3-7258-1784-9

Contents

applied sciences

MDPI

Editorial

Special Issue "International Conference Wood Science and Engineering in the Third Millennium—ICWSE 2023"

Lidia Gurau *, Mihaela Campean and Emilia-Adela Salca

Faculty of Furniture Design and Wood Engineering, Transilvania University of Brasov, 500068 Brasov, Romania; campean@unitbv.ro (M.C.); emilia.salca@unitbv.ro (E.-A.S.)
* Correspondence: lidiagurau@unitbv.ro

Citation: Gurau, L.; Campean, M.; Salca, E.-A. Special Issue "International Conference Wood Science and Engineering in the Third Millennium—ICWSE 2023". *Appl. Sci.* **2024**, *14*, 5800. https://doi.org/10.3390/app14135800

Received: 14 June 2024
Accepted: 1 July 2024
Published: 3 July 2024

1. Introduction

Every two years, the Faculty of Furniture Design and Wood Engineering of the Transilvania University of Brasov organises an international conference on wood science and engineering—ICWSE, which was held for the 13th in November 2023 and hosted over 100 participants from 30 countries worldwide. In this latest edition, some of participants opted to disseminate their work in the journal *Applied Sciences*, via a Special Issue dedicated to the conference. In order to cover a wide range of subjects of interests relevant to the wood industry, the selected topics were deliberately broad and were gathered in eight distinct sections: Wood Structure and Properties (6 papers); Wood Constructions (2 papers); Wood Drying and Heat Treatments (3 papers); Conservation and Restoration of Wooden Objects (1 paper); Wood-Based Materials (4 papers); Mechanical Wood Processing and Surface Quality (4 papers); Wood Preservation, Modification, Gluing and Coating (3 papers); Furniture Design (1 paper). There is a growing demand for wood products in the world today, but to meet future needs for these materials and overcome wood shortages, studies have focused on the best utilization of wood potential, either from less well-known species [1], fast-growing species [2], subfossil woods [3] or alternative and competitive composites [4,5], all deserving thorough investigation in order to promote the production of high-added-value products, without neglecting wood waste and its potential as a source of energy [6]. The domain of wood constructions has gained attention due to new developments in connectors that have the potential to meet all code requirements in terms of load-bearing capacity and dimensional stability [7–9]. The challenges of wood drying pose questions which must be considered by researchers, especially for species prone to drying defects [10]. At the same time thermal treatments continue to be interesting due to their potential to improve some of the properties of wood [11]. Manufacturers need to select the most suitable coating material for wood and wood-based panels in order to obtain finished products with the best performance and properties. Among the topics covered, the rise of digital printing in the wood industry [12] or the use of nanotechnologies and various treatments for wood modification [13] have drawn particular attention. The optimization of mechanical wood processing [14] and the flexibility in design offered by 3D printing [15], together with a reliable evaluation of the resulting surface quality [16], represent a subject intended to improve the marketing and appeal of finished wood products and their inspirational design [17]. The cultural heritage of wood was also not neglected, since the methods and techniques used to conserve works of art are part of the same scientific endeavours [18]. The following section provides an overview of the research papers contained in the Special Issue of the ICWSE conference, covering in their diversity the comprehensive range of topics outlined above.

2. An Overview of the Published Articles

2.1. Wood Structure and Properties

The work by Terzopoulou et al. (Paper 1) provides a review of the most important characteristics of *Ailanthus* wood, also called "tree-of-heaven", originating from China, but now spread to Europe and North America as environments become increasingly dry in the context of climate change. The paper presents a comprehensive review of this species in terms of its chemical properties, pharmacology applications, traditional and innovative uses, quality control, biological resistance, potential utilization in bioenergy, biofuels and wood-based products. The challenges and limitations of its uses are also included, in order to comprehensively outline the utilization potential of the "tree-of-heaven".

An overview of the current state of affairs of European poplar cultures, with a principal focus on Hungary, is offered by Komán et al. (Paper 2). Poplars (*Populus* spp.) from plantations represent a fast-growing wood species with a relatively short cutting age and rotation time meaning that they can produce maximum yields of timber or fibre. The authors provide an overview of the current data on poplars in Hungary, their quantity, species variety, age group composition, areas of use and other topics of interest. The work underlines the importance of this species which provides raw material to the wood industry, successfully replacing wood resources from natural forests.

The paper by Avram et al. (Paper 3) compares some physical and mechanical properties of three homogeneous species—linden, paulownia and balsa—in order to understand their characteristics in a comparative way, their report finally calling for paulownia wood as a replacement of linden wood in many applications.

Marttila et al. (Paper 4) draw attention to the material potential of Norway spruce and Scots pine, which remained standing and were harvested after a large forest fire in Finland. The authors found that the changes to the mechanical properties were mostly negligible, but the colour of the material had darkened in both species. Of the two species, spruce wood was more affected. The researchers recommend this material for mid-quality and low-quality sawn timber and selected end uses where the requirements for quality are moderate and aesthetic or hygienic criteria are not critical.

A less researched type of raw material, the subfossil oak, originating from the Romanian riverbanks, was studied by Nedelcu et al. (Paper 5) in comparison with aged oak recovered from existing constructions and new oak wood. Apart from the high variability of the properties of the subfossil oak from different trunks and its slightly lower swelling anisotropy, it was found that this material had a lower density, was more dimensionally unstable and had a lower compression strength parallel to the grain compared to the new oak. However, the fossil oak has important value for interior furniture with aesthetic requirements.

The last paper under the "wood structure and properties" topic investigated the tensile mechanical properties of oak (*Quercus robur* L. and *Quercus petraea* L.) sliced lamellae. The authors, Grubîi and Johansson (Paper 6), highlighted the correlation between the wood's check depth ratio, which is influenced by the lamella thickness, and the sliced lamellae strength and strain at failure. The conclusion was that, in the elastic domain, the stiffness was not significantly impacted by slicing check characteristic variations. Other findings indicate that slicing thickness substantially affects the mechanical properties of the modulus of rupture and strain at break. Digital image correlation (DIC) was used to analyse the failure mechanisms in the sliced lamellae, gaining insight into the fracture mechanisms of the tested samples. The study aimed to provide support for appropriate future product design, quality control and further investigations into the quality optimization of sliced lamellae.

2.2. Wood Constructions

The work of Agnantopoulou et al. (Paper 7) explores the windmills on the island of Sikinos and their mechanisms. Evidence of the utility, durability and simplicity of these wooden monuments of cultural heritage, the techniques employed in their construction and the proper application of different wood species ideally serving appropriate points in the construction, in accordance with their properties and characteristics, were recorded. The paper has a historical character, but it also reveals interesting data on the wood species used, which are more diverse than expected, involving red oak, juniper, olive wood, elm, chestnut and ash. Future work envisages comparison with windmills from other Greek islands.

The paper by Pomaroli and Kraler (Paper 8) has a strong innovative character as it presents a new type of wall–ceiling system connector made of wood for cross laminated timber (CLT) constructions. The proposed version is a three-dimensional, statically effective double-dovetail element, made of block-glued laminated veneer lumber (LVL) and manufactured in a CNC milling process. The results of the tests showed the high performance potential of this system connector, which can play an important role in future timber construction.

2.3. Wood Drying and Heat Treatments

The work of Deaconu et al. (Paper 9) approaches the drying of 50 mm thick Turkey oak (*Quercus cerris* L.) lumber. Both air drying and conventional kiln drying were applied in order to establish the optimum drying conditions for this particular wood species, which is denser than common oak (*Quercus robur* L.), and highly prone to cracks and deformations during the drying process. The influence of steaming prior to drying was also taken into consideration. Some recommendations for an optimised drying schedule were formulated.

Hajian et al. (Paper 10) compared the drying behaviour of western hemlock (*Tsuga heterophylla* (Raf.) Sarg.) with that of Norway spruce (*Picea abies* (L.) Karst.) and Scots pine (*Pinus sylvestris* L.) from the green to the oven-dried condition. A CT scanner was used instead of resistance-type pin electrodes in order to assess the moisture content evolution within the specimen, by calculating moisture content based on CT images. Consequently, the reason why hemlock dries much more slowly than the other two species could be identified, this being the presence of large wet pockets. An adapted drying schedule was developed for timber prone to the formation of such wet pockets.

Gomes et al. (Paper 11) investigated the thermal treatment of Teak (*Tectona grandis* L.f.) and the effects of this treatment on the colour and chemical composition of this valuable species. The heat treatment was carried out in two variants, in an autoclave (closed system) at 160 °C and in an electric oven (open system) at 185 °C and at 210 °C. The closed system demonstrated higher efficiency even though it operated with a lower temperature. This resulted in a higher mass loss, a stronger reduction of the hemicelluloses content and an increase of the extractive content for all solvents used.

2.4. Conservation and Restoration of Wooden Objects

This section was represented by the study by Moraru-Avram et al. (Paper 12), who investigated the effect of different solvents used in the process of conservation/restoration of polychrome linden wood panels degraded by xylophagous insects. Among the solvents used—water, acetone, and white spirit—acetone caused the greatest swelling and deformation of the treated panels. This study aimed to contribute to a better understanding of the material properties and their behaviour under restoration treatments.

2.5. Wood-Based Materials

The paper written by Chotikhun et al. (Paper 13) investigates the production and properties of wood plastic pellets (WPP) made from rubberwood sawdust and refuse-derived fuel (RDF). The samples were tested for chemical and physical properties and compared to standard wood pellets. Based on the findings of this research, in terms of heating value, density, mechanical durability, and moisture content, the mixed pellets can potentially be beneficial in alternative sustainable green energy as fuels and could to mitigate global warming.

In their paper, Barbu et al. (Paper 14) evaluated the physical and mechanical properties of single- and three-layered solid wood panels made of wood from European Paulownia plantations. A comparison with other similar lightweight commercially available solid wood panels (SWPs) was carried out. The findings of this work recommend Paulownia SWPs to be used as lightweight and sustainable core materials in sandwich structures, replacing balsa wood for various applications in the furniture and packaging industries, or as sport articles or non-load-bearing constructions.

Spirchez et al. (Paper 15) present the importance of the torrefaction process for poplar plywood and medium-density fibreboard, highlighting both the positive and negative aspects of the torrefaction process, such as, respectively, the decrease in water absorption, swelling and shrinkage, and colour on one side, and the mechanical resistance to static bending on the other. The obtained results emphasize that the mass loss increases at high temperatures, this being the main disadvantage of torrefaction.

The last paper of this section, by Zeleniuc et al. (Paper 16), evaluates the physical and mechanical performance of three types of hybrid composites made of date palm (*Phoenix dactylifera* L.), additional layers of cotton and Kevlar fibres. The fibres were formed into flat sheets and employed as reinforcement layers embedded in a polyester matrix. The influences of cotton (a natural and soft fibre) and Kevlar (a synthetic and strong fibre), were also investigated regarding the properties of bending strength and resistance to water and temperature changes in terms of freezing and drying. The results of this work are oriented towards finding an appropriate application for these composites, which are intended to be used in humid environments.

2.6. Mechanical Wood Processing and Surface Quality

In their work, Bedelean et al. (Paper 17) applied an artificial neural network modelling technique and response surface methodology to model and optimize the drilling process of plywood. Two artificial neural network models were developed to predict the thrust force and drilling torque based on drill tip angle, tooth bite, and drill type. It was found that the drill type significantly influenced the force during drilling, and the most significant factor affecting the drilling torque during the drilling of plywood was the tooth bite. The proposed method could be used as an optimization tool during the design phase of the furniture manufacturing process.

Paper 18, (Barboutis et al.) investigates the possibility of reproducing a precise copy of a carved wood object of medium size and considerable cultural value, which could continue to be used in its original location or could be exhibited in a museum space. 3D scanning, CNC machining and traditional wood carving handwork were applied and discussed. The current work transfers the experience of these reproduction techniques as applied to a 17th-century "Bema Door" element, sharing significant information and details on the tools, materials, process steps, etc., with the aim of being an inspiration for the preservation or reproduction of other highly important cultural heritage monuments.

Nicolau et al. (Paper 19) focus their work on 3D printed connectors for a chair leg-stretcher joint. Larch (*Larix decidua* Mill.) wood was used for the chair components and selective laser sintering (SLS) technology was employed for the 3D printed connector designed to assemble the chair elements. Diagonal tensile and compression tests were performed on the L-type joints with 3D printed connectors, along with the bending moments and micrographs taken of the broken areas of the 3D printed connectors. The results show that the SLS printing technology applied to build the connector does not ensure enough strength for the assembly, and further research on changes in printing parameters, in terms of laser power and scan speed, is needed.

The last paper of this section, written by Brenci and Gurau (Paper 20), is focused on the surface quality of beech wood obtained by profiled milling. In this study, the surface quality was analysed across stratified levels of irregularities differentiating between waviness, tool marks, fuzzy grain, and accidental surface gaps overlapped onto the deep anatomical cavities of the wood, such as pores from earlywood. The findings of the work show that the processing quality measured at different depths does not differ statistically, in spite of local differences in cutting speed. The statistical analysis helped the researchers understand the hierarchy of influencing factors and their significance. In this respect, the most important factor affecting the surface quality is the feed speed.

2.7. Wood Preservation, Modification, Gluing and Coating

Tokarczyk et al. (Paper 21) investigated the influence of selected technological parameters, such as the number of layers, energy doses distributed by the radiators and line speed, on the adhesion of varnish coatings formed in the process of varnishing with a roller. The objective of this paper was achieved by evaluating the phenomenon of wetting of coatings formed on HDF boards and taking into account the impact of surface roughness on this process.

The research performed by Henke et al. (Paper 22) sought to determine the impact of fibreboard (HDF) density, grinding parameters, the amount of basecoat and topcoat applied, the number of coats and different lamp powers on a combined industrial line using LED and mercury lamps at 50 m/min. It was assumed that these factors could significantly impact the coating's resistance and that simply learning about them and identifying the relationship could support entrepreneurs and manufacturing technologists and become a starting point for further work on varnish formulations and LED lamps.

The third paper under this topic, delivered by Torcatoru et al. (Paper 23), compared light-induced colour changes in uncoated and clear-coated wood surfaces with waterborne lacquers, for three hardwood species; namely, European maple (*Acer pseudoplatanus* L.), European ash (*Fraxinus excelsior* L.) and European walnut (*Juglans regia* L.). Highlighting the specific contribution of the wooden support and of the coating films to the total colour changes in the finished surfaces represented one of the main goals of this research.

2.8. Furniture Design

Zhach et al. (Paper 24) designed a versatile storage cube made of birch plywood that integrates magnetic structural components facilitating easy and a quick assembly. It is interesting, that the joint edges between the individual walls are covered with smooth plastic guards manufactured via additive 3D printing of corn fodder-based filament. The authors claim that this item of furniture is portable and can be converted into a table or a stool to serve children's gaming activities and other overall skills, at home or in the pre-primary and primary education spaces. The cubes were verified for their robustness and stability by applying the relevant standards.

3. Conclusions and Future Perspectives

With the publication of this Special Issue, we hope to contribute to a better understanding of the huge scientific potential of the natural and regenerable global reserve that is wood. To achieve this, we selected original works aiming at covering eight topics of research discussed across 24 research papers. The next conference will take place in November 2025, and will be, as usual, a forum for discussion and dissemination of the newest developments in the field of wood science and engineering.

Conflicts of Interest: The authors declare no conflicts of interest.

List of Contributions:

1. Terzopoulou, P.; Kamperidou, V.; Barboutis, I. Utilization Potential of Tree-of-Heaven Species Biomass—A Review. *Appl. Sci.* **2023**, *13*, 9185. https://doi.org/10.3390/app13169185.
2. Komán, S.; Németh, R.; Báder, M. An Overview of the Current Situation of European Poplar Cultures with a Main Focus on Hungary. *Appl. Sci.* **2023**, *13*, 12922. https://doi.org/10.3390/app132312922.
3. Avram, A.; Lunguleasa, A.; Spirchez, C.; Ionescu, C. Differences and Similarities between the Wood of Three Low-Density and Homogenous Species: Linden, Balsa, and Paulownia. *Appl. Sci.* **2023**, *13*, 10209. https://doi.org/10.3390/app131810209.
4. Marttila, J.; Möttönen, V.; Haapala, A.; Ylimäki, P.; Kilpeläinen, P.; Verkasalo, E. Wood Material Properties of Forest Fire-Damaged Norway Spruce and Scots Pine for Mechanical Wood Processing in Finland. *Appl. Sci.* **2024**, *14*, 238. https://doi.org/10.3390/app14010238.

5. Nedelcu, R.; Timar, M.; Porojan, M.; Beldean, E. Selected Physical and Mechanical Properties of Subfossil Oak (*Quercus* spp.) Compared to Aged Oak and Recent Oak. *Appl. Sci.* **2024**, *14*, 3294. https://doi.org/10.3390/app14083294.
6. Grubîi, V.; Johansson, J. The Influence of Slicing Thickness on the Perpendicular to Grain Tensile Properties of Oak (*Quercus robur* L. and *Quercus petraea* L.) Lamellae. *Appl. Sci.* **2023**, *13*, 12254. https://doi.org/10.3390/app132212254.
7. Agnantopoulou, E.; Barboutis, I.; Kamperidou, V. Wood Utilization in Windmill Mechanisms on Sikinos Island (Greece). *Appl. Sci.* **2023**, *13*, 9216. https://doi.org/10.3390/app13169216.
8. Pomaroli, A.; Kraler, A. System Connector "Double Dovetail Tenon"—A Wood–Wood Based Alternative for Connecting CLT-Elements. *Appl. Sci.* **2023**, *13*, 11727. https://doi.org/10.3390/app132111727.
9. Deaconu, I.; Georgescu, S.; Campean, M. Drying Behaviour of 50 mm Thick Turkey Oak Lumber. *Appl. Sci.* **2023**, *13*, 10676. https://doi.org/10.3390/app131910676.
10. Hajian, E.; Couceiro, J.; Hansson, L.; Sandberg, D. Drying Behaviour of Western Hemlock with Schedules Developed for Norway Spruce and Scots Pine. *Appl. Sci.* **2023**, *13*, 11083. https://doi.org/10.3390/app131911083.
11. Gomes, F.; Reis, M.; Rupf, A.; Silvares, P.; Ferreira, J.; Dias, L.; Madi, J.; Cademartori, P.; Neto, H.; Paes, J.; Batista, D. Chemical Composition and Color of Short-Rotation Teak Wood Thermally Modified in Closed and Open Systems. *Appl. Sci.* **2023**, *13*, 13050. https://doi.org/10.3390/app132413050.
12. Moraru-Avram, A.; Ionescu, C.; Lunguleasa, A. Effects of Solvents Used for Conservation/Restoration Treatments on Damaged Linden Panels of Cultural Heritage Assets. *Appl. Sci.* **2023**, *13*, 11148. https://doi.org/10.3390/app132011148.
13. Chotikhun, A.; Laosena, R.; Kittijaruwattana, J.; Lee, S.; Sae-Ueng, K.; Nakason, C.; Pianroj, Y.; Salca, E.A. Elemental Compositions of Wood Plastic Pellets Made from Sawdust and Refuse-Derived Fuel (RDF) Waste. *Appl. Sci.* **2023**, *13*, 11162. https://doi.org/10.3390/app132011162.
14. Barbu, M.; Radauer, H.; Petutschnigg, A.; Tudor, E.; Kathriner, M. Lightweight Solid Wood Panels Made of Paulownia Plantation Wood. *Appl. Sci.* **2023**, *13*, 11234. https://doi.org/10.3390/app132011234.
15. Spîrchez, C.; Lunguleasa, A.; Popescu, C.; Avram, A.; Ionescu, C. Properties of Un-Torrefied and Torrefied Poplar Plywood (PW) and Medium-Density Fiberboard (MDF). *Appl. Sci.* **2023**, *13*, 11950. https://doi.org/10.3390/app132111950.
16. Zeleniuc, O.; Mazaherifar, M.; Coșereanu, C.; Suciu, A. Date-Palm-Based Sustainable Hybrid Composite with Cotton and Kevlar Fibre Participation. *Appl. Sci.* **2024**, *14*, 1008. https://doi.org/10.3390/app14031008.
17. Bedelean, B.; Ispas, M.; Răcășan, S. Applying the Artificial Neural Network and Response Surface Methodology to Optimize the Drilling Process of Plywood. *Appl. Sci.* **2023**, *13*, 11343. https://doi.org/10.3390/app132011343.
18. Barboutis, I.; Kamperidou, V.; Economidis, G. Handcrafted Reproduction of a 17th Century Bema Door Supported by 3D Digitization and CNC Machining. *Appl. Sci.* **2023**, *13*, 11690. https://doi.org/10.3390/app132111690.
19. Nicolau, A.; Pop, M.; Georgescu, S.; Coșereanu, C. Application of Additive Manufacturing Technology for Chair Parts Connections. *Appl. Sci.* **2023**, *13*, 12044. https://doi.org/10.3390/app132112044.
20. Brenci, L.; Gurău, L. A Stratified Characterization of Surface Quality of Beech Processed by Profile Milling. *Appl. Sci.* **2024**, *14*, 129. https://doi.org/10.3390/app14010129.
21. Tokarczyk, M.; Lis, B.; Salca, E.A.; Krystofiak, T. Adhesion of Varnish Coatings as a Background for Analogue and Digital Printing Technologies. *Appl. Sci.* **2024**, *14*, 304. https://doi.org/10.3390/app14010304.
22. Henke, M.; Krystofiak, T.; Salca, E.A.; Lis, B. Gloss Level, Type of Paper and Colour of HDF Boards Finished with the Energy-Saving Acrylic Paint System. *Appl. Sci.* **2024**, *14*, 394. https://doi.org/10.3390/app14010394.
23. Torcătoru, M.; Timar, M. Light-Induced Colour Changes in Wood Surfaces in Indoor Conditions Determined by an Artificial Accelerated Test: Influence of Wood Species and Coating Materials. *Appl. Sci.* **2024**, *14*, 1226. https://doi.org/10.3390/app14031226.
24. Zach, M.; Tauber, J.; Dohnal, P.; Svoboda, J. Designing Multi-Functional Magnetic Storage Cubes for Use in Modern Homes and Schools. *Appl. Sci.* **2024**, *14*, 829. https://doi.org/10.3390/app14020829

References

1. Barboutis, I.; Kamperidou, V. Properties of two different thicknesses 3-ply plywood of Tree-Of-Heaven veneers. In Proceedings of the 22nd International Scientific Conference "Wood is Good-EU Preaccession Challenges of the Sector", Zagreb, Croatia, 21 October 2011.
2. Schiberna, E.; Borovics, A.; Benke, A. Economic Modelling of Poplar Short Rotation Coppice Plantations in Hungary. *Forests* **2021**, *12*, 623. [CrossRef]
3. Kolář, T.; Rybníček, M. Phisical and mechanical properties of subfossil oak (*Quercus* sp.) wood. *Acta Univ. Agric. Silvic. Mendel. Brun.* **2010**, *58*, 123–134. [CrossRef]
4. Copur, Y.; Guler, C.; Akgul, M.; Tascıoglu, C. Chemical properties of hazelnut husk and its suitability for particleboard production. *Build. Environ.* **2007**, *42*, 2568–2572. [CrossRef]
5. Nelis, P.A.; Michaelis, F.; Krause, K.C.; Mai, C. Kiri wood (*Paulownia tomentosa*): Can it improve the performance of particleboards? *Eur. J. Wood Prod.* **2018**, *76*, 445–453. [CrossRef]
6. Zawiślak, K.; Sobczak, P.; Kraszkiewicz, A.; Niedziółka, I.; Parafiniuk, S.; Kuna-Broniowska, I.; Tanaś, W.; Żukiewicz-Sobczak, W.; Obidziński, S. The Use of Lignocellulosic Waste in the Production of Pellets for Energy Purposes. *Renew. Energy* **2020**, *145*, 997–1003. [CrossRef]
7. Tannert, T. Structural Performance of Rounded Dovetail Connections. Ph.D. Thesis, University of British Columbia, Vancouver, BC, Canada, 2008.
8. Schickhofer, G. *Prüfung und Optimierung von CNC-Gefertigten Schwalbenschwanzzapfverbindungen*; Forschungsbericht Holz-Holz-Verbindungen: Graz, Austria, 2014.
9. Braun, M.; Pantscharowitsch, M.; Kromoser, B. Experimental investigations on the load-bearing behaviour of traditional and newly developed step joints for timber structures. *Constr. Build. Mater.* **2022**, *323*, 126557. [CrossRef]
10. Deaconu, I.; Campean, M. Survey concerning the challenges of industrial drying of Quercus lumber. *Pro Ligno* **2022**, *18*, 36–44.
11. Ferrari, S.; Allegretti, O.; Cuccui, I.; Moretti, N.; Marra, M.; Todaro, L. A revaluation of Turkey oak wood (*Quercus cerris* L.) through combined steaming and thermo-vacuum treatments. *BioResources* **2013**, *8*, 5051–5166. [CrossRef]
12. Pawlak, D.; Boruszewski, P. Digital printing in wood industry. Ann. WULS For. *Wood Technol.* **2020**, *109*, 109–115.
13. Spear, M.J.; Curling, S.F.; Dimitriou, A.; Ormondroyd, G.A. Review of Functional Treatments for Modified Wood. *Coatings* **2021**, *11*, 327. [CrossRef]
14. Hazir, E.; Koc, K.H. Optimization of wood machining parameters in CNC routers: Taguchi orthogonal array based simulated angling algorithm. *Maderas Cienc. Tecnol.* **2019**, *21*, 493–510. [CrossRef]
15. Ngo, T.D.; Kashani, A.; Imbalzano, G.; Nguyen, K.T.; Hui, D. Additive manufacturing (3D printing): A review of materials, methods, applications and challenges. *Compos. Part B Eng.* **2018**, *143*, 172–196. [CrossRef]
16. Kilic, M.; Hiziroglu, S.; Burdurlu, E. Effect of machining on surface roughness of wood. *Build. Environ.* **2006**, *41*, 1074–1078. [CrossRef]
17. Namichev, P.; Petrovski, M. Wood as a primary selection of material for furniture production. *J. Process Manag. New Technol.* **2019**, *7*, 6–12. [CrossRef]
18. Hunt, D. Properties of wood in the conservation of historical wooden artifacts. *J. Cult. Herit.* **2012**, *13S*, S10–S15. [CrossRef]

applied sciences

MDPI

Review

Utilization Potential of Tree-of-Heaven Species Biomass—A Review

Paschalina Terzopoulou, Vasiliki Kamperidou * and Ioannis Barboutis

Laboratory of Wood Technology, Department of Forestry and Natural Environment, Aristotle University of Thessaloniki, 54635 Thessaloniki, Greece; terzopouz@for.auth.gr (P.T.); jbarb@for.auth.gr (I.B.)
* Correspondence: vkamperi@for.auth.gr

Abstract: Climate change and the subsequent increase in global mean air temperature already present a significant impact on forest vegetation. Especially in the near future, several forest species are expected to be in danger of extinction or compelled to migrate to colder regions. Some common species will be replaced by highly climate-tolerant species, primarily exotic plants, among others. The tree of *Ailanthus*, or "tree-of-heaven", as it is widely known, constitutes a rapidly growing plant, considered to be native to parts of China, while since the middle of the 18th century, it gradually spread to Europe and North America. This species demonstrates a preference for warmer, drier environments, although it can also survive in a variety of habitats and endure pollution of urban areas. It is a species with several uses, such as for animal feed, fuel, timber, pharmaceutical applications, etc., while its suitability for specialized applications of high-added-value is constantly being investigated. Its wood has a desirable appearance and characteristics that are comparable to those of other hardwood species of similar densities/weight. This article discusses some of the most important characteristics of *Ailanthus* wood and presents a comprehensive and constructive review of the chemistry, pharmacology, traditional and innovative uses, quality control, biological resistance, potential utilization in bioenergy and biofuels and wood products (e.g., wood-based panels, other advanced structure materials, etc.), use challenges and limitations, in order to contribute to the utilization potential assessment of this species biomass.

Keywords: *Ailanthus*; biofuels; biomass; chemical composition; fast growing; invasive; mechanical strength; pulp; waste; wood

Citation: Terzopoulou, P.; Kamperidou, V.; Barboutis, I. Utilization Potential of Tree-of-Heaven Species Biomass—A Review. *Appl. Sci.* **2023**, *13*, 9185. https://doi.org/10.3390/app13169185

Academic Editors: Emilia-Adela Salca, Lidia Gurau and Mihaela Câmpean

Received: 1 August 2023
Revised: 8 August 2023
Accepted: 10 August 2023
Published: 12 August 2023

1. Introduction

Global warming is evidenced mainly by rising global mean air and ocean temperatures, extensive melting of snow and ice, and rising global mean sea levels. A layer of greenhouse gases mostly made of water vapor, in far lower amounts, carbon dioxide, methane, and nitrous oxide act as a thermal blanket around the Earth, absorbing heat and concurrently heating its surface to temperatures suitable for the growth of living creatures. Several experts declare that the "greenhouse effect" caused by human activity is mostly responsible for the current trend of global warming. As the main contributor to climate change, CO_2 is a rather small but significant part of the atmosphere. It is created by both natural processes, such as respiration and volcanic eruptions, and human activities, such as deforestation, land use changes, and fossil fuel burning [1,2].

Simulations of future vegetation distribution in the European area indicate that by the end of the century, at least one-third of the land surface area may be covered by different vegetation species [3]. Changes in the plant species are likely to be considerably more significant in southern Europe, with widespread replacement of forest by shrubland predicted, owing mostly to drier conditions and catastrophic impact of fire [4,5]. Only a few locations in Greece appear to have a process-based projection study of how forest ecosystems would adapt to climate change [6]. These simulations appear to support the overall trend of elevation changes in species distribution and the replacement of drought-sensitive species with drought-resistant ones, with favorable interactions with the local environment.

In parallel, the global demand for fibrous lignocellulosic materials coming from wood industries is constantly rising, while the production of industrial wood of high quality obtained from natural forests continues to decline [1]. The deficit of forest products in developing countries is mainly attributed to the depletion of forest resources due to deforestation, human intervention, institutional failure, etc. [2], while in developed countries, it is attributed to the reduction of forest areas intended for industrial wood production (versus other uses such as recreational areas, etc.) [7]. The remaining forests experience high pressures due to the higher demand for wood in the forest industry attributed to the ever-increasing population and new application areas [8].

In order to be able to meet future material and wood demands and overcome the existing wood shortage, studies have so far focused on the utilization of alternative and competitive materials in forest industries, especially as raw materials for the production of sawn wood, pulp, paper, biofuels, and other products [1,9,10]. The answer to the timber shortage problem could possibly be partly provided by the cultivation of some fast-growing, short-rotation species.

In recent years, many countries have focused their attention on the utilization of wood derived from species that are considered undesirable in natural ecosystems and are classified as invasive or pests (*Ailanthus altissima, Robinia pseudacacia, Ambrosia artemisiifolia,* etc.) because of their tendency to spread at such a rapid rate and to such an extent that they suppress native vegetation species. Such species have not been properly utilized in appropriate applications to date [11].

Ailanthus species, commonly known as tree-of-heaven (Figure 1), has concerned many studies around the globe, although so far, according to the literature, there is not enough information concerning its properties towards its extensive and successful use in a multitude of applications (wood-based products, composite materials, chemical products, biofuels, etc.). The properties and characteristics of this fast-growing species, which remains underestimated until now, should be thoroughly investigated in order to promote the production of high-added-value products utilizing such a low-quality and low-cost raw material.

Figure 1. *Ailanthus altissima* species.

A. altissima is a deciduous tree that is native to China, though it has become naturalized in many parts of the world, including North America, Europe, and Oceania. It constitutes a member of the *Simaroubaceae* family and is one of the five species of the genus *Ailanthus* Desf. The genus *Ailanthus* comprises about 10 species, which are native to Asia and northern Oceania. *A. altissima* was named by Swingle in 1916 [12]. Three varieties are recognized, vars. *Altissima, tanakai,* and *sutchuensis* [13]. Several other varieties have also been described. Apart from *A. altissima*, its congeners are *A. excelsa* Roxb., *A. Integrrifolia*

Lam., (including *A. calycina Pierre*), *A. triphysa* (Dennst.) Alston, [12,14]. In the research implemented by Saina et al. [15], the phylogenetic analysis using protein-coding of the genes revealed that *A. altissima* is genetically related to *Leitneria floridana*, as well as the family *Simaroubaceae* to the family *Rutaceae*. The word "*Ailanthus*" is derived from the Moluccan name "aylanto", which refers to the native *Ailanthus integrifolia* and means "tree of heaven". The word "*altissima*" means very high. Some other scientific names of the species are *Ailanthus cacodendron* (Ehrh.) Schinz and Thell, *Ailanthus giraldii* Dode, *Ailanthus glandulosa* Desf., *Ailanthus peregrina* (Buchoz) F. A. Barkley, *Ailanthus sutchuensis* Dode, *Ailanthus vilmoriniana* Dode, *Albonia peregrina* Buchoz, *Pongelion glandulosum* (Desf.) Pierre, *Rhus cacodendron* Ehrh., *Toxicodendron altissimum* Mill.

It is a highly adaptable species that is able to grow in a wide range of environmental conditions, including disturbed sites, urban areas, and forest edges. *A. altissima* presents both positive and negative ecological impacts. It can provide habitat for several wildlife species and has been used in traditional medicine, though it is also considered a highly invasive species that could displace native vegetation and reduce biodiversity. Its allelopathic effects can inhibit the growth of other plant species, and it can also release allelopathic compounds (allelochemicals, mainly quassinoids, including ailanthone, ailanthinone, and ailanthinol, chaparrinone, chaparrin, shinjudilactone, shinjulactone, etc.), contained mainly in the root bark and other parts of the plant, into the soil, affecting soil microbial communities and nutrient cycling. Management of *A. altissima* can be challenging, mainly due to its fast growth, ability to resprout from roots and cut stems, and its prolific seed production. Controlling methods include mechanical removal, chemical treatments, and the use of biological agents, although each has its own limitations and potential drawbacks [11].

The aim of this review is to identify, appraise, and synthesize all the evidence available so far on the biology, ecology, distribution, impacts, management, and utilization options, and potential of *A. altissima*. This review, dealing with this species' biomass, would undoubtedly provide a comprehensive and structured overview of the available literature and help identify research gaps and priorities for future research and management efforts.

2. Nature and Characteristics of *A. altissima*

2.1. Ecology and Distribution of A. altissima

A. altissima is a deciduous tree with a wide range of ecological tolerances, including tolerance to drought, pollution, and a variety of soil types. It is a fast-growing, climate-tolerant species that prefers warmer and drier environments, although it can also survive in a variety of habitats of a wide temperature- and air humidity-range, while it seems to withstand drought, using a highly effective mechanism for saving water with reduced water loss from leaves and reduced root hydraulic conductivity [12]. It is noticeable that in several regions of Asia with harsh and adverse climatic conditions and poor soils, this species holds a key role in the local economy and is characterized as a kind of multiple-purpose species.

It reproduces both sexually and asexually, with the latter being the main mechanism of spread [12]. This tree species can reach up to 25–30 m in height and has a distinctive appearance, with smooth gray-brown bark, large pinnate leaves, and clusters of small yellow-green flowers that develop into samara fruit. *A. altissima* has both dioecious and hermaphroditic forms, with male and female flowers produced on separate trees, while hermaphroditic flowers are produced on the same tree.

It can rapidly colonize disturbed areas and form dense thickets, competing with native vegetation. The tree has been found to exhibit allelopathic effects, releasing compounds that can inhibit the growth of other plant species. Allelopathy is the ability of a plant to release chemicals that inhibit the growth of other nearby plants. It also has a high level of insect herbivory, with several species of insects feeding on the tree's leaves, stems, and fruits [12,16,17].

Nowadays, it can be found in many parts of the world, including Europe, Asia, Africa, and North and South America. Distribution data for *A. altissima* can be collected from several global databases, such as (A) the Global Biodiversity Information Facility (GBIF),

(B) Centre for Agriculture and Biosciences International (CABI), (C) the Global Invasive Species Database (GISD), and (D) Delivering Alien Invasive Species Inventories for Europe (DAISIE). Espenschied-Reilly and Runkle [18] investigated the distribution and changes in abundance of *A. altissima* (Miller) Swingle, in a southwest Ohio woodlot. The study involved sampling of *A. altissima* population in a 12.6-hectare woodland from 1978 to 2006. It was concluded that this species was present in the study area throughout the study period, with a high initial abundance and a subsequent decline in population density over time. This decline in population density could be attributed to natural factors, such as competition with other species or mortality due to disease or insects. It was also revealed that it was most abundant in disturbed areas of the woodland, such as along the edges and in degraded areas that had been previously logged or cleared. Generally, this species is commonly found in disturbed habitats, such as roadsides, vacant lots, and abandoned fields. It can also invade natural ecosystems and outcompete native plant species [12]. The tree produces an extensive root system that can contribute to its ability to grow in a wide range of environmental conditions, including nutrient-poor soils and areas with high levels of pollutants [19]. It was suggested that the species has a high capacity for colonizing disturbed habitats, and its presence in the study area could be attributed to previously applied land-use practices [18].

Brusa and Holzapfel [20] investigated the population structure of the invasive tree species *A. altissima* and the role of land-use history and management practices in shaping its distribution. The study was conducted in an urban area of northern Italy and involved sampling 37 populations of this species across different land-use types. The importance of land-use history and management practices is highlighted by the researchers in shaping the distribution and population structure of *A. altissima* species. The results showed that this species was present in all land-use types, with a higher frequency in disturbed habitats such as abandoned areas and industrial sites. The genetic diversity of the populations was found to be high, indicating the high potential of this species for adaptation and evolution. This genetic diversity could be explained by its ability to rapidly colonize disturbed habitats and its high speed of growth. The seedlings often grow 1–2 m in their 1st year and the tree may reach a 20-m height and a 1.5-m diameter in 10 years [21,22].

It has a complex genome with an estimated size of 1.5 Gb. The genome has been sequenced and annotated, providing insights into the genetic basis of the tree's growth and development [12].

Overall, further research is necessary to fully comprehend the biology of *A. altissima*, including its genetic diversity, physiological adaptations to different environments, and interactions with other species in order to assist management strategies application concerning the control of this invasive species' spread and minimize its impact on native ecosystems [11].

2.2. Management of *A. altissima*

Various management strategies have been developed to control the spread of *A. altissima*, including physical, chemical, and biological control methods. Physical control methods include cutting, pulling, and girdling the tree. Chemical control methods involve the use of herbicides. Biological control methods involve the introduction of natural enemies, such as insects and fungi.

More specifically, Pisuttu et al. [23] investigated the physiological and biochemical responses of this species to the infection by the soilborne fungal pathogen, *Verticillium dahliae*. *V. dahliae* was found to cause significant damage to the leaves, roots, and stems of *A. altissima*, which resulted in reduced growth and biomass production. The infected plants also showed alterations in their antioxidant defense system and photosynthetic pigments, indicating oxidative stress and impaired photosynthesis. The study provided insights into the mechanisms underlying the decline of *A. altissima* under biotic stress and highlighted the potential of *V. dahliae* as a biological control agent against this invasive plant species.

Brusa and Holzapfel [20] revealed that management practices, such as cutting and herbicide application, were effective in reducing the density of this species populations, though the method did not eliminate the species. The researchers suggest that long-term management strategies that involve a combination of different control methods may be necessary to effectively manage its spread. Espenschied-Reilly and Runkle [18] declare that management practices that involve reducing the disturbance of the forest ecosystem and promoting the growth of native species could be effective in controlling its spread.

Nunes [24] discussed the creation of value chains to support sustainable control and eradication actions on *A. altissima*. They presented a theoretical framework for value chain creation and identified potential stakeholders and their roles in the process. They also highlighted the economic, environmental, and social benefits that can be derived from the sustainable management of this species, such as the production of biofuels [25–27], animal feed [28–30], and other value-added products [31]. Jat et al. [29] also characterized *Ailntahus* as a multiple-purpose, compatible species, stating that its leaves, bark, and residual biomass can be used, among other uses, as animal feed, combating drought and securing livelihood security of resource-poor people. Ganai et al. [30] concluded that harvested green *Ailanthus* leaves contained 18.22% CP, 4.40% EE, 15.79% CF, 50.04% NFE, 26.60% NDF, 18.10% ADF, 8.50% hemicellulose, 5.11% ADL, 11.41% TA, 1.98% Ca and 0.27% P in dry matter, are palatable, extremely delicious, and nutritious feed for sheep and goats that contribute to the health of animals. The creation of value chains could be a viable strategy for addressing the negative impacts of *A. altissima* on ecosystems and promoting sustainable development in affected regions.

2.3. *Ailanthus altissima Biomass Chemical Analysis*

A. altissima leaves and bark have been found to contain various extractives, mainly essential oils, that have antimicrobial and antioxidant properties. These extractives contain various chemical compounds, including sesquiterpenes, monoterpenes, and phenols, several alkaloids, including quassin, ailanthone, and ailtirone [30]. These alkaloids have been found to have insecticidal and antifungal properties [31]. Especially the bark contains a variety of bioactive compounds, including alkaloids, flavonoids, and terpenoids, which may contribute to its medicinal properties. The pharmacological properties of *A. altissima* bark, including its anti-inflammatory, analgesic, antioxidant, antimicrobial, and anticancer effects, have also been highlighted [32]. Kalaskar et al. [33] reported that the stem bark of *Ailanthus excelsa* Roxb is rich in phenolic compounds, known for their antioxidant and anti-inflammatory properties, and may have potential as a source of natural antioxidants and as a natural preservative in food and cosmetic industries. Extracts from leaves have even greater phenolic content than the other parts of this species tree, and a positive linear correlation between antioxidant activity and total phenolic content of all the extracts has been reported [34]. Further studies are necessary to evaluate the safety and efficacy of using this species in food and cosmetic products.

Ailanthone, which constitutes an extractive of this species, can be utilized for the production of various unique and valuable products (pharmaceuticals, herbicides, insecticides, and ecological products). Its leaves have been found to contain various flavonoids, including kaempferol and quercetin. These flavonoids have antioxidant and anti-inflammatory properties. *A. altissima* bark has been found to contain tannins, which have astringent properties, and coumarins that demonstrate anticoagulant properties and are employed in the treatment of various medical cases [32].

Caramelo et al. [35] described that the main extracts for analyzing the biological activities of *A. altissima* are essentially obtained from the leaves, seeds (e.g., essential oil), and bark. However, extracts from other plant parts, such as flowers and stems, have been rarely investigated.

Chuang et al. [36] investigated the biosynthesis pathway of the bioactive compounds of quassinoids in the invasive tree-of-heaven and confirmed their evolutionary origin from protolimonoids. This study contributed to the development of potential applications in the pharmaceutical and agricultural industries. Transcriptome sequencing and gene expression analysis were employed to identify the genes involved in the biosynthesis of quassinoids in this species. Thirty genes involved in the biosynthesis of quassinoids were identified, including several genes that are responsible for the formation of protolimonoids, which are the precursors to quassinoids. These findings could potentially lead to the production of new bioactive compounds with diverse biological activities, with applications in the pharmaceutical and agricultural industries, although further research is needed to explore these possibilities.

Pettersen [37], analyzing *Ailanthus* species from the Caucasus region, reported 46% cellulose, 14% lignin content, and 0.9% ash content. Demirbas [38] reported 26.6% hemicelluloses content, 46.7% cellulose, 26.2% lignin, and 0.5% ash content. Samariha and Kiaei [39] identified that in the wood trunk of this species, the cellulose content was 47.18%, which is suitable for pulp production, lignin 25.19%, extracts, 3.5%, and ash 1.25%, while in the branch wood, the percentages of cellulose were 44.12%, lignin 23.86%, extracts 3.2%, and ash 1.75%. Lignin was higher in the trunk compared to the branch by 5.3%. Furthermore, Hosseinzade et al. [40] found the lignin content in the trunk to be 25.19%, while Talaeipour et al. [41] found 23.33%, respectively. A difference in the amount of extract content was recorded between the stem and branch by 8.6% [41].

Saidur et al. [42] recorded 26.60% hemicelluloses content, 46.70% cellulose, 26.20% lignin, and 0.5% ash content, which chemical composition corresponds to a heating value of 18.93 MJ/kg. Rahman et al. [43] also found high levels of lignin (25.84%) and extractives (8.3%), such as phenolic compounds, flavonoids, and terpenes, and recorded quite low ash content, indicating that it could be suitable for use as a biofuel.

Bardak et al. [44], in the chemical analysis, carried out on the wood of *Ailanthus* grown in the Trabzon area, reported 49.8% cellulose, 29.1% hemicelluloses, 21.3% lignin, and 0.21% ash, while in the respective wood from Artvin area, the results showed 46.5% cellulose, 24.4% hemicelluloses, 24.9% lignin, and 0.35% ash. Terzopoulou and Kamperidou [32] chemically characterized the wood and bark of *A. altissima* originating from three different locations in North Greece, with a focus on the horizontal variability assessment of its chemical composition (Table 1). Significant differences were found in the chemical composition of biomass obtained from these different locations. Furthermore, bark presented higher extractive content, slightly higher lignin content, much higher ash, and slightly lower holocellulose content than wood. As the tree age increased, the wood and bark presented lower lignin, though a linear increase of holocellulose and extractives contents. Additionally, the heartwood demonstrated higher extractive content, lignin and ash, and lower holocellulose content than sapwood.

Küçük and Demirbaş [45] investigated the kinetics of the hydrolysis of *A. altissima* chips using an alkaline-glycerol solution and highlighted that the hydrolysis rate increased with increasing temperature, biomass concentration, and alkaline concentration.

Table 1. Findings of *A. altissima* chemical analysis.

Study/Source	Part	Extractives (%)	Lignin (%)	Holocellulose (%)	Ash (%)	Applications
Pettersen [37]	wood	-	14	46	0.9	
Khattak et al. [46]	wood	19.89 Alcohol-Benzene 18.6 hot water	22.19	80.12	-	pulp-paper
Demirbas [47]	wood	--	26.2	46.7	0.5	
Samariha and Kiaei [39]	stem branch	3.5 3.2	25.19 23.86	47.18 44.12	1.25 1.75	pulp-paper
Saidur et.al. [42]	biomass	-	26.2	46.7	0.25	-
Bardak et al. [44]	wood	--	21.3	49.8	0.21	-

Table 1. *Cont.*

Study/Source	Part	Extractives (%)	Lignin (%)	Holocellulose (%)	Ash (%)	Applications
Terzopoulou and Kamperidou [32]	latewood 10 y.o	7.2	33.2	73.3	0.6	-
	latewood 3 y.o	7.5	32.5	72	0.7	-
	latewood 30 y.o	8.7	41.8	63.3	0.8	-
	earlywood 10 y.o	10.9	36	75.3	0.8	-
	earlywood 3 y.o	8.1	41.3	73.9	0.8	-
	earlywood 30 y.o	12.6	40.7	59.6	0.8	-
	bark 10 y.o	17.7	43.5	56.4	4.3	-
	bark 3 y.o	17.3	39.2	62.8	4.7	
	bark 30 y.o	26.9	36.4	56.8	4.5	

2.4. Structural and Physical Properties of *A. altissima* Wood

Understanding the physical properties of *A. altissima* wood is also important for the determination of its suitability for different applications. The density of wood is reported to range from 0.46 to 0.72 g/cm^3, depending on the specific growth conditions and location [48,49]. This species is classified as a medium-weight woody species. The estimation of the density of a wood species constitutes valuable information for the evaluation of wood quality and its proper utilization.

The wood of *Ailanthus* is ring-shaped, white-yellowish, of smooth texture. The vessels of the late wood of this species are rather sparse, 170–250 μm in width, with simple perforations and smooth walls containing marginal cavities. Vessels in early wood are spiral and occur in clusters of 80–150 μm width [50]. The vessels are attached to the tracheids and are reinforced in spirals, where parenchyma also develops around the vessels. The rays of the wood core consist of several rows (5–7 cells). These cells have thin walls with many cavities, are 8–15 μm high, and the peripheral ones up to 32 μm. Arnaboldi et al. [51] reported that the wood of *Ailanthus* has a characteristic high porosity on the surface of the vessels, similar to oak and chestnut. In contrast to these two species, vessels in *Ailanthus* wood are grouped in 2–3 and only occur in the early wood. In the late wood, axial parenchyma cells are visible, where they tend to form bundles, and their arrangement does not depend on the vessels. Concerning the morphological characteristics of this species, studied by Kúdela and Mamoňová [52], the mean diameter of the early wood vessels was found to be 230 μm, which is in agreement with the value obtained by Balaban [53], and the length was found to be 259 μm, while in the late wood, the mean diameter of the vessels was 41 μm and the length was 369 μm, and the mean fiber length was 939 μm. The vessels in the early wood occur singly, but they are also found in two or three groups. The thin cell walls contain many cavities. The diameter of the vessels in the early wood is high, the vessels in the late wood appear in groups, and have thin walls. the axial parenchyma, is concentrated near the late wood cells, which are reinforced by threading, and finally, the core rays are most often composed of five rows. In general, the wood of *Ailanthus* is considered hard, of medium weight, difficult to split, and easy to bend. However, Demeter et al. [54] evaluated the physical properties of *Ailanthus* wood grown in different regions of India and reported that the wood had a low density, high porosity, and moderate strength properties, making it suitable for lightweight construction applications.

Porosity, however, is mainly associated with the quantity of vessels in the early wood, the number of which is about twice higher in the transverse and sapwood. Latewood cells are much less permeable to water than vessels. In addition to the quantity of vessels, their quality is an important factor. Based on cell anatomy, this species can contain a high amount of water and thus resist extremely dry conditions [52]. Li et al. [55] reported that the density, modulus of elasticity, and modulus of rupture of *Ailanthus* wood were found to be decreased as the moisture content increased.

The factors influencing the density of wood are the amount of water contained in the wood mass, its structure, and the main chemical composition of the wood (extractives, cellulose, hemicelluloses, lignin, etc.) in terms of linear and volumetric swelling. The mean swelling in the radial direction is 4.1% and classifies the wood as a wood species with low swelling. Based on the values of swelling in the tangential direction (10.8%) and volumetric swelling (16%), this species is classified as a wood species of moderate swelling [48]. Nevertheless, Li et al. [55] marked a high level of shrinkage and swelling and, therefore, poor dimensional stability.

Overall, the physical properties can vary depending on factors, such as location, moisture content, and processing. Danihelová et al. et al. [56] marked high sound velocity and acoustic stiffness, making *Ailanthus* species wood suitable for use in musical instruments, as well as a low density and moderate mechanical properties, making it suitable for lightweight construction applications. Merhar et al. [57] investigated the machinability of the most common invasive tree species in Slovenia, including *A. altissima*, *Robinia pseudoacacia*, and *Acer negundo*, which cause significant ecological damage and economic losses, and reported especially for *A. altissima* that it is relatively easy to be machined and can be processed using standard simple woodworking equipment.

2.5. Mechanical Properties of A. altissima

According to the literature, tree-of-heaven wood is characterized by satisfying mechanical properties, especially for its weight [50]. Merhar et al. [57] marked relatively low density and strength and considered this species less suitable for certain applications, those of high strength requirements. It presents a very high increase in strength after drying. Özçifçi and Hiziroglu [58] recorded lower modulus of elasticity (MOE) and modulus of rupture (MOR) than other hardwood species, such as oak and beech, as well as a lower density than these hardwoods. Li et al. [59] attributed the quite low MOE and MOR to the presence of extractives, such as tannins, although the presence of these extractives could improve the durability and resistance of wood to decay, as well as to the presence of lignin. In general, the MOE has been reported to range from 7.2 to 17.1 GPa, while its MOR wood has been reported to range from 50.5 to 91.2 MPa (Table 2). A more recent study by Olsson et al. [60] evaluated the mechanical properties of this wood species untreated and treated with a soybean oil-based resin. The results showed that the treated wood demonstrated improved MOE and MOR compared to untreated wood, indicating that the resin treatment had reinforced the wood. Of course, the stress and strain values of *A. altissima* wood vary depending on the specific growth conditions (age, growth rate, etc.), origin, and environmental conditions, as well as the testing methods used, and other factors such as density, structure, extractives and treatment applied [61]. Although its wood may not have similar or superior mechanical properties compared to other valuable hardwoods, it could still be considered suitable for certain applications where the weight, biological durability, or acoustic properties are of important consideration.

Table 2. Mechanical and physical properties of *A. altissima* wood.

Ref.	Density r10 (g/cm^3)	Shrinkage (%)	Axial Compression Strength (N/mm^2)	MOR/(Bend. Strength) (N/mm^2)	MOE/(Bend. Strength) (N/mm^2)	Impact Bending Strength (J/cm^2)	Hardness Radial (N)	Hardness Tangendial (N)
Požgaj et al. [62]	0.658	-	-	-	-	-	-	-
Moslemi Bhagwat [63]	0.531	10.81	36.3	81.5	10,492.41	3.94	7702	6839
Kúdela and Mamoňová [52]	0.545	16.00	-	-	-	-	-	-
Panayotov et al. [64]	0.62	14.85	55.83	71.11	11,906	-	4407	5407
Miao et al. [65]	0.638	16.2	110.9 axial	148.3	-	-	-	-
Barboutis and Kamperidou [66]	0.6	-	64.47	101.29	10,259.31	3.811	5687.8	5141.5
Medved et al. [67]	0.555	-	-	-	-	-	-	-
Szabolcs, and Varga [68]	0.653	-	53.4	111.07	11,665	6.11	-	-

2.6. Uses of A. altissima

This species could be used to meet many requirements of modern civilization. Widespread use in the past has demonstrated many successful applications in several fields. In China, it was mainly used in traditional folk medicine [69]. Extracts obtained from the bark of *Ailanthus* have been used in the past to fight diarrhea and dysentery. Extracts from the root of the species have been used to treat epilepsy, asthma, and fever. According to the literature, extracts of the fruit have been used to treat eye diseases and also as an astringent. In sprayed form, it seems to contribute to seborrhea and in the treatment of psoriasis [12]. Research has been carried out on its pharmacological effects [70], as well as on the detection of compounds with antiproliferative, cytotoxic, and antiviral action [70]. *A. altissima* seeds are a source of fatty oils and proteins. The oil is aromatic and bitter, though it can be used after refining [71]. Researchers believe that the extractives contained in *A. altissima* may have certain positive effects against cancer cells [70]. Additionally, Zhu et al. [72] reported the isolation and characterization of two new tetracyclic triterpenoids, called ailantinols A and B, obtained from the fresh bark of *A. altissima*. The authors also evaluated the potential cytotoxicity of the compounds against several human cancer cell lines and found that ailantinol B showed moderate activity against two of the tested cell lines.

Gürbüz and Kahramanoğlu [73] suggested that *A. altissima* leaf extracts can be used as a natural alternative to synthetic preservatives for maintaining the postharvest quality of fresh apricot fruits and that the leaf extracts favor the quality and shelf life of fresh apricot fruits by delaying the ripening process and reducing weight loss during storage. Leaf extracts were found to have a high content of phenolic and flavonoid compounds and exhibited significant antioxidant activity. However, further research is required to evaluate the safety and efficacy of using these extracts on a larger scale in the fruit industry. The study of Kim et al. [74] suggests that *A. altissima* leaves extract in lipopolysaccharide (LPS)-stimulated astrocytes has anti-inflammatory properties and may have therapeutic potential in treating neuroinflammatory diseases. Specifically, they found that the extract reduced the production of reactive oxygen species (ROS) and inhibited the activation of nuclear factor kappa-B (NF-κB) and mitogen-activated protein kinases (MAPKs) signaling pathways. However, the mechanisms of action should be further explored.

Andonova et al. [75] investigated the phenolic profile, antioxidant and DNA-protective capacity, and microscopic characteristics of aerial substances of *A. altissima* and showed that the aerial parts of this species are rich in phenolic compounds and have strong antioxidant and DNA-protective activities.

Its foliage has been used for silkworm breeding, and its wood as a material for various constructions [61] and traditionally as firewood, although the use as wood-based material applications will be analyzed in the following. Talaeipour et al. [41] revealed a high utilization potential of wood and branch material of this species for paper and wood pulp production.

The species has been introduced to various parts of the world as an ornamental one in urban environments, attributed to its resistance to pollution and its ability to act as a herbicide and repel native insects, it has replaced other, more sensitive, species such as ash (Tilia sp.) [12]. Limited literature has been detected concerning the use of *A. altissima* as a herbicide, although some research has been conducted on the allelopathic effects of the species on other plants [34].

It has been found that extracts of *A. altissima* leaves and roots inhibited the germination and growth of several plant species, including lettuce, radish, and tomato [76]. The researchers identified several chemical compounds in the extracts, including quassinoids, which are known for their allelopathic impact. Bajwa et al. [76] suggested that this species extract could be used as a natural herbicide to control the growth of weeds.

Galatowitsch et al. [77] found that the presence of *A. altissima* had a negative impact on the growth and survival of several native plant species in a riparian ecosystem. The authors investigated the invasiveness of wetland plants, including *A. altissima*, in temperate North

America and suggested that these findings could have implications for the management of wetland ecosystems and that efforts to control the spread of this invasive species may be necessary to maintain native plant biodiversity. This study indicated that the allelopathic effects of this species may contribute to its invasiveness and ability to outcompete native species.

Furthermore, Novak et al. [78] investigated this species' root extracts and isolated ailanthone allelopathic effects on the growth of three test species (lettuce, radish, and cress) in laboratory conditions. They used different concentrations of root extracts and isolated ailanthone to test their effects on the germination, root and shoot growth, and biomass accumulation of the test species. The results showed that both the root extracts and isolated ailanthone had allelopathic effects on the test species, but the effects varied depending on the concentration and the test species. The highest concentrations of root extracts and especially the isolated ailanthone significantly reduced the germination and growth of all test species, while lower concentrations had no significant effects or even promoted the growth of some species. The authors suggested that the allelopathic effects of tree-of-heaven on other plants could be attributed to various allelochemicals, including ailanthone.

Kozuharova et al. [79] explored the potential of the invasive plant *A. altissima* as a source of natural pesticides. They reported that the plant has a high content of biologically active compounds, including alkaloids, flavonoids, and tannins, which have shown insecticidal and fungicidal properties. The study included a comprehensive assessment of the phytochemical composition and biological activity of different parts of *A. altissima*, including leaves, stems, and seeds. They reported that the plant extracts showed strong insecticidal and fungicidal properties against various pests and pathogens, including aphids, spider mites, and powdery mildew, and also discussed the economic potential of this species as a source of natural pesticides. It was suggested that the development of a sustainable and eco-friendly production chain for this species' extracts could provide new opportunities for rural development and job creation.

Hopson et al. [80] examined the physiological and transcriptomic responses of *Arabidopsis thaliana* to ailanthone, as a potential bio-herbicide and analyzed the effect of different concentrations of this substance on the growth, chlorophyll content, and gene expression of *Arabidopsis*. The results showed that ailanthone had a dose-dependent effect on *Arabidopsis* growth and chlorophyll content. At higher concentrations, ailanthone inhibited the growth and chlorophyll content of *Arabidopsis*. Several differentially expressed genes in *Arabidopsis* were identified as a result of the exposure to ailanthone, and these genes were involved in various biological processes, including stress response, cell wall modification, and hormone signaling pathways. Light was shed on the action mechanisms of this compound on plant physiology and gene expression. However, the effects of ailanthone on other plant species have not been detected so far, or on the potential environmental impact of using this compound as an herbicide. Therefore, further research is needed to explore the potential efficacy and safety of using ailanthone as an herbicide and to assess its environmental impact.

Although these studies suggest that *A. altissima* may have potential as a natural herbicide, it is important to keep in mind that the species is still considered invasive and has negative impacts on native ecosystems. Using this species extracts as an herbicide or insecticide could potentially harm non-target plant species and contribute to the spread of this invasive species. Further research is needed to fully comprehend the potential risks and benefits of using *A. altissima* as an herbicide [81]. There are many alternative methods for pest control, including integrated pest management techniques that focus on prevention, monitoring, and the use of environmentally friendly methods [82].

Among several other uses, this species is also valued for the shade it provides and erosion control, particularly in cities where soils are poor, and the atmosphere is polluted [19,83]. In particular, since 1950 and onwards, several plantations of *A. altissima* have been established in different parts of Europe in order to protect barren, steeply sloping, and eroded soils on slopes and on soils disturbed after a mining activity. It has also been used in the rehabilitation of landfills [83]. In the Czech Republic and Hungary, it has been used towards afforestation and reforestation, and in some areas, it has also been used for honey production [12].

Mohebzadeh et al. [84] investigated the effectiveness of *A. altissima*, and organic amendments in remediating heavy metal polluted soil. The study was carried out in a greenhouse and involved four treatments: Control (untreated soil), organic treatments only, plant species only, and a combination of organic treatments. The concentrations of heavy metals (Cu, Zn, Pb, and Cd) in the soil were measured before and after the treatments. The results showed that this species was able to accumulate heavy metals in the tissues and that it is more effective in accumulating Cu and Zn. Organic treatments alone did not significantly affect heavy metal concentrations in the soil or plant. However, the combination of organic amendments and plant species resulted in a significant reduction of heavy metal concentrations in the soil, indicating that the two remediation strategies could work synergistically. Further research is needed to investigate the long-term effects of these treatments on soil health and plant growth.

According to Terzopoulou and Kamperidou [32], very low ash contents and high calorific values proved high potential for utilization in solid biofuel production. The optimum harvesting time for the highest possible utilization of extractives and holocellulose is after the 21st year of the tree, while for lignin, after the fifth year of the tree. Of course, for qualitative timber utilization, trees should be harvested before the age of 30 years since, in older trees, the wood begins to degrade [19].

2.7. Utilization of A. altissima Biomass in Wood-Based Products

This wood species has been used for a variety of purposes so far, including furniture, flooring, and construction. Comprehending the properties of this wood species is critical to the determination of its suitability for various applications.

The wood is light and soft, with a pale yellowish color in the sapwood and yellowish and pale green in the heartwood. According to the research of Barboutis and Vassiliou [50], who studied the physical and mechanical properties of its wood, it was found to be of medium density, while demonstrating good mechanical properties that resemble those of several medium-density broadleaf species, such as white Ash and Chestnut. It was also observed that this wood species grown in Greece was superior in terms of modulus of rupture and resistance to compressive forces compared to the values reported in the literature [63]. It was concluded that due to its excellent appearance and unique design, it could be successfully used as sawn timber in many timber constructions and also as wood veneers [85,86].

Its wood can be easily dried, machined, glued, and finished with ease in forming attractive designs on both radial and tangential surfaces. Additionally, it is quite resistant to wear/aging, therefore, it can be used extensively in furniture construction [50].

Its wood is frequently reported to be uneven and subject to staining. It is suitable for construction, packaging, furniture-making material, pulp production. *A. altissima* can be utilized in the production of paper, whose properties could be comparable to those of *Eucalyptus globulus*, one of the most common species used for paper production, especially in temperate climate regions [87]. There is an urgent need to find sources of alternative fibers due to the increasing demand for paper, which is about to reach 521 million t/year [87]. Due to its rapid growth and tolerance to poor soils, drying, and pollution, the utilization of this species in paper production would help recover degraded environments and, in parallel, could contribute to the conservation of native forests.

According to research by Vasileiou et al. [85], except for swelling after 24 h of water immersion, tree-of-heaven plywood of 3-mm thickness showed proportionately superior qualities and mechanical strength to the poplar plywood of 3-mm thickness, respectively. Furthermore, in *Ailanthus* veneers, no knots were observed, while their pattern was considered much more attractive. Divya et al. [88] evaluated the potential use of *Ailanthus excelsa* as an alternative tree species for plywood production raw material. Wood demonstrated similar properties to other commercially important plywood species, indicating its potential as an alternative source for plywood production. However, limitations were highlighted in terms of sample size and lack of evaluation of other crucial properties of plywood.

Additionally, tree-of-heaven wood can be used in the fiber industry, in the production of particleboards and fiberboards [63]. Moreover, Li et al. [89] investigated the potential of *A. altissima* to be used in the production of particleboard. They found that the particles present similar physical and mechanical properties as particles from other hardwoods commonly used in particleboard production. However, they noted that the species' high extractive content could potentially adversely affect the adhesion properties of the particleboard.

Medved et al. [67] investigated the use of non-native wood species (black locust, silver fir, tree-of-heaven) in wood-plastic composites (WPCs) and examined their mechanical and physical properties. They concluded that tree-of-heaven could participate in the feedstock materials of WPCs since it combines high growth rates with satisfying wood fibers' hygroscopic and mechanical properties. Evaluating the potential use of the various products and wood of the studied species is necessary to reduce the costs of extensive tree clearance [90] in a way that this removal would be economically viable [91].

Tree-of-heaven could benefit from a mild or medium intensity heat treatment process in order to be modified to an aesthetically pleasing wood with enhanced hygroscopic nature and properties, facilitating its use in cabinetry and in variable indoor and outdoor non-structural applications [61]. A recent study by Bardak et al. [44] evaluated the chemical properties of *A. altissima* wood treated with a combination of tannin and oil-based resins. The results showed that the dimensional stability and resistance to fungal decay were improved, as well as its mechanical properties.

Olek et al. [92] revealed that heat treatment caused a decrease in hemicelluloses content and an increase in cellulose crystallinity and lignin content. The study also noted that heat treatment improved its dimensional stability and decay resistance. Barboutis and Kamperidou [61] also reported that thermal treatment affected the dimensional stability and water-absorbing capacity of wood in a positive way, decreasing equilibrium moisture content, swelling (tangential–radial), and adsorption percent, compared to untreated wood. The anisotropy of wood was decreased only to a small extent, while the wood color changed towards darker tones. Only the most intensive treatment negatively influenced the modulus of rupture and impact bending strength of wood, while the elasticity and compression strength of treated wood were proved to be similar to those of untreated wood, while the described changes were attributed to permanent changes of chemical constituents associated to thermal degradation during treatment.

Qian et al. [93] investigated the application of ultrasonic-assisted treatments to affect the properties of *A. altissima* wood. The wood was subjected to ultrasound at 320 W and 25 kHz while being combined with various alkali solutions (1%, 4%, and 8% NaOH), and the resulting changes in wood density and pore structure were analyzed. The ultrasonic-assisted treatment resulted in the breakdown or conversion of cell-wall components, mainly hemicelluloses, resulting in changes in the density and pore structure of the wood. In comparison to water immersion samples, ultrasonic-assisted samples produced a consistent self-shrinking wood block with higher oven-dried and bulk density and reduced porosity, particularly applying 1% and 4% NaOH. In alkali-treated samples, the distribution of pores moved towards smaller sizes, notably in ultrasonic-assisted samples. The researchers also discovered that ultrasonic-assisted treatment enhanced the relative content of macropores (>0.5 m), with samples treated with ultrasonic-1% NaOH reaching 86.10%. Ultrasonic-assisted treatments can be utilized to build a self-sustaining substrate with changing densities, pore architectures, and characteristics. The materials produced presented a high density, tiny pores, and excellent drainage. These findings provide important information on how to enhance the quality and possible applications of *A. altissima* wood.

2.8. Utilization in Paper, Pulp- and Nanocellulose-Based Products

A. altissima wood, bark, and leaves have been used for papermaking in different parts of the world, especially in China. The bark of the tree contains long and strong fibers that are suitable for producing strong paper, while the leaves contain short fibers that can be used to produce softer and smoother paper [94]. The bark fibers can be obtained by scraping the external bark of the tree, boiling it, and then introducing it into the pulp, mixed with water, and formed into sheets, which are dried and finished into paper. *A. altissima* paper has a distinctive light brown color and a slightly rough texture. The paper is also known for its strength and durability, making it suitable for various applications, such as bookbinding and printing [94].

There is limited literature concerning the utilization of this species in paper and pulp production, although there are some studies conducted evaluating the potential of this species for use in the paper industry. Nevertheless, *A. altissima* has been found to have a high cellulose content, which is a key component in the production of paper and pulp [31]. Tang et al. [95] investigated the suitability of *A. altissima* as a raw material for kraft pulp production and found that this species presented a similar pulp yield and kappa number (a measure of pulp quality), as other hardwood species commonly used in the paper industry. Baptista et al. [87] examined the potential of using this species as an alternative fiber source for papermaking in terms of the physical and mechanical properties of the produced paper, as well as the chemical composition of the fibers. The paper made of this species' fibers demonstrated satisfying mechanical properties, including high tensile strength and burst resistance, attributed mainly to the high cellulose content and low lignin content, which are desirable properties for papermaking. It was noticed that *A. altissima* had a lower lignin content than other hardwood species, which could potentially make it easier to bleach the pulp to achieve the desired whiteness [95]. Terzopoulou and Kamperidou [32], who examined the impact of tree age on the wood and bark properties, reported that as the tree age increases, the wood and bark present lower lignin content, though a linear increase in holocellulose and extractives. It was found that the optimum harvesting time for the highest possible utilization of holocellulose (as well as extractives) is after the 21st year when the cellulose, hemicelluloses, and extractives content increase. Further research is undoubtedly necessary to explore the environmental implications of using this tree species as a fiber source, as well as to develop more sustainable methods of fiber harvesting. Overall, although there is some evidence to suggest that *A. altissima* could be used as a raw material in the paper and pulp industry, further research is needed to fully understand the potential of the species for this application.

Almeida et al. [96] produced nanocellulose using invasive tree species of *A. altissima* and *Acacia dealbata*, in order to create high-value-added materials. The process involved producing bleached pulps from the wood of these trees using kraft cooking, followed by pre-treatment through either TEMPO-mediated oxidation (*A. dealbata*) or enzymatic hydrolysis (*A. altissima*) and high-pressure homogenization. The resulting hydrogels were characterized in terms of physical and chemical properties, including rheology and surface energy. The study found that nano/microfibrils could be obtained from the wood of these invasive species and that TEMPO cellulose nanofibrils formed strong gels with high-yield stress points and viscosities (tensile strength of 79 MPa, Young's modulus of 7.9 GPa, and transparency of 88%, although the water vapor barrier was modest). Films produced from cellulose hydrogels were found to be characterized by good mechanical and optical properties.

2.9. Utilization in Bioenergy Production

There has been an increasing interest in the potential use of *A. altissima* as a biomass feedstock for energy production. The fact that this species is characterized by a high growth rate and can grow in a great range of conditions makes it a potentially valuable feedstock for the production of biofuels or other bio-based products.

A. altissima wood indeed seems appropriate to be used in the production of high-quality firewood, compared to that of white oak, walnut, and birch species. This quality, of course, refers to older trees since the wood of younger trees that has been grown quickly is less energetically dense and more brittle, can be easily split, while it may present slightly higher ash contents. It generally acquires a high calorific value (18,040.17 J·g^{-1}) and relatively low ash levels [66]. Kamperidou and Barboutis [66] concluded that *A. altissima* wood and bark marked high heating values and low ash content compared to *Robinia pseudoacacia* and other species, making those materials to be considered suitable for pellets production. Specifically, this wood species was found to contain 0.5–0.69% ash, which means that it can be used as a raw material for the production of pellets intended to be used for residential applications. The wood can be classified in the best quality class of pellets (ENplus-A1) and meets the requirements of EN14961-2 for very low ash values (<0.7%) [66]. *A. altissima* wood and bark pellets showed comparable physical and mechanical properties to pellets made from other commonly used wood species, indicating their potential for use as a feedstock in pellets production. Bučar et al. [97] also reported high potential of using *A. altissima* and *R. pseudoacacia*, as feedstock for pellet production, due to high energy density and low ash contents. It was noted that the use of invasive species as feedstock can provide a solution to the problem of invasive species management, as well as contribute to the development of sustainable energy systems.

It has also been used for coal production in many countries. Yang et al. [98] investigated the potential of *A. altissima* for use in co-firing with coal in power plants and found that it had similar combustion characteristics to other biomass feedstocks and could potentially be used as a renewable energy source in co-firing applications.

Jabeen et al. [27] focused on the production of biodiesel from *A. altissima* (Mill.) seed oil using a green and recyclable potassium hydroxide activated *A. altissima* cake and cadmium sulfide catalyst. The authors reported that the use of this species seed oil as a feedstock for biodiesel production is advantageous due to its non-edible nature and availability. It was mentioned that the use of green and recyclable potassium hydroxide-activated *Ailanthus* cake and cadmium sulfide catalyst is eco-friendly and sustainable. The optimal conditions for biodiesel production were found to be a molar ratio of 1:9 for oil to methanol, a catalyst loading of 1.0 wt.% based on oil weight, a reaction temperature of 60 °C, and a reaction time of 2 h. The produced biodiesel met the international standard specifications for biodiesel. A promising approach for the sustainable production of biodiesel using *A. altissima* (Mill.) seed oil and a green and recyclable potassium hydroxide activated *Ailanthus* cake and cadmium sulfide catalyst was presented, though it is crucial to note that cadmium sulfide is a toxic material, and its use should be carefully monitored to prevent any negative environmental impacts.

Additionally, the impact of graphene oxide nanoparticles, a novel fuel additive, on the engine performance and emission characteristics of a diesel engine was studied by Hoseini et al. [99]. Their results demonstrated that environmentally beneficial fuels may be produced using diesel fuel, graphene oxide (GO) nanoparticle additives, and *A. altissima* biodiesel.

Sahay and Rana [100] investigated the potential of hemicelluloses hydrolysate from *Ailanthus excelsa* wood as a feedstock for ethanol production, evaluating the chemical composition of the hydrolysate and its suitability for ethanol fermentation. They found that the hemicelluloses hydrolysate had a high sugar content and was potentially fermentable to ethanol. It had a low inhibitory effect on yeast growth, indicating its potential as a feedstock for ethanol production. Overall, the study highlighted the potential of *A. excelsa* wood as a feedstock for ethanol production and provided useful information for researchers and engineers interested in developing sustainable biofuels. However, further research is needed to explore the long-term sustainability and commercial viability of this process.

The study of Li and Chen [101] investigated the potential of *A. altissima* as a feedstock for the pyrolysis process and found that it had a high yield of bio-oil and a low ash content, which constitute desirable properties for a biofuel feedstock. However, the study indicated that this species presented a slightly lower calorific value compared to other biomass feedstocks, which could affect its economic viability. Ighalo and Adeniyi [102] investigated the in silico temperature sensitivity of the pyrolysis of tree-of-heaven, beech, and spruce, using the ASPEN Plus v8.8 software and simulated the pyrolysis process at temperatures ranging from 300 to 1000 °C, and analyzed the yields and properties of the pyrolysis products, including bio-oil, char, and gas. They found that the yield of bio-oil decreased with increasing temperature for all three species, with *Ailanthus* showing the highest bio-oil yield at all temperatures. The char yield was found to be increased with increasing temperature, while the gas yield decreased. The properties of the bio-oil produced varied with temperature and species, with *Ailanthus* showing the highest calorific value and the lowest acidity.

Liu et al. [103] investigated the potential of *A. altissima* for use in anaerobic digestion, a process that converts biomass into biogas, which can be used to generate electricity or heat, and reported that this species' biomass presented a high potential for methane production and could potentially be used as a feedstock for anaerobic digestion.

Generally, although there is significant evidence to suggest *A. altissima* to be used as a feedstock in energy production, further research is necessary to fully understand the potential of this species for such applications. In addition, it is important to consider the potential environmental impacts of using an invasive species in this way and to ensure that any use of *A. altissima* is in line with sustainable biomass production practices.

2.10. *Challenges and Limitations of Ailanthus Utilization*

Invasive alien species pose a significant threat to biodiversity conservation since they may disrupt ecosystem services, with subsequent socioeconomic and ecological implications that, in turn, affect human well-being. Global changes, such as rising temperature and anthropogenic disturbances, create favorable conditions for the introduction and expansion of invasive species, necessitating the proper management of this expansion and its impacts. Perez et al. [104] evaluated the potential impacts of four invasive alien species (*Ailanthus altissima*, *Baccharis halimifolia*, *Impatiens glandulifera*, and *Pueraria montana*) on three ecosystem services in Europe: Food provisioning, soil erosion regulation, and maintenance of biological diversity. Wild species diversity, erosion regulation, food provisioning (crops and livestock), harvested wild goods, and tourism or recreation were found to be adversely affected, according to the risk analysis protocol applied. Species distribution models indicated a shift in distribution ranges towards the north and east of Europe for these invasive species. Bivariate analysis revealed that the invasive species *I. glandulifera* and *A. altissima* were identified to have the highest potential impact on the contribution of natural ecosystems to crop provision and on soil erosion control, with specific mechanisms of impact, such as altering soil nutrient cycling, affecting soil microorganisms, or producing allelopathic compounds. According to the models, Western Europe and the British Isles were identified as the most affected regions, while the Mediterranean region showed lower projected impacts due to the northwards expansion of invasive species [104].

Because of its invasive nature, the constraints and limitations of *Ailanthus* use are triggered by a variety of socioeconomic concerns and ethical implications. To guarantee appropriate administration and usage, the following criteria must be carefully considered. First of all, the invasive habit of *Ailanthus* can have a severe impact on native ecosystems by outcompeting native vegetation, affecting soil chemistry, and disturbing normal ecological processes. To prevent aggravating ecological imbalances, utilization initiatives must properly examine and manage these repercussions. Secondly, the quick growth and abundant seed production of *Ailanthus* can result in the displacement of native vegetation species, and therefore, in a decrease in biodiversity and local food chain changes. This change of location has the potential to have a domino impact on ecosystem health and stability. Cross-breed between *Ailanthus* and native species can result in hybridization, which may cause genetic pollution and put at risk the inherited characteristics of local plant populations. This has the potential to have long-term implications for ecosystem dynamics and evolutionary processes [18]. Furthermore, harvesting *Ailanthus* for use may accidentally contribute to its spread since seeds can be dispersed during shipping and processing. This would raise the prospect of inadvertent introductions into new domains.

Its root system can cause damage to pavements, archaeological monuments, walls, and other structures because it causes erosion of substrates, and this presents one of its most important problems. Its effect can be chemical due to the secretion of organic acids from the roots decomposing calcium bicarbonate, which is the main component of the limestone used as a building material, or it can be mechanical, by the increase of the pressure of the roots, either above or near the buildings [105].

Positive impact has also been detected, as analyzed in this review article, which should be considered for inclusion in management cost-benefit evaluations. *A. altissima* stands out as a species that may provide wood of good quality and satisfying properties in accordance to its density and weight, medicinal ingredients, chemical products, biofuels, and has significant decorative appearance both as a standing tree and wood material produced [106]. The evaluation of beneficial effects should not be interpreted as an effort to counterbalance the negative effects but rather as a chance to contribute new information to scientists, managers, and policy-makers for the potential utilization of this species' biomass [107].

The possible influence of *Ailanthus* use in local economies and industries should be further and thoroughly studied. Depending on the extent of use, there may be resource competition between *Ailanthus*-based products and those derived from native/non-invasive plants. The decision to encourage Ailanthus use should be balanced against ethical concerns, such as the possible risk to natural ecosystems and traditional landscapes associated with our culture. It would be definitely critical to balance commercial interests with environmental and cultural values. Promoting the use of *Ailanthus* may unintentionally reinforce its existence and spread, complicating invasive species control efforts. Therefore, it is necessary to carefully examine if usage attempts may accidentally aid in its spread.

The need for management and mitigation strategies to address the potential impacts of invasive species on various ecosystem services seems to be of high urgency, especially in light of climate change and its interactions with invasive species spread.

In order to mitigate the potentially harmful impact, selective cultivation application would reduce the danger of biodiversity loss and genetic contamination, as well as the prioritization of the non-invasive, native, or less aggressive plant species cultivation. Rigorous measures implementation for *Ailanthus* production and trading would encourage the use of native species instead. Moreover, regular monitoring of *Ailanthus* and other alien species populations for evidence of hybridization and genetic contamination is of great importance, as well as the support of habitat restoration programs aimed at re-establishing native vegetation communities and increasing biodiversity in places impacted by Ailanthus invasion [104].

Since this species presents a preference for contaminated and disturbed soils and degraded habitats (along the roads on the outskirts and forest clearances, urban areas, and coastal zones of the Mediterranean islands [12,66,90]), it would be a good option to intensify our efforts towards the maintenance of good health of forest ecosystems, primarily avoiding severe degradation and applying rational and sustainable management.

3. Conclusions—Recommendation towards Future Research

Considering that *A. altissima* can be used for animal feed, bioenergy/biofuels, lumber, and wood-based products (plywood, particleboards, etc.) and composites, chemical industry products, pharmaceutical purposes, etc., it is frequently referred to as a species of several uses. When compared to other hardwood species of equivalent density and weight, its properties can be considered desirable, while it combines the advantages of pleasing appearance, high availability, fast growth, and low cost. It has been demonstrated that it is not appropriate for use in heavy structural applications, though it is acceptable for many other lighter applications, including furniture, boat construction, lining, decoration, pallets, matches, wood pulp, firewood, charcoal, etc., and its suitability for specialized applications of high added value is constantly being investigated. Despite the extensive research on the species of *A. altissima*, there are still several gaps in our knowledge about this invasive species. Here are a few areas where further research is needed.

There is a lack of genetic studies on *A. altissima* populations across its range. Understanding the genetic diversity and structure of these populations could provide insights into the species' evolutionary history and inform management strategies. There is limited research on the allelopathic effects of other plant species. Investigating the allelopathic compounds produced by this species could provide important information on its impact on plant communities and potential management options. Furthermore, there is limited research on the potential impacts of climate change on the distribution and spread of *A. altissima*. Taking into consideration the predicted changes in temperature and precipitation scenarios, it is important to understand how this invasive species may respond and adapt to these changes. There is also a lack of long-term studies on the effects of different management strategies for controlling *A. altissima*. Understanding the long-term ecological and economic impacts of different management approaches could help assist decision-making. There is limited research on the social and cultural impacts of *A. altissima* in areas where it is considered a valued or culturally significant species. This could include studies on the attitudes and perceptions of local communities towards the species and the impacts of its management on cultural practices and traditions. Although several studies have already been conducted in the field of potential application of this species in bio-based products and biofuels, they have been rather segmented, making it difficult to draw sound conclusions as regards the proper utilization of this species. Therefore, it would be of high importance to continue further examination of this species' biomass properties towards its rational utilization in an attempt to totally or partly substitute the respective conventional materials and products that are usually characterized by a high environmental impact, though without posing threats to the native ecosystem biodiversity.

Author Contributions: Conceptualization, I.B.; methodology, I.B., V.K. and P.T.; validation, P.T., V.K. and I.B.; formal analysis, V.K. and I.B.; investigation, P.T.; resources, I.B.; data curation, P.T.; writing—original draft preparation, P.T.; writing—review and editing, V.K. and I.B.; visualization, P.T. and V.K.; supervision, I.B. All authors have read and agreed to the published version of the manuscript.

Funding: This research received no external funding.

Institutional Review Board Statement: Not applicable.

Informed Consent Statement: Not applicable.

Data Availability Statement: No new data were created.

Conflicts of Interest: The authors declare no conflict of interest.

References

1. Ashori, A. Pulp and paper from kenaf bast fibers. *Fibers Polym. J.* **2006**, *7*, 26–29. [CrossRef]
2. Amoah, A.; Korle, K. Forest depletion in Ghana: The empirical evidence and associated driver intensities. *For. Econ. Rev.* **2020**, *2*, 61–80. [CrossRef]
3. Hickler, T.; Vohland, K.; Feehan, J.; Miller, P.A.; Smith, B.; Costa, L.; Giesecke, T.; Fronzek, S.; Carter, T.R.; Cramer, W.; et al. Projecting the Future Distribution of European Potential Natural Vegetation Zones with a Generalized, Tree Species-Based Dynamic Vegetation Model: Future Changes in European Vegetation Zones. *Glob. Ecol. Biogeogr.* **2012**, *21*, 50–63. [CrossRef]
4. Ruiz-Labourdette, D.; Schmitz, M.F.; Pineda, F.D. Changes in Tree Species Composition in Mediterranean Mountains under Climate Change: Indicators for Conservation Planning. *Ecol. Indic.* **2013**, *24*, 310–323. [CrossRef]
5. Santini, M.; Collalti, A.; Valentini, R. Climate Change Impacts on Vegetation Andwater Cycle in the Euro-Mediterranean Region, Studied by a Likelihood approach. *Reg. Environ. Chang.* **2014**, *14*, 1405–1418. [CrossRef]
6. Fyllas, N.M.; Koufaki, T.; Sazeides, C.I.; Spyroglou, G.; Theodorou, K. Potential Impacts of Climate Change on the Habitat Suitability of the Dominant Tree Species in Greece. *Plants* **2022**, *11*, 1616. [CrossRef]
7. Ashori, A. Non-wood fibers—A potential source of raw material in papermaking. *Polym-Plast Technol. Eng.* **2006**, *45*, 1133–1136. [CrossRef]
8. Giannotas, G.; Kamperidou, V.; Barboutis, I. Tree bark utilization in insulating bio-aggregates: A review. *Biofuels Bioprod. Biorefining* **2021**, *15*, 1989–1999. [CrossRef]

9. Copur, Y.; Guler, C.; Akgul, M.; Tascıoglu, C. Chemical properties of hazelnut husk and its suitability for particleboard production. *Build Environ.* **2007**, *42*, 2568–2572. [CrossRef]
10. Stefanidou, M.; Kamperidou, V.; Konstantinidis, A.; Koltsou, P.; Papadopoulos, S. Use of Posidonia oceanica fibres in lime mortars. *Constr. Build. Mater.* **2021**, *298*, 123881. [CrossRef]
11. Hytönen, J.; Nurmi, J. Heating value and ash content of intensively managed stands. *Wood Res.* **2015**, *60*, 71–82.
12. Kowarik, I.; Saumel, I. Biological flora of Central Europe: *Ailanthus altissima* (Mill.) Swingle. *Perspect. Plant Ecol. Evol. Syst.* **2007**, *8*, 207–237. [CrossRef]
13. Chen, S. *China Flora*; Science Press: Beijing, China, 1997; Volume 43, pp. 1–6.
14. Fu, L.; Hong, T. *Higher Plants of China; Beijing*; Book Gallery / / Mike Riley: Phoenix, AZ, USA, 2000; Volume 13.
15. Saina, J.K.; Li, Z.; Gichira, A.W.; Liao, W. The Complete Chloroplast Genome Sequence of Tree of Heaven (*Ailanthus altissima* (Mill.) (Sapindales: Simaroubaceae), an Important Pantropical Tree. *Int. J. Mol. Sci.* **2018**, *19*, 929. [CrossRef] [PubMed]
16. Bailly, C. Anticancer properties and mechanism of action of the quassinoid ailanthone. *Phytother. Res.* **2020**, *34*, 2203–2213. [CrossRef]
17. Wang, Z.; He, Z.; Zhao, Z.; Yi, S.; Mu, J. Influence of ultrasound-assisted extraction on the pyrolysis characteristics and kinetic parameters of eucalyptus. *Ultrason. Sonochemistry* **2017**, *37*, 47–55. [CrossRef]
18. Espenschied-Reilly, A.L.; Runkle, J.R. Distribution and changes in abundance of *Ailanthus altissima* (Miller) Swingle in a southwest Ohio woodlot. *Ohio J. Sci.* **2008**, *108*, 16–22. Available online: https://kb.osu.edu/bitstream/handle/1811/48375/V108N2_16.pdf?sequence=1 (accessed on 1 June 2023).
19. Hu, S.Y. Ailanthus. *Arnoldia* **1979**, *39*, 29–50.
20. Brusa, A.; Holzapfel, C. Population structure of *Ailanthus altissima* (Simaroubaceae): The role of land-use history and management1. *J. Torrey Bot. Soc.* **2018**, *145*, 55–68. [CrossRef]
21. Miller, J. *Ailanthus altissima* (Mill.) Swingle ailanthus. In *Silvics of North America: Hardwoods*; Agriculture Handbook 654; Burns, R.M., Honkala, B.H., tech. coords, Eds.; U.S. Department of Agriculture, Forest Service: Washington, DC, USA, 1990; Volume 2, pp. 101–104.
22. Evans, C.W.; Moorhead, D.J.; Bargeron, C.T.; Douce, G.K. *Invasive Plant Responses to Silvicultural Practices in the South*; Bugwood Network BW-2006-03; The University of Georgia Bugwood Network: Tifton, GA, USA, 2006; 52p. Available online: http://www.invasive.org/silvicsforinvasives.pdf (accessed on 15 May 2023).
23. Pisuttu, C.; Piccolo, E.L.; Paoli, L.; Cotrozzi, L.; Nali, C.; Pellegrini, E.; Lorenzini, G. Physiochemical responses of *Ailanthus altissima* under the challenge of *Verticillium dahliae*: Elucidating the decline of one of the world's worst invasive alien plant species. *Biol. Invasions* **2023**, *25*, 61–78. [CrossRef]
24. Nunes, L.J.R. Creation of Value Chains for the Sustainability of Control and Eradication Actions on *Ailanthus altissima* (Mill.) Swingle. *Environments* **2022**, *9*, 64. [CrossRef]
25. Brihaspati, S.; Srivastava, A.K.; Prakash, O. A Comprehensive Review on Rare Biodiesel Feedstock Availability, Fatty Acid Composition, Physical Properties, Production, Engine Performance and Emission. *Process Integr. Optim. Sustain.* **2023**, 1–36. [CrossRef]
26. Pal, V.; Sharma, V.; Gour, V.S. *Ailanthus excelsa* Roxb. in India: A multipurpose "tree of Heaven" for semi-arid regions. *For. Trees Livelihoods* **2023**, 1–16. [CrossRef]
27. Jabeen, M.; Munir, M.; Abbas, M.M.; Ahmad, M.; Waseem, A.; Saeed, M.; Kalam, M.A.; Zafar, M.; Sultana, S.; Mohamed, A.; et al. Sustainable Production of Biodiesel from Novel and Non-Edible *Ailanthus altissima* (Mill.) Seed Oil from Green and Recyclable Potassium Hydroxide Activated *Ailanthus* Cake and Cadmium Sulfide Catalyst. *Sustainability* **2022**, *14*, 10962. [CrossRef]
28. Musa Tibin, M.A.; Alsharifm, Z.A.; Mohammed, A.Y.I.; Jadalla, J.B.; Ebrahiem, M.A. Effects of feeding *Ailanthus excelsa* (Roxb.) leaves on desert lamb's feed intake, nutrients digestibility and growth performance. *Int. J. Vet. Sci. Res.* **2021**, *7*, 113–117. [CrossRef]
29. Jat, H.S.; Singh, R.K.; Mann, J.S. Ardu (*Ailanthus* sp.) in arid ecosystem: A compatiblespecies for combating with drought and securing livelihood security of resource poor people. *Indian J. Tradit. Knowl.* **2011**, *10*, 102–113.
30. Ganai, A.; Ahmad, S.B.; Bhat, J.I.A. Utilization of Ailanthus (*Ailanthus altissima*) Leaves in Sheep Feeding. 2010. Available online: https://www.researchgate.net/publication/314083647_Utilization_of_ailanthus_Ailanthus_altissima_leaves_in_sheep_feeding (accessed on 31 July 2023).
31. Heisey, R.M. Allelopathy and the secret life of *Ailanthus altissima*. *Arnoldia* **1997**, *57*, 28–36.
32. Terzopoulou, P.; Kamperidou, V. Chemical characterization of Wood and Bark biomass of the invasive species of Tree-of-heaven (*Ailanthus altissima* (Mill.) Swingle), focusing on its chemical composition horizontal variability assessment. *Wood Mater. Sci. Eng.* **2021**, *17*, 469–477. [CrossRef]
33. Kalaskar, M.G.; Sapkal, P.R.; Tatiya, A.U.; Jain, P.D.; Surana, S.J. Morphoanatomical and physicochemical studies on *Ailanthus excelsa* roxb. stem bark: A tree of heaven. *J. Drug Deliv. Ther.* **2019**, *9*, 128–131. [CrossRef]

34. Luís, Â.; Gil, N.; Amaral, M.E.; Domingues, F.; Duarte, A.P. *Ailanthus altissima* (Miller) Swingle: A source of bioactive compounds with antioxidant activity. *BioResources* **2012**, *7*, 2105–2120. Available online: https://bioresources.cnr.ncsu.edu/wp-content/uploads/2016/06/BioRes_07_2_2105_Luis_GADD_Ailanthus_Bioactiv_Cpds_Antioxidant_Activity_2555.pdf (accessed on 15 July 2023). [CrossRef]
35. Caramelo, D.; Pedro, S.I.; Marques, H.; Simão, A.Y.; Rosado, T.; Barroca, C.; Gominho, J.; Anjos, O.; Gallardo, E. Insights into the Bioactivities and Chemical Analysis of *Ailanthus altissima* (Mill.) Swingle. *Appl. Sci.* **2021**, *11*, 11331. [CrossRef]
36. Chuang, L.; Liu, S.; Biedermann, D.; Franke, J. Identification of early quassinoid biosynthesis in the invasive tree of heaven (*Ailanthus altissima*) confirms evolutionary origin from protolimonoids. *Front. Plant Sci.* **2022**, *13*, 958138. [CrossRef]
37. Pettersen, R.C. The chemical composition of wood. In *The Chemistry of Solid Wood*; Rowell, R.M., Ed.; Advances in Chemistry Series, 207; American Chemical Society: Washington, DC, USA, 1984; pp. 57–126.
38. Demirbas, A. Biodiesel from Wood Oils in Compressed Methanol. *Energy Sources Part A: Recovery Util. Environ. Eff.* **2009**, *31*, 1530–1536. [CrossRef]
39. Samariha, A.; Kiaei, M. Chemical Composition Properties of Stem and Branch in *Alianthus altissima* Wood. *Middle-East J. Sci. Res.* **2011**, *8*, 967–970.
40. Hosseinzade, A.; Arabtabar, H.; Golbabaei, F.; Familian, H.; Sadraei, N.; Habibi, M. Wood Properties of Eucalyptus steriaticalyx grown in southern region of Iran. *Res. Inst. For. Rangel.* **2001**, *243*, 62–123.
41. Talaeipour, M.; Hemmasi, A.H.; Ebrahimpour Kasmani, J.; Mirshokraie, A.; Khademieslam, H. Effects of fungal treatment on structural and chemical features of Hornbeam chips. *BioResources* **2010**, *5*, 477–487. [CrossRef]
42. Saidur, R.E.; Abdelaziza, A.; Demirbas, A.; Hossain, M.S.; Mekhilef, S. A review on biomass as a fuel for boiler. *Renew. Sustain. Energy Rev.* **2011**, *15*, 2262–2289. [CrossRef]
43. Rahman, H.M.A.; Javaid, S.; Ashraf, W.; Rasool, M.F.; Saleem, H.; Khan, S.A.; Ul-Haq, Z.; Anjum, S.M.; Ahmad, T.; Alqahtani, F. Effects of long-term *Ailanthus altissima* extract supplementation on fear, cognition and brain antioxidant levels. *Saudi Pharm. J.* **2023**, *31*, 191–206. [CrossRef]
44. Bardak, S.; Nemli, G.; Tiryaki, S. The Influence of Raw Material Growth Region, Anatomical Structure and Chemical Composition of Wood on the Quality Properties of Particleboards. *Maderas. Cienc. Tecnol.* **2017**, *19*, 363–372. [CrossRef]
45. Küçük, M.M.; Demirbaş, A. Kinetic study on hydrolysis of biomass (*Ailanthus altissima* chips) by using alkaline-glycerol solution. *Energy Convers. Manag.* **1999**, *40*, 1397–1403. [CrossRef]
46. Khattak, T.M.; Ghazi, J.Á.M.I.L.A. Suitability of some non-commercial fast growing wood yielding trees growing in Azad Kashmir for the production of pulp and paper. *Pak. J. Bot.* **2001**, *33*, 729–732.
47. Demirbas, M.F. Biorefineries for biofuel upgrading: A critical review. *Appl. Energy* **2009**, *86*, S151–S161. [CrossRef]
48. Wagenführ, R. *Holzatlas, Wood Atlas*, 6th ed.; Fachbuchverlag im Carl Hanser Verlag: Leipzig, Germany, 2007. (In German)
49. Hassan, K.; Tippner, J. Acoustic properties assessment of neem (*Azadirachta indica* A. Juss.) wood from trees irrigated with secondarily treated wastewater. *BioResources* **2019**, *14*, 2919–2930. Available online: https://bioresources.cnr.ncsu.edu/resources/acoustic-properties-assessment-of-neem-azadirachta-indica-a-juss-wood-from-trees-irrigated-with-secondarily-treated-wastewater/ (accessed on 10 January 2023). [CrossRef]
50. Barboutis, I.; Vassiliou, V. Physical and Mechanical Properties of the Wood of *Ailanthus altissima* (Miller) Swingle. In Proceedings of the 14th Pan-Hellenic Forestry Conference, Patra, Greece, 1–4 November 2009.
51. Arnaboldi, F.; Conedera, M.; Fonti, P. Caratteristiche anatomiche e auxometriche di Ailanthus altissima. Una specie arborea a carattere invasivo. *Sherwood* **2003**, *91*, 19–23. Available online: https://www.dora.lib4ri.ch/wsl/islandora/object/wsl:8663 (accessed on 31 July 2023).
52. Kúdela, J.; Mamoňová, M. Tree-of-heaven wood (*Ailanthus altissima*, Mill.)—Structure and properties. *Wood Struct. Prop.* **2006**, *6*, 275–287.
53. Balabán, K. *Nauka o Dřevě. 1. Část: Anatomie Dřeva*; Státní Zemědělské Nakladatelství: Praha, Czech Republic, 1955; 220p.
54. Demeter, A.; Saláta, D.; Tormáné Kovács, E.; Szirmai, O.; Trenyik, P.; Meinhardt, S.; Rusvai, K.; Verbényiné Neumann, K.; Schermann, B.; Szegleti, Z.; et al. Effects of the Invasive Tree Species Ailanthus altissima on the Floral Diversity and Soil Properties in the Pannonian Region. *Land* **2021**, *10*, 1155. [CrossRef]
55. Li, X.; Li, X.; Li, Y.; Li, C.; Li, J. Effect of moisture content on physical and mechanical properties of *Ailanthus altissima* wood. *BioResources* **2020**, *15*, 3142–3156.
56. Danihelová, A.; Čulík, M.; Danihelová, Z. Assessment of Properties of Ailanthus Wood for Xylophone Bars. *Akustika* **2015**, *24*, 42–49. Available online: https://www.akustikad.com/ (accessed on 31 July 2023).
57. Merhar, M.; Bučar, G.D.; Merela, M. Machinability Research of the Most Common Invasive Tree Species in Slovenia. *Forests* **2020**, *11*, 752. [CrossRef]
58. Özçifçi, A.; Hiziroglu, S. Physical and mechanical properties of *Ailanthus altissima* wood. *Eur. J. Wood Wood Prod.* **2013**, *71*, 179–181.
59. Li, X.; Li, Y.; Ma, S.; Zhao, Q.; Wu, J.; Duan, L.; Xie, Y.; Wang, S. Traditional uses, phytochemistry, and pharmacology of *Ailanthus altissima* (Mill.) Swingle bark: A comprehensive review. *J. Ethnopharmacol.* **2021**, *275*, 114121. [CrossRef]
60. Olsson, S.; Johansson, M.; Westin, M.; Ostmark, E. Reactive UV-absorber and epoxy functionalized soybean oil for enhanced UV-protection of clear coated wood. *Polym. Degrad. Stab.* **2014**, *110*, 405–414. [CrossRef]
61. Barboutis, I.; Kamperidou, V. Impact of Heat Treatment on the Quality of Tree-of-Heaven Wood. *Drv. Ind.* **2019**, *70*, 351–358. [CrossRef]

62. Požgaj, A.; Kurjatko, S.; Chovanec, D.; Babiak, M. *Štruktúra a Vlastnosti Dreva*, 2nd ed.; Príroda: Bratislava, Slovakia, 1997; 488p.
63. Moslemi, A.A.; Bhagwat, S.G. Physical and Mechanical Properties of the Tree of Heaven. *Wood Fiber* **1970**, *1*, 319–323.
64. Panayotov, P.; Kalmukov, K.; Panayotov, M. Biological and wood properties of *Ailanthus altissima* (mill.) swingle. *For. Ideas* **2011**, *17*, 122–130. Available online: https://forestry-ideas.info/files/issue/Forestry_Ideas_BG_2011_17_2_2.pdf (accessed on 5 May 2023).
65. Miao, X.; Chen, H.; Lang, Q.; Bi, Z.; Zheng, X.; Pu, J. Characterization of *Ailanthus altissima* Veneer Modified by Urea-formaldehyde Pre-polymer with Compression Drying. *BioResources* **2014**, *9*, 5928–5939. Available online: https://jtatm.textiles.ncsu.edu/index.php/BioRes/article/view/BioRes_09_4_5928_Miao_Ailanthus_altissima_Veneer (accessed on 1 January 2023). [CrossRef]
66. Barboutis, I.; Kamperidou, V. Utilization perspectives of wood and bark of the invasive species of *Ailanthus* and *Acacia* in the production of pellets. In Proceedings of the International Forestry and Environment Symposium, Trabzon, Turkey, 7–10 November 2017.
67. Medved, S.; Tomec, D.K.; Balzano, A.; Merela, M. Alien Wood Species as a Resource for Wood-Plastic Composites. *Appl. Sci.* **2021**, *11*, 44. [CrossRef]
68. Szabolcs, K.; Varga, D. Physical and mechanical properties of wood from invasive tree species. *Maderas Cienc. Tecnol.* **2021**, *23*, 1–8. [CrossRef]
69. Howard, J.L. Ailanthus altissima. In: Fire Effects Information System, [Online]. US Department of Agriculture, Forest Service, Rocky Mountain Research Station, Fire Sciences Laboratory (Producer). 2004. Available online: http://www.fs.fed.us/database/feis/S (accessed on 1 February 2023).
70. Kozuharova, E.; Lebanova, H.; Getov, I.; Benbassat, N.; Kochmarov, V. *Ailanthus altissima* (Mill.) Swingle a terrible invasive pest in Bulgaria or potential useful medicinal plant? Review paper. *Bothalia J.* **2014**, *44*, 213–230.
71. Cheng, J.Q.; Yang, J.J.; Liu, P. *Wood Science*; Chinese Forestry Publishing House: Beijing, China, 1992.
72. Zhu, T.H.; Zhang, W.N.; Gao, Y.Y.; Pei, Z.; Zhao, L.K.; Jiang, T.; Wu, Y.B.; Ni, Z.Y. Two New Tetracyclic Triterpenoids from the Fresh Bark of *Ailanthus altissima*. *Chem. Nat. Compd.* **2023**, *59*, 73–75. [CrossRef]
73. Gürbüz, R.; Kahramanoğlu, I. Possibility of using leaf extracts of tree-of-heaven (*Ailanthus altissima* (Mill.) Swingle) for the postharvest quality preservation of fresh apricot fruits. *Physiol. Mol. Plant Pathol.* **2021**, *113*, 101594. [CrossRef]
74. Kim, S.R.; Park, Y.; Li, M.; Kim, Y.K.; Lee, S.; Son, S.Y.; Lee, S.; Lee, J.S.; Lee, C.H.; Park, H.H.; et al. Anti-inflammatory effect of *Ailanthus altissima* (Mill.) Swingle leaves in lipopolysaccharide-stimulated astrocytes. *J. Ethnopharmacol.* **2022**, *286*, 114258. [CrossRef] [PubMed]
75. Andonova, T.; Muhovski, Y.; Slavov, I.; Vrancheva, R.; Georgiev, V.; Apostolova, E.; Naimov, S.; Mladenov, R.; Pavlov, A.; Dimitrova-Dyulgerova, I. Phenolic profile, antioxidant and DNA-protective capacity, and microscopic characters of *Ailanthus altissima* aerial substances. *Plants* **2023**, *12*, 920. [CrossRef] [PubMed]
76. Bajwa, R.; Farooq, M.; Cheema, Z.A.; Siddique, K.H.M. Allelopathic effects of *Ailanthus altissima* extracts on germination and growth of several crop and weed species. *Plant Growth Regul.* **2012**, *67*, 247–256.
77. Galatowitsch, S.M.; Anderson, N.O.; Ascher, P.D. Invasiveness in wetland plants in temperate North America. *Ecol. Eng.* **1999**, *13*, 145–167. [CrossRef]
78. Novak, M.; Novak, N.; Milinović, B. Differences in allelopathic effect of tree of heaven root extracts and isolated ailanthone on test-species. *J. Cent. Eur. Agric.* **2021**, *22*, 611–622. Available online: https://jcea.agr.hr/en/issues/article/3098 (accessed on 31 July 2023). [CrossRef]
79. Kozuharova, E.; Pasdaran, A.; Al Tawaha, A.R.; Todorova, T.; Naychov, Z.; Ionkova, I. Assessment of the Potential of the Invasive Arboreal Plant *Ailanthus altissima* (Simaroubaceae) as an Economically Prospective Source of Natural Pesticides. *Diversity* **2022**, *14*, 680. [CrossRef]
80. Hopson, C.A.; Natarajan, P.; Shinde, S.; Kshetry, A.O.; Challa, K.R.; Valenciana, A.P.; Nimmakayala, P.; Reddy, U.K. Physiological and Transcriptomic Analysis of *Arabidopsis thaliana* Responses to Ailanthone, a Potential Bio-Herbicide. *Int. J. Mol. Sci.* **2022**, *23*, 11854. [CrossRef]
81. Pavela, R.; Zabka, M.; Tylova, T.; Kresinova, Z. Insecticidal activity of compounds from *Ailanthus altissima* against spodoptera littoralis larvae. *Pak. J. Agric. Sci.* **2014**, *51*, 101–112. Available online: https://inis.iaea.org/search/searchsinglerecord.aspx?recordsFor=SingleRecord&RN=45082669 (accessed on 31 July 2023).
82. Tanaskovic, S.T.; Gvozdenac, S.; Kolarov, R.; Prvulovic, D. Antifeeding and Insecticidal Activity of *Ailanthus altissima* and Morus alba Extracts Against Gipsy Moth (*Lymantria dispar* (L.), Lepidoptera, Lymantridae) Larvae Under Laboratory Conditions. *J. Entomol. Res. Soc.* **2021**, *23*, 197–212. [CrossRef]
83. Lee, K.; Han, B.; Cho, W. The appropriate mounding height and selection of ornamental trees on consideration of the environmental characteristics in an apartment complex. In the case of Sanggyoi-Dong sanitary landfill. *Korean J. Environ. Ecol.* **1997**, *11*, 137–148.
84. Mohebzadeh, F.; Motesharezadeh, B.; Jafari, M.; Zare, S.; Aman, M.S. Remediation of heavy metal polluted soil by utilizing organic amendments and two plant species (*Ailanthus altissima* and *Melia azedarach*). *Arab. J. Geosci.* **2021**, *14*, 1211. [CrossRef]
85. Vasileiou, V.; Barboutis, I.; Kamperidou, V. Properties of Thin 3-PLY Plywood Constructed With Tree-of-Heaven and Poplar Wood. In Proceedings of the International Conference "Wood Science and Engineering in the Third Millennium"—ICWSE 2011, Brasov, Romania, 3–5 November 2011; pp. 323–329. Available online: http://users.auth.gr/jbarb/Publications/VASILEIOU%20ICWSE%202011%20.pdf (accessed on 31 July 2023).

86. Barboutis, I.; Kamperidou, V. Properties of two different thicknesses 3-ply plywood of Tree-Of-Heaven veneers. In Proceedings of the 22nd International Scientific Conference "Wood is Good-EU Preaccession Challenges of the Sector", Zagreb, Croatia, 21 October 2011.
87. Baptista, P.; Costa, A.P.; Simoes, R.; Amaral, M.E. *Ailanthus altissima*: An alternative fiber source for papermaking. *Ind. Crops Prod.* **2014**, *52*, 32–37. [CrossRef]
88. Divya, M.P.; Parthiban, K.T.; Manivasakan, S.; Packialakshmi, M. *Ailanthus excelsa*: An alternate tree species for plywood production. *Pharma Innov. J.* **2022**, *SP-11*, 1184–1187.
89. Li, H.; Fang, G.; Cai, L.; Li, J. Utilization of *Ailanthus altissima* (Mill.) Swingle for particleboard production. *BioResources* **2018**, *13*, 1038–1050.
90. VDOF. *Control and Utilization of Tree of Heaven: A Guide for Virginia Landowners*; Virginia Department of Forestry: Charlottesville, VA, USA, 2009. Available online: https://www.dof.virginia.gov/trees/tree-of-heaven.htm (accessed on 1 June 2022).
91. Zhu, S.; Fang, X.; Wu, Q. Characterization of *Ailanthus altissima* Fiber as a Potential Reinforcement in Polymer Composites. *J. Nat. Fibers* **2017**, *14*, 462–474.
92. Olek, W.; Bonarski, J.T. Texture changes in ther-mally modified wood. *Arch. Metall. Mater.* **2008**, *53*, 207–211. Available online: http://journals.pan.pl/dlibra/jour-nal/97808 (accessed on 31 July 2023).
93. Qian, J.; Gao, J.; He, Z.; Yi, S. Self-shrinking *Ailanthus altissima* substrate obtained by ultrasonic-assisted treatment: Density and pore structure characteristics. *Ind. Crops Prod.* **2022**, *175*, 114221. [CrossRef]
94. Ferreira, P.J.; Gamelas, J.A.; Carvalho, M.G.; Duarte, G.V.; Canhoto, J.M.; Passas, R. Evaluation of the papermaking potential of *Ailanthus altissima*. *Ind. Crops Prod.* **2013**, *42*, 538–542. [CrossRef]
95. Tang, X.; Guo, J.; Yan, Z.; Wu, X.; Wang, S. Kraft pulp production from *Ailanthus altissima*. *J. Wood Sci.* **2015**, *61*, 501–507.
96. Almeida, R.O.; Ramos, A.; Alves, L.; Potsi, E.; Ferreira, P.J.T.; Carvalho, M.G.V.S.; Rasteiro, M.G.; Gamelas, J.A.F. Production of nanocellulose gels and films from invasive tree species. *Int. J. Biol. Macromol.* **2021**, *188*, 1003–1011. [CrossRef]
97. Bučar, G.D.; Gospodarič, B.; Smolnikar, P.; Stare, D.; Kranjc, N.; Prislan, P. Invasive Species as Raw Material for Pellets Production. In Proceedings of the 30th International Conference on Wood Science and Technology—ICWST 2019, Zagreb, Croatia, 12–13 December 2019; "Implementation of wood science in woodworking sector" & 70th Anniversary of Drvna Industrija Journal. pp. 60–68. Available online: https://www.sumfak.unizg.hr/site/assets/files/3845/proceedings_icwst-2019.pdf (accessed on 1 February 2022).
98. Yang, H.; Chen, Q.; Chen, H. Co-firing of *Ailanthus altissima* with coal in a circulating fluidized bed combustor. *Fuel Process. Technol.* **2015**, *131*, 387–393.
99. Hoseini, S.S.; Najafi, G.; Ghobadian, B.; Ebadi, M.T.; Mamat, R.; Yusaf, T. Novel environmentally friendly fuel: The effects of nanographene oxide additives on the performance and emission characteristics of diesel engines fuelled with *Ailanthus altissima* biodiesel. *Renew. Energy* **2018**, *125*, 283–294. [CrossRef]
100. Sahay, S.; Rana, R.S. Hemicellulose hydrolysate from *Ailanthus excelsa* wood potentially fermentable to ethanol. *J. Trop. For. Sci.* **2017**, *29*, 172–178.
101. Li, X.; Chen, Y. Pyrolysis of *Ailanthus altissima* for bio-oil production: Effect of pyrolysis conditions on product yields and characteristics. *Biomass Bioenergy* **2016**, *85*, 167–176.
102. Ighalo, J.O.; Adeniyi, A.G. An In Silico Temperature Sensitivity Study of the Pyrolysis of Beech, *Ailanthus* and *Spruce*. *Eur. J. Sustain. Dev. Res.* **2020**, *4*, em0137. [CrossRef] [PubMed]
103. Liu, Z.; Li, L.; Yuan, X. Anaerobic digestion of *Ailanthus altissima* for biogas production: Kinetics, methane yield, and energy balance. *Energy Sources Part A Recovery Util. Environ. Eff.* **2017**, *39*, 667–672.
104. Pérez, G.; Vilà, M.; Gallardo, B. Potential impact of four invasive alien plants on the provision of ecosystem services in Europe under present and future climatic scenarios. *Ecosyst. Serv.* **2022**, *56*, 101459. [CrossRef]
105. Almeida, M.T.; Mouga, T.; Barracosa, P. The weathering ability of higher plants. The case of *Ailanthus altissima* (Miller) Swingle. *Int. Biodeterior. Biodegrad.* **1994**, *33*, 333–343. [CrossRef]
106. Sladonja, B.; Susek, M.; Guillermic, J. Review on invasive tree of heaven (*Ailanthus altissima* (Mill.) Swingle) conflicting values: Assessment of its ecosystem services and potential biological threat. *Environ. Manag.* **2015**, *56*, 1009–1034. [CrossRef]
107. Vimercati, G.; Kumschick, S.; Probert, A.F.; Volery, L.; Bacher, S. The importance of assessing positive and beneficial impacts of alien species. *NeoBiota* **2020**, *62*, 525–545. [CrossRef]

applied sciences

MDPI

Article

An Overview of the Current Situation of European Poplar Cultures with a Main Focus on Hungary

Szabolcs Komán, Róbert Németh and Mátyás Báder *

Institute of Wood Technology and Technical Sciences, Faculty of Wood Engineering and Creative Industries, University of Sopron, Bajcsy-Zsilinszky 4, 9400 Sopron, Hungary; koman.szabolcs@uni-sopron.hu (S.K.); nemeth.robert@uni-sopron.hu (R.N.)
* Correspondence: bader.matyas@uni-sopron.hu; Tel.: +36-99-518-187

Featured Application: As poplar is a fast-growing tree that can be found all over the world, it is an important raw material for the timber industry and is becoming increasingly important as its use expands and is the subject of ongoing research in many countries. For this reason, it is essential to know the amount of living stock available, which is detailed in this study.

Abstract: Among fast-growing wood species, poplars (*Populus* spp.) can be grown well. Thanks to their valuable wood, they also play an important role in the world. In Europe, regarding poplar forest areas, Hungary is ranked second, where the territorial proportion of poplars has not changed in the past ten years, and currently, two-thirds of them are constituted by two species. The proportion of grey poplar (*P.* × *canescens*) is 42%, while the proportion of 'Pannonia' hybrid (*P.* × *euramericana cv. Pannonia*) is 22%. With regard to gross wood production, they are second after black locust (*Robinia pseudoacacia*), and their standing tree volume is significant for up to 30 years of age. The average age of hybrid poplars is 21 years, while that of the natives is 26 years. Its wood is used by the packaging industry (pallets and boxes) and the furniture industry (700,000 m^3), the plywood industry (200,000 m^3), and the fiber industry (300,000 m^3), but it is also an important source for energetic purposes (150,000 m^3).

Keywords: diameter at breast height; hybrid; native; seedling; standing tree volume; utilization; tree felling volume

Citation: Komán, S.; Németh, R.; Báder, M. An Overview of the Current Situation of European Poplar Cultures with a Main Focus on Hungary. *Appl. Sci.* **2023**, *13*, 12922. https://doi.org/10.3390/app132312922

Academic Editors: Emilia-Adela Salca, Lidia Gurau and Mihaela Câmpean

Received: 15 October 2023
Revised: 25 November 2023
Accepted: 28 November 2023
Published: 2 December 2023

1. Introduction

After World War II, due to the rapid growth of poplars (*Populus* spp.), there was growing interest in this species, and over the past few decades, the poplar genus has gained a unique position in ecology, commercial application, and science [1,2]. The formation of the Hungarian poplar growing situation (Figure 1) is the result of several poplar tree-planting programs following World War II [3]. Poplar cultivation research in the second half of the 20th century clarified the site conditions for the effective cultivation of hybrid poplars, the advisable and optimal cultivation technologies in the conditions in Hungary, and set up a selection of hybrid poplar species corresponding to their abilities. All of these, together with the generally safe timber sales opportunities, made the production of hybrid poplars economical and profitable [4,5]. In addition, breeding has been going on all over Europe ever since [6,7], and poplars seem to be a good model for tree physiology and genomics research [8,9]. This shows that this issue is important across Europe and is still being addressed today.

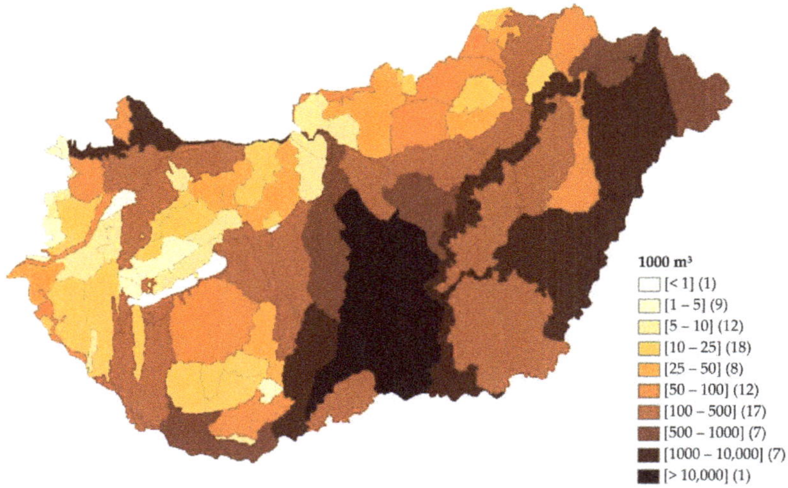

Figure 1. The spread of poplars (*Populus* spp.) in Hungary. The quantities indicated by the colors refer to the whole landscape unit, regardless of its area. The 92 official forest stands in the country shown were defined and used by the National Land Centre. The number of occurrences of the quantity categories is shown in parentheses (source of data: [10]).

Fast-growing plantation tree species, such as poplars, provide raw materials to sectors of the wood industry as substitutes for harvesting from natural forests or imported wood products [5,11]. The sub-sectors are industrial processing (pulp, paper, engineered wood products, plywood, veneer and other boards, saw timber, packing crates, pallets, furniture, and increasingly, bio-energy) and valuable non-wood products (e.g., livestock fodder, medicinal extracts, and food products) [12]. As described by Hafner et al. [13], the use of wood as a raw material will slowly increase in the future, while the use of wood for energetic purposes will increase dramatically. As evidenced, wood used for energetic purposes in Germany already exceeded the wood used for material purposes in 2012, which includes poplars, too. Besides all of these, it is environmentally important that it plays a role in the protection of soil, the regeneration of waste land, the restoration of natural river-bank environments, phytoremediation, and the reduction of the effects of climate change and air pollution [11,14–16]. Their disadvantage is that poplar trees are allergenic, but in the case of industrial plants far from cities, this is not a real problem. One major advantage of poplars and willows (*Salix* spp.) is that they have a relatively short cutting age, i.e., rotation time, to produce maximum yields of timber or fiber on a sustainable basis [8,9]. Taking into account their large volume, future poplar utilization should be based on an integrated and total use concept, i.e., produce the highest-value products or product combinations from each given raw material input so that nothing is wasted [17]. In this case, energetic use is also considered waste because value creation is based on the manufacturing of products.

This article provides an overview of the current data on poplars in Hungary, including the quantity, available in Europe. Among others, it analyzes features such as species variety, age group composition, and areas of use. Although most of these data are widely available, it would require more research to identify them individually, and they are often difficult to find in a comparable format. Thus, this article fills a gap in the information on poplar wood species and is therefore necessary to provide a fresh overview for professionals working in the field.

2. Materials and Methods

The Hungarian data are from the database published by the National Land Centre of the Hungarian Government (Nemzeti Földügyi Központ; NFK) in 2022 [10]. This source includes data for the years between 2011 and 2021. The raw data were classified; averages were calculated, and charts, graphs, and tables were created in Excel (Microsoft Corporation, Redmond, WA, USA). The Hungarian statistics contain data on 119,423 forest stands with poplar trees (exact species, age, diameter, stand area, standing tree volume, ownership, soil type, soil thickness, etc.), which were divided into 92 official forest stands in the country.

The National Statistical Data Collection Programme (Országos Statisztikai Adatfelvételi Program; OSAP) was also used to obtain annual data (standing tree volume and wood consumption) for different tree species [18]. This allowed us to verify, among other things, the accuracy of our results extracted from the detailed database. The results therefore show the statistical status of the forests in Hungary in 2021. The European data are from the International Poplar Commission (IPC) report [12], published officially by the Food and Agriculture Organization (FAO, Rome, Italy). This kind of data separation made it possible to provide the most timely and spatially accurate results, as outlined in the Results and Discussion section.

3. Results and Discussion

Taking into account the poplar stocks in Europe (about 1,064,000 ha), Hungary has a significant proportion, as shown in Figure 2 using the IPC data [12]. With 198,520 hectares of poplar tree-growing areas, it is outperformed only by France [10,12]. Nearly 200,000 hectares alone are greater than the total quantities in other Central and Eastern European countries. Here, the native and hybrid species are present in almost the same quantity, similarly in France. While in Spain we almost find only hybrid poplars, in Bulgaria or Slovenia, the situation is just the opposite.

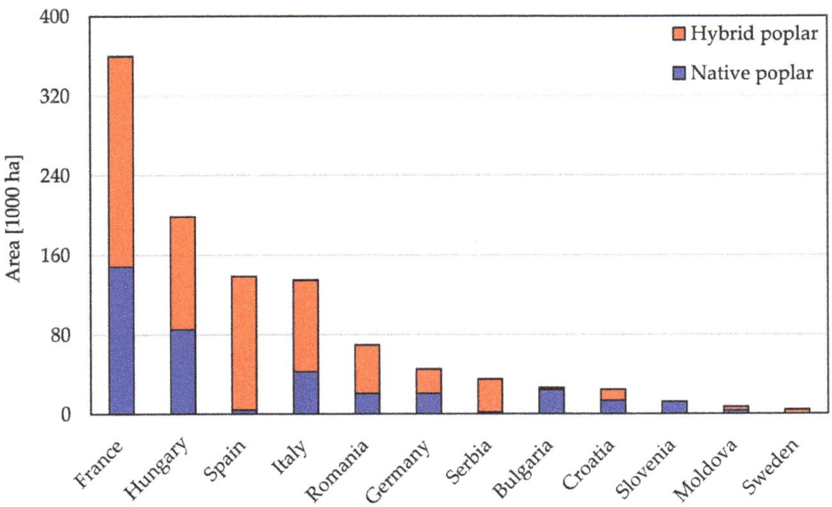

Figure 2. Area of native and hybrid poplar trees (*Populus* spp.) in Europe (based on [10,12]).

The area occupied by poplars in Hungary has remained practically unchanged over the past 10 years. Following sites of oaks and black locust, it has a 10.6% area compared to the total forest area. Considering the area of the countries, the area occupied by poplar trees in Hungary is even more remarkable. Hungary is in first place with 2.18%, followed by Slovenia with 0.59% and France with 0.57%. This can be explained by history, as Hungary lost about two-thirds of its territory in World War I. This largely meant mountainous and forested areas, so the proportion of flat land in the country has increased greatly. Poplar, among other species, was considered to be a suitable tree for these areas, so considerable planting was carried out to reduce the continuing shortage of timber [4].

Looking at the standing tree volume, oaks (*Quercus* spp.), black locust, beech (*Fagus sylvatica*), and pine (*Pinus* spp.) trees outperform the poplars, but their 9.4% volume is still significant in Hungarian forests (Figure 3), which resulted in 37,904,662 m^3 of poplar in 2021.

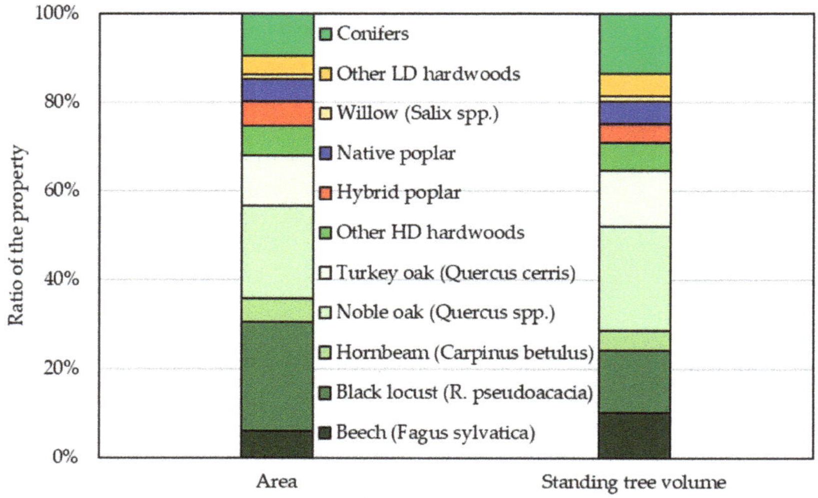

Figure 3. Area and standing tree volume of the tree species groups in Hungary (based on [10]). Abbreviations: HD—high density; LD—low density.

First of all, because, by analyzing the age of the tree species in 10-year intervals in Figure 4, black locust and poplar have the largest area in the first 30 growing years. In the later age groups, however, their ratio significantly decreases, which is due to their short cutting cycle.

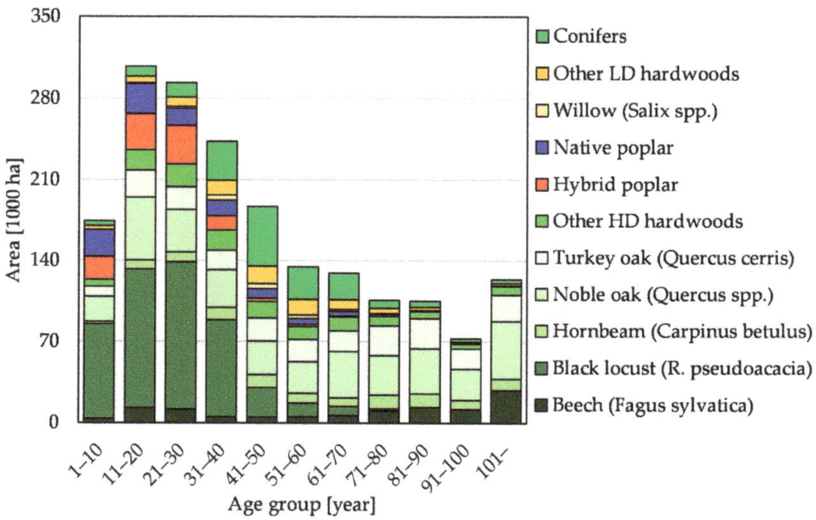

Figure 4. Area of species groups by age classes in Hungary (based on [10]). Abbreviations: HD—high density; LD—low density.

Examining the size of both native and hybrid poplar area (white poplar (*P. alba*), black poplar (*P. nigra*), grey poplar (*P.* × *canescens*), aspen (*P. tremula*)) in Hungary in the last 10 years, it can be seen that while the area of the hybrid poplars decreases by 1–2% per year, the area of the native poplars increases by nearly the same ratio (Figure 5). For the standing tree volume, the amount of hybrid poplars has remained practically unchanged, while the native poplars in the country have seen a 4–5% yearly increase.

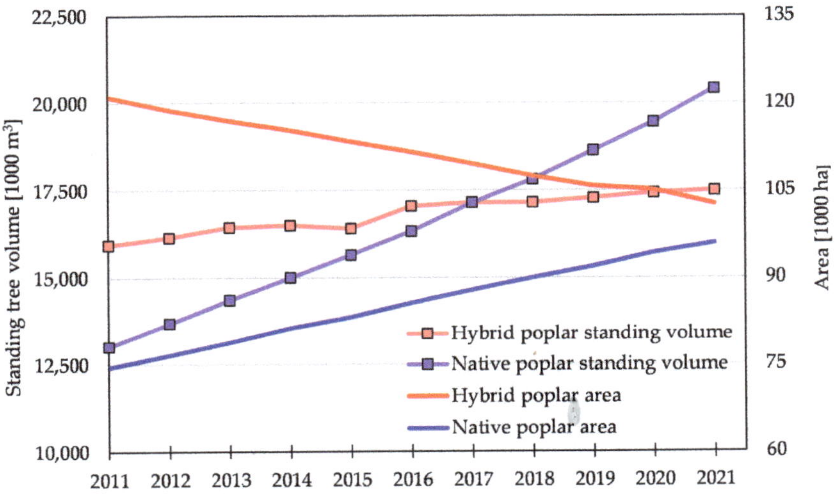

Figure 5. Changes in area and standing tree volume of poplars (*Populus* spp.) in Hungary (based on [10]).

In the last three years of annual production of seedlings, native poplars are in first place (Figure 6). The proportion of all hybrid poplar seedlings is approximately 20%. Although the number of hybrid poplar species recognized by the state is over 20, the number of both 'I-214' and 'Pannonia' hybrid poplar species (*P.* × *euramericana cv. I-214* and *P.* × *euramericana cv. Pannonia*, respectively) is outstanding, with a share of more than 80% among the hybrids.

In accordance with the proportion of the standing tree volume, poplars represent a significant amount of the harvested volume in Hungary (averagely 1400 gross thousand m^3 yearly) since they have been harvested in almost the same amount as black locust and conifers in recent years (Table 1). For a one-hectare annual increment, the native poplars are first, while the hybrid poplars are ranked second.

Considering the area proportions of the different poplars, practically two species constitute two-thirds of the total area (Figure 7). Grey poplar is present in the highest volume, with 42% (76,282 ha), followed by 'Pannonia' (22%—39,975 ha). It is interesting to note that despite being a hybrid (albeit a natural hybrid), grey poplar is widespread over a very large area and, in many cases, is treated as native, while it is an inter-specific combination of *P. alba* and *P. tremula* [6].

Mainly, the native poplar species have the highest average age, as well as some hybrid poplar species that have been available for a longer period of time (Figure 8). With regard to the average diameter at breast height, the situation is not so obvious. This feature is influenced by several factors, such as the conditions at the site, the target range to be produced, the specific growth characteristics of the poplar species, etc. According to Tóth [19], the optimal ecological cutting age of the hybrid poplars in Hungary is 18–20 years, which can, of course, be modified by the aforementioned factors. Presently (data are from 2021), it is about 24 years for the hybrid poplars, while it is about 34 years for the native poplars. It can be seen that some hybrid species, such as the 'Pannonia' and the 'Kopecky'

P. × *euramericana cv. Pannonia* and *P.* × *euramericana cv. Koltay*, respectively, despite the lower average age, have similar average diameters at breast height as some older native or hybrid poplar species. Overall, the average age of the hybrid poplars is 21 years, while the average age of the native poplars is 26 years.

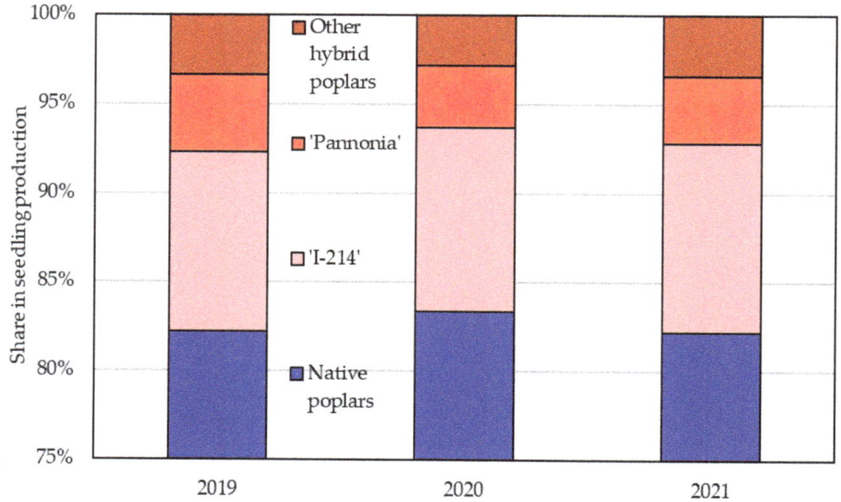

Figure 6. Annual production of poplar (*Populus* spp.) seedlings in Hungary. Note that the diagram shows only the top quartile. The two specified hybrids are *P.* × *euramericana cv. I-214* and *P.* × *euramericana cv. Pannonia* (based on [10]).

Table 1. Total tree felling volume by species groups in Hungary. The bold rows show the summary for the three main categories and finally the total amount (based on [10]).

Species Groups	2019	2020 Gross Thousand m^3	2021	Average Yearly Ratio
Beech (*Fagus sylvatica*)	670	674	683	9.5%
Black locust (*Robinia pseudoacacia*)	1502	1303	1421	19.7%
Hornbeam (*Carpinus betulus*)	248	223	234	3.3%
Noble oak (*Quercus* spp.)	935	892	972	13.1%
Turkey oak (*Quercus cerris*)	751	721	801	10.6%
Other high-density hardwoods	329	319	341	4.6%
High-density hardwoods total	**4435**	**4132**	**4452**	**60.8%**
Hybrid poplar	1174	952	1306	16.0%
Native poplar	250	239	287	3.6%
Willow (*Salix* spp.)	46	28	39	0.5%
Other low-density hardwoods	264	201	241	3.3%
Low-density hardwoods total	**1734**	**1420**	**1873**	**23.5%**
Conifers	**1145**	**1029**	**1198**	**15.7%**
Altogether	**7314**	**6581**	**7523**	**100.0%**

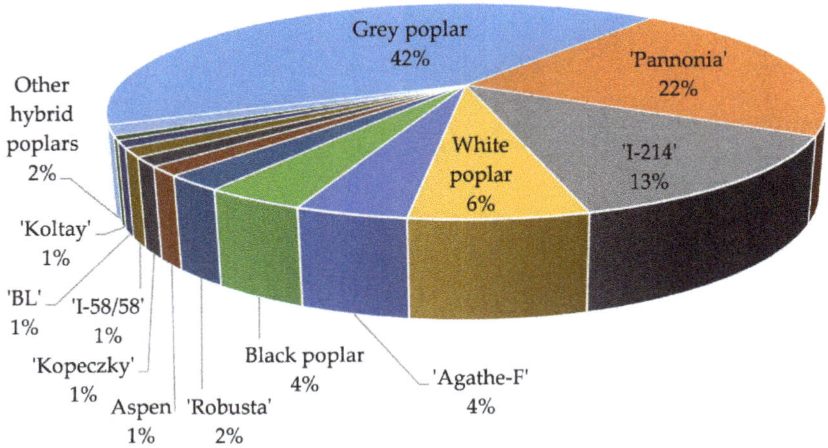

Figure 7. Poplar (*Populus* spp.) species in the Hungarian forests. Scientific names of the species in the same descending order: *P. × canescens; P. × euramericana cv. Pannonia; P. × euramericana cv. I-214; P. alba; P. × euramericana cv. Agathe-F; P. nigra; P. × canadensis cv. Robusta; P. tremula; P. × euramericana cv. Kopecky; P. × euramericana cv. I-58/58; P. × euramericana cv. BL-Costanzo; P. × euramericana cv. Koltay* (based on [10]).

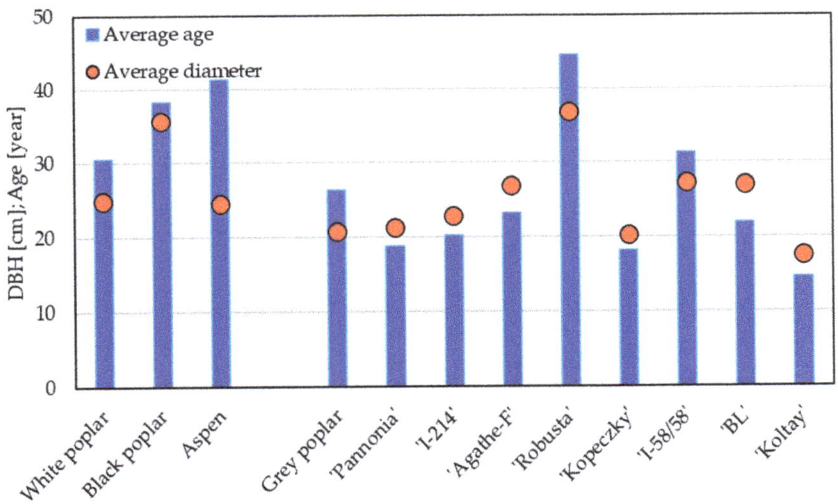

Figure 8. Average age and diameter at breast height (DBH) of Hungarian poplars (*Populus* spp). The species are shown on the diagram in order of their abundance: natives in the first group, and hybrids in the second group. Latin names of the species listed in the same order: *P. alba; P. nigra; P. tremula; P. × canescens; P. × euramericana cv. Pannonia; P. × euramericana cv. I-214; P. × euramericana cv. Agathe-F; P. × canadensis cv. Robusta; P. × euramericana cv. Kopecky; P. × euramericana cv. I-58/58; P. × euramericana cv. BL-Costanzo; P. × euramericana cv. Koltay* (based on [10]).

Use of Poplar

The use of the wood of poplars can be grouped into the following professional areas. In parentheses are the amount of hybrid poplar, the amount of native poplar, and the ratio of these two compared to the total amount of logs harvested in 2022 in Hungary, respectively, based on forestry data [10]:

- Veneer log (205,967 m^3; 8667 m^3; 95%).

- Sawlog (512,458 m^3; 78,119 m^3; 50%).
- Other logs used by sawmills (118,961 m^3; 27,413 m^3; 40%).
- Logs for fiberboard production (137,967 m^3; 41,870 m^3; 29%).
- Logs for paper production (63,385 m^3; 47,575 m^3; 49%).
- Industrial wood chips and their raw material (roundwood equivalent) (57,610 m^3; 90 m^3; 79%).
- All other industrial log assortments (50,492 m^3; 7640 m^3; 28%).
- Logs for energetic purposes (107,562 m^3; 47,103 m^3; 3%).

The poplars, whose wood can be used in various ways, provide the most important raw material for mass wood production (boxes, pallets, plywood, etc.). The largest part of the sawmill industry's purpose is the production of pallet elements. In the furniture industry, upholstery frames, panels, and other hidden components are produced from poplar, but they are also used to make log furniture. The poplars provide in Hungary the most important raw material for match production.

The possibility of the use of hybrid poplars is the same as that of the native poplars, but their advantage is that their properties can be improved by genetic modification [20,21]. Due to their poor mechanical strength, they are used less as building materials. For their broader use, it is important to improve their physical and mechanical properties [22]. Carrying out such procedures, e.g., heat treatment, several researchers [23–26] have also achieved encouraging results. In most cases, the aim of breeding is to increase the wood yield, but in many cases, the shape of the trunk is an important aspect. Unfortunately, in general, breeders usually pay little attention to improving the physical and mechanical properties. For this reason, the density, compressive strength, tensile strength, and bending modulus of elasticity of native poplars are typically among the highest of all poplars, and in most cases, they are definitely higher than those of hybrids [27].

From the point of view of biomass utilization, poplar trees are widely grown in many countries and used by applying different technologies. Wood from the short-rotation plantation of poplars has economic and environmental benefits compared to coal and other fossil fuels [28]. Poplar short-rotation plantations are one of the most important biomass resources for energetic purposes [29], which occupy approximately 50,000 hectares of land in Europe [30]. In the last decade, in Hungary, short rotations (10–15 years) and even more intensive cultivation hybrid poplars have appeared in industrial wood plantations; their territory is currently approaching 5000 hectares.

The utilization of poplars as biomass is their least value-added use. It would be much better to give more opportunities to this species in industrial processing. This would solve a number of problems, because on the forestry side, poplar logs can be produced with significantly higher value, and on the timber industry side, it would be easier to manage the growing demand for timber worldwide. In addition, plantation wood production methods and technologies can not only provide a solution to the growing demand for timber but also reduce the economic pressure on natural forests. Traditional forestry seeks to maintain near-natural conditions, and essentially, native species provide the biological basis for timber production. Plantation tree growing, on the other hand, focuses on the use of improved varieties that can be used in a cultivation technology environment adapted from intensive agriculture. For cultivated varieties, it is also necessary to assess the production values to determine the possible locations and production technologies for the new varieties and to determine the economic yield (wood yield) that can be expected while the technological conditions are maintained.

In summary, the timber industry needs poplar plantations with longer cutting cycles and trunks with a good shape to reduce the proportion of juvenile wood and wood defects, which are the basic requirements for industrial timber. After a lot of breeding and material testing work, substitutes for conifers could even be produced, as the area of the latter is predicted to decline significantly in Europe over the next century [31].

4. Conclusions

There is a growing demand for wood in Europe. Besides its use as an industrial raw material for wood products, its use for energetic purposes is also increasing (e.g., 50,000 hectares of land in Europe). The wide spread of poplars (*Populus* spp.) is mainly due to their rapid growth, short rotation period, and the various usability of their wood. Thanks to these characteristic features, they have become an important source of raw material for the forestry and wood industry in Hungary; therefore, there is already a significant standing tree volume quantity in Hungary (37,904,662 m^3 on 198,520 ha) regarding European parameters (1,064,000 ha). Over the last decade, their territorial share has practically remained unchanged, and the proportion of hybrid poplars is somewhat higher than that of the native poplars. Two species constitute two-thirds of the total area: the grey poplar (76,282 ha) and the 'Pannonia' hybrid with 39,975 ha (*P.* × *canescens* and *P.* × *euramericana cv. Pannonia*, respectively). Most of the available quantity is 10–30 years old due to the short rotation period of the poplars. Regarding their use, they provide materials in large quantities to produce pallets and plywood. Considering the examined period, the last decade, it can be stated that poplar trees are important raw materials for the Hungarian wood industry. This statement is also true for Europe, as evidenced by several ongoing breeding programs.

Author Contributions: Conceptualization, S.K.; methodology, S.K.; software, S.K. and M.B.; validation, S.K., M.B. and R.N.; formal analysis, S.K.; investigation, S.K. and M.B.; resources, R.N.; data curation, S.K.; writing—original draft preparation, S.K.; writing—review and editing, M.B.; visualization, S.K. and M.B.; supervision, M.B. and R.N.; project administration, R.N.; funding acquisition, R.N. All authors have read and agreed to the published version of the manuscript.

Funding: This article was made in frame of the project TKP2021-NKTA-43, which has been implemented with the support of the Ministry of Culture and Innovation of Hungary from the National Research, Development, and Innovation Fund, financed under the TKP2021-NKTA funding scheme.

Institutional Review Board Statement: Not applicable.

Informed Consent Statement: Not applicable.

Data Availability Statement: Restrictions apply to the availability of these data. Data was obtained from National Land Centre of Hungary and are available on individual request from the National Land Centre of Hungary.

Acknowledgments: We express our thanks to Maximilián Cziczer for preparing the first figure.

Conflicts of Interest: The authors declare no conflict of interest. The funders had no role in the design of the study; in the collection, analyses, or interpretation of data; in the writing of the manuscript; or in the decision to publish the results.

References

1. du Cros, E.T. Breeding Strategies with Poplars in Europe. *For. Ecol. Manag.* **1984**, *8*, 23–39. [CrossRef]
2. Stobrawa, K. Poplars (*Populus* spp.): Ecological Role, Applications and Scientific Perspectives in the 21st Century. *Balt. For.* **2014**, *20*, 204–213.
3. Schiberna, E.; Borovics, A.; Benke, A. Economic Modelling of Poplar Short Rotation Coppice Plantations in Hungary. *Forests* **2021**, *12*, 623. [CrossRef]
4. Tóth, B. *Description on the Hybrid Poplar Species (Nemesnyár-Fajták Ismertetője)*; Agroinform Kiadó: Budapest, Hungary, 2006.
5. Marogel-Popa, T.; Marcu, M.V.; Borz, S.A. Postural Risk in Manual Planting Operations of Poplar: Two Options Compared. *Sustainability* **2020**, *12*, 5531. [CrossRef]
6. Stanton, B.J.; Neale, D.B.; Li, S. Populus Breeding: From the Classical to the Genomic Approach. In *Genetics and Genomics of Populus*; Jansson, S., Bhalerao, R., Groover, A., Eds.; Springer: New York, NY, USA, 2010; pp. 309–348, ISBN 978-1-4419-1540-5.
7. Biselli, C.; Vietto, L.; Rosso, L.; Cattivelli, L.; Nervo, G.; Fricano, A. Advanced Breeding for Biotic Stress Resistance in Poplar. *Plants* **2022**, *11*, 2032. [CrossRef] [PubMed]
8. Rosso, L.; Cantamessa, S.; Bergante, S.; Biselli, C.; Fricano, A.; Chiarabaglio, P.M.; Gennaro, M.; Nervo, G.; Secchi, F.; Carra, A. Responses to Drought Stress in Poplar: What Do We Know and What Can We Learn? *Life* **2023**, *13*, 533. [CrossRef]
9. Shi, Y.-J.; Mi, J.-X.; Huang, J.-L.; He, F.; Chen, L.-H.; Wan, X.-Q. Study on the Origin and Classification of Two Poplar Species on the Qinghai–Tibet Plateau. *Forests* **2023**, *14*, 1003. [CrossRef]

10. NFK (National Land Centre). *Hungarian Forest Related Data (Magyarország Erdeivel Kapcsolatos Adatok)*; Nemzeti Földügyi Központ (National Land Centre): Budapest, Hungary, 2022.
11. Zhang, Y.; Tian, Y.; Ding, S.; Lv, Y.; Samjhana, W.; Fang, S. Growth, Carbon Storage, and Optimal Rotation in Poplar Plantations: A Case Study on Clone and Planting Spacing Effects. *Forests* **2020**, *11*, 842. [CrossRef]
12. FAO (Food and Agriculture Organization). *Poplars and Other Fast-Growing Trees—Renewable Resources for Future Green Economies. Synthesis of Country Progress Reports*; 25th Session of the International Poplar Commission; Forestry Policy and Resources Division, Food and Agriculture Organization: Rome, Italy, 2016.
13. Hafner, A.; Ott, S.; Bodemer, E.; Winter, S. A Case Study for End of Life Reuse and Recycling Survey Methodologies: The Höllentalanger Cottage. *J. Civ. Eng. Arch.* **2014**, *8*, 1211–1220. [CrossRef]
14. Nervo, G.; Coaloa, D.; Vietto, L.; Giorcelli, A.; Allegro, G. Current Situation and Prospects for European Poplar Culture: The Role of Research; Actas del Tercer Congreso Internacional de las Salicáceas en Argentina 'Los álamos y lossaucesjunto al paisaje y eldesarrolloproductivo de la Patagonia'; Neuquen, Argentina, 2011, p. 9. Available online: https://www.academia.edu/7390321/Tercer_Congreso_Internacional_de_Salic%25C3%25A1ceas_en_Argentina (accessed on 21 November 2022).
15. Cantamessa, S.; Rosso, L.; Giorcelli, A.; Chiarabaglio, P. The Environmental Impact of Poplar Stand Management: A Life Cycle Assessment Study of Different Scenarios. *Forests* **2022**, *13*, 464. [CrossRef]
16. Pucher, C.; Erber, G.; Hasenauer, H. Europe's Potential Wood Supply by Harvesting System. *Forests* **2023**, *14*, 398. [CrossRef]
17. Balatinecz, J.; Mertens, P.; de Boever, L.; Yukun, H.; Jin, J.; Van Acker, J. Properties, Processing and Utilization. In *Poplars and Willows: Trees for Society and the Environment*; Isebrands, J., Richardson, J., Eds.; CABI Books: Wallingford, UK, 2014; pp. 527–561. [CrossRef]
18. KSH (Hungarian Central Statistical Office) National Statistical Data Collection Programme (Országos Statisztikai Adatfelvételi Program; OSAP). Available online: https://www.ksh.hu/osap (accessed on 21 November 2022).
19. Tóth, B. Plantation-grown of hybrid poplars (Nemesnyárasok ültetvényszerű fatermesztése). In *Ültetvényszerű Fatermesztés I.*; Führer, E., Rédei, K., Tóth, B., Eds.; Mezőgazda Kiadó—ERTI: Budapest, Hungary, 2003; pp. 19–96, ISBN 963-286-013-6.
20. Rathke, J.; Sinn, G.; Harm, M.; Teischinger, A.; Weigl, M.; Müller, U. Effects of Alternative Raw Materials and Varying Resin Content on Mechanical and Fracture Mechanical Properties of Particle Board. *BioResources* **2012**, *7*, 2970–2985. [CrossRef]
21. Benke, A.; Cseke, K.; Borovics, A. Benke A, Cseke K, Borovics A Dunántúli Leuce nyár populációk genetikai vizsgálata RAPD és cpDNS markerekkel. *Erdészettudományi Közlemények* **2011**, *1*, 83–94.
22. Wang, J.; Guo, X.; Zhong, W.; Wang, H.; Cao, P. Evaluation of Mechanical Properties of Reinforced Poplar Laminated Veneer Lumber. *BioResources* **2015**, *12*, 43–55. [CrossRef]
23. Bak, M. The Effect of Oil-Heat-Treatment on Some Major Properties of Poplar Wood (Növényi Olajokban Hőkezelt Nyár Faanyag Tulajdonságainak Vizsgálata). Ph.D. Thesis, University of West-Hungary, Sopron, Hungary, 2012.
24. Cai, J.B.; Ding, T.; Yang, L. Dimensional Stability of Poplar Wood after Densification Combined with Heat Treatment. *Appl. Mech. Mater.* **2012**, *152–154*, 112–116. [CrossRef]
25. Gao, H.; Sun, M.-Y.; Cheng, H.-Y.; Gao, W.-L.; Ding, X.-L. Effects of Heat Treatment under Vacuum on Properties of Poplar. *BioResources* **2015**, *11*, 1031–1043. [CrossRef]
26. Marcon, B.; Goli, G.; Matsuo-Ueda, M.; Denaud, L.; Umemura, K.; Gril, J.; Kawai, S. Kinetic Analysis of Poplar Wood Properties by Thermal Modification in Conventional Oven. *iForest* **2018**, *11*, 131–139. [CrossRef]
27. Molnár, S.; Farkas, P. Hybrid poplars (Euramerican poplars)—Populus x euramericana (Nemes nyár (euramerikai nyárak)—Populus x euramericana). In *The World's Industrial Wood Species (Földünk Ipari Fái)*; Molnár, S., Farkas, P., Börcsök, Z., Zoltán, G., Eds.; Erfaret Nonprofit Kft.: Sopron, Hungary, 2016; pp. 98–99, ISBN 978-963-12-5239-2.
28. Klasa, A.; Karlen, D. Poplar. In *Cellulosic Energy Cropping Systems*; Karlen, D., Ed.; U.S. Department of Agriculture: Agricultural Research Service: Lincoln, NE, USA, 2014; pp. 183–199.
29. Berndes, G.; Hoogwijk, M.; van den Broek, R. The Contribution of Biomass in the Future Global Energy Supply: A Review of 17 Studies. *Biomass Bioenergy* **2003**, *25*, 1–28. [CrossRef]
30. Lindegaard, K.N.; Adams, P.W.R.; Holley, M.; Lamley, A.; Henriksson, A.; Larsson, S.; von Engelbrechten, H.-G.; Esteban Lopez, G.; Pisarek, M. Short Rotation Plantations Policy History in Europe: Lessons from the Past and Recommendations for the Future. *Food Energy Secur.* **2016**, *5*, 125–152. [CrossRef] [PubMed]
31. Lindner, M.; Maroschek, M.; Netherer, S.; Kremer, A.; Barbati, A.; Garcia-Gonzalo, J.; Seidl, R.; Delzon, S.; Corona, P.; Kolström, M.; et al. Climate Change Impacts, Adaptive Capacity, and Vulnerability of European Forest Ecosystems. *For. Ecol. Manag.* **2010**, *259*, 698–709. [CrossRef]

applied sciences

MDPI

Article

Differences and Similarities between the Wood of Three Low-Density and Homogenous Species: Linden, Balsa, and Paulownia

Anamaria Avram, Aurel Lunguleasa *, Cosmin Spirchez and Constantin Stefan Ionescu

Department of Wood Processing and Design of Wooden Products, Transilvania University of Brasov, 500036 Brasov, Romania; anamaria.avram@unitbv.ro (A.A.); cosmin.spirchez@unitbv.ro (C.S.); ionescu.constantin.stefan@unitbv.ro (C.S.I.)
* Correspondence: lunga@unitbv.ro

Abstract: The use of woods with a low density and good structural uniformity has intensified in recent decades, paving new ways for their superior valorization. This research study aimed to examine the wood of three deciduous wood species with low densities and great uniformity of structure, namely linden wood, with an oven-dried density of 461 kg/m^3, paulownia wood, with an oven-dried density of 304 kg/m^3, and balsa wood, with an oven-dried density of 118 kg/m^3. The physical and mechanical properties of these species were studied using various methodologies. The obtained results show that, although they have significant differences in terms of densities and mechanical characteristics, the coloristic properties of the three analyzed species obtained using the CIELab are almost similar. As a general conclusion, based on all the properties found in this research, it can be concluded that paulownia wood is more appropriate than linden or balsa for use, being able to replace linden in its uses.

Keywords: linden; paulownia; balsa; low density; structural uniformity

Citation: Avram, A.; Lunguleasa, A.; Spirchez, C.; Ionescu, C.S. Differences and Similarities between the Wood of Three Low-Density and Homogenous Species: Linden, Balsa, and Paulownia. *Appl. Sci.* **2023**, *13*, 10209. https://doi.org/10.3390/app131810209

Academic Editor: Victor Franco Correia

Received: 8 August 2023
Revised: 22 August 2023
Accepted: 24 August 2023
Published: 11 September 2023

1. Introduction

The wood of different wood species differs greatly from one species to another. Even within the same genus and family, there is a diversity of different wood species. Going further, even within the same wood species, its wood can differ depending on the type of soil, climatic conditions, geographical area, etc. For the majority of wood-processing technologies, and in finished or semifinished products, a homogeneity of the wood species used is needed [1,2]; otherwise, the produced products will have major deficiencies or will have reduced marketability. This homogeneity can be characterized from a structural point of view, including density [3,4], reaction to atmospheric humidity or water, color [5] or the color difference between the sapwood and heartwood (mature wood) [6], the presence/absence of defects, resistances, and surface roughness [3], with all of these depending on the processing technology used and the final product that is obtained. It is known that the wood of the same wood species has a different structure. There are differences between the sapwood and heartwood, the annual rings of early and late wood, or the type of deciduous wood species. From this point of view, the most uniform species are those with small and uniform distributed pores, such as linden wood. There are fields in which only homogeneous woods are used due to variation in their density and structure [1,7], such as when casting molds in the shoe industry, in the textile industry, and in manual sculptures, and when casting molds of alloys of small shapes in the foundry industry.

In the field of manufacturing wood-based composite boards, such as LVL [8], wood species with appropriate densities are needed; only in this way can chips be compatible with and compacted according to the technology used in this field. Also, the color of the wood species used in mixture chips, flakes, or chops must be similar and uniform; otherwise, the boards obtained will have uneven colors or a reduced esthetic appearance. In the field of restoration of cultural heritage objects with wooden supports, it is recommended to use the same wood species, especially species of a density equivalent to that found in the degraded object to be restored [9–11]. Historically, cultural heritage objects were mainly made of linden wood due to the structural homogeneity of this species, as well as the homogeneity of its shrinkage/swelling in the two structural directions (the ratio of tangential shrinkage to radial shrinkage). When it is restored, the wood in degraded heritage objects has a low density, with the mass losses being 25–50% depending on the degree of degradation. This is why the replaced wood material must have a density equivalent to that of the object to be restored (much smaller than the initial one); otherwise, the product would be heavier and additional stresses would be induced. Consequently, if linden wood is used for restoration, it should be artificially degraded or torrefied to reduce its density to that appropriate for the wood in the restored object, or to use another species with a lower density, such as balsa wood or paulownia wood [12,13]. In the field of veneers and timbers, species that are very closed in terms of density, texture, pattern, and color are considered uniform, even with similarities inside the annual rings [4,8]. The investigation methods of wood homogeneity were different from one author to another, and included the use of X-rays [4] or a combination of light transmittance and mechanical stiffness [8]. The analyzed materials included bamboo veneers and LVL [8], white fir, Aleppo pine, European beech, walnut, and white oak [3].

Objectives. This paper aims to analyze the uniformity of the wood of three homogeneous wood species, namely linden, balsa, and paulownia, from a structural and coloristic point of view, with respect to their physical–mechanical properties, and their relationship with water (water absorption and shrinkage). In this study, linden wood is taken as the element of comparison, emphasizing the specific properties of paulownia wood, which is promoted as the replacement for linden wood in all possible uses. This paper begins with the premise that linden and balsa woods are unsuitable species for replacing very old wood from heritage objects, and that paulownia wood could be a much more advantageous solution.

2. Materials and Methods

Anatomical structure. The first element of the analysis of the homogeneity or differences between the three wood species (linden, balsa, and paulownia) was their anatomical structure, including a study of the cross-sections of these three deciduous species, and their inclusion in the category of species with typically annular or uniformly scattered pores [1,6,7]. For this purpose, three cross-sections of 20 mm × 20 mm of the analyzed species were taken, as shown in Figure 1.

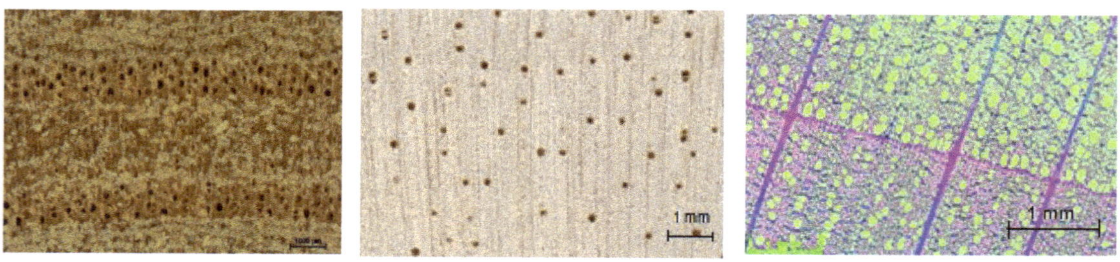

(a) (b) (c)

Figure 1. Microscopic cross-sectional images of paulownia (**a**), balsa (**b**), and linden (**c**).

Unit density. The density of the three wood species was determined as mass per unit volume, as defined by international standards in the field [14,15]. Because wood is found at different moisture content levels, for a good comparison of the three wood species, both the mass and the volume of 30 mm × 20 mm × 20 mm samples were determined at the same moisture content of 0% (oven-dried state). For this purpose, the samples were dried for 24 hours in an oven with a 105 °C temperature. Taking into account the volume of the rectangular parallelepiped, under which the samples' density was determined, the wood density was calculated as follows (Equation (1)):

$$\rho_0 = \frac{m}{l \cdot b \cdot g} \cdot 10^6 \left[\mathrm{kg/m^3}\right] \quad (1)$$

where ρ_0 is the density at a wood moisture content of 0%; m is the mass of the wooden specimen, in g; l is the length of the specimen, in mm; b is the width of the specimen, in mm; and g is the thickness of the specimen, in mm.

Ten samples were used for this test.

The CIEL*a*b* color of the 3 species. Immediately after cutting its log into lumber, paulownia wood darkens in color to a light gray. This coloring is observable on a small thickness of about 0.7–0.8 mm due to the oxidation of wood sap when it comes into contact with atmospheric air [16–18]. This blackish color is kept as long as the wood lumber is not processed. The dark color will disappear when the timber surface is planned to a thickness of at least 0.7–0.8 mm, obtaining the real and beautiful color of paulownia wood surface [19–21].

A portable colorimeter of type Tes-135A, manufactured by Tess Electrical, Electronic Corp. (Taipei, Taiwan), was used for the color determination, using the CIEL*a*b* system. The CIEL*a*b* color space is one of the multiple color spaces (alongside CIEXYZ, CMYK, NCS, RGB, etc.) that was defined by the International Commission on Illumination (abbreviated as CIE) in 1976, which makes possible the transition from visual appreciation of colors to numerical quantification. It is based on human perception of colors as opposing colors, in which the group of red with green and the group of blue with yellow are usually opposing colors (Figure 2).

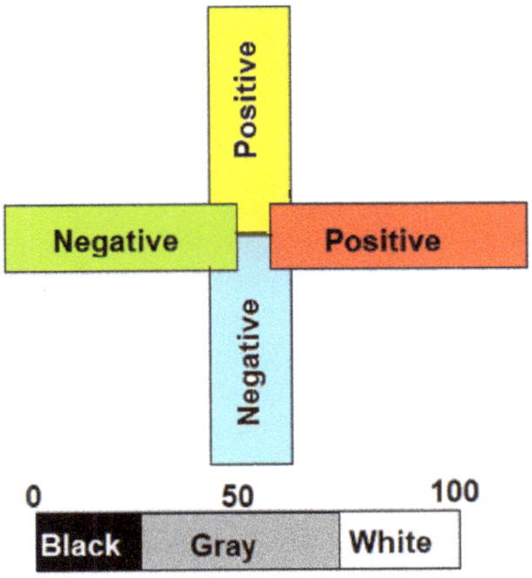

Figure 2. Coloristic space CIEL*a*b*.

This is why the CIEL*a*b* color space is quantified by three distinct parameters: L*, a*, and b*. In this color space, the letter "L*" represents lightness and has a value of 0 for black and 100 for white, with values in between being a series of shades of gray. The "a*" axis refers to the green–red opposition, with negative values toward green and positive values toward red. The "b*" axis quantifies the blue–yellow opposition, with negative values toward the blue zone and positive values toward the yellow zone [5,18].

For the color study, the three radial sections shown below were used (Figure 3).

(a) (b) (c)

Figure 3. Radial sections for color determination: (**a**) *Paulownia tomentosa*; (**b**) balsa (*Ochroma pyramidale*); and (**c**) linden (*Tilia platyphyllos*).

From Figure 3, it can be seen visually that, roughly, the three species are similar in color, with paulownia wood being a little toward brown, and linden wood being a little toward reddish.

Water absorption after two hours of immersion. Water absorption after immersion in water for 2 h was determined for the absolutely dry samples, which were kept in the oven at 105 °C for 6 h [22–25]. The linden, balsa, and paulownia samples had the average dimensions of 20 mm × 20 mm × 30 mm, where the dimension of 30 mm was their length. The mass after immersion and the mass after drying in the oven were determined for the three wood species. Based on these masses, water absorption (*WA*) was determined according to the following relationship (Equation (2)):

$$WA = \frac{m_w - m_0}{m_0} \cdot 100\ [\%] \quad (2)$$

where m_w is the mass of the immersed specimen, in g, and m_0 is the mass of the oven-dried test piece, in g.

Seven specimens were used for this test.

Shrinkage in the radial/tangential direction and volume. Wood shrinkage was performed on the samples of the same types and dimensions as in the case of absorption. These samples were cut in such a way so that the radial and tangential directions were clearly identified [26,27]. In this test, with the help of an electronic caliper, the initial dimensions in the wet state and the final ones in the absolutely dry state were determined. Based on the masses of the samples in the wet and dry state, the wood moisture content was also determined.verifying whether the requirement of a certain moisture content above the fiber saturation point (over 30% moisture content) was fulfilled. If the moisture content of the samples did not touch 30% (Figure 4), the necessary correction was performed; the value for 1% moisture content was determined, and then, this value was multiplied by 30%.

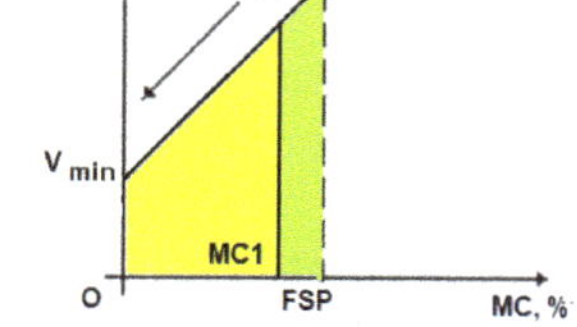

Figure 4. Wood shrinkage diagram: MC—moisture content, in %; FSP—fiber saturation point, in %; V_{max}—the maximum volume of wood, in cm^3; V_{min}—the minimum volume of wood, in cm^3; MC1—moisture content under FSP, in %; V1—volume of wood for moisture content MC1, in cm^3.

The total shrinkage (throughout the saturation moisture range) in the radial, tangential, longitudinal, and volumetric directions was determined based on the dimensions before and after drying according to the following specific relationships (Equations (3) and (4)):

$$S_{tl} = \frac{l_{max} - l_{min}}{l_{max}} \cdot 100\ [\%]; S_{tt} = \frac{t_{max} - t_{min}}{t_{max}} \cdot 100\ [\%] \tag{3}$$

$$S_{tr} = \frac{r_{max} - r_{min}}{r_{max}} \cdot 100\ [\%]\ S_{tv} = \frac{V_{max} - V_{min}}{V_{max}} \cdot 100\ [\%] \tag{4}$$

where S_{tl} is the total longitudinal shrinkage, in %; l_{max} is the maximum length of the specimen, after immersion in water, in mm; l_{min} is the minimum length of the sample, after drying in the oven, in mm; S_{tt} is the total tangential shrinkage, in %; t_{max} is the maximum thickness (on thickness) of the specimen, after immersion in water, in mm; t_{min} is the minimum thickness of the sample, after drying in the oven, in mm; S_{tr} is the total radial shrinkage, in %; r_{max} is the maximum radial of the specimen, after immersion in water, in mm; r_{min} is the minimum radial of the sample, after drying in the oven, in mm; S_{tv} is the total volumetric shrinkage, in %; V_{max} is the maximum volume of the specimen, after immersion in water, in mm; and l_{min} is the minimum volume of the sample, after drying in the oven, in mm.

The partial volumetric shrinkage (only in a portion of the moisture fiber saturation), such as from the saturation point of the fiber to a certain MC1 value (Figure 4), can be determined according to the following equation (Equation (5)):

$$S_{pv} = \frac{V_{max} - V_1}{V_{max}} \cdot 100\ [\%] \tag{5}$$

where S_{pv} is the partial volumetric shrinkage, in %, and V_1 is the volume of wood corresponding to the MC1.

Modulus of resistance and modulus of elasticity to static bending conforming to ISO 13061-4: 2014 [28]. Before testing, the samples were conditioned to obtain an average moisture content of 12%. The modulus of resistance (MOR) and modulus of elasticity (MOE) to static bending were determined according to the following general relationships (Equations (6) and (7)):

$$\sigma_i = \frac{3 \cdot F_{max} \cdot l}{2 \cdot b \cdot t^2} \left[\frac{\mathrm{N}}{\mathrm{mm}^2}\right] \tag{6}$$

$$E_m = \frac{l_1^3 \cdot (F_1 - F_2)}{4 \cdot b \cdot t^3 \cdot (a_2 - a_1)} \left[\frac{\mathrm{N}}{\mathrm{mm}^2}\right] \tag{7}$$

where F_{max} is the maximum force of breaking, in N; l is the distance between supports, in mm; b is the width of the specimen, in mm; t is the thickness of the specimen, in mm; F_1 is the value of force at 10% from the maximum force, in N; F_2 is the value of force at 40% from the maximum force, in N; a_1 is deformation for force F_1, in mm; and a_2 is deformation for force F_2.

Ten samples were tested in order to obtain statistically significant values.

Compressive strength conforming to ISO 13061-17 [28,29]. Before testing, the samples were conditioned in order to have up to 12% moisture content. The compressive strength parallel to the grain used in the work was obtained in the moment of breaking when the layers of wood in the samples were sheared or crushed, and the force dropped suddenly.

Therefore, the resistance to compression parallel to the wood grain was determined as the ratio between the maximum breaking force and the area of the breaking surface (Equation (8)):

$$\sigma_c = \frac{F_{max}}{b_1 \times b_2} \left[\frac{\mathrm{N}}{\mathrm{mm}^2}\right] \tag{8}$$

where σ_c is the compressive strength, in N/mm^2; F_{max} is the maximum force of breakage, in N; and b_1 and b_2 are the widths of the specimen in tangential and radial directions, in mm.

Ten replicates of each type of specimen were used for this test.

Brinell hardness conforming to EN 1534:2003 [30]. Brinell hardness quantifies the resistance of a wood surface when a metal ball tries to penetrate and destroy it. The punch that is pressed onto the surface of the wood is created by a ball with a diameter of 10 mm, and, to highlight the diameter of the mark, it is necessary to arrange a copy paper between the punch and the wooden surface. This resistance is higher on the transverse section, compared to the longitudinal–radial and longitudinal–tangential sections, but, due to the fact that the transverse section is less used in industrial practice of wood processing, only the radial and tangential sections are used in research. Brinell hardness was calculated as follows (Equation (9)):

$$HB = \frac{2 \times P}{\pi \times D(D - \sqrt{D^2 - d^2})} \left[\frac{N}{mm^2}\right] \quad (9)$$

where P is the applied force, related to wood density, in N; D is the diameter of the punch ball, in mm; and d is the diameter of the trace left on the wood, in mm.

Ten valid samples were used for testing each wood species.

Ecological aspects of the three species. All forests, through the trees that they have in their area, have several functions, among which are hydrological, soil protection, esthetic, recreational, ecological, and production functions [31]. The ecological function of forests is related to physiology, namely the fact that humans need a certain amount of oxygen for breathing. Oxygen is released into the atmosphere by forest vegetation (about 1.3 t of oxygen per ton of woody mass, in a day) through its chlorophyll assimilation process. A mature tree produces about 1.7 kg of oxygen/day, which means the oxygen requirement of a human for three days [32]. Also, in the same process of tree photosynthesis, a large amount of carbon dioxide is consumed, approximatively 1.8 t per one cubic meter of wood. On hot days, a hectare of a highly productive forest consumes an average of 220–280 kg of carbon dioxide per day. Wood production of the three forestry species in this study is different: linden at 350 m^3/ha, balsa at 550 m^3/ha, and paulownia at 950 m^3/ha. The volume of timber obtained after the first 3–4 years of paulownia is 0.3–0.8 $m^3/tree$, i.e., 200–500 m^3/ha. The medium age of exploitation is 75 years for linden, 50 years for balsa, and 15 years for paulownia [33,34]. To establish the ecological factor for all three species, the unitary coefficient for linden forests was considered, which was used as a reference for comparison with the other two forest species (paulownia and balsa).

The comparative analysis of the three species. The method of comparative analysis of the 3 wood species took linden wood as the element of comparison because it is currently considered the most homogeneous species, especially from the point of view of the ratio of tangential shrinkage to radial shrinkage. Several analysis criteria were used, namely anatomical structure, density, resistance, color, shrinkage and swelling, and the ratio of tangential shrinkage to radial shrinkage. All the analyzed characteristics for paulownia and balsa woods were compared to those of linden to obtain positive/negative percentages, using the following relationship (Equation (10)):

$$R = \frac{C_{b,p} - C_l}{C_l} \cdot 100\ [\%] \quad (10)$$

where $C_{b,p}$ is the characteristic of balsa/paulownia wood, and C_l is the characteristic of linden wood.

Based on the all obtained values of characteristics, it was cumulatively determined which of the two species (balsa and paulownia) is more appropriate as a possible replacement for linden wood in its uses.

Statistical analysis. Standard deviation and arithmetic mean were calculated for each group of tests. Also, the regression tendences and the Pearson coefficients R^2 were identified on the graphs obtained using Microsoft Excel 2019 (Microsoft Corp., Redmond, WA, USA). With the help of the statistical program Minitab 18 (Penn State University, State College, PA, USA) and its generated graphs, the mean, standard deviation, *p*-value, and Anderson–Darling coefficient of all tested values were found, with a confidence interval of 95% or an error of 0.05.

3. Results

3.1. Physical and Mechanical Wood Properties (Base on the Data from Reference [1])

Table 1 shows a comparison between linden, balsa, and paulownia species, with the first two species being often used in restoration, consolidation, and conservation processes [12], and the last one being a possible species to be used in this field.

Table 1. Comparison between the wood species of linden, balsa, and paulownia [1].

No.	Wood Species		Linden	Balsa	Paulownia
1.	Distribution		Europe	America	Asia and North America
2.	Tree size—max height and diameter, in m		20–40; 1.5–2	18–28; 1–1.2 m	10–20 m; 0.6–1.2
3.	Density kg/m^3	0%	420	120	250
4.		12%	535	150	280
5.	MOR, MPa		85.4	19.6	37.8
6.	MOE, GPa		11.71	3.71	4.38
7.	Compressive strength, MPa		44.8	11.6	20.7
8.	Shrinkage, %	Radial	5.0	2.3	2.5
9.		Tangential	7.5	6.2	3.9
10.		Volumetric	12.6	8.6	6.5
11.		T/R ratio	1.5	2.6	1.6
12.	Quality coefficient, MOR/ρ		159.6	130.6	135.0
13.	Quality coefficient, σ_c/ρ		83.7	77.3	73.9

Even if the three species have different distribution areas (linden in Europe, balsa in America, and paulownia in Asia and North America), they are found together in all corners of the world. Linden is considered a uniform species, even if it has a high density, while balsa has a density reduction of 71.5%, and paulownia has a density reduction of 47.6% (for densities at 12% moisture content) (Table 1, Figures 5 and 6). Thus, from the point of view of density, paulownia is closer to linden than balsa.

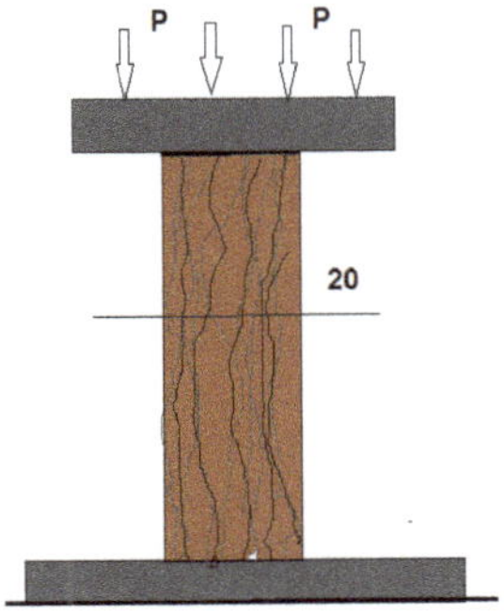

Figure 5. Compressive strength of a wood specimen.

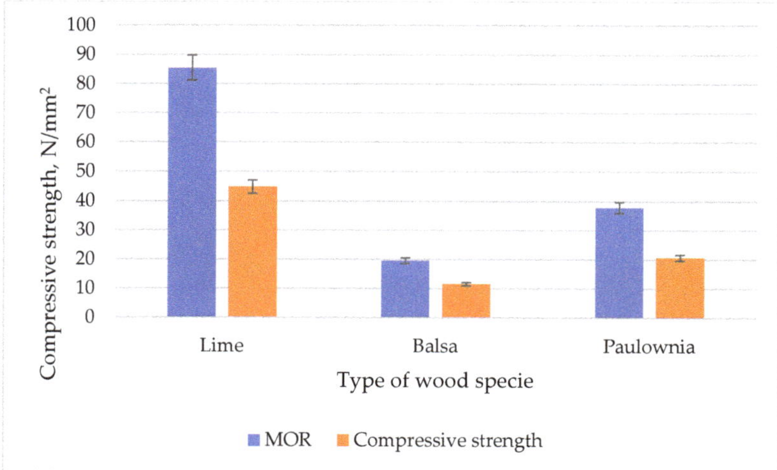

Figure 6. Resistance to compression of lime, balsa, and paulownia woods.

From the point of view of resistance to static bending (MOR), linden wood had the highest value due to its maximum density, with the decrease in resistance being 77% in the case of balsa wood and 55.7% in the case of paulownia wood. These decreases in bending strength (MOR) were greater than those in density, meaning that the dependence relationship was not proportional. From the point of view of the modulus of elasticity, linden wood was the most elastic, with the decrease in elasticity being 68.3% in the case of balsa wood and 62.5% in the case of paulownia wood. From the point of view of compressive strength, linden wood had the highest resistance at 44.8 N/mm^2, with the strength of balsa wood decreasing by 74.1%, and that of paulownia wood by 53.7%. Thus, from the point of view of compression resistance, paulownia wood is very close to that of linden, especially after its lower density has been taken into account.

Following the analysis shown in Table 1, from the point of view of shrinkage, linden wood had the highest values with values of 5% in the radial direction, 7.5% in the tangential direction, and 12.6% in volume, and the lowest values were identified for paulownia wood with values of 2.5% in the radial direction, 3.9% in the tangential direction, and 6.5% in volume. There were very large decreases in the shrinkage of paulownia wood (compared to that of linden wood) in the radial direction by 50%, in the tangential direction by 48%, and in volume by 48.4%. The best factor of homogeneity in relation to humidity and water is the ratio between tangential and radial shrinkage; this ratio was very good in the case of linden and paulownia wood, but it was also above the average value of 2.0 for all species, in the case of balsa wood (Figures 7 and 8).

Figure 8 and Table 2 show that the density of each wood species is different depending on its moisture content, taking as the reference values the moisture content values of absolutely dry wood (MC = 0%), at 12%, and above the saturation point of fiber (MC > 30%). The first two values (at 0% and 12% moisture content) are common ones, but the third value (over the fiber saturation point) is also important because, when the value is above the saturation point of fiber, wood no longer swells and shrinks (it does not change its dimensions, with the volume remaining constant), but it still absorbs water (increases its mass), thus significantly increasing the density. It was observed that, although the immersion time was the same, the moisture content increased more in balsa wood, which is a less dense species (with more free holes for water to penetrate).

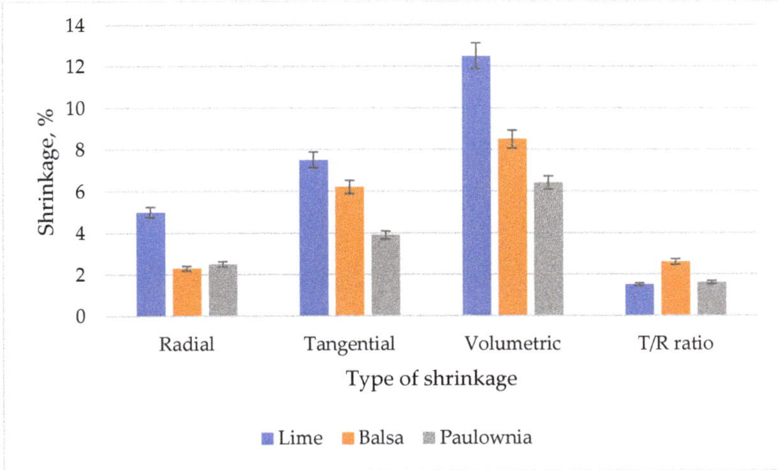

Figure 7. Shrinkage of lime, balsa, and paulownia woods.

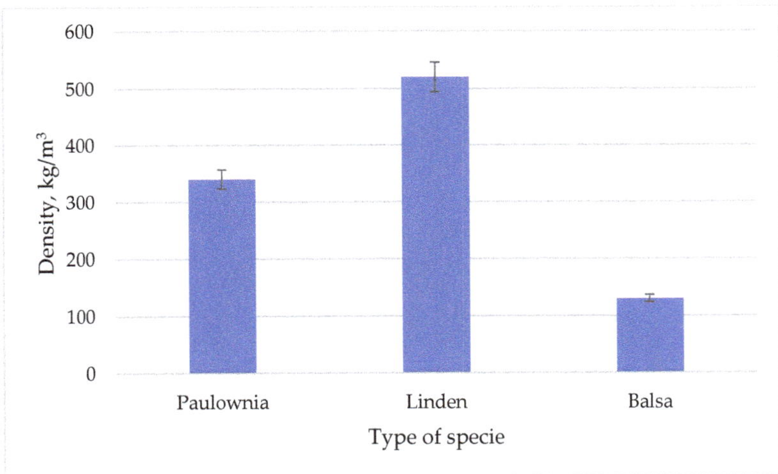

Figure 8. Density of the three wood species.

Table 2. Results regarding the density of the three wood species.

Species	Density, kg/m^3, Related to Moisture Content (MC)		
	MC = 0%	MC = 12 %	Over the Fiber Saturation Point
Paulownia	310	340	380, MC = 30%
Linden	460	520	580, MC = 40%
Balsa	120	130	210, MC = 85%

3.2. *Experimental Results for Wood Density*

For comparison, the density of the samples in an absolutely dry state was examined (Figure 9). It was observed that the density of linden wood was the highest (461 kg/m^3), and that of balsa wood was the lowest (118 kg/m^3).

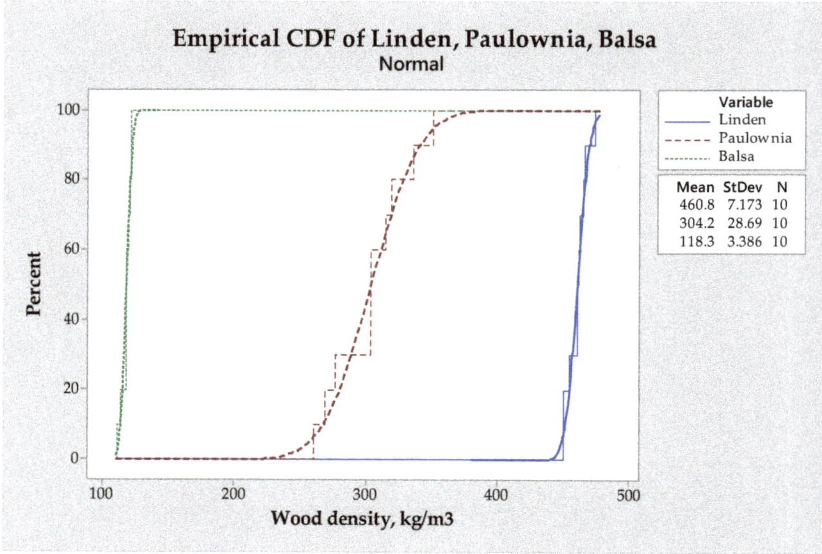

Figure 9. Empirical cumulative distribution function (CDF) of wood density.

Bearing in mind that density is greatly influenced by moisture content, to find the density values depending on the moisture content, the following relationship was used (Equation (11)):

$$\rho_{MC} = \rho_0 \cdot \frac{1 + MC}{1 + \rho_0 \cdot MC} \quad (11)$$

where ρ_0 is the oven-dried wood density, in g/cm^3, and MC is the moisture content, in decimal.

Also, to find the maximum moisture content (MC_{max}) of each wood species, the following relationship was used (Equation (12)):

$$MC_{max} = \frac{1.5 - \rho_0}{1.5 \cdot \rho_0} \quad (12)$$

The obtained value of wood densities are presented in Table 3.

Table 3. Density of wood species at different moisture content.

Wood Species	Density at Different Moisture Content, ρ_u, kg/m^3							
	0%	12%	30%	60%	90%	120%	150%	MC_{max}
Linden	461	489	526	578	619	672	681	681 (MC_{max} = 150%)
Paulownia	304	328	362	411	454	490	522	612 (MC_{max} = 262%)
Balsa	118	130	148	176	202	227	250	540 (MCmax = 780%)

*3.3. Colorimetry of the Three Wood Species Using CIEL*a*b* Space*

3.3.1. The Color of Paulownia Wood before and after Planning

Immediately after cutting the log into lumber, the paulownia wood darkened in color to a light gray. This coloring was observed on a small thickness of about 0.5–0.7 mm, due to the oxidation of wood sap when it came into contact with atmospheric air. This color was kept as long as the wood was not processed via planning. The dark color disappeared when the timber was planned to a thickness of at least 0.7 mm, obtaining the real beautiful color of paulownia wood surface (Table 4).

Table 4. The color changes of paulownia timber wood before and after planning.

No.	Timber Surface before Planning			Timber Surface after Planning		
	L*	a*	b*	L*	a*	b*
1.	4.725	−3.36	−12.86	63.92	−23.41	21.13
2.	2.435	−15.03	−10.16	64.49	−23.36	21.18
3.	3.293	−1.45	−6.68	64.29	−23.66	21.17
4.	5.779	−7.65	−10.47	63.4	−22.87	20.41
5.	8.369	−19.41	−11.78	55.52	−22.94	20.61
6.	7.734	−5.66	−7.43	58.43	−22.28	20.89
7.	6.451	−12.71	−9.87	61.72	−21.89	21.75
8.	3.478	−6.59	−12.66	59.39	−24.86	22.02
9.	4.673	−2.99	−8.64	64.15	−22.89	21.17
10.	7.184	−13.67	−9.17	62.63	−21.64	20.97
Mean	5.412	−8.852	−9.97	61.794	−22.98	21.13
SD	1.913	−0.569	0.197	2.886	−0.884	0.453

Figure 10 shows the degree of black and white of the paulownia timber before and after planning, clearly demonstrating an increase in color from white to black (from 5.41, to 61.79 units).

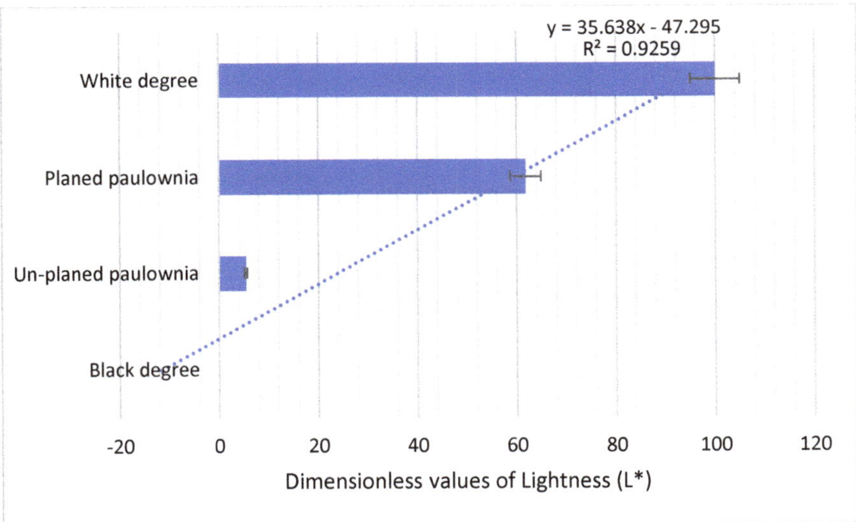

Figure 10. The degree of white/black (L*) of paulownia timber before and after planning.

Figures 11 and 12 show the green–red and blue–yellow degrees of paulownia timber before and after planning. There was a decrease in the degree of green–red from −8.8 to −22.9 units (14.1 units) and a large increase in the degree of blue–yellow from −9.9 to 21.1 units (an increase of 31 units).

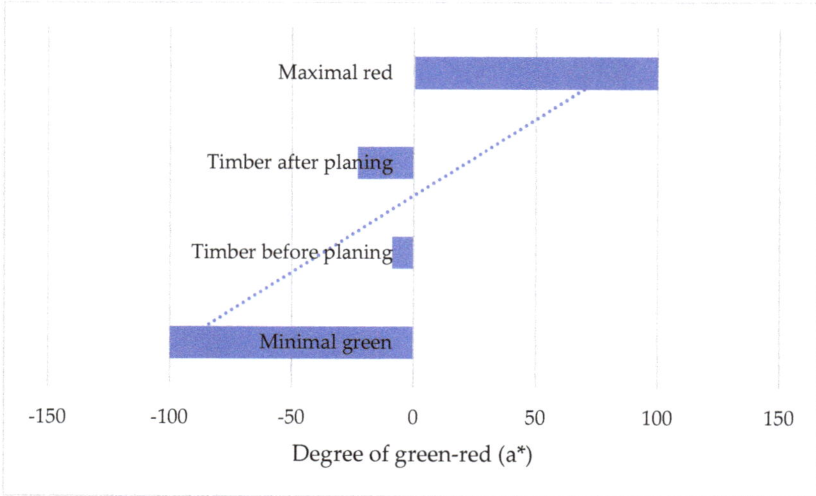

Figure 11. Degree of green–red (a*) of paulownia timber before and after planning.

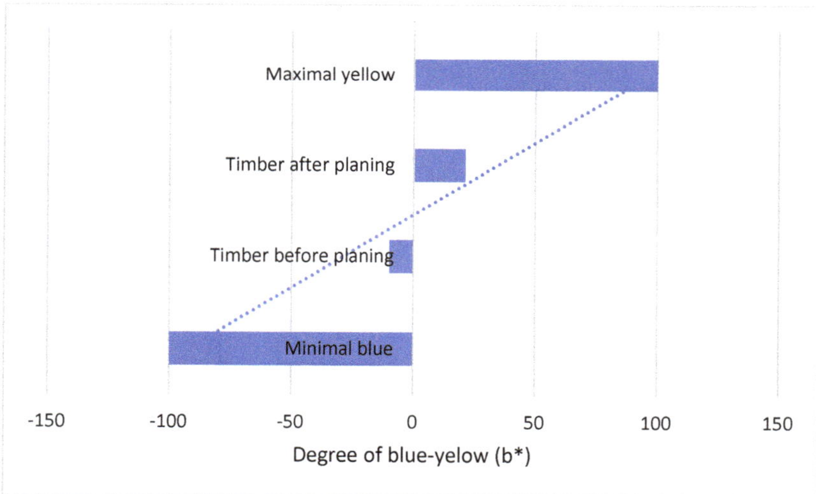

Figure 12. Degree of yellow–blue (b*) of paulownia timber before and after planning.

3.3.2. Results of the Color Analysis of the Three Species

The comparative color analysis of the three species is highlighted in Figure 13, indicating a higher luminance for balsa (Figure 10), a degree of red that is accentuated negatively (35.28–37.62) for all analyzed wood species, and a very low degree of yellowness (3.04–3.66).

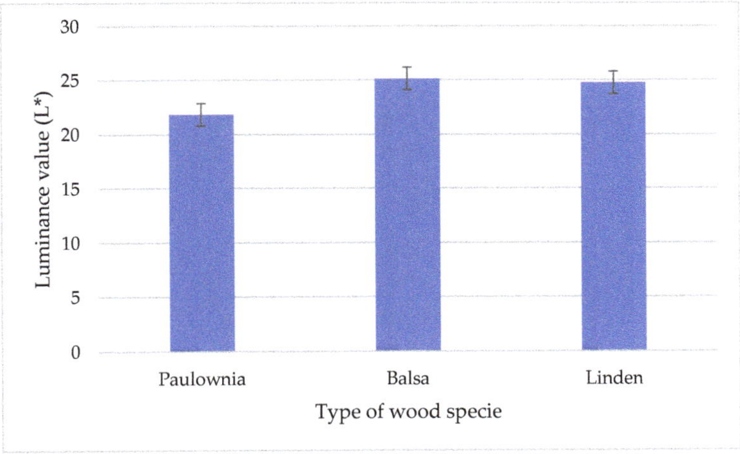

(a)

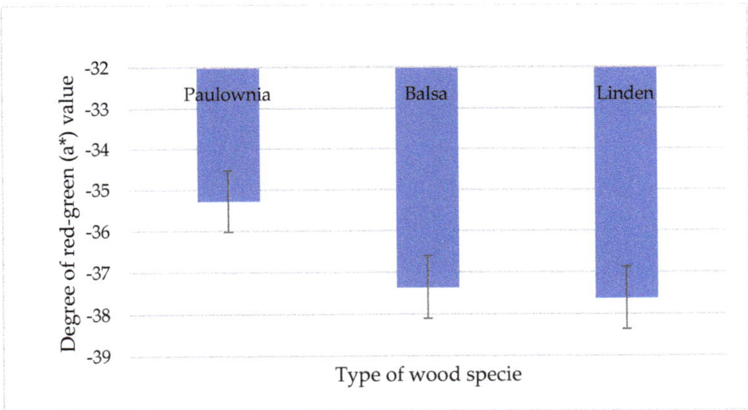

(b)

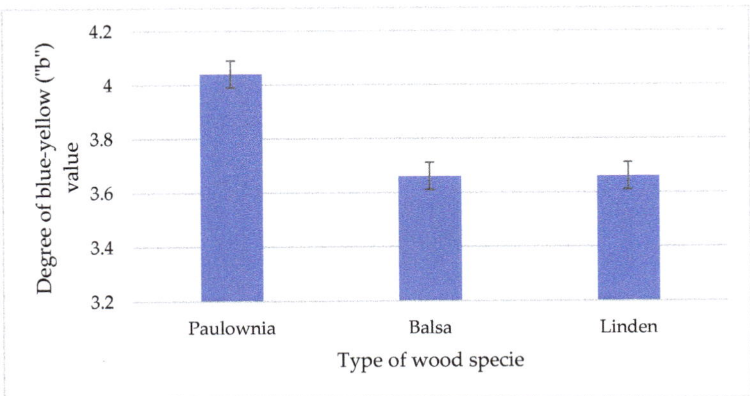

(c)

Figure 13. Values of the CIELab color space for luminance (L*) (**a**), for the degree of red–green (a*) (**b**), and for the degree of blue–yellow (b*) (**c**).

It could be observed that both luminance (L*), red–green contrast (a*), and blue–yellow contrast (b*) had some very close values, which means that from a coloristic point of view, the three analyzed species are appropriate and homogeneous (Figure 13). Furthermore, when the "a*" and "b*" values of the three species were plotted in a two-dimensional plane (Figure 14), it could be noticed that all three points are located in the fourth quadrant of the plane, and the points are very close to each other. Thus, from a color point of view, the three species are very similar.

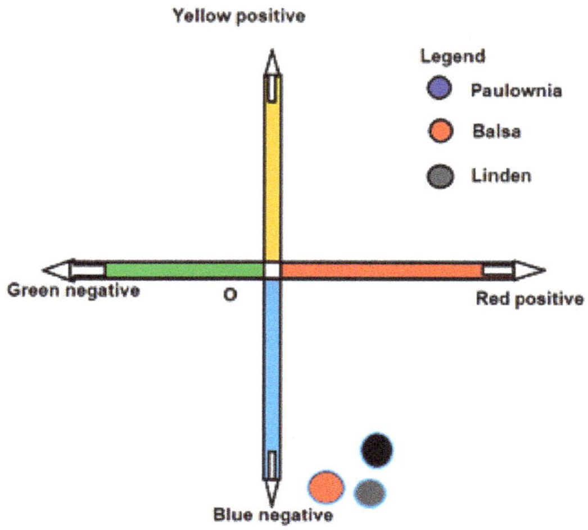

Figure 14. Positioning the color points of linden, balsa, and paulownia wood in a two-dimensional plane (positive–negative).

3.4. Results for Water Absorption

The results for water absorption of the three wood species are presented in Table 5 and Figure 15. It is clearly demonstrated that paulownia wood had the lowest water absorption (27.3%) and balsa wood had the highest one (85.3%).

Table 5. Results for water absorption of the three wood species.

No.	Paulownia Mass, g		WA, %	Linden Mass, g		WA, %	Balsa Mass, g		WA, %
	Final	Initial		Final	Initial		Final	Initial	
1	6.31	4.98	26.70	8.61	6.11	40.91	2.86	1.53	86.92
2	5.51	4.28	28.73	8.66	6.26	38.33	2.93	1.58	85.44
3	6.1	4.82	26.55	8.84	6.43	37.48	2.71	1.5	80.66
4	6.46	5.07	27.41	8.78	6.42	36.76	2.91	1.58	84.17
5	6.56	5.15	27.37	8.41	6.1	37.86	2.82	1.53	84.31
6	6.41	5.05	26.93	8.5	6.13	38.66	2.89	1.51	91.39
7	6.6	5.18	27.41	8.83	6.34	39.27	2.62	1.39	88.48
Mean	6.279	4.932	27.305	8.66	6.26	38.47	2.82	1.52	85.92
St. Dev	0.38	0.311	0.723	0.17	0.14	1.35	0.114	0.064	3.440

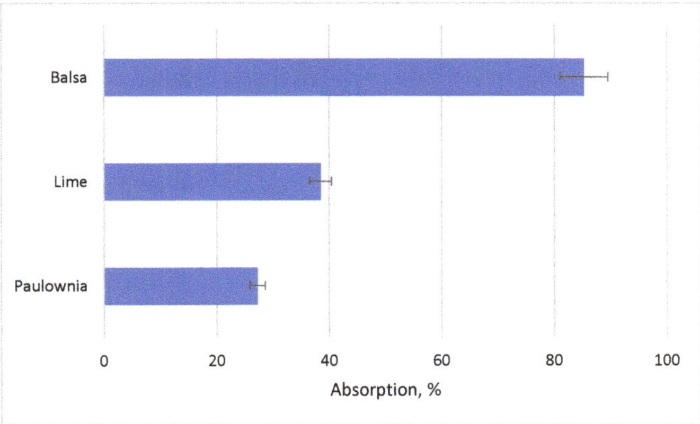

Figure 15. Water absorption of the three wood species.

3.5. Results for Wood Shrinkage

The results for wood shrinkage are presented in Table 6 and Figure 16. It can be seen that the largest volumetric shrinkage was found in linden wood (8.29%), and a smaller volumetric shrinkage was found in paulownia and balsa wood (4.13% and 3.39%, respectively).

Table 6. Wood shrinkage for paulownia, linden, and balsa.

No.	Paulownia, mm					Linden, mm					Balsa, mm				
	T	R	L	V	T/R	T	R	L	V	T/R	R	T	L	V	T/R
1	2.50	1.61	0.034	4.15	1.55	4.30	3.37	0.14	7.81	1.27	1.08	1.53	0.40	3.01	1.41
2	1.64	0.88	0.034	2.55	1.86	4.46	3.18	4.13	11.7	1.40	1.84	1.95	0.48	4.27	1.05
3	2.57	1.31	0.209	4.09	1.96	3.81	3.69	0.00	7.50	1.03	1.61	1.64	0.27	3.52	1.01
4	2.78	0.87	0.138	3.78	3.19	4.54	3.45	0.28	8.27	1.31	1.55	1.66	0.20	3.42	1.07
5	2.87	1.44	0.172	4.48	1.99	4.82	3.05	0.17	8.04	1.58	0.66	1.23	0.27	2.16	1.86
6	2.74	1.69	0.550	4.98	1.62	4.97	2.21	0.49	7.67	2.24	1.58	1.86	0.17	3.60	1.17
7	2.81	1.70	0.345	4.86	1.65	3.99	2.80	0.21	6.99	1.42	1.57	1.86	0.34	3.77	1.18
M	2.56	1.36	0.212	4.13	1.88	4.41	3.11	0.77	8.29	1.4	1.42	1.67	0.31	3.39	1.25
ST	0.43	0.36	0.184	0.81	0.52	0.42	0.49	1.49	1.59	0.38	0.4	0.24	0.1	0.66	0.16

T—tangential; R—radial; L—longitudinal; M—mean; ST—standard deviation.

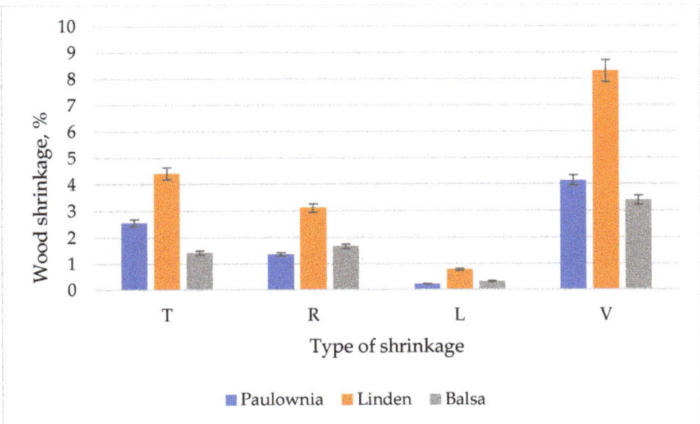

Figure 16. Shrinkage of the wood species. T—tangential; R—radial; L—longitudinal; V—volumetric.

3.6. Brinell Hardness

The Brinell hardness (BH) results are presented in Figure 17, which highlights the maximum value for linden wood (20.4 N/mm^2) and the minimum value for balsa wood (11.1 N/mm^2). It is also noticeable the value of Brinell hardness is higher in the radial section for all three analyzed species.

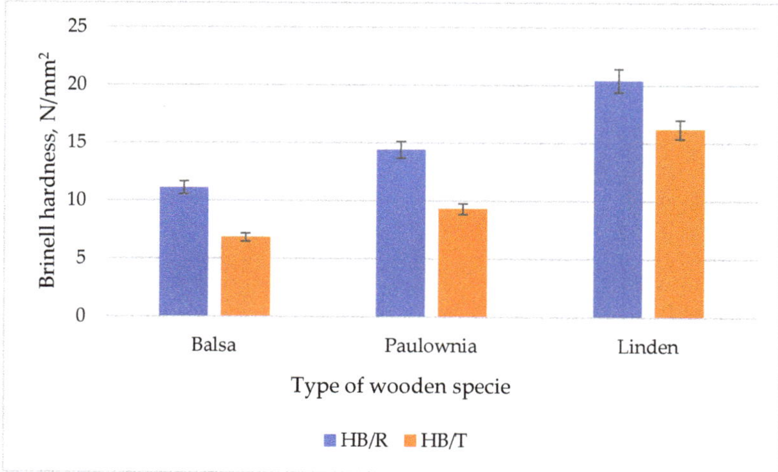

Figure 17. Brinell hardness for balsa, paulownia, and linden wood in the radial (HB/R) and tangential (HB/T) sections.

3.7. Results for Compressive Strength

The results regarding compressive strength are presented in Figure 18, where it can be seen that the maximum compression resistance was obtained for linden wood (52.9 N/mm^2), and the minimum value was obtained for balsa wood (9.1 N/mm^2).

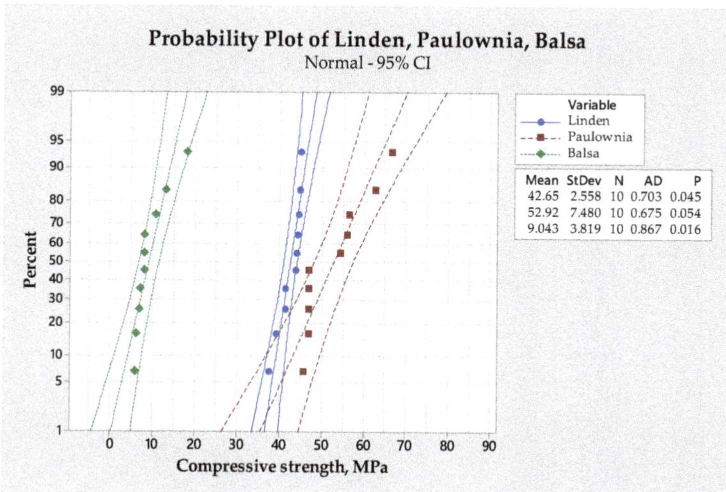

Figure 18. The compressive strength of all three wood species.

3.8. Results for the Modulus of Resistance (MOR) to Static Bending

The results obtained for the modulus of resistance (MOR) upon static bending for the three wood species corresponded with their density, showing a value of 92.3 N/mm^2 in the case of linden wood, 62.1 N/mm^2 in the case of paulownia wood, and 14.1 N/mm^2 in the case of balsa wood (Figure 19). The influence of wood density on the modulus of resistance (MOR) is also demonstrated in Figure 20, with paulownia wood as an example.

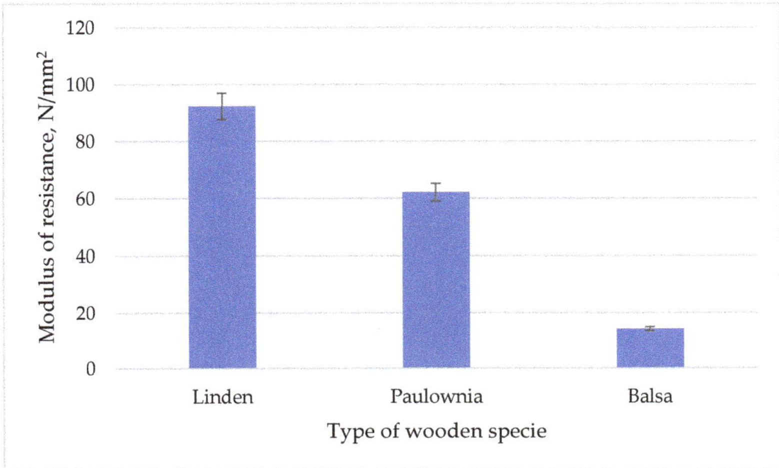

Figure 19. Modulus of resistance (MOR) obtained for the three wood species.

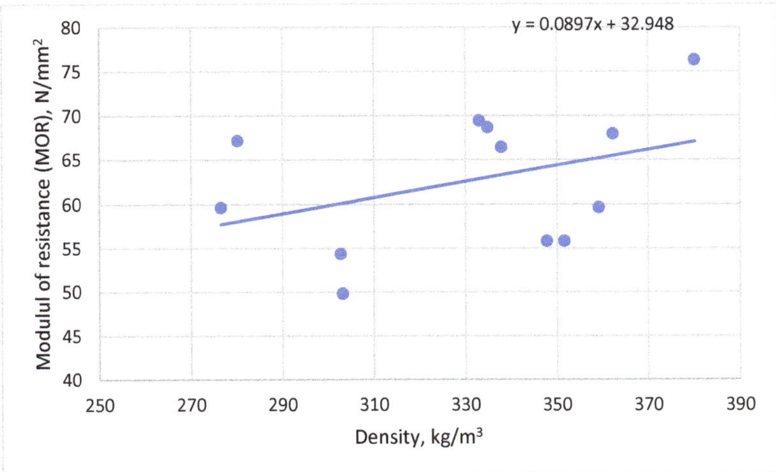

Figure 20. Influence of density on the modulus of resistance (MOR) to static bending strength for paulownia.

3.9. Results for the Modulus of Elasticity (MOE) to Static Bending

The results of the modulus of elasticity (MOE) to static bending are presented in Figure 21, where it is demonstrated (as in the case of the modulus of resistance) that the maximum value was obtained for linden wood (6782 N/mm^2), and the minimum value was obtained for balsa wood (927 N/mm^2). A direct proportionality was observed between wood density and the modulus of elasticity (MOE) for paulownia wood (because the graphs of the influence of density on the modulus of elasticity for linden and balsa were similar) (Figure 22).

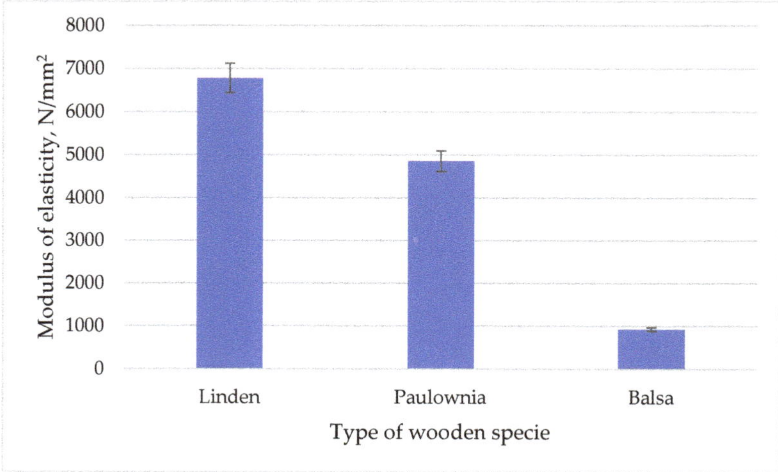

Figure 21. Modulus of elasticity obtained for the three wood species.

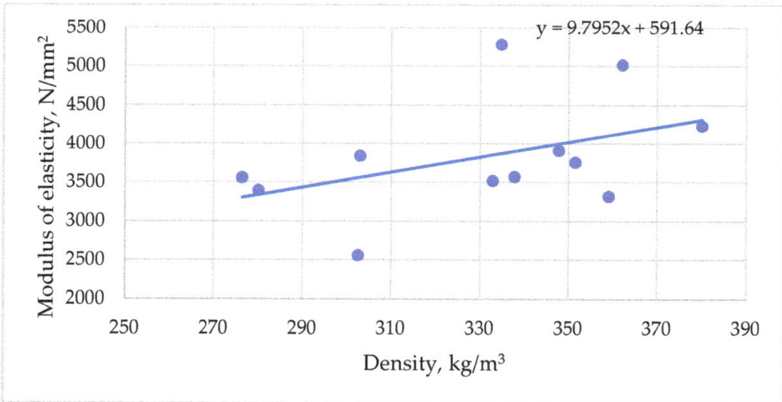

Figure 22. Influence of density on the modulus of elasticity (MOE) for paulownia wood.

3.10. Ecological Influences

The results related to the ecology of the three forest species started from the absorption of carbon dioxide of 1.8 t/t wood and removal of oxygen of 1.3 t/t wood, but also considering the different growth rates of the three species, which are 350 m^3/ha in the case of linden tree, 550 m^3/ha in the case of balsa tree, and 850 m^3/ha in the case of paulownia tree. Taking into account the densities of these wood species, a quantity of mass of 650 t/ha was obtained in the case of linden, 3600 t/ha in the case of balsa, and 3400 t/ha in the case of paulownia. In this way, 845 t of O_2/t wood and 1170 t of CO_2/ha in the case of linden, 4680 t of O_2/wood and 6480 t CO_2/ha in the case of balsa, and 4420 t O_2/t wood and 6120 t CO_2/ha in the case of paulownia were obtained. If the ecological factor was a unitary one in the case of linden forests, then a factor of 5.5 was obtained for balsa forests and a factor of 5.2 was obtained for paulownia forests.

4. Discussion

To be able to have a conclusive discussion regarding the results obtained in this work, all results are grouped in Table 7. Also, this table includes the comparison values obtained for paulownia and balsa wood to those obtained for linden wood, corresponding to one of the relationships (Equation (10)).

Table 7. Comparison of the characteristics of the balsa and paulownia wood species related to linden one.

No.	Characteristics	Linden	Paulownia		Balsa	
		Value	Value	Eq. 11%	Value	Eq. 11%
1.	Unit density, ρ_0, kg/m^3	461	304	−34.0	118	−74.4
2.	Luminance	21.8	25.1	+15.1	24.7	+13.3
3.	a* value	−35.2	−37.3	−5.9	−37.6	−6.8
4.	b* value	4.0	3.6	−10.0	3.6	−10.0
5.	Absorption, %	38	27	+7.8	85	−123.6
6.	Tangential shrinkage (TS), %	4.4	2.5	+43.1	1.4	+68.1
7.	Radial shrinkage (RS), %	3.1	1.3	+58.1	1.6	+48.3
8.	Ratio, TS/RS	1.4	1.8	−28.5	1.2	+14.2
9.	Compressive strength, N/mm^2	52.9	42.5	−2.3	9.1	−82.7
10.	Modulus of resistance, N/mm^2	92.3	62.1	−32.7	14.1	−84.7
11.	Modulus of elasticity, N/mm^2	6782	4852	−28.4	927	−86.3
12.	Brinell hardness (R), N/mm^2	20.4	14.4	−29.4	11.1	−45.8
13.	Ecological coefficient O_2; CO_2	1	5.5	+550	5.2	+520
14.	Quality coefficient, MOR/$\rho_{0(g/cm^3)}$	200.2	204.2	+1.9	119.4	+0.8
15.	Quality coefficient, $\sigma_c/\rho_{0(g/cm^3)}$	114.7	139.8	+21.8	77.1	−32.7
	Total points	----	----	+526.6	-----	+117.7

The first analysis of the results shown in Table 7 was a comparison between the characteristics of the three species, namely how close the characteristics of paulownia and balsa are to those of linden wood. It was clearly observed that paulownia wood was much more appropriate as a replacement for linden wood with a total of 526.6 points, thus being 4.4 times more appropriate than balsa wood.

Paulownia wood had a density lower than that of linden by 34%, but higher than that of balsa wood by 1.62 times. It was observed that the experimental value of balsa wood density was closest to that obtained by other researchers [1], with the difference only 1.6% lower. In the case of linden wood, the difference between the two values was higher than 9.7%, and in the case of paulownia wood, the difference was higher than 21.6%. From the point of view of density, the three analyzed species are different as the differences are greater than 5–10% between them, paulownia wood being more appropriate as a replacement for linden [35,36]. Some authors [17] found a wood density of about 300 kg/m^3 for paulownia, which was very close to the value of 304 kg/m^3 of this research.

The analysis of color parameters with regard to the natural color of planed wood showed that from a color point of view, the three wood species are very appropriate (L = 21.8–25.1; a = −(35.2–37.6); and b = 3.6–4), with the points of all three species being positioned in the fourth trigonometric quadrant. Appropriate values were also found by other authors [5,18], and a similar value of paulownia was obtained by others [17].

The water absorption of these three wood species was dependent on their porosity; it increased with a decrease in the density of these species. The exception to this rule was paulownia wood, which had a lower absorption than linden wood. The explanation for this phenomenon could be attributed to the microscopic structure of this wood species

(Figure 1) and the fact that it has pores filled with cellulosic formations that prevent the penetration of water into the wood.

As a general trend, the wood shrinkage of these three wood species decreased with a decrease in density (regardless of whether the direction of analysis was radial or tangential), with balsa wood having the lowest shrinkage (1.4% in the tangential direction and 1.6% in the radial direction) [37]. The values of the ratio of tangential shrinkage to radial shrinkage showed that these three wood species are uniform, with balsa wood and linden wood having the lowest shrinkage coefficients.

The Brinell hardness of the three wood species had values corresponding to the values reported in other research studies in the field [38,39]), but there were also small differences between them. For example, the Brinell hardness of balsa wood was only 22.8% lower than that of paulownia wood, although the density difference between the two wood species was 74.4%.

The compressive strength of these three species (Figure 17) was different and depended on density, with a higher compressive resistance (52.9 N/mm^2) being obtained for linden, which had a higher wood density (461 kg/m^3). As a general trend, from the point of view of compression strength, paulownia wood has a value closer to linden wood. The values obtained in this research were slightly higher than those reported in the literature [1,40], being higher by 15.3% for linden, 8.4% for paulownia, and 8.2% for balsa wood.

The bending strength (MOR) was in accordance with the density of the wood species, with higher densities corresponding to higher resistances. The dependence relationship between the densities and resistances of these three species (Figure 19) was approximated as linear; from this point of view, paulownia wood has superior characteristics compared to the other two species. Differences between the values obtained in this research and those reported in the literature [1,41] were also identified; in this case the difference was about 7.4% for linden wood. The modulus of elasticity (MOE) maintained the same trend as the modulus of resistance (MOR), showing a decrease along with a decrease in density. However, the differences between the values of the three species have increased, as the modulus of elasticity of paulownia wood (4852 N/mm^2) becoming closer to that of linden wood (6782 N/mm^2), while the modulus of elasticity of balsa wood (927 N/mm^2) moving further away. The values obtained in this work differed slightly from those reported in the literature [1]. For instance, there was a large difference of about 4.72 times in the case of balsa wood. In the direction of what was mentioned above, in the reference [23], an MOR value of 111 N/mm^2 and an MOE value of 11,479 N/mm^2 were found for beech wood, but with a much higher density than the species analyzed in this study, of 727 kg/m^3.

Based on the results shown in Table 7, due to the different rates of growth, paulownia and balsa wood had superior ecologically characteristics compared to linden wood, being about five times higher [30,33]. From an ecological point of view, the results showed that paulownia and balsa wood are superior to linden wood but are very similar to each other, with the difference between balsa and paulownia of only 5.4%.

The quality coefficient (related to both compressive strength and modulus of resistance) of paulownia wood was better than balsa and linden. From this point of view, paulownia wood could be considered a strong species that is comparable to the strongest wood species in the word.

5. Conclusions

Generally speaking, the three deciduous woody species are as appropriate as possible, having well-observed similarities, and, because that, they can be considered homogenous.

The closest to linden is paulownia wood, which can replace linden in its uses in modeling, restoration, conservation, sculpture, etc.

Ecologically, balsa wood and paulownia wood are better than linden wood due to their faster growth.

The mechanical properties (compressive strength, MOR, MOE, and Brinell hardness) of the three species are consistent with their density.

From the point of view of quality coefficient, paulownia wood is the best species, compared to linden and balsa woods.

In a broader context, this research study falls into the area of knowledge of wood species with a low density and a good structural homogeneity.

Among the limitations of this research study and possible new areas of investigation there are the behavior of these species during thermal treatments at temperatures above 200 °C, in a nitrogen environment, or when there is no access to oxygen.

Author Contributions: Conceptualization, A.L. and A.A.; methodology, C.S.; software, C.S.I.; validation, A.L., A.A. and C.S.; formal analysis, C.S.I.; investigation, A.A.; resources, A.A.; data curation, A.L.; writing—original draft preparation, A.A.; writing—review and editing, AL.; visualization, C.S.I.; supervision, C.S.; project administration, A.L.; funding acquisition, A.L. All authors have read and agreed to the published version of the manuscript.

Funding: This research received no external funding.

Institutional Review Board Statement: Not applicable.

Informed Consent Statement: Not applicable.

Data Availability Statement: Not applicable.

Acknowledgments: The authors thank the Transilvania University of Brasov for the technical and logistical support provided.

Conflicts of Interest: The authors declare no conflict of interest.

References

1. Wood Database. Available online: https://www.wood-database.com/ (accessed on 12 June 2023).
2. Lark-Horovitz, K. A Simple Method for Testing Homogeneity of Wood. *Nature* **1936**, *137*, 663. [CrossRef]
3. Hektor, T.; Leonidha, P.; Entela, L. Evaluation of wood surface roughness depending on species characteristics. *Maderas. Cienc. Tecnol.* **2015**, *17*, 285–292. [CrossRef]
4. Echols, R.M. Uniformity of wood density assessed from X-rays of increment cores. *Wood Sci. Technol.* **1973**, *7*, 34–44. [CrossRef]
5. Grekin, M. Color and color uniformity variation of Scots pine wood in the air-dry condition. *Wood Fiber Sci.* **2007**, *39*, 279–290.
6. Identification of Soft or Hardwood. Available online: https://extension.tennessee.edu/publications/documents/PB1692.pdf (accessed on 3 March 2023).
7. Basic Guide to Identification of Hardwoods and Softwoods Using Anatomical Characteristics. Available online: https://extension.msstate.edu/sites/default/files/publications/publications/p2606.pdf (accessed on 6 May 2023).
8. Zhou, H.; Wei, X.; Smith, L.M.; Wang, G.; Chen, F. Evaluation of Uniformity of Bamboo Bundle Veneer and Bamboo Bundle Laminated Veneer Lumber (BLVL). *Forests* **2019**, *10*, 921. [CrossRef]
9. Avram, A.; Ionescu, C.S.; Lunguleasa, A. A consolidation of degraded lime wooden support from heritage objects using two types of consolidant. *BioResources* **2023**, *18*, 4580–4597. [CrossRef]
10. Fierascu, R.C.; Doni, M.; Fierascu, I. Selected aspects regarding the restoration/conservation of traditional wood and masonry building materials: A short overview of the last decade findings. *Appl. Sci.* **2020**, *10*, 1164. [CrossRef]
11. Popescu, C.-M.; Popescu, M.-C.; Vasile, C. Characterization of fungal degraded lime wood by FT-IR and 2D IR correlation spectroscopy. *Micro. J.* **2010**, *95*, 377–387. [CrossRef]
12. Avram, A.; Ionescu, C.Ş.; Lunguleasa, A. Some Methods for the Degradation-Fragility Degree Determination and for the Consolidation of Treatments with Paraloid B72 of Wood Panels from Icon-Type Heritage Objects. *Forests* **2022**, *13*, 801. [CrossRef]
13. Wang, Y.; Schniewind, A.P. Consolidation of deteriorated wood with soluble resins. *J. Am. Inst. Conserv.* **1985**, *24*, 77. [CrossRef]
14. *ISO 13061-2: 2014*; Physical and Mechanical Properties of Wood—Test Methods for Small Clear Wood Specimens—Part 2: Determination of Density for Physical and Mechanical Tests. International Organization for Standardization (ISO): Geneva, Switzerland, 2014.
15. *ASTM D2395-17*; Standard Test Methods for Density and Specific Gravity (Relative Density) of Wood and Wood-Based Materials. ASTM International: West Conshohocken, PA, USA, 2017.
16. Suri, L.F.; Purusatama, B.D.; Kim, J.H.; Yang, G.U.; Prasetia, D.; Kwo, G.J.; Hidayat, W.; Lee, S.H.; Febrianto, F.; Kim, N.H. Comparison of physical and mechanical properties of *Paulownia tomentosa* and *Pinus koraiensis* wood heat-treated in oil and air. *Eur. J. Wood Wood Prod.* **2022**, *80*, 1389–1399. [CrossRef]
17. Suri, I.F.; Kim, J.H.; Purusatama, B.D.; Yang, G.U.; Prasetia, D.; Lee, S.H.; Hidayat, W.; Febrianto, F.; Park, B.H.; Kim, N.H. Comparison of the color and weight change in *Paulownia tomentosa* and *Pinus koraiensis* wood heat-treated in hot oil and hot air. *BioResources* **2021**, *16*, 5574–5585. [CrossRef]

18. Liu, X.Y.; Timar, M.C.; Varodi, A.M.; Yi, S.L. Effects of ageing on the color and surface chemistry of Paulownia wood (*P. elongata*) from fast growing crops. *BioResources* **2016**, *11*, 9400–9420. [CrossRef]
19. Akyildiz, M.H.; Kol, H.S. Some technological properties and uses of paulownia (*Paulownia tomentosa* Steud.) wood. *J. Environ. Biol.* **2010**, *31*, 351–355. [PubMed]
20. Akyildiz, M.H. Screw-nail withdrawal and bonding strength of paulownia (*Paulownia tomentosa* Steud.) wood. *J. Wood Sci.* **2014**, *60*, 201–206. [CrossRef]
21. Ashori, A.; Nourbakhsh, A. Studies on Iranian cultivated paulownia—A potential source of fibrous raw material for paper industry. *Eur. J. Wood Prod.* **2009**, *67*, 323–327. [CrossRef]
22. *ISO 13061-15: 2017*; Physical and Mechanical Properties of Wood—Test Methods for Small Clear Wood Specimens—Part 15: Determination of Radial and Tangential Swelling. International Organization for Standardization (ISO): Geneva, Switzerland, 2017.
23. Borůvka, V.; Novák, D.; Šedivka, P. Comparison and Analysis of Radial and Tangential Bending of Softwood and Hardwood at Static and Dynamic Loading. *Forests* **2020**, *11*, 896. [CrossRef]
24. Reinprecht, L. *Wood Deterioration, Protection and Maintenance*, 1st ed.; John Wileys & Sons: Chichester, UK, 2016; p. 376; ISBN 978111910653.
25. Wagenführ, R.; Wagenführ, A. *Holzatlas (E-Book)*, 2021 ed.; Carl Hanser Fachbuchverlag: Munchen, Germany, 2021.
26. Côté, W.A. Chemical Composition of Wood. In *Principles of Wood Science and Technology*; Springer: Berlin/Heidelberg, Germany, 1968. [CrossRef]
27. Ates, S.; Yonghao, N.; Akgul, M.; Tozluoglu, A. Characterization and evaluation of *Paulownia elongota* as a raw material for paper production. *Afr. J. Biotec.* **2008**, *7*, 4153–4158. Available online: http://www.academicjournals.org/AJB (accessed on 5 May 2023).
28. *ISO 13061-4: 2014*; Physical and Mechanical Properties of Wood—Test Methods for Small Clear Wood Specimens—Part 4: Determination of Modulus of Elasticity in Static Bending. International Organization for Standardization: Geneva, Switzerland, 2014.
29. *ISO 13061-17: 2017*; Physical and Mechanical Properties of Wood—Test Methods for Small Clear Wood Specimens—Part 17: Determination of Ultimate Stress in Compression Parallel to Grain. International Organization for Standardization: Geneva, Switzerland, 2017.
30. *EN 1534: 2003*; Wood and Parquet Flooring. Determination of Resistance to Indentation (Brinell)—Test Method. European Committee for Standardization: Brussels, Belgium, 2003.
31. Ilie, D.M.; Păduri din România. *Elemente de Dendrologie şi Ecologie (Forests in Romania Elements of Dendrology and Ecology)*; Sibiu: Print House of "Lucian Blaga" University: Sibiu, Romania, 2015; ISBN 978-606-12-1156-2. Available online: https://www.infolife.ro/wp-content/uploads/2016/11/Paduri-din-Romania.pdf (accessed on 20 June 2023).
32. Clapa, D.; Fira, A.; Simu, M.; Vasu, L.B.; Buduroi, D. Improved In Vitro Propagation of *Paulownia elongata*, *P. fortunei* and its Interspecific Hybrid *P. elongata* x *P. fortune*. *Bull. UASVM Hortic* **2014**, *71*, 6–14. [CrossRef]
33. Ecomatcher. Available online: https://www.ecomatcher.com/how-to-calculate-co2-sequestration/ (accessed on 25 November 2022).
34. Borrega, M.; Gibson, L. Mechanics of balsa (*Ochroma pyramidale*) wood. *Mech. Mater.* **2015**, *84*, 75–90. [CrossRef]
35. Cáceres-Hidalgo, E. Caracterización Físico-Mecánica de la Madera de Paulownia Elongate: Physico-mechanical Characterization of Paulownia Elongata Wood. Master's Thesis, Escuela Técnica Superior de Ingenierías Agrarias, Universidad de Valladolid, Valladolid, Spain, 2016. Available online: http://uvadoc.uva.es/handle/10324/18787 (accessed on 5 May 2023).
36. Candan, Z.; Korkut, S.; Unsal, O. Effect of thermal modification by hot pressing on performance properties of paulownia wood boards. *Ind. Crops Prod.* **2013**, *45*, 461–464. [CrossRef]
37. Budakçı, M.; Pelit, H.; Sönmez, A.; Korkmaz, M. The effects of densification and heat post-treatment on hardness and morphological properties of wood materials. *BioResources* **2016**, *11*, 7822–7838. [CrossRef]
38. Wu, C.; Vahedi, N.; Vassilopoulos, A.; Keller, T. Mechanical properties of a balsa wood veneer structural sandwich core material. *Constr. Build. Mater.* **2020**, *265*, 120193. [CrossRef]
39. TranVan, L.; Legrand, V.; Jacquemin, F. Thermal decomposition kinetics of balsa wood: Kinetics and degradation mechanisms comparison between dry and moisturized materials. *Polym. Degrad. Stabil.* **2014**, *110*, 208–215. [CrossRef]
40. Da Silva, A.; Kyriakides, S. Compressive response and failure of balsa wood. *Int. J. Solids Struct.* **2007**, *44*, 8685–8717. [CrossRef]
41. Mitani, A.; Barboutis, I. Changes Caused by Heat Treatment in Color and Dimensional Stability of Beech (*Fagus sylvatica* L.) Wood. *Drvna Industrija* **2014**, *65*, 225–232. [CrossRef]

applied sciences

MDPI

Article

Wood Material Properties of Forest Fire-Damaged Norway Spruce and Scots Pine for Mechanical Wood Processing in Finland

Juhani Marttila [1,*], Veikko Möttönen [1], Antti Haapala [2], Pekka Ylimäki [3], Petri Kilpeläinen [4] and Erkki Verkasalo [1]

1 Natural Resources Institute Finland (Luke), Production Systems, Yliopistokatu 6B, FI-80100 Joensuu, Finland; veikko.mottonen@luke.fi (V.M.); erkki.verkasalo@luke.fi (E.V.)
2 Department of Chemistry, Faculty of Science, Forestry and Technology, University of Eastern Finland (UEF), Yliopistokatu 7 (Futura), FI-80100 Joensuu, Finland; antti.haapala@uef.fi
3 UPM Forest, Kaarinantie 700, FI-20540 Turku, Finland; pekka.ylimaki@upm.com
4 Natural Resources Institute Finland (Luke), Production Systems, Viikinkaari 9, FI-00790 Helsinki, Finland; petri.kilpelainen@luke.fi
* Correspondence: juhani.marttila@luke.fi; Tel.: +358-29-532-2164

Abstract: Due to climate change, the risk of forest fires has increased in Europe, resulting in challenges in the allocation of salvaged wood. We studied the raw material potential for wood products of Norway spruce and Scots pine sawn log trees that remained standing after a large forest fire in Kalajoki, Finland, in July 2021. Eight burned trees, with four reference trees per species, were sampled as standard specimens, and measurements were analyzed with linear mixed models. The effects of fire on the modulus of elasticity and rupture, Brinell hardness, moisture gradient, and color were measured on clear wood specimens of sapwood and heartwood. The wood density, level of fire damage, and height location of a tree were used as additional predictors. The results show some changes in the sapwood material. Spruce wood underwent stronger changes after the fire than pine wood, probably due to spruce wood having a thinner bark and a longer crown. The moisture content decreased in spruce, and the color darkened in both spruce and pine. Changes in the mechanical properties were mostly negligible, but a small increase in the Brinell hardness in spruce and a small decrease in the modulus of rupture in pine were observed. Fresh salvaged wood can be a suitable material for middle-quality and lower-quality wood products. The spread of char and soot into wood and wood processing machinery still limits its usage, especially for spruce.

Keywords: *Picea abies*; *Pinus sylvestris*; wildfire; forest fire; salvaged wood; sawn wood; wood products; clear wood properties; moisture gradient; density; mechanical properties; wood color

Citation: Marttila, J.; Möttönen, V.; Haapala, A.; Ylimäki, P.; Kilpeläinen, P.; Verkasalo, E. Wood Material Properties of Forest Fire-Damaged Norway Spruce and Scots Pine for Mechanical Wood Processing in Finland. *Appl. Sci.* **2024**, *14*, 238. https://doi.org/10.3390/app14010238

Academic Editors: Stefano Invernizzi, Emilia-Adela Salca, Lidia Gurau and Mihaela Câmpean

Received: 15 November 2023
Revised: 7 December 2023
Accepted: 14 December 2023
Published: 27 December 2023

1. Introduction

Wildfires and resultant forest disturbances have become common globally, and their scale and severity have increased since the beginning of the 2000s [1–4]. This has mostly been attributed to climate change and sudden regional weather phenomena, such as droughts and heat loads during summer seasons, but partly due to human behavior and machine work activities in nature and the built environment [5–10]. Most wildfires, with respect to the overwhelmingly largest areas, have been in the Russian Federation, Canada, the U.S. West, Australia, and Mediterranean countries [2].

In Finland, the forest fire incidence level is very low [3]. For example, 470 hectares of land damaged by forest fires was reported in 2017, and the benchmark of over 1000 hectares was destroyed only in 2006 [11]. There is a short span of seasons when forests are vulnerable to fire, ranging from May to September, with June and July remaining as the peak periods; relatively cool and rainy summers, high forest and second-home road densities, and efficient aerial fire control and prevention have contributed to the low incidence rate [12]. However, with climate change causing drought periods during summer, more trees suffering and drying from bark beetle damage, and the practice of leaving standing and wind-fallen trees for biodiversity management after final harvesting, vulnerability to fires is predicted to increase in the coming decades [13].

The forest stand structure and site, tree species, and fire intensity largely determine the consequences for trees and their probability of surviving a forest fire [7,14,15]. A dense stand, dry site, small diameter, thin bark, and large crown have been observed to lead to a high mortality of trees [5–10]. Owing to fire, drying that results in checking may occur in sapwood if the bark deteriorates first; additionally, notable charring and grayish discoloration occur because of char and soot, but the heartwood should remain essentially unchanged [6,8,12,15,16]. In wood material, drying starts slowly due to direct heating at a temperature as low as 20–100 °C; thereafter, the rate increases, and the intensity multiplies [17]. Chemical transformation affecting the physico-mechanical properties starts at 50–100 °C with the decomposition of certain extractives, and the major changes begin at 160–180 °C with the decomposition of hemicelluloses and above 250 °C with lignin [18]. Physical changes start above 150 °C with a decrease in the equilibrium moisture content and a reduction in hygroscopicity, absorbability, and surface activity; these changes continue and later cause an increase in dimensional stability, color changes, and a decrease in some mechanical properties [17–19].

Few studies have considered the effects of forest fires on the physico-mechanical properties of wood or wood products. No effects on density or shrinkage properties have been found in the wood of loblolly pine in Brazil [20] or maritime pine in Portugal [21]. However, significant reductions in the bending properties (the modulus of rupture, MOR, and the modulus of elasticity, MOE), compression strength, and toughness values have been observed for loblolly pine when flames consume the total crown, and the compression strength has been shown to be reduced when the crown is partially consumed [20]. Maritime pine wood showed a similar response in a severe forest fire, but the reduction in the properties was traced more to biological degradation due to prolonged wood harvesting than the fire [21]. Fire effects on physico-mechanical properties and chemical composition do not cause sufficient chemical degradation or strength reduction, which would cause the rejection of these woods for commodity uses of mechanical wood processing; however, in structural uses, caution should be taken for wood from trees that have been fully or partially consumed by fire. In Uganda, Caribbean pine wood from 10-year-old burned trees has a significantly lower density, MOE, and MOR than wood from unburned trees, and it should not be used for high-strength structural purposes but rather in low-strength construction works, such as shuttering and ceilings [22].

Both the direct and indirect effects of forest fire are present in wood salvaged from fire areas, and their nature and severity depend on the time between the fire and harvesting [5,14,16]. Trees with lowered vitality are susceptible to insect and microbial attacks and wind falls; the resultant wood defects lower the basic quality of sawn timber and veneer-based products but do not ruin their potential use. The susceptibility varies a lot between individual forest stands, but the following generalizations have been presented [16,23]: (1) blue stain occurs in the first year after the fire, lowering the visual quality; (2) heartwood starts to decay, and the deterioration of sapwood accelerates in the second year, reducing the quality for high-strength products; and (3) the sapwood of most coniferous species is unworthy for any construction purpose three years after the fire. The types and rates of deterioration of burned wood after forest fires in North America are illustrated in Figure 1.

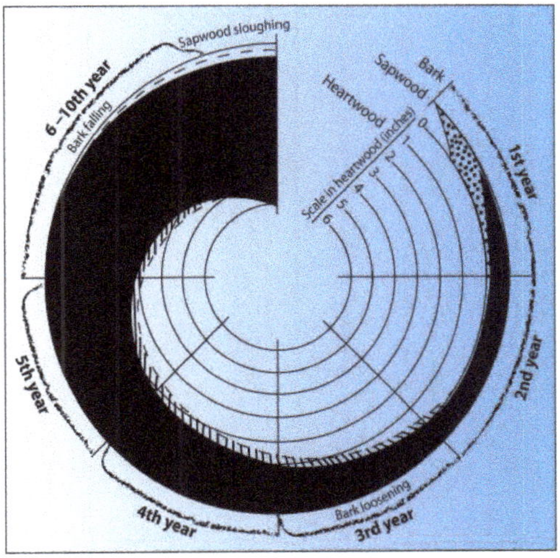

Figure 1. Effects of forest fire on the deterioration of softwood timber according to time from harvesting in the northern and western regions of the United States and Canada [16].

In practice, wood salvaged from forest fire areas has been allocated to either bioenergy or low-grade sawn timber; in North America, some wood has also been allocated to laminated veneer lumber (LVL) [16,24–26]. The suitability of burned wood has also been studied for oriented strand board (OSB) and medium-density fiberboard (MDF) [27,28]. Burned wood is not accepted for pulp, paper, or paperboard processes or products due to char and soot [28]. In North America, burned timber has been posited to still be usable for sawn timber if the logs are harvested within two years after the fire [29], but other sources claim that harvesting should be done at the latest one year after the fire to recover the full economic benefits of such logging [30,31]. In Finland, the Forest Damage Prevention Act suggests that forest-damaged trees should be harvested from the forest if their volume exceeds 10 or 20 m^3/ha for Norway spruce or Scots pine, respectively, counting trees with a stump diameter of more than 10 cm [32].

The objective of this study was to investigate the raw material potential of sawn wood and to select further processed products of Norway spruce and Scots pine sawn log trees from salvage harvesting in forest fire areas with northern European climatic conditions by comparing selected physical and mechanical properties of fire-affected trees and virgin reference trees. The effects of fire on the MOE, MOR, Brinell hardness (HB), moisture gradient, and wood color (CIE L*a*b*) were studied on clear wood specimens of outer sapwood and inner heartwood, and linear mixed models were constructed to analyze the sources of variation. Recommendations are provided for sawmill managers, further processors, and forest owners on wood allocation and potential products from forest fire areas in Finland.

2. Materials and Methods

Wood samples of forest fire trees were collected from burned stands of Norway spruce and Scots pine in an area of 227 hectares in Kalajoki, Western Finland (64°00′15.0″ N 24°08′09.0″ E), in November 2021, four months after the fire in July 2021. In Kalajoki, flammability was high before the fire. The local temperature was 27 °C, and the air humidity was 44% on the first day of the fire, 26 July 2021. The wind speed was 3.5 m/s, and in gusts, it was 9–10 m/s. In the beginning, the fire spread 2–3 m/min, and at its highest, the rate of spread was 15–20 m/min. The fire reached the entire area in two and a half days (until 28 July 2021), and in its most intense phases, it spread into forests as a crown fire [33].

The studied stands are presented in Figures 2 and 3. The samples of the reference trees were collected simultaneously from stands located near the fire area in Kalajoki and supplemented with trees in February 2022 from Raahe (64°23′36.8″ N 25°03′01.8″ E) to obtain comparable reference data in terms of tree age and diameter.

Figure 2. The burned Norway spruce stands studied in Kalajoki, Finland, at the time of sample collection in November 2021.

Figure 3. The burned Scots pine stands studied in Kalajoki, Finland, at the time of sample collection in November 2021.

The stems of the burned trees and reference trees were cut in the forest into 4 m logs up to a top diameter of about 10 cm, which is close to the commercial minimum diameter of small-diameter logs. Four normal-sized log burned trees (diameter at breast height, DBH, about 25–30 cm) and four small-log burned trees (DBH, about 14–18 cm) of both Norway spruce and Scots pine were selected, and the total number of logs with references was 24 (Table 1). Before the cut, the length of the area damaged by the fire was measured on the burned trees based on the signs visible on the bark in two ways: (1) by considering the entire circumference of the stem, i.e., the shell surface burned, and (2) by considering all clear signs of the effects of the fire on the bark of the stem. From each 4 m log, a block of approximately 60 cm was further cut as the material to be used in physical and mechanical tests. The blocks (total number of 72) represented the butt, middle, and top logs.

In the case of spruce, the diameters and lengths of the burned and reference trees were comparable, but the normal-sized log burned trees were younger than the reference trees (Table 2). For pine, the lengths and diameters of the burned and reference trees were comparable, but the burned trees were slightly older than the reference trees.

Table 1. Framework of tree sampling by species and size of burned trees in Kalajoki and reference trees in Kalajoki and Raahe.

Species	Stem Class	DBH [1] Class, mm	N, Burned Kalajoki	N, Reference Kalajoki	N, Reference Raahe
Norway spruce	Normal-log tree	250–300	4	0	2
Norway spruce	Small-log tree	140–180	4	2	0
Scots pine	Normal-log tree	250–300	4	0	2
Scots pine	Small-log tree	140–180	4	0	2

[1] DBH = diameter at breast height.

Table 2. Age (at stump), dimensions, and length of the burned area of forest fire trees and reference trees by species and stem class: means and ranges of variation. The burned area is the vertical extent from the stump upwards: Burned 1 = complete surface mantle of the tree is burned. Burned 2 = burned parts are visible in the surface mantle area of the tree.

Fire Class	Species	Stem Class	Age, Years	DBH, mm	Height, m	Burned 1, m	Burned 2, m
Burned	Spruce	Normal-log tree	52 (50–56)	268 (259–284)	18.0 (16.9–19.4)	2.0 (0.5–3.1)	5.2 (2.3–6.7)
Burned	Spruce	Small-log tree	52 (45–56)	169 (145–183)	14.0 (10.2–16.6)	1.0 (0.7–1.3)	7.0 (1.7–11.5)
Burned	Pine	Normal-log tree	82 (72–90)	267 (252–280)	19.4 (16.7–20.7)	2.6 (1.8–3.0)	4.3 (4.0–4.6)
Burned	Pine	Small-log tree	69 (56–82)	156 (135–174)	16.5 (15.5–17.7)	1.8 (1.1–2.7)	3.1 (2.0–3.6)
Reference	Spruce	Normal-log tree	74 (68–79)	263 (255–271)	20.1 (20.0–20.1)	–	–
Reference	Spruce	Small-log tree	54 (53–55)	159 (151–167)	13.9 (13.5–14.3)	–	–
Reference	Pine	Normal-log tree	76 (74–77)	261 (252–271)	21.4 (20.4–22.4)	–	–
Reference	Pine	Small-log tree	62 (53–71)	158 (147–168)	15.6 (15.2–16.0)	–	–

Sixty-centimeter blocks were through and through sawn with a band saw in Joensuu in February 2022. A 60 mm thick pith-centered balk from each block was sawn towards the surface, totaling 72 balks for further tests. If there were visually noticeable differences in the intensity of the burning on different sides of the block, the direction of sawing was set in such a way that the most heavily burned area (the more damaged side) of the stem was placed on one edge of the pith-centered balk. The slightly burned area (the less damaged side) was placed on the opposite edge of the balk. The sawing direction was set for the reference blocks and the blocks that were burned entirely around the circumference to minimize knots and other defects in the final balks.

The balks were stored at room temperature (around 20 °C) for one week. The top end was trimmed with a blade saw from each balk (length of 60 cm); one cross-sectional slice was sawn to determine bark thickness, basic density, and moisture content; and two slices (thickness of 10 mm) were sawn for annual ring and chemical analyses.

The ring width was, on average, larger in the burned trees than in the reference trees, excluding the top and middle logs of Scots pine, indicating the somewhat younger age of the burned trees. The bark was generally thin in all materials, partly because of the multiple handling of logs and balk during sampling. As expected, the bark was thinner in the burned trees than in the reference trees for both species. Among the butt logs, the bark was thicker for Scots pine than for Norway spruce but thinner in the middle and top logs. A large ring width may indicate tree vitality and recovery ability after a fire. Thick bark before the fire may indicate a high protective capacity against the fire, while thin bark after the fire may indicate its large reductive and devastating effects. The variations in ring width and bark thickness in the data are described in Table 3. Before the preparation of the mechanical test specimens, the bark and outer branches were removed with a precision circular table saw from the outer faces of the remaining part of each balk (approximate length of 55 cm). Subsequently, the balks were planed on flat sides to a 60 mm thickness, and bars of 60 × 30 mm (tangential × radial) were sawn in the longitudinal direction from the outer faces (sapwood section) and the core of 10–15 annual rings from the pith (heartwood section). It was hypothesized that the effect of the fire was at its highest in the sapwood pieces, which were located immediately close to the bark. The nominal number of bars was 288.

Table 3. Mean bark thickness (mm) and annual ring width (mm) on the more damaged and less damaged sides of forest fire trees and on the two opposite sides of reference trees by species and vertical log location in the balks after being sawn. Middle logs and top logs have common means of bark thickness.

	Norway Spruce			Scots Pine		
Burned Trees	**Butt Log**	**Middle Log**	**Top Log**	**Butt Log**	**Middle Log**	**Top Log**
Bark thickness, more damaged side	3.30	3.02	3.02	4.56	1.71	1.71
Bark thickness, less damaged side	2.89	3.42	3.42	4.46	1.76	1.76
Ring width, more damaged side	2.71	2.69	2.55	2.06	1.73	1.24
Ring width, less damaged side	2.80	2.75	2.53	2.06	1.75	1.26
Reference Trees	**Butt log**	**Middle log**	**Top log**	**Butt log**	**Middle log**	**Top log**
Bark thickness, side of maximum bark	4.33	3.87	3.87	5.58	1.89	1.89
Bark thickness, opposite side	4.19	3.93	3.93	5.42	1.73	1.73
Ring width, side of maximum bark	1.88	2.12	2.50	1.68	1.84	1.80
Ring width, opposite side	1.73	1.90	2.40	1.70	1.81	2.13

The bars, coded with marks on the outer faces, were conditioned for 7–8 days in a weather chamber (60 °C, RH 45%). Subsequently, the bars were slightly planed on both sides. A piece of 340 mm was cut and planed from the pith side to a thickness of 20 mm. The samples were finished with a dual-blade circular saw to standard test specimens (20 × 20 × 340 mm) for determining the bending properties of clear wood (Table 4). The specimens were conditioned under standard conditions (20 °C, RH 65%) to reach a moisture content of 12% (dry weight basis). From the remaining parts of the bars, specimens for the HB test (60 × 28.5 × 90 mm) and CIE L*a*b* color analysis (30 × 20 × 60 mm) were prepared.

Table 4. The number of samples used for testing air-dry density, modulus of elasticity (MOE), flexural strain, and modulus or rupture (MOR) by species, damage class, radial position, and log type.

	Norway Spruce			Scots Pine		
Wood Material	**Butt Log**	**Middle Log**	**Top Log**	**Butt Log**	**Middle Log**	**Top Log**
Sapwood, heavily burned side	8	12	6	15	6	5
Sapwood, slightly burned side	8	12	4	15	6	5
Heartwood, burned trees	20	12	–	20	7	–
Sapwood, reference trees	12	8	8	16	8	6
Heartwood, reference trees	12	6	–	11	4	4

For the mechanical tests, the air-dry density was determined for each specimen according to ISO standard 13061-2 [34]. The MOE (E_{12}, GPa) and the static bending strength, the MOR ($\sigma_{b,12}$, MPa), were determined using a three-point bending test in the tangential direction according to standards ISO 13061-4 and ISO 13061-3, respectively [35,36]. Flexural stress (σ_f, MPa) and strain (ε_f, %) were calculated according to standard ISO 178 [37]:

$$E_{12} = \frac{Pl^3}{4bh^3 f} \tag{1}$$

$$\sigma_{b,12} = \frac{3P_{max}l}{2bh^2} \tag{2}$$

$$\sigma_f = \frac{3P_{appl}l}{2bh^2} \tag{3}$$

$$\varepsilon_f = \frac{600fh}{l^2} \quad (4)$$

where P is the difference between the higher load and the lower load (N); l is the span (mm); b is the specimen width (mm); h is the specimen height (mm); f is the deflection (mm); P_{max} is the maximum load (N); and P_{appl} is the applied load (N).

Stored energy (U, MJ/m^3) at maximum stress was calculated using the following formula:

$$U = \int_0^{\varepsilon_{max}} \sigma_f d\varepsilon \quad (5)$$

where ε_{max} is the strain at maximum stress.

The moisture content of the samples was determined after each mechanical test and oven drying, and it was calculated on a dry weight basis, according to ISO standard 13061-1 [38]. The samples were weighed, dried at 103 °C for 24 h, and weighed dry. On average, the moisture content was 9.9% in pine samples and 10.7% in spruce samples. An approximate adjustment of the MOE and MOR to 12% moisture content was performed according to the given ISO standard.

HB was determined under air-dry conditions following standard EN 1534 [39]. A force of 1 kN was applied to a steel ball with a contact radius of 10 mm. In the test, defect-free spots from the outer edge of the sample were chosen.

$$HB = \frac{2F}{\pi D\left(D - \sqrt{D^2 - d^2}\right)} \quad (6)$$

$$d = 2\sqrt{(10h - h^2)} \quad (7)$$

where HB is the Brinell hardness (N/mm^2); F is the maximum load force (N); D is the diameter of the ball (mm); d is the diameter of the residual indentation (mm); and h is the depth of the residual indentation (mm). The MOE, MOR, and HB were determined using the universal material testing machine Zwick Z050 (ZwickRoell GmbH, Ulm, Germany; maximum load of 50 kN).

Variations in basic density and moisture content variation, called the moisture gradient, were determined starting from the 72 battens, which were sawn in the radial direction from the cross-sectional slices (see before). The samples were stored in airproof plastic bags. From each batten, 2–4 cubes (dimensions of 20 × 20 × 20 mm) were sawn between the pith and bark, with the number of cubes depending on the diameter of the original log. The first cube was always taken immediately under the bark, and the following cubes were taken towards the pith at a distance of 20 mm each. If the log diameter did not allow for a 20 mm distance between the third and fourth cubes, the fourth cube was taken with a smaller distance from the third cube. The basic density and moisture gradient were determined from the same samples. The basic density of wood (kg/m^3) was determined using the oven-dry method, according to standard ISO 13061-2 [34].

A color analysis was performed on HB specimens according to standard ISO/CIE 11664-4 [40] with a portable spectrophotometer (Konica Minolta 2600d, Osaka, Japan). In the CIELAB color space, L* indicates lightness from black to white, a* represents the red/green coordinate, and b* represents the yellow/blue coordinate. Two measurements were taken from each specimen on its outer edge. The specimens were conditioned in a weather chamber approximately 1 week before the measurements. The color difference between the reference trees and burned trees was calculated using the CIE76 color difference formula:

$$\Delta E^*_{ab} = \sqrt{(\Delta L^*)^2 + (\Delta a^*)^2 + (\Delta b^*)^2} \quad (8)$$

where ΔE^*_{ab} is the color difference, and ΔL^*, Δa^*, and Δb^* are the differences in the corresponding CIELAB coordinates.

Descriptive statistics of the density and mechanical properties and the Pearson correlation between them are reported. Linear mixed-effect models (LMMs) were used to determine the variability in the mechanical properties, moisture gradient, color, and effects of selected factors on them:

$$A_i = b_0 + dmg_j + h_k + r + \rho_a + r * \rho_a + u_s + u_{ts} + \varepsilon, \tag{9}$$

$$MC = b_0 + dmg_j + h_k + r + \rho_b + dmg_j * r + r * \rho_b + u_s + u_{ts} + \varepsilon, \tag{10}$$

$$C_m = b_0 + dmg_j + h_k + r + dmg_j * r + u_s + u_{ts} + \varepsilon, \tag{11}$$

where A_i is the modulus of elasticity (GPa), modulus of rupture (MPa), or Brinell hardness (N/mm^2); MC is the moisture content (%); C_m is the CIELAB color coordinate L^*, a^*, or b^*; b_0 is the intercept; dmg_j is the fire damage class; h_k is the height class; r is the dummy term of the radial position (sapwood); ρ_a is the air-dry density (kg/m^3); ρ_b is the basic density (kg/m^3); $r * \rho_a$ is the interaction term of the radial position and air-dry density; $dmg_j * r$ is the interaction term of the fire damage class and radial position; $r * \rho_b$ is the interaction term of the radial position and basic density; u_s is the random stand effect; u_{ts} is the random tree effect; and ε is the random error term.

The LMM allows for the inclusion of dependent fixed effects (such as fire damage and radial position) and more than one source of random variation (such as tree and stand-level effect) in the model [41]. The fire damage class was included as a fixed factor in all models. The linear mixed-effects model (MIXED) procedure in IBM SPSS Statistics fits the model to the data. Normal probability distributions were tested and therefore assumed in the models. The accuracy of the predictions of the linear model was determined by calculating the proportion of explained variance (R^2), bias, and root mean square error (RMSE) using only the fixed effects. The relative bias (bias, %) and relative error (RMSE, %) were calculated by dividing the bias and RMSE by the mean of the measured values.

3. Results

3.1. Mechanical Properties

3.1.1. Density

Descriptive statistics of the air-dry density are presented in Table 5. The highest density was observed in sapwood in both spruce and pine. In pine sapwood, the density in the butt logs was higher than in the middle or top logs. In spruce, the density variation between log types was small, especially in the burned trees. The highest variation was observed in the butt logs of pine, where the variation decreased in the middle and top logs. The variation in the density of spruce was smaller than that of pine, and the differences in the variation between the log types were modest. Due to the young cambial age of the top logs, heartwood samples did not occur in the burned trees or spruce reference trees. The results indicate that the spruce reference trees and burned pine trees had higher density values than the respective burned spruce trees and pine reference trees. The low density in the burned Norway spruce trees indicates rapid growth in young stands. Pine had more variation in size and age than spruce, which partly explained the higher variation in pine density.

Table 5. Air-dry density by species, damage class, radial position, and log type (kg/m^3). Values are presented as the means, with the standard deviation in parentheses.

Wood Material	Norway Spruce			Scots Pine		
	Butt Log	Middle Log	Top Log	Butt Log	Middle Log	Top Log
Sapwood, heavily burned side	446.5 (42.6)	448.7 (32.5)	442.0 (18.4)	581.3 (71.8)	489.9 (41.3)	489.2 (33.4)
Sapwood, slightly burned side	448.9 (44.1)	455.5 (32.5)	446.1 (34.4)	585.4 (67.4)	476.2 (31.3)	489.0 (38.8)
Heartwood, burned trees	388.3 (32.3)	402.0 (25.1)	–	544.2 (95.5)	478.3 (28.4)	–
Sapwood, reference trees	474.8 (40.7)	448.9 (55.5)	448.0 (47.3)	527.7 (78.1)	476.3 (63.5)	446.2 (45.9)
Heartwood, reference trees	467.5 (38.8)	446.6 (33.4)	–	468.6 (62.2)	447.1 (22.4)	450.0 (60.6)

3.1.2. Modulus of Elasticity

Descriptive statistics of the MOE are presented in Table 6. The MOE and air-dry density were strongly correlated in spruce (0.761) and pine (0.752). The differences in variation between the species and between the log types were more minor than in density. The differences in the MOE between the burned trees and reference trees in both species corresponded to the differences in density. The highest MOE was observed in the pine sapwood of the burned trees in the butt logs and the smallest in the spruce heartwood of the burned trees in the butt logs.

Table 6. Modulus of elasticity (MOE) by species, damage class, radial position, and log type (GPa). Values are presented as the means, with the standard deviation in parentheses.

Wood Material	Norway Spruce			Scots Pine		
	Butt Log	Middle Log	Top Log	Butt Log	Middle Log	Top Log
Sapwood, heavily burned side	8.0 (1.6)	9.2 (1.6)	7.7 (1.6)	12.3 (1.7)	9.1 (1.4)	8.9 (1.2)
Sapwood, slightly burned side	8.5 (1.9)	9.3 (1.0)	8.0 (0.4)	11.6 (2.0)	9.7 (1.6)	9.3 (1.4)
Heartwood, burned trees	5.4 (1.3)	7.0 (0.8)	–	9.4 (1.4)	9.1 (1.2)	–
Sapwood, reference trees	10.4 (2.7)	9.8 (2.2)	9.6 (2.4)	10.6 (2.9)	9.5 (2.4)	7.7 (2.4)
Heartwood, reference trees	9.8 (2.5)	10.0 (2.1)	–	8.3 (2.6)	7.9 (2.1)	6.7 (1.6)

In the linear mixed model for the MOE, significant fixed factors differed between the species. Density was the most significant factor for both species ($p < 0.01$) (Table 7). Height class was very significant for spruce and significant for pine ($p < 0.01$). The radial position was a significant factor for pine but had no statistical significance for spruce. Fire damage class was not a significant factor for either species, although the significance was close to 0.05 in spruce. The interaction between the radial position and density was very significant for pine ($p < 0.001$) but did not have any significance for spruce.

The fixed coefficients of the models for the MOE are presented in Table 8. A higher density indicates a higher MOE. In both spruce and pine, the MOE was the lowest in the top logs and the highest in the middle logs. In pine, the MOE was higher in sapwood than in heartwood. The positive interaction between the radial position and density in pine indicated that density had a more substantial positive effect on the sapwood MOE than on heartwood. However, fire damage did not have any significant effect on the MOE. In spruce, the RMSE and proportion of stand-level variation (83.0%) of the total random variation were high. Most of the random variation in pine was residual (51.8%), and the RMSE was smaller.

Table 7. Fixed effects in the linear mixed model (Equation (9)) for the modulus of elasticity (MOE) and the variation in random effects.

	Norway Spruce				Scots Pine			
Source	**F**	**df1**	**df2**	**Sig.**	**F**	**df1**	**df2**	**Sig.**
Fire damage class	2.782	2	120	0.066	0.760	2	119	0.470
Height class	10.522	2	120	<0.001	5.644	2	119	0.005
Radial position	0.062	1	120	0.804	7.100	1	119	0.009
Air-dry density	17.392	1	120	<0.001	66.026	1	119	<0.001
Radial position * air-dry density	0.287	1	120	0.593	14.004	1	119	<0.001
Random effect	**Variance**				**Variance**			
Stand	6.866				0.068			
Tree	0.269				0.983			
Residual	1.135				1.130			
Fitting statistics								
R^2	36.7%				64.5%			
Bias	−1.01 (−11.9%)				0.00 (0.0%)			
RMSE	2.14 (25.4%)				1.47 (15.0%)			

* refers to interaction.

Table 8. Fixed coefficients in the linear mixed model (Equation (9)) for the modulus of elasticity (MOE).

	Norway Spruce				Scots Pine			
Variable	**Estimate**	**Std. Error**	**95% CI**		**Estimate**	**Std. Error**	**95% CI**	
			Lower	**Upper**			**Lower**	**Upper**
Intercept	1.717	2.866	−3.956	7.391	1.314	1.375	−1.408	4.036
Fire damage class (ref: no damage)								
More damaged side	0.869	0.442	−0.007	1.745	−0.020	0.443	−0.897	0.857
Less damaged side	1.056	0.450	0.165	1.947	−0.206	0.442	−1.082	0.669
Height class (ref: top log)								
Butt log	0.034	0.313	−0.586	0.654	0.729	0.401	−0.065	1.524
Middle log	0.938	0.311	0.323	1.552	0.864	0.415	0.042	1.686
Radial position (ref: heartwood)								
Sapwood	−0.584	2.353	−5.243	4.074	−4.899	1.764	−8.391	−1.407
Air-dry density	0.015	0.005	0.005	0.025	0.013	0.003	0.008	0.019
Interaction of radial position with air-dry density (ref: heartwood * density)								
Sapwood * density	0.003	0.005	−0.007	0.013	0.012	0.003	0.005	0.019

* refers to interaction.

3.1.3. Modulus of Rupture

The descriptive statistics of the MOR are presented in Table 9. The MOR had a strong correlation with air-dry density (0.720 and 0.770) and a very strong correlation with the MOE (0.919 and 0.900) in spruce and pine, respectively. Therefore, the MOR in each group mainly followed the MOE. Although the MOR in heartwood in the burned butt logs of spruce was the smallest of all groups, the relative difference from the other groups was smaller than in the case of the MOE.

Table 9. Modulus of rupture (MOR) by species, damage class, radial position, and log type (MPa). Values are presented as means, with the standard deviation in parentheses.

	Norway Spruce			Scots Pine		
Wood Material	**Butt Log**	**Middle Log**	**Top Log**	**Butt Log**	**Middle Log**	**Top Log**
Sapwood, heavily burned side	74.3 (13.5)	75.2 (11.8)	62.4 (13.5)	104.2 (19.6)	81.2 (11.6)	81.2 (9.8)
Sapwood, slightly burned side	75.3 (12.3)	77.1 (7.8)	62.8 (7.5)	108.3 (19.1)	82.9 (11.8)	81.9 (17.5)
Heartwood, burned trees	51.7 (10.6)	58.7 (9.3)	–	84.4 (12.5)	82.6 (9.0)	–
Sapwood, reference trees	86.4 (16.9)	77.2 (12.8)	77.5 (14.8)	92.8 (20.7)	83.8 (18.7)	71.8 (15.5)
Heartwood, reference trees	77.3 (20.5)	77.4 (12.8)	–	67.2 (16.8)	72.1 (13.9)	67.3 (13.2)

No significant differences were observed in the MOR of the spruce sapwood of the butt or middle logs, but the MOR of the sapwood of the burned trees in the top logs was lower than in the reference trees (Table 10). As for the MOE, density was the most significant factor affecting the MOR of spruce and pine ($p < 0.001$). The radial position was a very significant factor in pine ($p < 0.001$) but did not have any significance in spruce. Height class was a significant factor in spruce ($p < 0.05$), but it did not have any significance in pine. In contrast to the MOE, fire damage class was a significant factor in pine (<0.05). The interaction between the radial position and density was a significant factor in pine ($p < 0.01$) but not in spruce. The RMSE values were lower in pine than in spruce. As in the case of the MOE, most of the random variation was on the stand level (69.0%) for spruce and on the residual level for pine (55.3%).

Table 10. Fixed effects in the linear mixed model (Equation (9)) for the modulus of rupture (MOR) and the variation in random effects.

	Norway Spruce				Scots Pine			
Source	**F**	**df1**	**df2**	**Sig.**	**F**	**df1**	**df2**	**Sig.**
Fire damage class	0.981	2	120	0.378	3.593	2	118	0.031
Height class	4.478	2	120	0.013	0.649	2	118	0.525
Radial position	0.028	1	120	0.867	20.600	1	118	<0.001
Air-dry density	18.419	1	120	<0.001	81.054	1	118	<0.001
Radial position * air-dry density	0.399	1	120	0.529	31.767	1	118	<0.001
Random effect	**Variance**				**Variance**			
Stand	215.127				53.704			
Tree	0.000				22.311			
Residual	96.850				94.116			
Fitting statistics								
R^2	42.8%				67.1%			
Bias	−5.76 (−8.1%)				0.86 (1.0%)			
RMSE	14.0 (19.8%)				11.6 (13.3%)			

* refers to interaction.

Table 11 shows the fixed coefficients of the models for the MOR. The effect of density was positive for the MOR. The sole fixed coefficient of the dummy factor of the sapwood position was negative in pine. However, because of the observed positive interaction between sapwood and density, the density indicated an increase in the MOR, and the effect of density on sapwood was larger than on heartwood. In spruce, the top logs had a lower MOR than the butt logs or middle logs. The only significant effect of fire damage was observed in pine, where the burned wood had a lower MOR than the wood from the reference trees.

Table 11. Fixed coefficients in the linear mixed model (Equation (9)) for the modulus or rupture (MOR).

Variable	Norway Spruce Estimate	Std. Error	95% CI Lower	Upper	Scots Pine Estimate	Std. Error	95% CI Lower	Upper
Intercept	15.511	19.585	−23.266	54.288	42.146	11.974	−23.266	54.288
Fire damage class (ref: no damage)								
More damaged side	3.505	3.198	−2.827	9.838	−11.485	4.285	−2.827	9.838
Less damaged side	4.496	3.288	−2.013	11.006	−9.319	4.316	−2.013	11.006
Height class (ref: top log)								
Butt log	6.391	2.865	0.719	12.064	2.830	2.894	0.719	12.064
Middle log	8.474	2.834	2.864	14.084	3.137	2.899	2.864	14.084
Radial position (ref: heartwood)								
Sapwood	−3.182	19.017	−40.833	34.470	−57.080	12.576	−40.833	34.470
Air-dry density	0.108	0.038	0.033	0.183	0.065	0.021	0.033	0.183
Interaction of radial position with air-dry density (ref: heartwood * density)								
Sapwood * density	0.027	0.043	−0.058	0.112	0.144	0.026	−0.058	0.112

* refers to interaction.

3.1.4. Flexural Strain

Typical stress–strain curves in the three-point bending test are illustrated in Figure 4. The higher force values at the same deformation (deflection) in Scots pine than in Norway spruce indicated higher MOE and MOR values. The corresponding stored energy values at maximum stress for the curves presented in Figure 3 were 0.81 MJ/m^3 (spruce, burned), 0.58 MJ/m^3 (spruce, reference), 1.16 MJ/m^3 (pine, burned), and 0.73 MJ/m^3 (pine, reference).

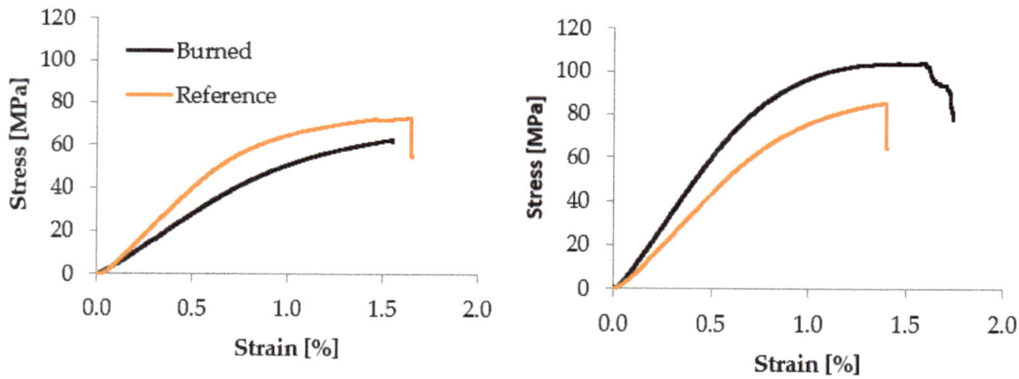

Figure 4. Typical flexural stress–strain curves of sapwood in burned wood (heavily burned side) and reference wood of Norway spruce (**left**) and Scots pine (**right**) in butt logs.

The flexural strain at maximum force during the three-point bending test is presented in Table 12. Strain and air-dry density had a modest negative correlation in spruce (−0.266) and a negligible correlation in pine (−0.071). The highest strain values at maximum force were observed in the reference trees in pine sapwood in the top logs and the lowest in the slightly burned trees in spruce heartwood in the butt logs. In Norway spruce, strain generally decreased towards the top, but in Scots pine, the lowest values were observed in most cases in the middle logs. In the middle and top logs, the strain values were, on average, higher in pine than in spruce in all groups.

Table 12. Flexural strain at maximum force by species, damage class, radial position, and log type (%). Values are presented as means, with the standard deviation in percentage points in parentheses.

	Norway Spruce			Scots Pine		
Wood Material	**Butt Log**	**Middle Log**	**Top Log**	**Butt Log**	**Middle Log**	**Top Log**
Sapwood, heavily burned side	1.47 (0.11)	1.34 (0.11)	1.29 (0.18)	1.42 (0.19)	1.39 (0.11)	1.43 (0.08)
Sapwood, slightly burned side	1.48 (0.15)	1.28 (0.15)	1.19 (0.32)	1.52 (0.19)	1.40 (0.11)	1.47 (0.07)
Heartwood, burned trees	1.53 (0.37)	1.24 (0.32)	–	1.37 (0.21)	1.44 (0.09)	–
Sapwood, reference trees	1.42 (0.18)	1.31 (0.21)	1.35 (0.14)	1.59 (0.25)	1.46 (0.13)	1.60 (0.23)
Heartwood, reference trees	1.23 (0.18)	1.32 (0.16)	–	1.47 (0.22)	1.36 (0.21)	1.46 (0.10)

3.1.5. Brinell Hardness

Descriptive statistics of HB are presented in Table 13. HB was strongly correlated with air-dry density in spruce (0.629) and pine (0.701). The correlations of HB with the MOE and MOR were moderate in spruce (0.465 and 0.476, respectively) but higher in pine (0.678 and 0.721, respectively). This was consistent with the more significant effects of density on the MOE and MOR in pine in the linear mixed model analysis than in spruce.

Table 13. Brinell hardness (HB) by species, damage class, radial position, and log type (MPa). Values are presented as means, with the standard deviation in parentheses.

	Norway Spruce			Scots Pine		
Wood Material	**Butt Log**	**Middle Log**	**Top Log**	**Butt Log**	**Middle Log**	**Top Log**
Sapwood, heavily burned side	12.4 (1.9)	12.0 (2.7)	11.8 (2.0)	15.3 (2.4)	12.4 (1.0)	12.6 (2.0)
Sapwood, slightly burned side	11.9 (1.8)	11.4 (1.6)	10.1 (1.8)	14.9 (2.6)	12.3 (1.1)	12.6 (0.7)
Heartwood, burned trees	9.9 (1.8)	9.6 (1.3)	–	14.7 (2.2)	12.5 (1.8)	–
Sapwood, reference trees	11.4 (1.1)	10.6 (2.1)	10.2 (2.0)	13.9 (4.3)	12.2 (2.9)	10.3 (2.1)
Heartwood, reference trees	11.2 (2.1)	11.2 (2.6)	–	11.7 (2.1)	10.7 (1.3)	9.9 (2.2)

Similar to other mechanical properties, density was the most significant factor for HB (Table 14). In pine, the radial position and the interaction between the radial position and density had some effects, similar to the MOE and MOR. In spruce, these factors were not significant. Instead, fire damage class proved significant ($p < 0.01$). The RMSE value was slightly smaller in spruce than in pine. Most of the random variation in both species was residual (74.8% for spruce and 53.6% for pine). Density had a positive effect on HB in both spruce and pine (Table 15). In spruce, the sapwood position was a positive predictor of HB. Although the estimate for the dummy term of sapwood was negative in pine, because of the positive interaction between the dummy term or sapwood position and density, the higher density indicated a higher HB in pine sapwood. Similar to the MOR, density had a higher effect on hardness in pine sapwood than in pine heartwood. It was notable that fire damage positively affected HB in spruce.

Table 14. Fixed effects in the linear mixed model (Equation (9)) for Brinell hardness (HB) and variation in random effects.

	Norway Spruce				Scots Pine			
Source	**F**	**df1**	**df2**	**Sig.**	**F**	**df1**	**df2**	**Sig.**
Fire damage class	5.093	2	120	0.008	2.661	2	119	0.074
Height class	1.800	2	120	0.170	1.914	2	119	0.152
Radial position	0.037	1	120	0.848	4.549	1	119	0.035
Air-dry density	25.152	1	120	<0.001	31.115	1	119	<0.001
Radial position * density	0.068	1	120	0.794	6.071	1	119	0.015

Table 14. *Cont.*

	Norway Spruce				Scots Pine			
Source	**F**	**df1**	**df2**	**Sig.**	**F**	**df1**	**df2**	**Sig.**
Random effect	**Variance**				**Variance**			
Stand	0.000				1.782			
Tree	0.600				1.034			
Residual	1.779				3.254			
Fitting statistics								
R^2	46.9%				42.9%			
Bias	0.00 (0.0%)				0.01 (0.0%)			
RMSE	1.49 (13.6%)				2.21 (16.6%)			

* refers to interaction.

Table 15. Fixed coefficients in the linear mixed model (Equation (9)) for Brinell hardness.

	Norway Spruce				Scots Pine			
Variable	**Estimate**	**Std. Error**	**95% CI**		**Estimate**	**Std. Error**	**95% CI**	
			Lower	**Upper**			**Lower**	**Upper**
Intercept	0.163	2.357	−4.505	4.830	7.452	2.2330	3.030	11.873
Fire damage class (ref: no damage)								
More damaged side	1.509	0.519	0.483	2.536	−1.568	0.7980	−3.148	0.012
Less damaged side	0.659	0.525	−0.380	1.699	−1.832	0.7973	−3.411	−0.254
Height class (ref: top log)								
Butt log	0.697	0.392	−0.080	1.474	0.879	0.5381	−0.187	1.944
Middle log	0.364	0.389	−0.407	1.134	0.139	0.5398	−0.929	1.208
Radial position (ref: heartwood)								
Sapwood	0.568	2.958	−5.288	6.424	−4.992	2.3406	−9.626	−0.357
Air-dry density	0.023	0.006	0.012	0.034	0.001	0.010	0.0039	0.002
Interaction of radial position with air-dry density (ref: heartwood * density)								
Sapwood * density	−0.002	0.006	−0.015	0.011	0.012	0.0048	0.002	0.021

* refers to interaction.

3.2. Moisture Gradient

The moisture content at different distances from the log surface is illustrated for the burned wood and the wood of the reference trees of spruce and pine by log type in Figure 5. A clear trend of decreasing moisture content from the surface to the pith was observed in all cases. The highest moisture content was observed in the surface layer of the reference trees for all log types (140–160%), and the lowest moisture content was observed in the core of the butt logs and middle logs of the burned and reference trees (30–40%). Due to the smaller diameter of the top logs, no samples were obtained there at 12–14 cm from the log surface.

Because the logs had varying thicknesses and the moisture content was expected to vary by distance from the log surface, the fixed effect of the radial position in the linear mixed models was divided into two classes. The specimen at 0–2 cm from the log surface represented the tree surface (outer sapwood), and the specimen at the farthest distance from the log surface represented the core (inner heartwood).

The fixed effects are illustrated in Table 16. In spruce, height class was highly significant for the moisture gradient ($p < 0.001$); other significant factors included fire damage class and the interaction between fire damage class and the radial position. In pine, the radial position and the interaction between the radial position and basic density were highly significant factors ($p < 0.001$), and basic density and height class were significant ($p < 0.01$). Most of the random variation was residual in spruce (53.5%) and pine (50.0%). The RMSE values were lower for pine than for spruce. For both spruce and pine, the moisture content was the lowest in the butt logs and the highest in the top logs (Table 17). In spruce, the burned wood had a lower moisture gradient than the wood from the reference trees because the forest fire decreased the moisture content in sapwood. The moisture content was higher in sapwood than in heartwood in pine and spruce. In pine, the fixed coefficient for basic density was positive, but due to the interaction between the radial position and basic density, the higher the basic density, the lower the moisture content in sapwood.

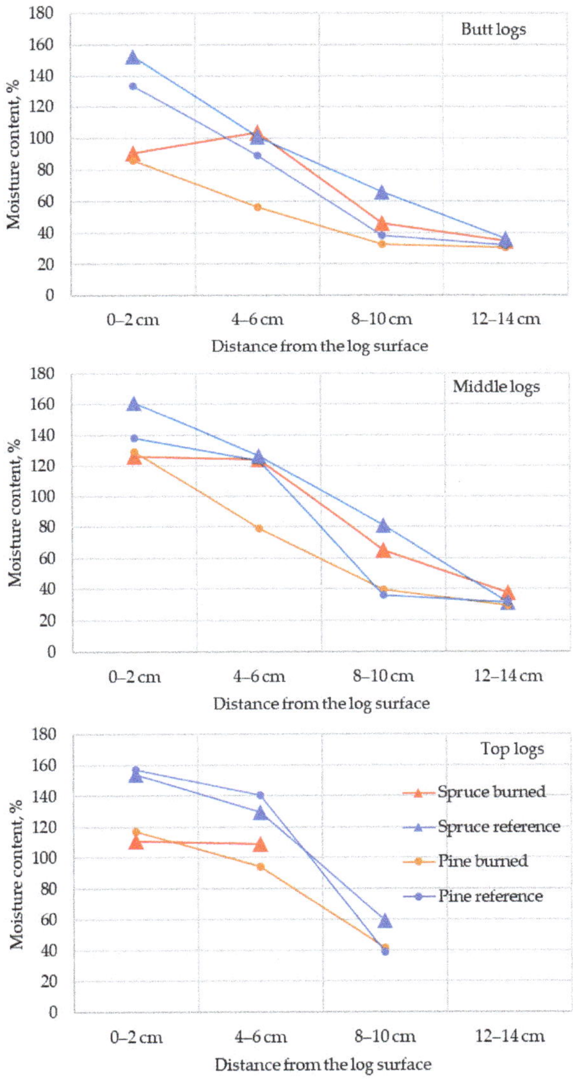

Figure 5. The moisture content of wood at different distances from the log surface in the butt, middle, and top logs.

Table 16. Fixed effects in the linear mixed model (Equation (10)) for moisture gradient and variation in random effects.

	Norway Spruce				Scots Pine			
Source	**F**	**df1**	**df2**	**Sig.**	**F**	**df1**	**df2**	**Sig.**
Fire damage class	5.943	2	133	0.003	0.410	2	150	0.665
Height class	9.215	2	133	<0.001	4.974	2	150	0.008
Radial position	1.400	1	133	0.239	93.330	1	150	<0.001
Basic density	1.563	1	133	0.213	8.400	1	150	0.004
Fire damage class * radial position	5.367	2	133	0.006	0.671	2	150	0.513
Radial position * basic density	0.040	1	133	0.841	54.860	1	150	<0.001
Random effect	**Variance**				**Variance**			
Stand	654.551				373.526			
Tree	275.985				151.485			
Residual	1070.354				525.179			
Fitting statistics								
R^2	42.4%				79.5%			
Bias	1.55 (1.6%)				0.00 (0.0%)			
RMSE	42.4 (44.9%)				21.6 (26.0%)			

* refers to interaction.

Table 17. Fixed coefficients in the linear mixed model (Equation (10)) for moisture content.

	Norway Spruce				Scots Pine			
Variable	**Estimate**	**Std. Error**	**95% CI**		**Estimate**	**Std. Error**	**95% CI**	
			Lower	**Upper**			**Lower**	**Upper**
Intercept	134.068	51.752	31.704	236.431	17.895	31.324	−43.998	79.788
Fire damage class (ref: no damage)								
More damaged side	−30.077	16.500	−62.713	2.559	−18.665	29.080	−76.124	38.793
Less damaged side	−30.108	16.946	−63.627	3.411	−20.975	29.098	−78.471	36.520
Height class (ref: top log)								
Butt log	−29.964	7.100	−44.007	−15.922	−14.387	5.682	−25.614	−3.160
Middle log	−14.362	8.178	−30.537	1.813	−2.399	5.890	−14.037	9.239
Radial position (ref: heartwood)								
Sapwood	92.055	59.173	−24.986	209.097	292.496	28.989	235.217	349.775
Basic density	−0.107	0.126	−0.356	0.142	0.136	0.060	0.016	0.255
Interaction of fire damage class * radial position (ref: no damage or heartwood)								
More damaged side * sapwood	−44.223	13.530	−70.985	−17.462	−9.415	9.096	−27.387	8.557
Less damaged side * sapwood	−24.310	13.809	−51.624	3.003	−0.641	9.064	−18.550	17.268
Interaction of radial position with basic density (ref: heartwood * basic density)								
Sapwood * basic density	−0.032	0.157	−0.343	0.280	−0.530	0.072	−0.672	−0.389

* refers to interaction.

3.3. Wood Color

The CIE L*a*b* color values differed between the burned and reference spruce sapwood at all tree heights. In pine sapwood, the difference in color between the burned and reference trees was evident in the butt logs but smaller in the middle and top logs. The L*, a*, and b* values are presented in Figure 6, Figure 7, and Figure 8, respectively.

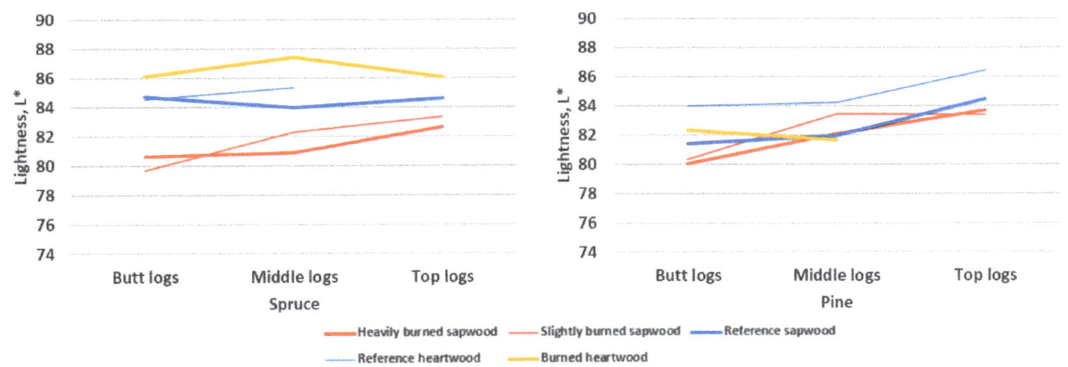

Figure 6. Lightness (L*) of spruce and pine wood by log type: sapwood of burned trees on the opposite sides (concerning the intensity of forest fire) and heartwood of burned trees, sapwood, and heartwood of reference trees. Moisture content is 12%.

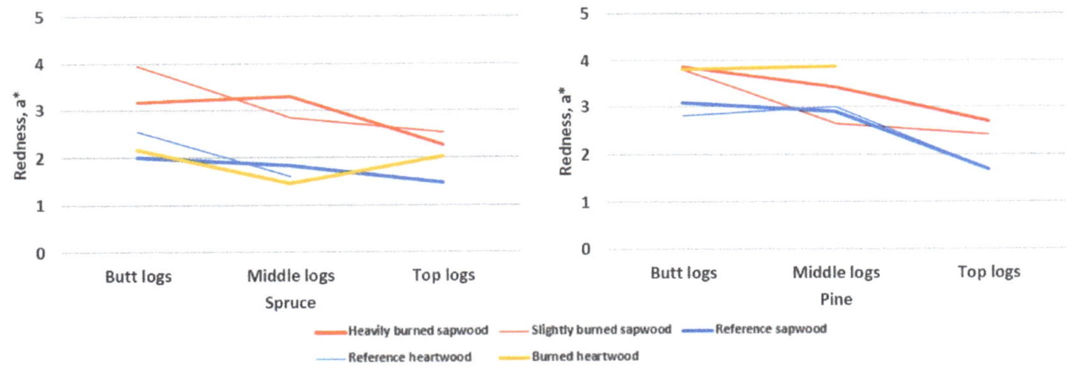

Figure 7. Redness (a*) of spruce and pine wood by log type: sapwood of burned trees on the opposite sides (concerning the intensity of forest fire) and heartwood of burned trees, sapwood, and heartwood of reference trees. Moisture content is 12%.

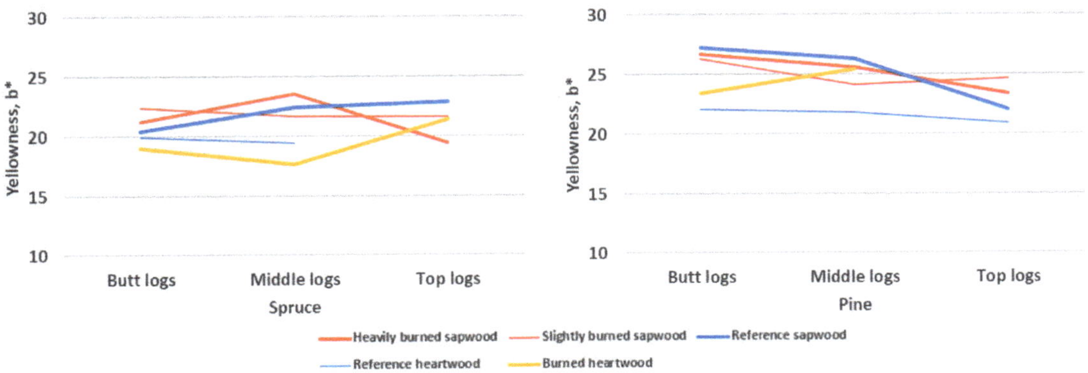

Figure 8. Yellowness (b*) of spruce and pine wood by log type: sapwood of burned trees on the opposite sides (concerning the intensity of forest fire) and heartwood of burned trees, sapwood, and heartwood of reference trees. Moisture content is 12%.

The fixed effects for the CIE L*, a*, and b* coordinates are presented separately for spruce wood in Table 18. The radial position was highly significant ($p < 0.001$) regarding all color coordinates. Height class was significant for the L* and a* coordinates ($p < 0.05$). The interaction between fire damage class and the radial position was very significant ($p < 0.001$) for the L* coordinate and significant ($p < 0.05$) for the a* coordinate. Most of the random variation was residual for all color coordinates (59.3–62.1%). The RMSE value was at its lowest, and the R^2 value was the highest for the L* coordinate. The fixed coefficients of the models for the color coordinates of spruce wood are presented in Table 19. Sapwood was darker and closer to green and yellow than heartwood, possibly because of the higher content of darker latewood in sapwood than in heartwood. Sapwood in the butt logs and middle logs was also darker than in the top logs. In the top logs, the a* coordinate was closer to green than in the butt and middle logs. Due to the fire damage class and radial position interaction, sapwood was darker and closer to green in the burned trees. This was explained by the most substantial effects of forest fires on the outer layer of the wood.

Table 18. Fixed effects in the linear mixed model (Equation (11)) for the color of Norway spruce wood and the variation in random effects.

	L*	a*	b*			L*	a*	b*
Source	F	F	F	df1	df2	Sig.	Sig.	Sig.
Fire damage class	0.084	0.286	0.093	2	234	0.919	0.751	0.911
Height class	4.202	5.974	0.156	2	234	0.016	0.003	0.856
Radial position	75.455	16.096	23.851	1	234	<0.001	<0.001	<0.001
Fire damage class * radial position	12.827	6.100	0.385	2	234	<0.001	0.003	0.681
Random effect	**Variance**	**Variance**	**Variance**					
Stand	6.677	1.481	9.158					
Tree	0.654	0.046	0.064					
Residual	10.704	2.427	15.125					
Fitting statistics								
R^2	31.9%	16.8%	9.4%					
Bias	−0.01 (0.0%)	0.00 (0.2%)	0.06 (0.3%)					
RMSE	3.29 (3.9%)	1.54 (62.7%)	3.91 (18.9%)					

* refers to interaction.

Table 19. Fixed coefficients in the linear mixed model (Equation (11)) for the color of Norway spruce and the variation in random effects.

	L		a*		b*	
Variable	**Estimate**	**Std. Error**	**Estimate**	**Std. Error**	**Estimate**	**Std. Error**
Intercept	86.309	2.750	1.446	1.287	19.544	3.191
Fire damage class (ref: no damage)						
More damaged side	1.725	3.781	−0.266	1.770	−0.931	4.393
Less damaged side	1.243	3.780	−0.219	1.770	−0.861	4.393
Height class (ref: top log)						
Butt log	−1.727	0.649	0.950	0.308	0.101	0.765
Middle log	−0.810	0.691	0.395	0.327	0.380	0.815
Radial position (ref: heartwood)						
Sapwood	−0.872	0.777	−0.142	0.369	2.045	0.921
Interaction of fire damage class * radial position (ref: no damage or heartwood)						
More damaged side * sapwood	−4.968	1.066	1.395	0.506	0.709	1.261
Less damaged side * sapwood	−4.364	1.068	1.650	0.507	1.095	1.265

* refers to interaction.

The fixed effects for the CIE L*, a*, and b* coordinates are presented separately for pine wood in Table 20. Similar to spruce, height class was highly significant ($p < 0.001$) for all color coordinates. The radial position was very significant ($p < 0.001$) for the L* and b* coordinates, but not for the a* coordinate, albeit the naturally red-brown color of pine heartwood. The fire damage class was significant for the a* coordinate. The fixed coefficients of the models for the color coordinates are presented in Table 21. As for pine, most of the random variation was residual, primarily for all color coordinates (61.0–32.4%). The R^2 values were, in general, lower than in spruce. Due to the apparent color difference between earlywood and latewood, the variation in the share of earlywood and latewood in the measurement spots probably explains most of the color variation. All R^2 values were relatively low.

Table 20. Fixed effects in the linear mixed model (Equation (11)) for the color of Scots pine and the variation in random effects.

	L*	a*	b*			L*	a*	b*
Source	**F**	**F**	**F**	**df1**	**df2**	**Sig.**	**Sig.**	**Sig.**
Fire damage class	0.719	3.541	0.952	2	252	0.488	0.030	0.387
Height class	12.434	7.621	7.885	2	252	<0.001	<0.001	<0.001
Radial position	13.129	0.233	31.437	1	252	<0.001	0.629	<0.001
Fire damage class * radial position	1.120	1.673	2.068	2	252	0.328	0.190	0.129
Random effect	**Variance**	**Variance**	**Variance**					
Stand	6.451	1.860	9.699					
Tree	0.000	0.079	0.446					
Residual	10.692	3.041	15.849					
Fitting statistics								
R^2	17.0%	13.4%	16.9%					
Bias	0.00 (0.0%)	0.00 (0.1%)	0.04 (0.2%)					
RMSE	3.32 (4.0%)	1.74 (53.9%)	3.96 (16.0%)					

* refers to interaction.

Table 21. Fixed coefficients in the linear mixed model (Equation (11)) for color coordinates of Scots pine and variation in random effects.

	L		a*		b*	
Variable	**Estimate**	**Std. Error**	**Estimate**	**Std. Error**	**Estimate**	**Std. Error**
Intercept	86.663	2.633	1.744	1.423	19.463	3.251
Fire damage class (ref: no damage)						
More damaged side	−2.359	3.687	1.543	1.988	2.333	4.542
Less damaged side	−1.517	3.680	0.280	1.984	0.810	4.533
Height class (ref: top log)						
Butt log	−2.925	0.595	1.255	0.322	2.906	0.735
Middle log	−1.947	0.667	0.960	0.358	2.507	0.818
Radial position (ref: heartwood)						
Sapwood	−2.403	0.664	0.108	0.359	4.309	0.819
Interaction of fire damage class * radial position (ref: no damage or heartwood)						
More damaged side * sapwood	1.476	1.034	−0.828	0.557	−2.573	1.272
Less damaged side * sapwood	1.039	1.013	0.165	0.546	−1.354	1.247

* refers to interaction.

The color differences between the wood from the burned trees and reference trees is presented in Table 22 as ΔE^*_{ab} values. The values were calculated based on the L, a*, and b* estimates in Tables 19 and 21. In the sapwood layer, a noticeable color difference was observed on the more damaged and less damaged sides compared with the wood from the reference trees in Norway spruce and Scots pine. In heartwood, the color change was less perceptible, especially in Norway spruce and on the less damaged side. However, due to the relatively high standard error in the estimated L, a*, and b*, caution is needed when interpreting single ΔE^*_{ab} values.

Table 22. ΔE^*_{ab} (Equation (8)) values between damaged wood and Norway spruce and Scots pine reference wood.

Fire Damage Class	Radial Position	Norway Spruce	Scots Pine
More damaged side	Sapwood	4.61	5.29
More damaged side	Heartwood	1.98	3.66
Less damaged side	Sapwood	4.77	4.77
Less damaged side	Heartwood	1.53	1.74

4. Discussion

Due to the global increase in forest fires during the 2000s, the direct and indirect damage that deteriorates raw wood materials has become more frequent in tree stock, resulting in challenges in the allocation of salvaged wood to the roundwood market and profitable utilization in forest industries [3,16,24]. The risk increase is not only commonly traced to climate change, regional droughts, and heat loads but also to human behavior and working with machines in nature and the built environment [5–10].

The data from this study represent a typical forest fire in Finland in terms of vastness, intensity, and season. According to our visual observations in the studied stands, the mortality of trees and fire patterns were consistent with previous findings in practical forestry. Based on the crown's status, most spruces were judged as dead or dying, whereas almost all pines remained living. The vertical length of the fire effect, starting from the base of the tree, was relatively short, with values of 9% for spruce and 12% for pine when the bark was required to have burned for the entire circumference, but the values were as high as 39% for spruce and 20% for pine when any signs of the fire effect were considered on bark. In a study on two forest stands in northern Europe, 20% of Scots pine trees survived in a heavy forest fire, with the proportion reaching 50% for trees with a DBH of 20 cm or more if the fire remained at a low height and below the living crown, whereas 99% of Norway spruce trees and 98% of broadleaved trees died [42].

Our results show that fire damage had only minor or no significant effects on the wood properties studied, evident only in sapwood. The changes in wood properties were more substantial in spruce than in pine, presumably because of the thinner bark, especially in the lower parts of the trees, that protects wood less from deterioration, and the longer crown that accelerates the spread of fire in the middle and top parts of trees [12,15,23,42].

In the mechanical tests, the physical properties and position of a tree explained most of the variation in wood properties. As expected, wood density proved to be the strongest predictor by far, followed by height location and the radial position. The fire damage class of trees at respective heights was identified as a significant factor only for the MOR in Scots pine (decrease) and for HB in Norway spruce (increase). In general, the models obtained were more accurate in predicting the mechanical properties of Scots pine. In Norway spruce, the bias and variation by random effects were relatively high, and the R^2 values were relatively low. This indicates that only some of the essential factors in modeling the bending properties of spruce are recognized. In flexural strain at maximum force, no clear differences were detected between the burned and the reference trees. In spruce logs, flexural strain decreased towards the top, but in pine, the lowest values were observed in the middle logs.

In studies on Brazilian loblolly pine and Portuguese maritime pine, where the forest fire intensities were heavier than in our study, more substantial effects on wood properties were observed. The consumption of the full crown of loblolly pine by fire led to a decrease in the MOE and MOR in static bending by 17% and 12%, respectively, in addition to decreases in ultimate stress and stress at the proportional limit in parallel compression by 19% and 34% and the resilience coefficient for toughness by 17% [20]. Similar reductions in these properties have been observed for maritime pine, albeit associated with reductions in biological degradation because of harvest 5–7 months after the fire [21]. The effect of the biological and chemical degradation of wood translates into changes in viscoelastic behavior [43].

Clear positive correlations between density and mechanical properties were noted in our study, as in numerous previous studies [44,45]. In sapwood (mature wood), the mechanical properties were also higher than in heartwood (juvenile wood), particularly in pine. Structural differences between the mature wood and juvenile wood of softwoods have been reported in several compilation works [45–47].

Forest fire caused a decrease in the moisture content in spruce sapwood due to its thin bark and partly dead wood (high mortality of trees). The moisture gradient between sapwood and heartwood after the fire was larger in pine wood than in spruce. The radial position and basic density had significant effects on the moisture content. Density and the moisture content were partially negatively correlated for pine but not for spruce. In both species, the moisture content decreased from the butt to the top. When modeling the moisture content, the RMSE values were relatively high in both species, and the model's explanatory power was higher in Scots pine. The slight difference in harvesting time may have influenced the moisture content difference between the burned trees and reference trees, with the former being harvested in November and the latter in February (except for small-log spruce trees). In Northern Europe, the moisture content of spruce and pine is the highest in the middle of winter and usually 2–4% lower in November than in February [48,49].

In Norway spruce and Scots pine sapwood, a perceptible color difference due to the fire was observed. In heartwood, the ΔE values indicated slight changes in color but were mostly below the perceptibility level of $\Delta E^*_{ab} = 2$ for wooden surfaces [50]. The wood color was the darkest in the burned Norway spruce wood of all studied wood materials.

Changes in wood color might indicate some chemical changes in the wood at high temperatures during the forest fire or some staining due to soot and char. The increase in the relative proportion of darker latewood with an increased cambial age explains the difference between heartwood and sapwood. In Scots pine, the color difference due to heartwood formation is naturally visible between sapwood and heartwood. A sharp boundary between heartwood and sapwood occurs naturally in Scots pine, while in Norway spruce, yellowish heartwood is not clearly distinguished from sapwood [51,52]. The differences in color parameters between heartwood and sapwood can mainly be explained by their different chemical compositions. The color of Scots pine heartwood is usually brown with a hue of red. The color of sapwood, especially earlywood, is typically very light, with a hue of yellow in both species. In addition, the proportion of darker latewood is higher in heartwood than in earlywood.

It is well known that the temperature level and burning time of the exposed area during a fire are critical factors for changes in the physico-mechanical and chemical properties of wood [12,23]. We had no information available on the thermal circumstances during the forest fire, but we assumed that the temperature elevation stayed lower than in the prescribed burning. The following temperatures were measured in experiments in the forest: 400–500 °C at ground level, 220 °C at the 3 m level, 150 °C at the 6 m level, and 45 °C at the 9 m level [15].

In a forest fire, bark provides protective support for wood, slowing or inhibiting the rise in temperature and the chemical and physical alteration of wood. Accordingly, in the loblolly pine study in Brazil, only the most severe burning level, where the flames consumed the total crown, had any effect on the chemical composition of the wood, decreasing the content of cold water-soluble extracts (tannins, gums, sugars, pigments, or colorants) due to the movement of water within the wood tissue [20]. Similarly, no chemical effects of fire on wood were observed in an Aleppo pine study conducted in Turkey [53].

The mortality rate of the burned trees and the high deterioration rate of fire-affected wood were evident for Norway spruce but less severe for Scots pine. Similar to North American softwoods [16], rapid wood procurement after a fire is essential to prevent excessive drying and further biological–physical degradation of the material. In Finland, the Forest Damage Prevention Act effectively prevents the advancement of secondary defects to a harmful degree in salvaged wood [32].

The partial suitability of burned wood has also been found for wood panel applications. Jack pine wood in Wisconsin, USA, had a similar engineering performance for oriented strand board (OSB) at four fire damage levels and met the standard mechanical properties when all char was removed from the flakes [27]. In a parallel study on red pine wood from fire-killed, fire-affected, and virgin trees, the addition of 20% charred bark among flakes reduced bending strength, still providing OSB mechanical properties meeting the standard requirements for all fire levels, but bark addition of up to 20% improved the stability of the boards. Scots pine fiber material from burned wood was tested as a mixture of virgin beech and oak to manufacture a medium-density fiberboard (MDF) in Turkey. All test panels met the mechanical properties of general-purpose MDF panels, but surface roughness remained higher and dimensional stability remained weaker than control panels from unburned beech and oak [28].

5. Conclusions

This study, based on wood materials from a forest fire area and a neighboring forest area in Western Finland, showed fire-induced changes in the physical and mechanical properties of the sapwood of Norway spruce and Scots pine in Northern Europe. The changes in the properties were more substantial in spruce than in pine, presumably because the thinner bark, especially in the lower parts of trees, protects wood less from deterioration, and the longer crown accelerates the spread of fire in the middle and top parts of trees.

In spruce, the moisture content decreased, and the color darkened compared to the reference trees. In pine, color changes were observed. Spruce wood underwent greater changes, probably due to the thinner bark and the longer crown being more susceptible to fire. The changes in mechanical properties were mostly unessential. In spruce, a slight increase in HB was observed, and in pine, a slight decrease in bending strength was observed.

Due to the minor deterioration, and from a mechanical point of view, fresh wood salvaged from forest fire areas, especially Scots pine but also Norway spruce, is a potential raw material for middle-quality and low-quality sawn timber and selected end-uses where the requirements for quality are moderate and aesthetic or hygienic criteria are not critical, including the following: varieties of construction, non-food packaging and pallets, and other less sensitive uses of wood. However, most physico-mechanical properties of virgin wood change only a little in a forest fire, especially in spruce char and soot, making use difficult in many product applications and manufacturing processes. Sapwood is prone to excessive drying, which may initiate checking in wood after being sawn, sliced, or rotary cut but reduces the need for drying wood products. We recommend further studies on the following to create and develop the use of wood from forest fire areas: (1) chemical composition and surface activity of burned wood, (2) performance and usability in different types of production and products (mechanical, chemical, and hybrid), (3) potential for diversified biorefining processes and chemical treatments, and (4) technical procedures to remove or reduce char and soot in wood material before processing.

Author Contributions: Conceptualization, V.M., A.H., P.K. and E.V.; Data curation, V.M., P.Y. and E.V.; Formal analysis, J.M., V.M., P.Y. and E.V.; Funding acquisition, J.M. and V.M.; Methodology, J.M., V.M. and E.V.; Project administration, J.M., P.K. and E.V.; Software, J.M. and V.M.; Supervision, V.M., A.H. and E.V.; Validation, J.M., V.M., A.H. and E.V.; Writing—original draft, J.M., V.M., P.Y. and E.V.; Writing—review and editing, J.M., V.M. and E.V. All authors have read and agreed to the published version of the manuscript.

Funding: This research was mainly funded by Puumiesten Ammattikasvatussäätiö (The Foundation for Finnish Wood Professionals' Education), project Wood properties and product potential of forest fire damaged timber in wood product uses, project no. 41007-00246600. Additional funding was provided for working hours by Luke's strategic project PosFibes, project no. 41007-00212300, and as a grant for the M.Sc. thesis of Pekka Ylimäki by Pyhäjoen Puu Oy, a wood product company located near the forest fire area in Pyhäjoki, Northern Ostrobothnia.

Institutional Review Board Statement: Not applicable.

Informed Consent Statement: Not applicable.

Data Availability Statement: Data are available on request from the corresponding author.

Acknowledgments: The authors want to thank the Natural Resources Institute Finland Luke technical staff in the Joensuu Research Unit for their contribution to preparing and trimming the wood laboratory samples. Päiviö Nikki, the owner of Pyhäjoen Puu Oy, participated in the preliminary discussion on the research problem and provided the site for the pre-processing of sample blocks and wood discs. The Forestry Association Pyhä-Kala from Kalajoki assisted in selecting the study stands. Two local forest entrepreneurs from Kalajoki and Raahe cut the sample trees and transported the logs to the pre-processing site. Luke Joensuu Research Unit and UEF, Department of Environmental and Biological Sciences (Joensuu), provided the laboratory infrastructure and methods for the project.

Conflicts of Interest: The authors declare no conflicts of interest.

References

1. ESA Climate Office. Multi-Decade Global Fire Dataset Set to Support Trend Analysis. The European Space Agency, 28 January 2021. Available online: https://climate.esa.int/de/news-events/multi-decade-global-fire-dataset-set-support-trend-analysis/ (accessed on 25 September 2023).
2. Burgueño Salas, E. Global Forest Cover Loss by Wildfires 2001–2022. Statista 2023. Available online: https://www.statista.com/statistics/1401539/forest-loss-by-wildfires/ (accessed on 25 September 2023).
3. Camia, A.; Durrant, T.; San-Miguel-Ayanz, J. *The European Fire Database: Technical Specifications and Data Submission*; EUR—Scientific and Technical Research series; Publications Office of the European Union: Luxembourg, 2014. [CrossRef]
4. San-Miguel-Ayanz, J.; Durrant, T.; Boca, R.; Maianti, P.; Liberta, G.; Artes Vivancos, T.; Jacome Felix Oom, D.; Branco, A.; De Rigo, D.; Ferrari, D.; et al. *Forest Fires in Europe, Middle East and North Africa 2021*; EUR—Scientific and Technical Research Series; Publications Office of the European Union: Luxembourg, 2022. [CrossRef]
5. Flannigan, M.D.; Amiro, B.B.; Logan, K.A.; Stocks, B.J.; Wotton, B.M. Forest Fires and Climate Change in the 21st Century. *Mitig. Adapt. Strateg. Glob. Chang.* **2005**, *11*, 847–859. [CrossRef]
6. Roshani, S.H.; Kumar, P.; Masroor, M.; Rahaman, M.H.; Rehman, S.; Ahmed, R.; Sahana, M. Forest Vulnerability to Climate Change: A Review for Future Research Framework. *Forests* **2022**, *13*, 917. [CrossRef]
7. Dupuy, J.; Fargeon, H.; Martin-St. Paul, N.; Pimont, F.; Ruffault, J.; Guijjaro, M.; Madrigal, H.J.; Fernandes, P. Climate change on future wildfire danger and activity in southern Europe: A review. *Ann. For. Sci.* **2020**, *77*, 35. [CrossRef]
8. De Rigo, D.; Libertà, G.; Houston Durrant, T.; Artés Vivancos, T.; San-Miguel-Ayanz, J. *Forest Fire Danger Extremes in Europe under Climate Change: Variability and Uncertainty*; JRC Technical Reports; Publication Office of the European Union: Luxembourg, 2017. Available online: https://data.europa.eu/doi/10.2760/13180 (accessed on 13 December 2023).
9. Tang, Y.; Zhong, S.Y.; Luo, L.F.; Bian, X.D.; Heilman, W.E.; Winkler, J. The potential impact of regional climate change on fire weather in the United States. *Ann. Am. Assoc. Geogr.* **2015**, *105*, 1–21. [CrossRef]
10. DiVirgilio, G.; Evans, J.P.; Blake, S.A.P.; Armstrong, M.; Dowdy, A.J.; Sharples, J.; McRae, R. Climate change increases the potential of extreme wildfires. *Geophys. Res. Lett.* **2019**, *46*, 8517–8526. [CrossRef]
11. Natural Resources Institute Finland. Suomen Metsätilastot 2021—Finnish Forest Statistics 2021: Helsinki. 2022. Available online: https://stat.luke.fi/sites/default/files/suomen_metsatilastot_2021_verkko.pdf (accessed on 18 September 2023).
12. Wallenius, T. Major decline in fires in coniferous forests—reconstructing the phenomenon and seeking for the cause. *Silva Fenn.* **2011**, *45*, 36. [CrossRef]
13. Kilpeläinen, A.; Kellomäki, S.; Strandman, H.; Venäläinen, A. Climate change impacts on forest fire potential in boreal conditions in Finland. *Clim. Chang.* **2010**, *103*, 383–398. [CrossRef]
14. Budi, A.S. Effects of Forest Fire on Wood: A Biological (Anatomical Study). In *Rehabilitation of Degraded Tropical Forest Ecosystems: Workshop Proceedings*; Kobayashi, S., Turnbull, J.W., Toma, T., Mori, T., Majid, M.N.M.A., Eds.; Center for International Forest Research (CIFOR): Bogor, Indonesia, 2001; pp. 57–68.
15. Karjalainen, J. *Tuli Pohjoisissa Havumetsissä ja Metsänhoidollinen Kulotus [Fire in Northern Coniferous Forests and Silvicultural Burning]*; Tiedote 5/94; Metsähallitus, kehittämisyksikkö: Rovaniemi, Finland, 1994. Available online: https://julkaisut.metsa.fi/assets/pdf/mt/tuli_pohjoisissa_havumetsissa.pdf (accessed on 12 January 2022).

16. Barkley, Y. *Salvage Logging after a Wildfire*; Forestry Extension Series, Fire 11; Station Bulletin No. 96. University of Idaho Extension, Idaho Forestry Wildlife and Range Experiment Station: Moscow, ID, USA, 2020. Available online: https://www.uidaho.edu/search?q=Salvage%20Logging%20after%20a%20Wildfire (accessed on 12 January 2023).
17. Hill, C.A.S. *Wood Modification: Chemical, Thermal and Other Processes*; Wiley Series in Renewable Resources; Wiley: Hoboken, NJ, USA, 2006. [CrossRef]
18. Fengel, D.; Wegener, G. *Wood: Chemistry, Ultrastructure, Reactions*; Walter de Gruyter: Berlin, Germany; New York, NY, USA, 1984. [CrossRef]
19. Piernik, M.; Woz'niak, M.; Pinkowski, G.; Szentner, K.; Ratajczak, I.; Krauss, A. Impact of the Heat Treatment Duration on Colour and Selected Mechanical and Chemical Properties of Scots Pine Wood. *Materials* **2022**, *15*, 5425. [CrossRef]
20. Bortoletto Júnior, G.; Moreschi, J.C. Physical-mechanical properties and chemical composition of Pinus taeda mature wood following a forest fire. *Bioresources* **2003**, *87*, 231–238. [CrossRef]
21. Carvalho, A. *Madeiras "Salvadas" de Fogos Florestais [Woods Saved from Forest Fires]*; Departamento de Tecnologia dos Produtos Florestais, Estac~ao Nacional de Tecnologia dos Produtos Agrarios/INIA: Alcobaca, Portugal, 1986.
22. Zziwa, A.; Mukasa, J.; Kizito, S. Structural suitability of 10-year old Pinus caribaea timber with a forest fire history in farm buildings. *AgricEngInt CIGR J.* **2020**, *22*, 49–58.
23. Lowell, E.C.; Cahill, J.M. Deterioration of Fire Killed Timber in Southern Oregon and Northern California. *West. J. Appl. For.* **1996**, *11*, 125–131. [CrossRef]
24. Cole, R. Selling Burned Timber. Los Angeles Times. 24 August 2014. Available online: https://northwestmanagement.com/selling-burned-timber/ (accessed on 16 November 2022).
25. Reader, T. After the Fire: Finding Value in Burned Timber 2-3-2. Colorado State University, Center for Collaborative Conservation. 2018. Available online: https://collaborativeconservation.org/2019/02/12/after-the-fire-finding-value-in-burned-timber-within-the-2-3-2/ (accessed on 12 January 2023).
26. West Fraser. Salvaging Wood Products from Charred Trees. 2022. Available online: https://www.westfraser.com/responsibility/stories-responsibility/salvaging-wood-products-charred-trees (accessed on 14 December 2022).
27. Moya, L.M.; Winandy, J.E.; Tze, W.T.Y.; Ramaswamy, S. Use of fire-impacted trees for oriented strand board. *For. Prod. J.* **2008**, *58*, 45–52.
28. Akgül, M.; Ayrilmis, N.; Camlibel, O.; Korkut, S. Potential utilization of burned wood in manufacture of medium density fiberboard. *J. Mater. Cycles Waste Manag.* **2012**, *15*, 195–201. [CrossRef]
29. Watson, P.; Potter, S. Burned wood in the pulp and paper industry: A literature review. *For. Chron.* **2004**, *80*, 473–477. Available online: https://pubs.cif-ifc.org/doi/pdf/10.5558/tfc80473-4 (accessed on 14 December 2022). [CrossRef]
30. Johnson, J. California Wildfire Fallout: Timber Industry Confronted with Too Many Dead Trees, Warns of Damaged Forest. San Francisco Chronicle, 28 November 2021, Updated 29 November 2021. Available online: https://www.sfchronicle.com/bayarea/article/California-wildfire-fallout-Timber-industry-16655624.php (accessed on 12 January 2023).
31. Militz, H. Georg-August-Universität Göttingen, Department of Wood Biology and Wood Products, Göttingen, Germany. Personal Communication, 4 September 2023.
32. Finlex. Laki Metsätuhojen Torjunnasta [Finnish Law on Preventing Forest Damage]. Finlex 1087/2013. Available online: https://www.finlex.fi/fi/laki/alkup/2013/20131087 (accessed on 18 January 2022).
33. Saarela, M. Kalajoen Metsäpalo 2021. Tapaustutkimus Palotapahtumasta ja Suojeluratkaisusta. [Kalajoki Forest Fire 2021: A Case Study of a Fire Event and a Conservation Solution.]. Bachelor's Thesis, Tampere University of Applied Sciences, Tampere, Finland, 2022.
34. *ISO 13061-2:2014*; Physical and Mechanical Properties of Wood—Test Methods for Small Clear Wood Specimens—Part 2: Determination of Density for Physical and Mechanical Tests. International Organization for Standardization: Geneva, Switzerland, 2014.
35. *ISO 13061-4:2014*; Physical and Mechanical Properties of Wood—Test Methods for Small Clear Wood Specimens—Part 4: Determination of Modulus of Elasticity in Static Bending. International Organization for Standardization: Geneva, Switzerland, 2014.
36. *ISO 13061-3:2014*; Physical and Mechanical Properties of Wood—Test Methods for Small Clear Wood Specimens—Part 3: Determination of Ultimate Strength in Static Bending. International Organization for Standardization: Geneva, Switzerland, 2014.
37. *ISO 178:2019*; Plastics—Determination of Flexural Properties. International Organization for Standardization: Geneva, Switzerland, 2019.
38. *ISO 13061-1:2014*; Physical and Mechanical Properties of Wood—Test Methods for Small Clear Wood Specimens—Part 1: Determination of Moisture Content for Physical and Mechanical Tests. International Organization for Standardization: Geneva, Switzerland, 2014.
39. *CEN-EN 1534:2020*; Wood Flooring and Parquet—Determination of Resistance to Indentation—Test Method. European Committee for Standardization (CEN): Brussels, Belgium, 2020.
40. *ISO/CIE 11664-4:2019*; Colorimetry—Part 4: CIE 1976 L*a*b* Colour Space. International Commission on Illumination (CIE). International Organization for Standardization: Geneva, Switzerland, 2019.
41. Fox, J. *Applied Regression Analysis and Generalized Linear Models*, 3rd ed.; Sage Publications, Inc.: New York, NY, USA, 2016.
42. Kolström, T.; Kellomäki, S. Tree survival in wildfires. *Silva Fenn.* **1993**, *27*, 5521. [CrossRef]
43. Broda, M.; Spear, M.; Curling, S.; Dimitriou, A. Effects of Biological and Chemical Degradation on the Properties of Scots Pine—Part II: Wood-Moisture Relations and Viscoelastic Behaviour. *Forests.* **2022**, *13*, 1390. [CrossRef]

44. Bodig, J.; Jayne, B.A. *Mechanics of Wood and Wood Composites*, 2nd ed.; Krieger Publishing Company: Malabar, FL, USA, 1993.
45. Kretschmann, D.E. Mechanical Properties of Wood. In *Wood Handbook: Wood as an Engineering Material;* General Technical Report FPL-GTR282; Department of Agriculture, Forest Service, Forest Products Laboratory: Madison, WI, USA, 2021. Available online: https://www.fpl.fs.usda.gov/documnts/fplgtr/fplgtr282/fpl_gtr282.pdf (accessed on 30 March 2023).
46. Zobel, B.J.; Sprague, J.R. *Juvenile Wood in Forest Trees;* Springer: Berlin/Heidelberg, Germany, 1998.
47. Moore, J.R.; Cown, D.J. Corewood (Juvenile Wood) and Its Impact on Wood Utilisation. *Curr. For. Rep.* **2017**, *3*, 107–118. [CrossRef]
48. Hakkila, P. Polttohakepuun kuivuminen metsässä [Forest seasoning of wood intended for fuel chips]. *Commun. Inst. For. Fenn.* **1962**, *54*, 1–82.
49. Repola, J.; Heikkinen, J.; Lindblad, J. Pulpwood green density prediction models and sampling-based calibration. *Silva Fenn.* **2021**, *55*, 19. [CrossRef]
50. Buchelt, B.; Wagenführ, A. Evaluation of colour differences on wood surfaces. *Eur. J. Wood Wood Prod.* **2012**, *70*, 389–391. [CrossRef]
51. Grekin, M. Color and color uniformity variation of Scots pine wood in the air-dry condition. *Wood Fiber Sci.* **2007**, *39*, 279–290.
52. Torniainen, P.; Popescu, C.-M.; Jones, D.; Scharf, A.; Sandberg, D. Correlation of Studies between Colour, Structure and Mechanical Properties of Commercially Produced ThermoWood® Treated Norway Spruce and Scots Pine. *Forests* **2021**, *12*, 1165. [CrossRef]
53. Antonović, A.; Barčić, D.; Kljak, J.; Ištvanić, J.; Podvorec, T.; Stanešić, J. The Quality of Fired Aleppo Pine Wood (*Pinus Halepensis* Mill.) Biomass for Biorefinery Products. *Croat. J. For. Eng.* **2018**, *39*, 313–324.

applied sciences

MDPI

Article

Selected Physical and Mechanical Properties of Subfossil Oak (*Quercus* spp.) Compared to Aged Oak and Recent Oak

Ruxandra Nedelcu *, Maria Cristina Timar *, Mihaela Porojan and Emanuela Carmen Beldean

Faculty of Furniture Design and Wood Engineering (FDMIL), Transilvania University of Brașov, 500036 Brașov, Romania; mporojan@unitbv.ro (M.P.); ebeldean@unitbv.ro (E.C.B.)
* Correspondence: ruxandra.nedelcu@unitbv.ro (R.N.); cristinatimar@unitbv.ro (M.C.T.)

Abstract: Subfossil oak (SO) wood material, originating from three different buried trunks discovered in recent years by excavations in riverbanks on Romanian territory, was analysed in this research. Aged oak recovered from constructions (AO_C) and recent/new oak wood material (NO) were also investigated to provide comparative data for the SO. The oven-dry density and the basic density, the total volumetric and linear swelling and shrinkage coefficients and the compression strength parallel to the grain were the selected physical and mechanical properties considered. The experimental results showed a lower density of SO compared to NO and AO_C tested by up to about 19–20%, alongside a trend of increased dimensional instability, with variability among the tested assortments. The compression strength parallel to the grain was reduced by 19–31% compared to NO. The properties of AO-C were closer to those of NO, but differences between wood materials from different sources and of different ages were registered. A positive linear correlation was found between compression strength parallel to grain and the basic density for all types of material and assortments tested. These comparative results have to be considered by designers and engineers in the valorisation of SO in furniture design and other applications.

Keywords: oak; subfossil wood; aged wood; recent wood; density; swelling; shrinkage; compression strength

Citation: Nedelcu, R.; Timar, M.C.; Porojan, M.; Beldean, E.C. Selected Physical and Mechanical Properties of Subfossil Oak (*Quercus* spp.) Compared to Aged Oak and Recent Oak. *Appl. Sci.* **2024**, *13*, 3294. https://doi.org/10.3390/app14083294

Academic Editors: Stefano Invernizzi and Giuseppe Lazzara

Received: 6 March 2024
Revised: 28 March 2024
Accepted: 9 April 2024
Published: 13 April 2024

1. Introduction

Subfossil wood is a material that has come under the attention of researchers more seriously in the last decade due to both scientific reasons and its economic potential as an unconventional natural resource. Subfossil wood is an unfossilized wood which has been deposited in anaerobic conditions in rivers, swamps, or moraine sediments for hundreds to thousands of years [1–4]. Subfossil wood is also known as *bog wood* and *abonos wood* [5]. The sub-fossilisation process in burial conditions is accompanied by microstructural and chemical changes. Studies in the field revealed an apparent increase in lignin content due to the prevalent degradation of polysaccharides, especially hydrolysis of hemicelluloses. A decrease or even a slight increase in cellulose, alongside a decrease in its crystallinity index, were reported by different authors, depending on the subfossil wood analysed, conditions of burial and age, though this was found to be not a statistically significant factor. Sub-fossilisation also results in a significant increase in ash content due to the minerals penetration into the wood structure during the long-term anoxic ageing process in specific burial conditions [1–6].

Due to its natural durability related to the high content of tannins, oak (*Quercus* spp.) is one of the most common wood species that has been preserved in such conditions, rendering it a grey-to-black colour as a result of the formation of iron–tannin complexes, and a very beautiful appearance that could be valorised in design projects [5,6].

Aged wood is considered the wood exposed to aerobic conditions for long periods of time, prevalent in wooden buildings or wooden objects [1,7]. The conditions for wood storage have a significant effect on the ageing process, generally defined as a slow, gradual and irreversible alteration of the chemical and physical structure of the material under the action of environmental factors. When wood has been stored indoors, in dry air, where no significant variation in temperature, UV radiation or water contact were present, the ageing effect on the structure and wood properties is minimal if no biodegradation was present. Contrarily, if the wood has been exposed outdoors, complex chemical degradation by UV radiation, thermal degradation, erosion and biological deterioration may occur [7]. Different chemical processes are responsible for the ageing of external surfaces and subsurfaces, with an important contribution of light, especially UV radiation causing lignin degradation, compared to inner bulk wood in the absence of light, where hemicelluloses are the most affected [8]. Nevertheless, chemical and properties changes are not limited to the surface of wood (e.g., colour, roughness), but mechanical properties can also be affected by ageing in oxic conditions under variable conditions of temperature and relative humidity. Interest in recovering and valorisation of aged wood from constructions in innovative furniture design has grown in recent years, so acknowledging the changes in properties compared to recent wood is important.

Discovering subfossil buried wood during excavation work in riverbeds or riverbanks, or archaeological work in various sites, has raised scientific interest, leading to intensified research in this field [9]. Subfossil wood might be also found as archaeological wood, raising the scientific interest of both restorers in the field of Cultural Heritage and wood scientists in direct relation to its historical significance, specific properties and conservation challenges [6,9–12].

Regarding the study of subfossil wood worldwide, a rather limited number of papers were published before 2000. Some of them focused on defining and understanding the phenomena of ageing and fossilization of wood or material dating by dendrochronology [1], while others referred to the physical and mechanical properties of the material [13,14]. The interest in subfossil wood has continuously grown in conjunction with the progress of investigation methods and equipment. The vast majority of studies after 2000 focused on the dating of subfossil wood by dendrochronology and other methods [15,16] or dating in conjunction with material characterisation [17–19].

The most investigated species of subfossil wood were as follows: oak, elm, ash, pine, fir, etc. Oak (*Quercus* spp.) seems to be the prevalent subfossil wood discovered and studied in Europe, especially in East and Central Europe [2,4,9–11,17,18,20–23]. A particular interest in subfossil oak has been evident in research carried out in the Czech Republic [2–4,17,18,24], Poland [9,15], Serbia [21], Bosnia Herzegovina [5], Italy [22,23], Austria [16], and Germany [14].

The complex physical and chemical modifications, in terms of loss of structural chemicals or tissue integrity [24], occurring in wood during the long sub-fossilisation process are reflected in changes in the physical and mechanical properties compared to recent wood [2–6].

Density is a physical property of wood largely investigated in correlation with anatomical structure and mechanical properties due to its practical importance for processing and treating technologies. Several researchers reported slightly reduced values of the basic density of subfossil oak compared to recent wood (e.g., 91% of the value of recent wood according to [22]). It was also observed that density does not necessarily decrease with the age of subfossil oak, but it depends on the location of burial, condition of deposition and degree of degradation of wood components [2,13]. When archaeological oak excavated from fluvial terraces was investigated, a higher density than contemporary wood was obtained, and the author explained that this result is due to a strong supersaturation of wood with mineral compounds [20]. Values of basic density almost equal to those for recent oak, without noticeable differences between internal and external areas of the subfossil trunk, were reported for subfossil oak from Italy [23]. The oven-dry density reported by different authors varied in the ranges of 544–667 kg/m^3 for new wood [25] and 526–668 kg/m^3 for subfossil oak [3].

Increased dimensional instability was reported for subfossil oak. Double values of shrinkage and swelling compared to recent oak and the fact that these values did not change with the age of subfossil wood were reported [2]. Similarly, an increased affinity for water and almost double swelling values compared with recent oak were found by other researchers [4,24]. The same increasing trend of wood shrinkage with double radial values or even triple tangential values compared to recent wood was registered in another research [23].

In terms of mechanical properties, it was stated that compression strength parallel to the grain of subfossil oak does not exceed the value of 40 MPa, being approx. 25–50% lower than the strength of recent wood [2,3,14,20]. A similar trend, but with a lower decrease of about 10–23%, was also reported [23].

Despite all these general trends, the properties of subfossil oak wood are highly variable, depending on the burial conditions, composition of the environmental sediments and age of the material, though this seems to be not the most relevant influencing factor. Therefore, chemical investigation of subfossil wood is necessary to explain the changes in properties [2,4,24].

In Romania, subfossil wood is almost unknown to users. However, researchers have recently studied subfossil wood of various species discovered on Romanian territory, focusing mainly on dendrochronological evaluation [26–28] or some specific chemistry-related characteristics [6,10,11,29]. Subfossil oak from the Suceava region was investigated by dendrochronology [28], while chemometric characteristics of historical/archaeological oak from the 14th century, wood–water relationship and biological degradation of subfossil oak were also research topics [6,10,11].

The physical and mechanical properties and their relationship were less studied on subfossil oak/wood material from Romania. Therefore, the present study aims to fill some lack of information in this field, contributing to the enrichment of the available database on subfossil oak and its properties.

The aim of the research presented in this paper was to comparatively investigate some relevant physical and mechanical properties of subfossil oak originating from three buried trunks excavated from riverbanks in different locations in Romania. Aged wood recovered from construction and new wood were included in this research as comparison reference samples. Density, swelling, shrinkage and compression parallel to grain were selected in this research as relevant properties for furniture and interior design elements.

2. Materials and Methods

2.1. Wood Material and Test Samples

Three types of oak wood material, namely subfossil oak (SO), aged oak (AO) and recent (new) oak (NO) were employed in this research. The SO material originated from three buried trunks (S1, S2, S3) accidentally discovered in riverbeds in three locations in Romania, as presented in Figure 1, alongside their GPS coordinates. The SO wood material provided to our faculty for testing was in various forms: air-dried timber pieces (S1, S3) or crosscut log sections (S2). Two such pieces (100 mm thick discs), one cut from the base of the trunk (S2a) and the second cut at 6 m height from the base (S2b), were investigated. Wood elements recovered from constructions on the occasion of their demolition or rehabilitation were used as sources of AO material. There were two elements (C1, C2), namely a beam and a pillar from the traditional porch of a house built around 1930 in Valea Mare-Bratia, Argeș County and one beam (C3) from the roof structure of a house built around 1970 in Codlea, Brasov County. An approximate age of 90 years in the conditions of use class 3 (UC 3) according to [30] could be considered for AO_C1-C2, while 30 years in service in UC 2, followed by 25 years storage in conditions of UC 3+ (outdoors, covered, near soil contact), should be acknowledged for AO_C3. Four pieces of recent oak timber (50 mm thick) from the workshop of the Faculty of Furniture Design and Wood Engineering were employed as NO material for comparison.

Figure 1. Types of wood material and phases of test samples preparation: (**a**) log cross sections (SO_S2a, SO_S2b), (**b**) timber pieces (SO_S1, SO_S3) and (**c**) elements of aged wood recovered from constructions (AO_C).

Prismatic test samples with dimensions of (30 × 20 × 20) in (mm × mm × mm) on the (L × Ra × Tg) directions were prepared by careful mechanical processing from the air-dried wood material of all the presented assortments. This type of samples was employed in this research to determine all the selected physical and mechanical properties, namely density, coefficients of swelling and shrinkage, and the compression parallel to grain.

Depending on the form of the available material (timber pieces, wooden log cross sections, elements recovered from constructions), adequate processing phases and cutting schemes were applied (Figure 1). Special attention was paid to preliminary cuttings to remove all obviously degraded material (AO) and/or sapwoodby-case basis. Then, cuttings in the radial direction were carried out to ensure obtainment by further adequate processing of intermediary baguettes (scantlings) with cross sections of 20 × 20 (mm × mm) in the Ra and Tg directions, respectively. The annual rings were parallel to one of the cross section sides. The final test samples resulted from crosscutting at 30 mm lengthin the longitudinal direction.

All the samples were coded and numbered during the whole mechanical processing steps so that their location in the original material could be clearly understood and traceability ensured. Numbering of the intermediary baguettes/scantlings started from the bark/outer part of un-edged timber inwards. The first two scantlings were considered sapwood, while heartwood was considered from scantling four inwards. Core wood (pith) was eliminated (when present) in all cases. Only samples following standard conditions were used for testing.

The codes of the sample series corresponding to the three types of oak wood material (SO, AO, NO), respectively the seven different assortments, alongside the total number of samples investigated for each assortment (mentioned as figures in brackets) were as follows: SO_S1 (37), SO_S2a (35), SO_S2b (37), SO_S3 (32), AO_C1_C2 (63), AO_C3 (43), NO (64).

2.2. Experimental Methods

2.2.1. Density and FSP

The oven-dried density (ρ_0) and basic density (ρ_c) were calculated considering the ratio between the oven-dried mass of the sample (m_0) and its oven-dried volume (V_0), respectively, and maximum volume ($V_{\max}$) at fibre saturation point (FSP) or above, according to Equations (1) and (2).

Drying of samples was achieved in a laboratory oven at 103 ± 2 °C until constant weight. They were weighed with an analytical balance with precision 0.001 g and measured by calliper with an accuracy of 0.01 mm.

$$\rho_0 = m_0/V_0, \left[\text{kg/m}^3\right] \tag{1}$$

$$\rho_c = m_0/V_{\max}, \left[\text{kg/m}^3\right] \tag{2}$$

The experimentally determined values of oven-dried and basic densities were then employed to calculate the FSP based on their mathematical correlation [31,32], according to Equation (3).

$$\text{FSP} = \left(\frac{1000}{\rho_c} - \frac{1000}{\rho_0}\right) \times 100, [\%] \tag{3}$$

2.2.2. Swelling and Shrinkage

Methods adapted from relevant standards in force were applied [33–36] to determine swelling and shrinkage properties.

For the total swelling coefficients, $\alpha_{\max}$, the samples were oven-dried until a constant mass, and the dimensions (l) were measured on the three directions L, Ra, Tg with 0.01 mm accuracy. The specimens were then completely immersed in distilled water and reweighed and measured after 72, 96 and 324 h, when they reached maximum constant dimensions.

The total swelling coefficients on the longitudinal, radial, tangential directions and volumetric swelling were calculated according to Equations (4) and (5).

$$\alpha_{(L,Ra,Tg)\max} = \frac{l_{(L,Ra,Tg)\max} - l_{(L,Ra,Tg)0}}{l_{(L,Ra,Tg)0}} \cdot 100, [\%] \tag{4}$$

$$\alpha_{V\max} = \frac{V_{\max} - V_0}{V_0} \cdot 100, [\%] \tag{5}$$

where

$l_{(L,Ra,Tg)\max}$ = the maximum dimensions (at FSP or above) in longitudinal, radial, and tangential directions, respectively, [mm];

$l_{(L,Ra,Tg)0}$ = the oven-dry dimensions in the longitudinal, radial, and tangential directions, respectively, [mm];

$V_{\max}$ = the maximum volume (at FSP or above), [mm^3];

V_0 = oven-dry volume, [mm^3].

The same samples were then employed to determine the total shrinkage coefficients $\beta_{\max}$. For this purpose, the wet samples from immersion phase were air conditioned for a few days and then dried in a laboratory oven at 103° ± 2 °C until a constant mass. The dimensions were measured after 30 min of cooling in desiccator and the total linear and volumetric shrinking coefficients were calculated according to Equations (6) and (7).

$$\beta_{(L,Ra,Tg)\max} = \frac{l_{(L,Ra,Tg)\max} - l_{(L,Ra,Tg)0}}{l_{(L,Ra,Tg)\max}} \cdot 100, [\%] \tag{6}$$

$$\beta_{V\max} = \frac{V_{\max} - V_0}{V_{\max}} \cdot 100, [\%] \tag{7}$$

where

$l_{(L,Ra,Tg)\max}$ = the maximum dimensions (at FSP or above) in longitudinal, radial, and tangential directions, respectively, [mm];

$l_{(L,Ra,Tg)0}$ = the oven-dried dimensions in the longitudinal, radial, and tangential directions, respectively, [mm];

$V_{\max}$ = the maximum volume at FSP or above [mm^3];

V_0 = the oven-dried volume, [mm^3].

The linear and volumetric swelling and shrinkage coefficients for 1% moisture content variation (K_α and K_β) were then calculated with Equations (8) and (9), respectively.

$$K_\alpha = \frac{\alpha_{\max}}{\text{FSP}} \tag{8}$$

$$K_\beta = \frac{\beta_{\max}}{\text{FSP}} \tag{9}$$

2.2.3. Compression Strength (Ultimate Stress in Compression Parallel to Grain)

The ultimate stress in compression (compression strength) parallel to grain was determined according to [37]. The test was performed by means of a ZWICK/ROELL universal testing machine (type BT1FB050TN.D30/2007, produced in Ulm Germany), endowed with a device (support) for positioning the specimen and a device for applying the load. The compression test (Figure 2) consisted of measuring the maximum load required to break the specimen under a static load applied parallel to grain. The machine software automatically recorded and displayed all measured data and the calculated value of compression strength. A data correction depending on the real moisture content of tested samples (W, %) was performed in order to assess compression strength at a 12% moisture content ($\sigma_{c,12}$).

$$\sigma_{c,W} = \frac{P_{\max}}{b \times h}, \left[\text{N/mm}^2\right] \tag{10}$$

where

σ_c = ultimate stress in compression parallel to grain at the actual moisture content (W) of tested sample, [N/mm^2];

$P_{\max}$ = maximum breaking force of the sample, [N];

$b \times h$ = cross section area of the test sample ($Ra \times Tg$), [mm^2].

$$\sigma_{c,12} = \sigma_{c,W} \cdot [1 + \alpha \cdot (W - 12)], \left[\text{N/mm}^2\right] \tag{11}$$

where

α = correction coefficient of compression strength for 1% MC difference, equal to 0.04

W = moisture content of the tested sample [%].

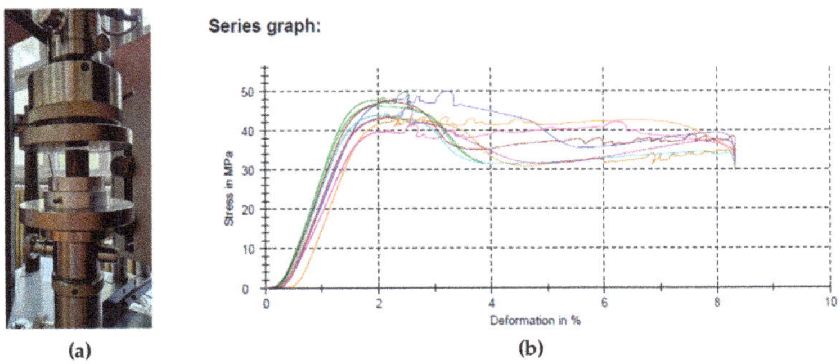

Figure 2. Determination of compression strength: (**a**) detail of ZwickRoell universal machine equipped with compression device; specimen positioned for testing; (**b**) examples of stress–deformation diagrams automatically registered (different colours represent different samples).

2.3. Statistical Analysis

Experimental data were statistically analysed employing Minitab 17.1.0 statistic software. Statistical analysis consisted of calculation of the mean, median and standard deviation values of each set of data. Box plot graphs were employed to better highlight the variation in the studied properties among the types of investigated samples and variability of experimental data for each assortment. Null hypothesis tests were applied to check if the differences between the data sets/mean values of multiple samples (one-way ANOVA) were significantly different or not at a confidence level of 95% (α = 0.05).

3. Experimental Results

3.1. Physical Properties of Subfossil Oak in Comparation with Aged Oak from Construction and Recent Oak

3.1.1. Wood Density

The oven-dry and basic densities of all the seven assortments from the three types of oak wood material (SO, AO, NO) are summarized in Table 1 and graphically illustrated by the box plots in Figure 3.

Table 1. Oven-dry density (ρ_0) and basic density (ρ_c) of subfossil oak (SO) from different sources compared to aged oak recovered from constructions (AO_C) and recent oak (NO)—average values and standard deviations.

Wood Material	NOR	ρ_0, kg/m^3	ρ_c, kg/m^3
SO_S1	15	609 ± 28	504 ± 19
SO_S2a	13	658 ± 48	508 ± 36
SO_S2b	12	752 ± 47	568 ± 32
SO_S3	10	690 ± 29	558 ± 19
AO_C1_C2	25	733 ± 47	623 ± 34
AO_C3	20	685 ± 39	585 ± 32
NO	27	755 ± 63	629 ± 49

Notes: NOR = Number of replicates.

It can be observed that average values of oven-dry density, ρ_0, for the SO assortments varied between 609 ± 28 kg/m^3 (SO_S1) and 752 ± 47 kg/m^3 (SO_S2b). For the two AO assortments, the average values were 685 ± 39 kg/m^3 (AO_C3) and 733 ± 47 kg/m^3 (AO_C1_C2). All these values were slightly lower than the experimental average value 755 ± 63 kg/m^3 for NO.

The obtained results are supported by the literature data. According to [38], the oven-dry density of *Quercus Robur* and *Petraea* (NO) can vary between 390 and 930 kg/m^3 with an average of 650 kg/m^3, while values of about 618–696 kg/m^3 were reported by [2]. The reported values for SO varied in the range of about 526–734 kg/m^3 [3]. Also, lower density values for SO compared to NO were previously reported in the literature, e.g., [14,22].

Statistical analysis of our experimental data showed that there were no statistically significant differences between the SO_S2a and SO_S3 assortments and between NO and SO_S2b and AO_C1-C2 assortments in terms of oven-dry density. On the other hand, there was a statistically significant difference in oven-dry density of about 93 kg/m^3 between SO_S2a and SO_S2b samples originating from the same log but from different positions on the trunk length (base and 6m from the base), with a 14% higher density in the upper position. This higher value was statistically similar to that determined for NO.

Usually, wood in the trunks of oak (ring-porous) trees shows a maximum density at the base of the trunk, then decreases to remain constant over a relatively large height, and then increases again under the crown. Also, the density of ring-porous wood material is correlated positively with the width of the annual rings respective to the higher proportion of latewood [31,39]. Our results showed, however, an axial increase in oven-dry density for SO_S2 over the value at the base at approximately 2/3 height of the 9 m trunk. Presence of tension wood might be an explanation.

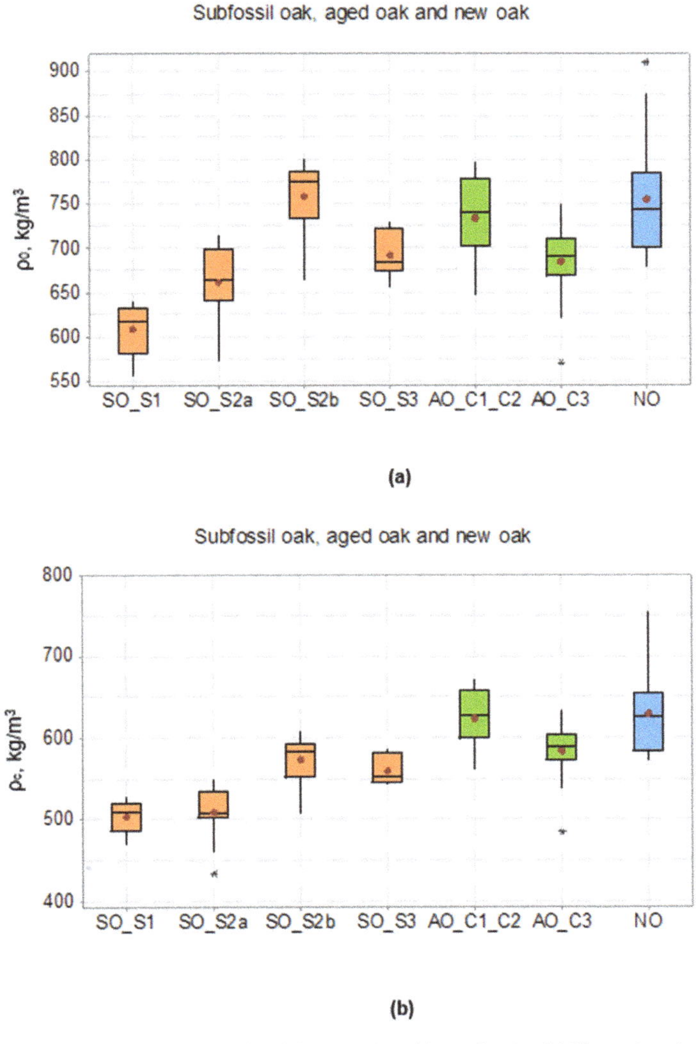

Figure 3. Box plots of oven-dried density (**a**) and basic density (**b**) illustrating the variation in these values among the different assortments of oak wood and dispersion of data in each group. Note: Box plots include median value line, mean (average) value (red dot), whiskers for minimum and maximum values and outliers (*).

In terms of basic density (ρ_c), it can be observed that the average value for NO was the highest (629 $\pm$ 49 kg/m^3), followed by the average values of AO (585–623 kg/m^3) and lower average values for the SO assortments (504–568 kg/m^3). Similar values for basic densities of SO were previously reported: 521.7 kg/m^3 [22] and 530 kg/m^3 [23].

To conclude the density investigation, the experimental data showed, in good accordance with the literature, that the density of sub-fossil oak varied among assortments from different locations and in the same log, being generally lower (with up to 19.3–19.9%) compared to new oak, while the density of aged oak may be comparable or slightly lower.

A slight decrease in basic density could be attributed partially to a minimal decomposition of the cell walls and partially to the extraction of water-soluble substances as a result of leaching in a burial environment [40]. Also, the variability of density among the wood material from various assortments and the data dispersion for each assortment, due to several particularities of wood structure (e.g., the width of the annual rings, other anatomical features, wood extractives content, absorption of other components), as reflected by the boxplots in Figure 3, should be acknowledged.

3.1.2. Swelling and Shrinkage, Anisotropy and FSP

The total linear and volumetric swelling and shrinkage coefficients are presented in Table 2 and graphically illustrated by the box plots in Figure 4.

Table 2. Comparative values of total swelling and total shrinkage coefficients of subfossil oak (SO), aged oak (AO_C), and new oak (NO)—average values and standard deviations.

Wood Material	Total Swelling				Total Shrinkage			
	$\sigma_{L\,max}$, %	$\alpha_{Ra\,max}$, %	$\alpha_{Tg\,max}$, %	$\alpha_{V\,max}$, %	$\beta_{L\,max}$, %	$\beta_{Ra\,max}$, %	$\beta_{Tg\,max}$, %	$\beta_{V\,max}$, %
SO_S1	0.7 ± 0.3	8.4 ± 1.5	10.7 ± 1.3	20.9 ± 1.4	0.7 ± 0.4	7.6 ± 1.2	9.0 ± 1.2	16.4 ± 1.0
SO_S2a	1.0 ± 0.3	8.5 ± 1.0	18.4 ± 1.2	29.6 ± 1.8	0.7 ± 0.3	8.0 ± 0.7	14.3 ± 0.8	21.7 ± 1.1
SO_S2b	1.2 ± 0.7	9.4 ± 0.7	19.5 ± 0.8	32.4 ± 1.7	0.9 ± 0.6	8.9 ± 0.5	14.4 ± 0.6	22.7 ± 1.0
SO_S3	0.6 ± 0.3	6.4 ± 0.6	15.4 ± 2.1	23.6 ± 1.8	0.7 ± 0.2	6.4 ± 0.6	11.9 ± 0.5	18.0 ± 0.9
AO_C1_C2	0.3 ± 0.1	6.1 ± 1.3	10.5 ± 1.5	17.6 ± 1.4	0.2 ± 0.1	6.5 ± 0.9	9.7 ± 0.9	15.6 ± 0.7
AO_C3	0.7 ± 0.2	5.5 ± 0.4	10.3 ± 0.6	17.1 ± 0.8	0.5 ± 0.3	5.6 ± 0.4	9.7 ± 0.7	15.1 ± 0.8
NO	0.5 ± 0.3	5.4 ± 0.9	13.4 ± 1.0	20.0 ± 1.4	0.4 ± 0.4	5.4 ± 0.8	10.2 ± 1.3	15.3 ± 1.5

It could be observed that, generally, higher total swelling and shrinking coefficients were determined for SO compared to NO. The average values of the total swelling coefficients determined for the different SO assortments varied in the ranges 20.9–32.4% (*V*), 6.4–9.4% (*Ra*) and 10.7–19.5% (*Tg*), with maximum values for the SO_S2b samples, which also presented the higher density. The corresponding values for NO were 20.0% (*V*), 5.4% (*Ra*) and 13.4% (*Tg*), respectively. The smaller tangential swelling and shrinkage coefficients for the SO_S1 assortment compared to recent oak NO must be also noted, which might be correlated with the higher density of the NO wood specimens investigated in this research. The total swelling and shrinkage coefficients of AO samples were close to those for recent oak.

The values of total shrinkage obtained in this research, higher for SO compared to NO, as similarly reported in the literature, e.g., [2–4,13], were situated in the characteristic intervals presented in [38] for subfossil oak and new oak. Some exceptions could be noted: slightly lower values of radial and tangential shrinkage for new oak and a single higher value of longitudinal shrinkage for SO.

A remark has to be made on the high longitudinal swelling and shrinking coefficients of SO, especially the assortments SO_S2b ($\alpha_{L\,\max} = 1.2$, $\beta_{L\,\max} = 0.9$) and SO_S2a ($\alpha_{L\,\max} = 1.0$, $\beta_{L\,\max} = 0.7$), compared to NO ($\alpha_{L\,\max} = 0.5$, $\beta_{L\,\max} = 0.4$). They might indicate the presence of tension wood [41]. The direct relationships between anatomical pattern, density and local swelling of oak wood were demonstrated [42].

The increased swelling of SO compared to recent oak could probably be due to the "leaching of degradation products or inorganic compounds stored in wood during the fossilization process" and chemical analysis of leaching water would be necessary in order to find out which compounds are responsible for the distinct swelling if subfossil oak is immersed in water [4].

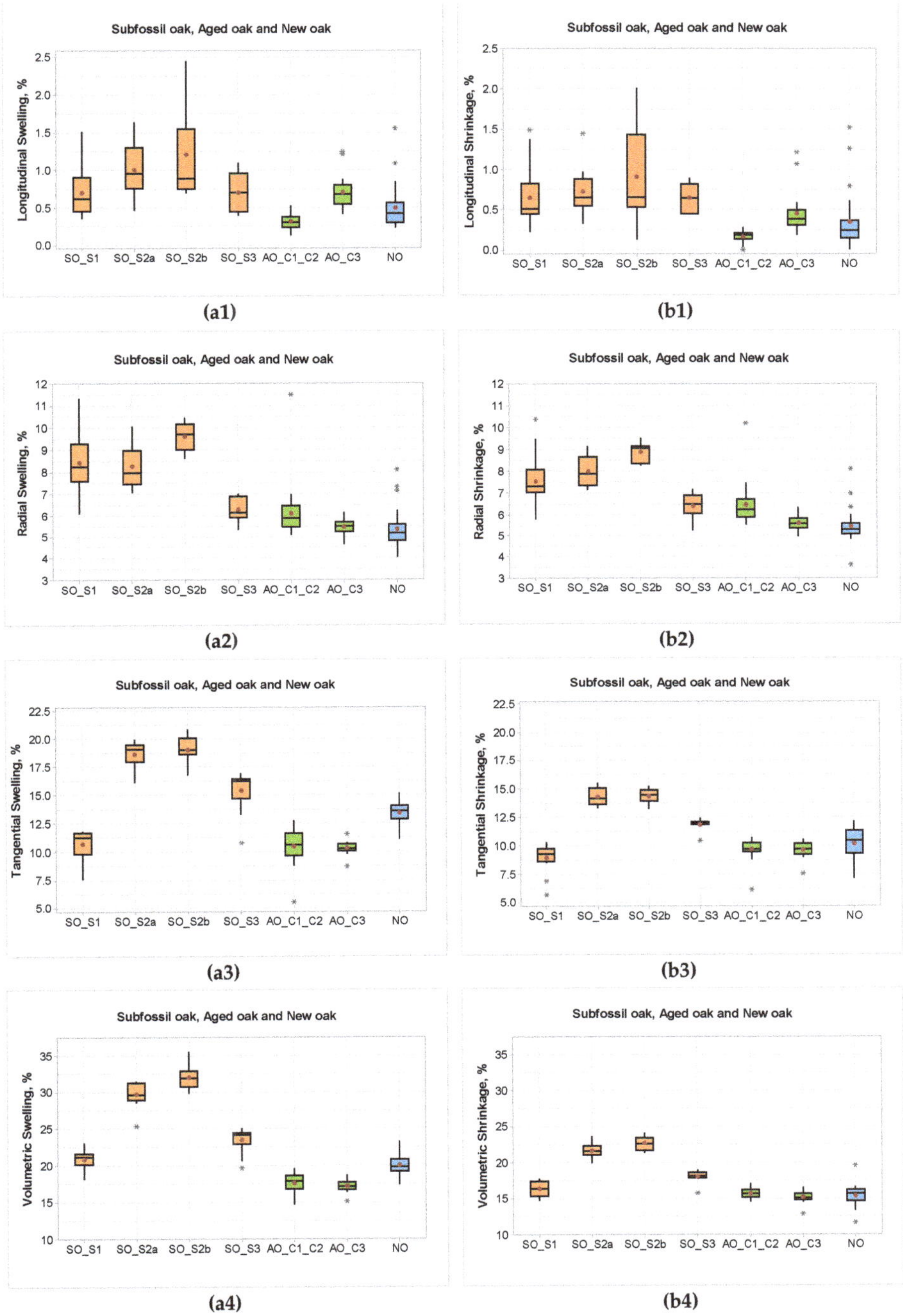

Figure 4. Box plots of total swelling (**a1**–**a4**) and total shrinkage coefficients (**b1**–**b4**), illustrating the variation in these values among the different assortments of oak wood and dispersion of data in each group. Note: Box plots include median value line, mean (average) value (red point), whiskers for minimum and maximum values and outliers (*).

The coefficient of anisotropy calculated as ratios between the total tangential and radial swelling were situated in the range of 1.3–2.4 for the SO assortments and 1.7–1.9 for AO_C assortments, being slightly lower than the value of 2.5 calculated for the NO employed in this research. The same trend was observed for shrinking. To conclude, the experimental data in this research indicate that increased swelling and shrinking could be expected for SO compared to NO, though the anisotropy of dimensional changes might be slightly reduced. These aspects are important in the design of furniture and other wood products.

The values of FSP calculated based on Equation (3) and swelling and shrinking coefficients for 1% variation in the MC, calculated based on the total swelling and shrinking coefficients and corresponding FSP values, are cumulated in Table 3.

Table 3. Fibre saturation point, swelling and shrinkage coefficients for 1% moisture content (*W*) variation, K_α, K_β.

Wood Material	FSP, %	Swelling Coefficients for 1% *W* Variation			Shrinkage Coefficients for 1% *W* Variation			
		$K_{\alpha,Ra}$, %	$K_{\alpha,Tg}$, %	$K_{\alpha,V}$, %	$K_{\beta,L}$, %	$K_{\beta,Ra}$, %	$K_{\beta,Tg}$, %	$K_{\beta,V}$, %
SO_S1	34	0.25	0.31	0.61	0.02	0.22	0.26	0.48
SO_S2a	45	0.19	0.41	0.66	0.02	0.18	0.32	0.48
SO_S2b	43	0.22	0.45	0.75	0.02	0.21	0.33	0.53
SO_S3	34	0.19	0.45	0.69	0.02	0.19	0.35	0.53
AO_C1_C2	24	0.25	0.44	0.73	0.01	0.27	0.40	0.65
AO_C3	25	0.22	0.41	0.68	0.02	0.22	0.38	0.60
NO	27	0.20	0.50	0.75	0.01	0.20	0.38	0.58

It can be observed that SO presents higher FSP values (34–45%) compared to NO (27) and AO (24–25). The variation in this property for all the studied assortments and variability and dispersion of individual data within each group is highlighted by the boxplots in Figure 5.

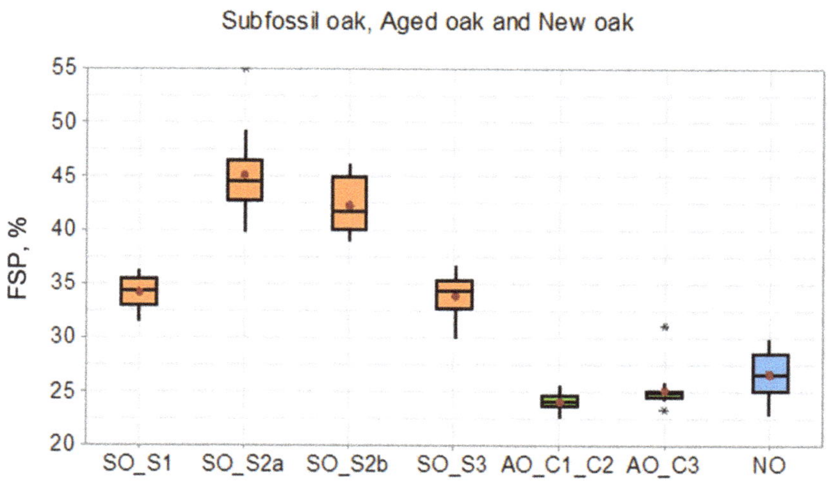

Figure 5. Box plots of FSP values, illustrating the variation in this property among the different assortments of oak wood and dispersion of data in each group. Note: Box plots include median value line, mean (average) value (brown point), whiskers for minimum and maximum values and outliers (*).

According to [3,4] and literature cited herein, an increase in affinity to water might be attributed to a higher amount of the amorphous part of cellulose in subfossil oak and "freer bonding sites in cellulose" due to hydrolysis of hemicelluloses. FTIR investigations of our samples (not included in this paper) highlighted the prevalent degradation of hemicelluloses in subfossil wood so the above explanation might apply. However, it has to be highlighted that the calculation method employed was conceived for recent wood and might not be fully applicable for subfossil or aged woods due to the chemical changes in the cell walls.

3.2. Mechanical Properties of Subfossil Oak Compared with Aged Oak from Construction and Recent Oak

The results of compression tests parallel to grain are summarised in Table 4, while the variation in this mechanical property among the different assortment of oak wood and its correlation with the basic density is illustrated in Figure 6.

Table 4. Compression parallel to grain for sub-fossil oak (SO) assortments compared to recent oak (RO) and aged oak recovered from constructions (AO_C): ultimate compression stress $\sigma_{c,12}$ (average values with standard deviations), strain at break point ε (range) and Modulus of elasticity *Ec* (minimum, mean and average values).

Wood Material	NOR	$\sigma_{c,12}$ MPa	Strain at Break Point, mm/mm	Modulus of Elasticity, E_c Min MPa	Mean MPa	Max MPa
SO_S1	15	46 ± 3	0.020–0.023	3806	4144	4808
SO_S2a	13	46 ± 8	0.020–0.023	2937	3765	4494
SO_S2b	13	54 ± 5	0.023–0.027	3432	3523	3832
SO_S3	12	50 ± 3	0.020–0.023	3197	3585	3708
AO_C1_C2	31	69 ± 7	0.023–0.030	2236	4706	5334
AO_C3	20	54 ± 7	0.020–0.033	1594	3854	4579
NO_N	29	66 ± 6	0.020–0.033	4680	4887	5114

Notes: NOR = Number of replicates; $\sigma_{c,12}$ = ultimate compression stress parallel to grain corrected for W 12%; E_c modulus of elasticity calculated for Min = minimum; Mean and Max = maximum values of $\sigma_{c,12}$ employing the experimental stress–strain diagrams on the elastic domain (between 10% and 40% of the maximum breaking load), according to [43].

An average ultimate compression stress value of 66 ± 6 MPa was obtained for the tested recent oak material (NO). This value is in good accordance with the reference literature data indicating an interval of variation of 54–76 MPa for *Quercus robur*, with an average of 61 MPa, while the corresponding values for *Quercus petraea* are 48–65–70 MPa [38]. The different assortments of subfossil oak presented compression strengths significantly lower (by 20–30%) compared to NO, in the range 46–54 MPa, which agrees with reference literature 33–38–52 MPa [38] The values for the two assortments of aged oak recovered from construction were significantly different, being either by 18% lower compared to NO (54 ± 7 MPa for AO_C3), or similar with NO (AO-C1-C2 with average 69 ± 7 MPa, statistically not different to NO). Despite this variability of the average values of compression strength parallel to grain among the studied oak wood assortments (Figure 6a), all values correlated well with the basic density of the wood material (Figure 6b). A linear positive correlation with a high correlation factor (R^2 = 0.8955) was obtained for the experimental results of this research.

The average values of modulus of elasticity varied in the range 3523–4144 MPa for the assortments of SO, being 15–30% lower than the average value of 4887 MPa determined for NO. Average values of modulus of elasticity varying in the range of 4490–10,342 MPa for subfossil oak from various sources and of different ages compared to 12,351 MPa for recent oak were previously reported [3].

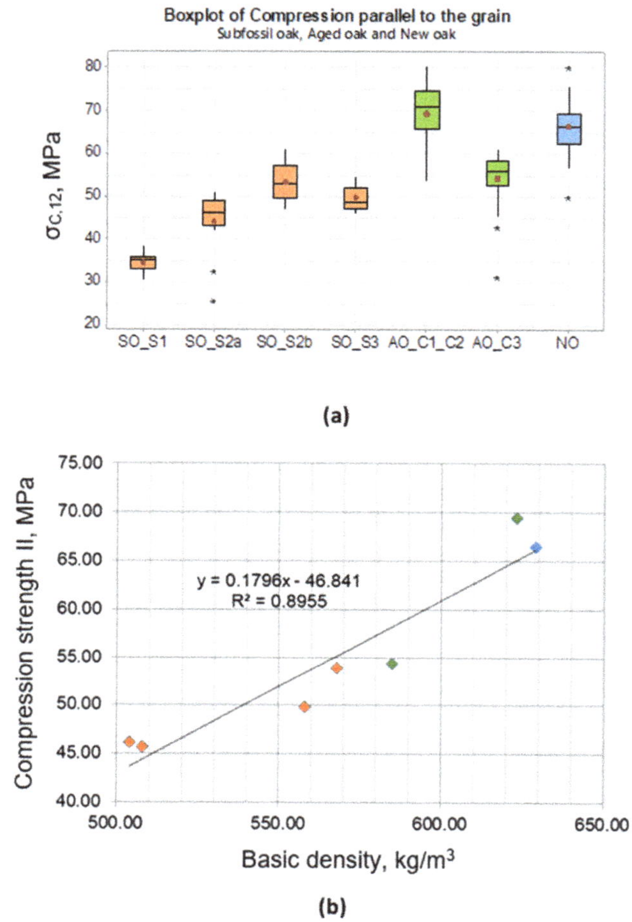

Figure 6. Ultimate compression stress parallel to grain: (**a**) Box plots illustrating variation among the different assortments of oak wood and dispersion of data in each group; (**b**) linear correlation with basic density. Note: Box plots include median value line, mean (average) value (red point), whiskers for minimum and maximum values and outliers (*); SO—orange colour, AO—green colour and NO—blue colour.

The fact that the mechanical properties of subfossil wood are lower than those of recent wood has been a common conclusion of several researchers, though differences in the actual values reported were present. Compression strength parallel to the grain for subfossil oak was found to vary in the range of 50–80% of that of recent oak according to [3] and the literature cited therein. Variations in density and decreases in ultimate compression stress of subfossil oak compared to recent wood were associated with chemical changes occurring during the sub-fossilisation process, including some degradation of the main wood components and settlements of some minerals into the wood structure. These changes seem to depend more on the environment during fossilisation than on time. Linear positive correlations between compression strength and density were determined for both subfossil and recent oak, with a higher correlation of variables for the latter [3]. However, it was also highlighted that the chemical changes occurring in subfossil oak are the key factors determining its properties: a lower content of cellulose, a relatively low content of lignin and a high proportion of ash may explain low compression strength for samples with high density [3].

The samples were also analysed to assess the failure patterns by compression parallel to the grain. Under the action of the compression force, the anatomical elements tend to separate from each other and buckle individually. The specimens break along a radial plane of minimum resistance by sudden detachment or by sliding and shearing along a plane oblique to the axis or by combining these two phenomena. For most of the samples, the break occurred along one or several planes of minimum resistance, sometimes accompanied by longitudinal cracks, as exemplified in Figure 7b–e. There were no notable differences among the subfossil oak, recent oak, and aged oak samples in terms of failure patterns or prevalence of a certain pattern, except for failure by fibre sliding (Figure 7a) observed only for SO, especially SO_S2.

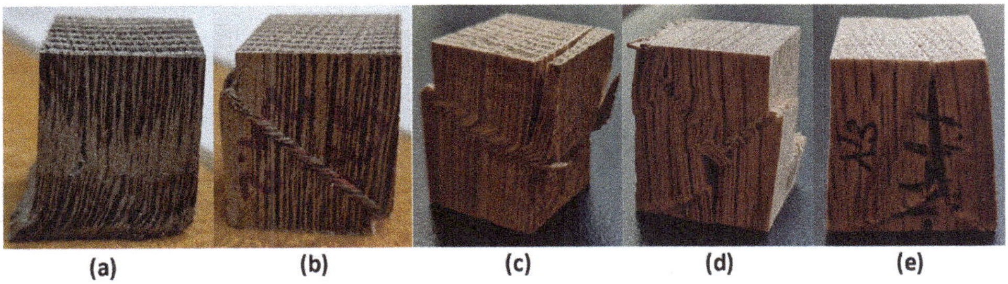

Figure 7. Exemplification of the main types of wood failure occurring by compression parallel to grain: (**a**) sliding of fibers; (**b**,**c**) break along one plane of minimum resistance, with a longitudinal crack (**c**); (**d**,**e**) break along two or more planes of minimum resistance with longitudinal cracks.

4. Conclusions

Subfossil oak (SO) wood material originating from three different buried trunks recently excavated from riverbanks on Romanian territory was analysed in this research in comparison with aged oak recovered from constructions (AO_C) and recent/new oak wood material (NO).

The experimental results showed a high variability of the properties of SO from different trunks and from the same trunk as a function of axial position and significant differences when compared with NO and AO. A lower density (by up to 20%), increased hygroscopicity and dimensional instability, but slightly lower swelling anisotropy were the main physical characteristics of SO compared to NO. At the same time, ultimate compression stress parallel to the grain and the corresponding modulus of elasticity were also reduced (by 20–30%). By comparison, only slight changes in the tested physical and mechanical properties were determined for the assortments of AO when compared to NO, though some significant differences were registered among the two assortments as a function of age and service/storage conditions.

The physical and mechanical properties tested and the differentiating features of NO, SO and AO are important for interior furniture with artistic valences and decorative woodwork and contribute to understanding the scientific and commercial value of this material.

Author Contributions: Conceptualization, R.N. and M.C.T.; methodology, R.N. and M.C.T.; supervision, M.C.T.; preparation of test samples, R.N.; investigations: R.N., M.C.T., M.P. and E.C.B., software/statistics, R.N.; validation, M.C.T. and M.P.; writing—original draft preparation, R.N., M.C.T., M.P. and E.C.B.; writing—review and editing, R.N., M.C.T., M.P. and E.C.B.; visualization, M.C.T. All authors have read and agreed to the published version of the manuscript.

Funding: This research received no external funding.

Institutional Review Board Statement: Not applicable.

Informed Consent Statement: Not applicable.

Data Availability Statement: The original contributions presented in the study are included in the article, further inquiries can be directed to the corresponding authors.

Acknowledgments: This work is part of a budget-funded PhD research program undertaken by the first author within the Interdisciplinary Doctoral School of the Transilvania University of Brasov, Romania. Special thanks are addressed to Eng Adrian Baban for making this research possible by providing the subfossil oak material SO_S1 and SO_S2 analyzed within this research. The authors are extremely grateful to the technical staff of the workshop of the Faculty of Furniture Design and Wood Engineering of Transilvania University, especially to Eng. Valeriu Budulan and Eng. Martin Szakacs for their dedicated effort in the preparation of test samples. The authors also acknowledge the PRO-DD structural funds project (POS-CCE, O.2.2.1., ID 123, SMIS 2637, no. 11/2009) for providing the infrastructure used in this work at the ICDT Research Institute of the Transilvania University of Brasov-Romania.

Conflicts of Interest: The authors declare no conflicts of interest. The founders had no role in the design of the study in the collection analysis or interpretation of data in the righting of the manuscript or in the decision to publish the results.

References

1. Fengel, D. Aging and fossilisation of wood and its components. *Wood Sci. Tehnol.* **1991**, *25*, 153–177. [CrossRef]
2. Kolář, T.; Rybníček, M. Phisical and mechanical properties of subfossil oak (*Quercus*, sp.) wood. *Acta Univ. Agric. Silvic. Mendel. Brun.* **2010**, *58*, 123–134. [CrossRef]
3. Kolar, T.; Rybnícek, M.; Strelcová, M.; Hedbávný, J.; Vít, J. The changes in chemical composition and properties of subfossil oak deposited in holocene sediments. *Wood Res.* **2014**, *59*, 149–166.
4. Baar, J.; Nevrly, O. Swelling of subfossil oak under different conditions. In Proceedings of the Iaws Plenary Meeting—Eco-Efficient Resource Wood with Special Focus on Hardwoods, Sopron, Hungary, 15–18 September 2014. [CrossRef]
5. Rede, V.; Sara Essert, S.; Kocijan, M.; Dubravac, T. Influence of Ageing on Abrasion Volume Loss, Density, and Structural Components of Subfossil Oak. *Appl. Sci.* **2022**, *12*, 1814. [CrossRef]
6. Ghavidel, A.; Hofmann, T.; Bak, M.; Sandu, I.; Vasilache, V. Comparative archaeometric characterization of recent and historical oak (*Quercus* spp.) wood. *Wood Sci. Technol.* **2020**, *54*, 1121–1137. [CrossRef]
7. Kránitz, K.; Sonderegger, W.; Bues, C.T.; Niemz, P. Effects of aging on wood: A literature review. *Wood Sci. Technol.* **2016**, *50*, 7–22. [CrossRef]
8. Liu, X.Y.; Timar, M.C.; Varodi, A.M.; Sawyer, G. An investigation of accelerated temperature induced ageing of four wood species: Colour and FTIR. *Wood Sci. Technol.* **2017**, *51*, 357–378. [CrossRef]
9. Zisi, A. Forest Wood through the Eyes of a Cultural Conservator. *Forests* **2021**, *12*, 1001. [CrossRef]
10. Ghavidel, A.; Bak, M.; Hofmann, T.; Vasilache, V.; Sandu, I. Evaluation of some wood-water relations and chemometric characteristics of recent oak and archaeological oak wood (*Quercus robur*) with archaeometric value. *J. Cult. Herit.* **2021**, *51*, 21–28. [CrossRef]
11. Ghavidel, A.; Jorbandian, A.; Bak, M.; Gelbrich, J.; Morrell, J.J.; Sandu, I.; Hosseinpourpia, R. Degradation assessment of archaeological oak (*Quercus* spp.) buried under oxygen-limited condition. *Holzforschung* **2023**, *77*, 198–207. [CrossRef]
12. Lisuzzo, L.; Hueckel, T.; Cavallaro, G.; Sacanna, S.; Lazzara, G. Pickering Emulsions based on Wax and Halloysite Nanotubes: An Ecofriendly Protocol for the Treatment of Archaeological Woods. *CS Appl. Mater. Interfaces* **2021**, *13*, 1651–1661. [CrossRef] [PubMed]
13. Borgin, K.; Tsoumis, G.; Passialis, G. Density and shrinkage of old wood. *Wood Sci. Tehnol.* **1979**, *13*, 49–57. [CrossRef]
14. Schwab, E. Holzeigenschaften einer Subfossile Eiche. *Holz Als Roh-Und Werkst.* **1994**, *52*, 210. [CrossRef]
15. Krapiek, M. Holocene Dendrochronological standards for subfossil oaks from the area of Southern Poland. *Stud. Quat.* **2001**, *18*, 47–63.
16. Klusek, M.; Pavelczyk, S. Stable carbon isotope analysis of subfossil wood from Austrian Alps. *Geochronometria* **2014**, *41*, 400–408. [CrossRef] [PubMed]
17. Kolář, T. Oak Wood Properties Change on Ana Example of Subfossil Trunks. Dissertation, Mendel University, Brno, Czech Republic, 31 January 2012.
18. Kolář, T.; Gryc, V.; Rybníček, M.; Vavrčík, H. Anatomical analysis and species identification of subfossil oak wood. *Wood Res.* **2012**, *57*, 251–264.
19. Baar, J.; Paschová, Z.; Hofmann, T.; Kolář, T.; Koch, G.; Saake, B.; Rademacher, P. Natural durability of subfossil oak: Wood chemical composition changes through the ages. *Holzforschung* **2020**, *74*, 47–59. [CrossRef]
20. Mañkowski, P.; Kozakiewicz, P.; Drodek, M. The selected properties of fossil oak wood from medieval burgh in P£oñsk. *Wood Res.* **2016**, *61*, 287–298.

21. Veizović, M.; Popović, Z.; Todorović, N.; Milić, G. Drying quality and colour of subfossil oak from Central Serbia. *Glasnik Sumarskog fakulteta* **2018**, *2018*, 157–172. [CrossRef]
22. Urso, T.; Martinelli, N.; Pignatelli, O.; Scortegagna, T. The subfossil trunk of Chiarano (TV). *Conserv. Sci. Cult. Herit.* **2018**, *15*, 41–52. [CrossRef]
23. Meneghello, C. Bog Oak: Physico-mechanical and Chemical Characterization of Two Logs from Venetian Plan. In Proceedings of the XI Congresso Nazionale AIAr, Napoli, Italy, 28–30 July 2021. [CrossRef]
24. Sinkovic, T.; Govorcin, S.; Dubravac, T.; Roth, V.; Sedlar, T. Comparison some physical and mechanical properties of abonos and recent oak (*Quercus robur* L.). *Šumarski List* **2009**, *891*, 605–611.
25. Giagli, K.; Baar, J.; Fajstavr, M.; Gryc, V.; Vavrcik, H. Tree-ring width variation of wood density in Fraxinus excelsior L. and Quercus robur L. growing in Floodplain Forests. *BioResources* **2018**, *13*, 804–819. [CrossRef]
26. Arvai, M.; Popa, I.; Mîndrescu, M.; Nagy, B.; Kern, Z. Dendrochronological assessment and radiocarbon dating of subfossil coniferous macroremains excavated from a peat bog, Maramures Mts, Romania. In Proceedings of the Late Pleistocene and Holocene Climatic Variability in the Carpathian Balkan Region, Cluj/Kolozsvár, Romania, 6–9 November 2014; Stefan cel Mare University Press: Suceava, Romania, 2014.
27. Radoane, M.; Nechita, C.; Chiriloaei, F.; Rădoane, N.; Popa, I.; Roibu, C.; Robu, D. Late Holocene fluvial activity and correlations with dendrochronology of subfossil trunks: Case studies of north-eastern Romania. *Geomorphology* **2015**, *239*, 142–159. [CrossRef]
28. Kern, Z.; Popa, I. Dendrochronological and radiocarbon analyses of subfossil oaks from the foothills of the Romanian Carpathian. *Geochronometria* **2016**, *543*, 113–120. [CrossRef]
29. Ghavidel, A.; Gelbrich, J.; Ku, A.; Vasilache, V.; Sandu, I. Investigation of Archaeological European White Elm (Ulmus laevis) for Identifying and Characterizing the Kind of Biological Degradation. *Heritage* **2023**, *3*, 60. [CrossRef]
30. *EN 335:2013*; Durability of Wood and Wood-Based Products—Use Classes: Definitions, Application to Solid Wood and Wood-Based Products. BSI Standards Publication: London, UK, 2013.
31. Filipovici, J. *Study of Wood, Vol II*; Didactic Publishing House: Bucharest, Romania, 1965. (In Romanian)
32. Deaconu, I.; Porojan, M.; Timar, M.C.; Bedelean, B.; Campean, M. Comparative Research on the Structure, Chemistry, and Physical Properties of Turkey Oak and Sessile Oak Wood. *BioResources* **2023**, *18*, 5724–5749. [CrossRef]
33. *ISO 13061-13:2016*; Physical and Mechanical Properties of Wood. Test Methods for Small Clear Wood Specimens. Part 13: Deter. Determination of Radial and Tangential Shrinkage. ISO: Geneva, Switzerland, 2016.
34. *ISO 13061-14:2016*; Physical and Mechanical Properties of Wood. Test Methods for Small Clear Wood Specimens. Part 14: Determination of Volumetric Shrinkage. ISO: Geneva, Switzerland, 2016.
35. *ISO 13061-15:2017*; Physical and Mechanical Properties of Wood. Test Methods for Small Clear Wood Specimens. Part 15: Determination of Radial and Tangential Swelling. ISO: Geneva, Switzerland, 2016.
36. *ISO 13061-16:2017*; Physical and Mechanical Properties of Wood. Test Methods for Small Clear Wood Specimens. Part 16: Determination of Volumetric Swelling. ISO: Geneva, Switzerland, 2016.
37. *ISO 13061-17:2017*; Physical and Mechanical Properties of Wood. Test Methods for Small Clear Wood Specimens. Part 17: Determination of Ultimate Stress in Compression Parallel to Grain. ISO: Geneva, Switzerland, 2016.
38. Wagenführ, R. *Holzatlas (Wood atlas)*; Fachbuchverlag Leipzig im Carl Hanser Verlag: München, Germany, 2007; pp. 255–264.
39. Kollmann, F.F.P.; Cote, W.A., Jr. *Principles of Wood Science and Technology, Vol I. Solid Wood*; Springer: Berlin/Heidelberg, Germany; GmbH & Co. KG: Hamburg, Germany, 2012; 592p, ISBN 9783642879302.
40. Lucejko, J.J.; Tamburini, D.; Zborowska, M.; Babiński, L.; Modugno, F.; Colombini, M.P. Oak wood degradation processes induced by the burial environment in the archaeological site of Biskupin (Poland). *Herit. Sci.* **2020**, *8*, 44. [CrossRef]
41. Clair, B.; Thibaut, B. *Physical and Mechanical Properties of Reaction Wood. The Biology of Reaction Wood*; Springer Series in Wood Science; Springer: Berlin/Heidelberg, Germany, 2014; 249p, ISBN 978-3-642-10813-6. [CrossRef]
42. Badel, E.; Bakour, R.; Perre, P. Investigation of the relationships between anatomical pattern, density and local swelling of oak wood. *IAWA J.* **2006**, *27*, 55–71. [CrossRef]
43. Jiang, J.; Lu, J.; Zhou, Y.; Zhao, Y.; Zhao, L. Compression strength and modulus of elasticity parallel to the grain of oak wood at ultra-low and high temperatures. *BioResources* **2014**, *9*, 3571–3579. [CrossRef]

applied sciences

MDPI

Article

The Influence of Slicing Thickness on the Perpendicular to Grain Tensile Properties of Oak (*Quercus robur* L. and *Quercus petraea* L.) Lamellae

Victor Grubîi * and Jimmy Johansson

Department of Forestry and Wood Technology, Linnaeus University, 351 95 Växjö, Sweden; jimmy.johansson@lnu.se
* Correspondence: victor.grubii@lnu.se; Tel.: +46-724539121

Abstract: The mechanical properties of sliced lamellae are critical for structural and decorative engineered wood products. This study evaluates the impact of slicing thickness on the tensile mechanical properties of plain oak (*Quercus robur* L. and *Quercus petraea* L.) sliced lamellae, perpendicular to the grain direction. The mechanical performance parameters in terms of the modulus of rupture (MOR), strain at break, and modulus of elasticity (MOE), were analysed using a one-way analysis of variance (ANOVA) and contrast analysis. Our findings indicate that slicing thickness substantially affects the mechanical properties of the modulus of rupture and strain at break, whilst the modulus of elasticity was somewhat independent of the slicing thickness. The mean MOR value increased from 0.8 to 1.43 N mm^{-2} for an increase in the sliced lamellae thickness from 1.5 to 4.5 mm. The strain at break increased, on average, from 1.37 to 2.64% for an increase in the sliced lamellae thickness from 1.5 to 4.5 mm. The MOE was approximately 100–120 N mm^{-2}, indicating a substantially reduced stiffness compared to other sliced lamellae species and solid oak reported values. The slicing check depth ratio diminished from approximately 69% to 50% for an increase in the sliced lamellae thickness from 1.5 mm to 4.5 mm. These findings indicate a negative correlation between the slicing check depth ratio and the tensile performance perpendicular to the grain, suggesting the importance of obtaining an optimal slicing quality. This study employs digital image correlation (DIC) analysis to gain insights into the fracture mechanisms of the tested sliced lamellae and provides an alternative method for strain and stress measuring. The DIC analysis highlighted the role of slicing checks in the stress concentration and ultimate failure areas. This research provides insights into the fracture behaviour of sliced lamellae that are perpendicular to the grain, which is critical for the performance of both structural and decorative products.

Keywords: veneer mechanical properties; tensile strength; slicing checks; digital image correlation (DIC); slicing thickness

Citation: Grubîi, V.; Johansson, J. The Influence of Slicing Thickness on the Perpendicular to Grain Tensile Properties of Oak (*Quercus robur* L. and *Quercus petraea* L.) Lamellae. *Appl. Sci.* **2023**, *13*, 12254. https://doi.org/10.3390/app132212254

Academic Editors: Giangiacomo Minak, Emilia-Adela Salca, Lidia Gurau and Mihaela Câmpean

Received: 25 September 2023
Revised: 18 October 2023
Accepted: 31 October 2023
Published: 13 November 2023

1. Introduction

Sliced lamellae, or veneer sheets produced via flat slicing, are thin wood elements obtained via compression cutting. Veneers are produced through rotary cutting or peeling, i.e., cutting in the wood's tangential direction, and sliced lamellae are produced through parallel to the grain flat slicing. The leading product segment for the veneering industry is plywood, with a global production of 90–100 mln. m^3, according to the European Panel Federation estimates from 2015 [1]. Another significant veneer-based product is laminated veneer lumber (LVL), with approximately 3 mln. m^3 global production [2] and a somewhat high compound annual growth rate (CAGR) of 13.5% [1]. Decorative veneers in the panel industry are also estimated to be a significant market. Flat-sliced lamellae are primarily used in decorative panel lamination, where a crown-cut or timber appearance is desirable. Due to raw material supply challenges, sliced lamellae in product realisation could be extended to a series of multi-layered products such as engineered wood flooring, interior panels, and furniture. The potential impact of extensively using sliced lamellae for joinery products is high; for example, the engineered wood flooring market is estimated at nearly 90 mln. m^2 in Europe alone [3].

The main advantage of the slicing technique over traditional sawing is an increased yield due to no kerf being required when cutting. Due to the kerfless cutting aspect, slicing produces fewer combustion-use by-products than sawing, allowing for the sustainable and cascaded use of the wood raw material. Moreover, slicing technology may increase the fire and health safety of production units compared to traditional sawmills. Hence, the main advantage of the slicing technique resides in producing thin wood elements without compromising the surface quality; therefore, a high volume-to-area conversion ratio is achieved.

Plain and lengthwise slicing are the most common flat slicing techniques [4]. Plain slicing is performed on vertical slicers where the cutting is perpendicular to the material direction, similar to a 0–90 cutting situation [5]. Lengthwise slicing is performed on horizontal slicers using a highly oblique slicing knife and occurs when cutting parallel to the grain, following the specimen feed direction. The advantage of plain slicing over lengthwise slicing is the increased processing speed [6]. Material cutting in slicing occurs via wood compression in front of the knife's cutting edge, assisted by a pressure bar displaced on the back-side of the cut veneer. The role of the pressure bar is to counter excessive tensile forces perpendicular to the grain that would lead to lamellae collapse [4,5]. Hence, parts of the cutting energy may not be countered in the slicing process, and it is released by generating small cracks, known as lathe or slicing checks [5]. Slicing checks are a significant quality issue of sliced lamellae and sliced-lamellae-based products. For instance, slicing checks were shown to negatively influence the shear strength in plywood [7,8] and LVL panels [9].

Considerable research studies on veneer quality have considered slicing check characteristics to be a principal focus area. The depth of the lathe checks in relation to the veneer thickness, known as the check depth ratio and check frequency, were used to characterise slicing checks [4,10,11]. The impact of some production parameters on slicing check characteristics was evaluated, e.g., the pre-treatment temperature, cutting speed and pressure bar geometry, the pre-compression ratio, and species [4]. Wood structure and density significantly impact the slicing quality; therefore, medium- and high-density wood species, i.e., above 500 kg/m^3, require a thermo-hydro pre-treatment before slicing [4]. The pre-treatment temperature was positively correlated with check depth ratio measurements in birch (*Betula pendula* Roth) and beech (*Fagus sylvatica* L.) species [12–14]. In rotary-cut beech veneers, increasing the pre-compression ratio reduced the check depth ratio [14]. A too-high pre-compression may result in significant damage on the pressure bar veneer side as surface roughness and small cracks [15]. The slicing quality depends on the wood species, mainly due to macro-anatomical features such as rays [14]. Maple, birch, and beech sliced lamellae and rotary-cut veneers were found to be tight-cut, i.e., with a check depth ratio below 50% [10,12,16,17], while oak species were found to be loose-cut, i.e., with a check depth ratio above 50% [9,16,18].

Hence, slicing- or sliced lamellae thickness is one of the most critical production variables impacting the processing efficiency and sliced lamellae quality. Sliced lamellae thickness can impact the sustainability aspects of sliced-lamellae-based products, i.e., reduced adhesive consumption with increasing sliced lamellae thickness due to fewer adhesive layers. The check depth ratio was positively correlated with the veneer thickness for rotary-cut beech veneers [9,10], but a negative correlation was found in plain-sliced oak lamellae [18]. Such discrepancies in results may be due to the significant wood fracture mechanisms when cutting in different wood growth directions, i.e., the tangential direction in rotary cutting and a mixture of the tangential and radial direction in plain slicing.

Besides decreasing the mechanical performance of veneer-based products, slicing checks are assumed to impact the surface-checking of veneer and sliced-lamellae-based products [19,20]. Surface-checking is an extensive defect represented by small surface cracks appearing on the veneer's surface and sliced-lamellae-based products. This phenomenon occurs due to residual tensile stresses developed on the surface of a multi-layered structure subjected to a moisture gradient [21]. A significant role in surface-checking is attributed to the low mechanical properties of sliced lamellae. For example, while sliced-lamellae-based engineered wood flooring has exhibited significant surface-checking, the sawn-lamellae-based one showed no surface checking at all [22]. The material edge-chipping was another drawback reported in sliced-lamellae-based engineered wood flooring production [22].

Consequently, knowledge of sliced lamellae's mechanical properties is essential for appropriate product design, quality control, and further investigations on sliced lamellae quality optimisation. Due to the occurrence of slicing checks being mainly in the wood radial direction and surface checking occurring when the material is under surface tension loading, it is assumed that tensile strength perpendicular to the grain is a critical parameter that could describe veneer or sliced lamellae mechanical performance. Regarding this aspect, a relevant study represents the investigation of the mechanical properties of plain- and lengthwise-sliced beech lamellae [6] and birch rotary-cut veneers [23]. Hence, these studies evaluated low slicing thicknesses, i.e., 0.5–1.5 mm. Such lamella thicknesses are uncommon in solid-wood-based flooring product realisation, while the studied species, unlike oak, are generally considered to be suitable for slicing.

Digital image correlation (DIC) is a gradient displacement tracking method used primarily in wood mechanics to analyse fracture occurrence [24,25], characterise various loading cases [26,27], or measure deformations [28]. DIC can also be applied to detect small defects, such as surface checks in veneer-based products [24,29,30]. Therefore, the analysis of sliced lamellae's fracture behaviour could benefit from the use of DIC to gain insights into the role of slicing checks on the failure mechanisms.

Given the importance of veneers and sliced lamellae in various product applications, it is crucial to understand the material mechanical properties and the impact of slicing checks to ensure optimal product performance. Therefore, the primary objective of the current work was to evaluate the effect of slicing thickness on the tensile mechanical properties of oak (*Quercus robur* L. and *Quercus petraea* L.) sliced lamellae. The investigated sliced lamellae thickness was chosen concerning the potential application for engineered wood flooring top-layer manufacturing, i.e., 1.5, 2.5, 3.5, and 4.5 mm. Another objective was to evaluate the correlation between the lamellae slicing check depth ratio and their mechanical performances. The final objective of this study was to analyse the failure mechanisms in sliced lamellae using digital image correlation (DIC). The formulated objectives aimed to provide comprehensive and practical recommendations to guide enhanced product design and application in sliced-lamellae-based products.

2. Materials and Methods

2.1. Materials

The material part of this study's investigations was the sliced lamellae of oak grown in Northern Europe. The sliced lamellae were produced from 32 freshly sawn timber blocks with nominal dimensions of 2500 × 220 × 85 mm. Before slicing, the timber blocks were pre-treated via water soaking at 90 °C for 58 h. Lamellae slicing was carried out on two vertical slicing machines, depending on the nominal thickness. The 1.5 mm sliced lamellae were produced on a typical (low-thickness) veneer slicing of type Cremona, while the 2.5–4.5 mm groups were sliced on a particular Babcock-BSH vertical slicer designed for thick sliced lamellae slicing (Merscha mill in Celje). The slicing parameters, such as the cutting speed, lead (vertical opening between pressure bar and knife cutting edge), gap (horizontal opening between pressure bar and knife cutting edge), and pressure, were adjusted for the cutting thickness based on experienced operator knowledge for optimal quality. Both vertical slicing machines were equipped with a slicing knife with an 18° knife bevel angle. The knives were inspected and resharpened after each block processing. The drying of the sliced lamellae was performed according to manufacturer recommendations; i.e., the 1.5 and 2.5 mm thick groups were dried on a multi-deck automatic veneer dryer at

80 °C for approx. 8–13 min, while the 3.5 and 4.5 mm groups were dried in a conventional dryer at 40 °C for 4–6 days, including a 24 h indoor climate seasoning step. The resultant moisture content (MC) was measured to be 5.4–6.8% MC for the 1.5 and 2.5 mm thick sliced lamellae and 8.2–8.6% MC for the 3.5 and 4.5 mm thick sliced lamellae after being transported from the production site in Slovenia to the laboratory in Sweden. All sliced lamellae were stored in a climate chamber at room temperature and a relative humidity (RH) of 60% for approximately 24 months, corresponding to approx. 11% equilibrium MC (EMC). Specimen testing was performed at a climate of 30% RH and room temperature for 0 to 28 h.

The tensile mechanical properties of the oak sliced lamellae were determined based on a non-standardised tensile test procedure. The specimen requirements depicted for the perpendicular to the grain tensile testing procedures—such as for the rotary veneers, DIN 52188-1979 [31], or those aimed at solid wood specimens, such as JIS Z 2101, ASTM D143, or ISO 13910 [32]—could not be fulfilled in the case of the sliced lamellae. Due to the characteristic brittleness and potential damage during machining, the specimens were obtained with a rectangular shape, similar to in [6], with nominal dimensions of 150 × 20 mm (length × width) (Figure 1A), and were retrieved from the central part of a sliced lamella. A carpentry band-saw was used to produce the specimens used in this study. Special care was taken to limit induced vibrations in the specimens that could alter their mechanical properties. Six specimens per lamella were cut (from its middle part, i.e., produced with consecutive cuts). The specimens were obtained from straight-grained wood, free of knots, cracks, pith, and sapwood. An additional two specimens per lamella were cut adjacent to the testing specimens to perform slicing check measurements and the DIC analysis. The specimens were obtained from one lamella per sliced flitch (the bundle of sliced lamellae from one timber block). Five flitches per thickness group were randomly selected, i.e., 20 sliced lamellae, resulting in 120 specimens.

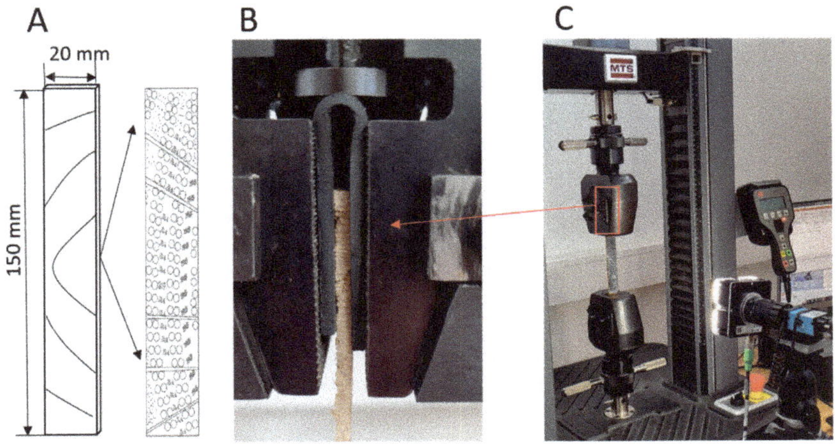

Figure 1. Testing specimens' representation (**A**); specimen fixing in the testing machine's grips (**B**); experimental setup for the DIC acquisition (**C**).

2.2. Methodology

The slicing check characteristics, check depth ratio, and check frequency were measured using a semi-automated detection method based on the surface-staining of the sliced lamellae specimens using a black ink dye, image acquisition of the sliced lamellae cross-section, and the deployment of a slicing check detection algorithm, according to the procedures described in [18]. The algorithm is based on image processing and analysis, i.e., a series of image enhancement and morphological filtering steps to determine the shape of the slicing checks. The method's outputs are the check depth ratio and check frequency.

2.2.1. Mechanical Tensile Tests

Mechanical testing was performed on an MTS Exceed Model E43.104 (MTS Systems Corporation, Eden Prairie, MN, USA) mechanical test machine equipped with a 10 kN load force [33]. During testing, a rubber band was placed between the grips of the testing machine and the tested specimens (Figure 1B). The rubber band's role was to limit potential damages induced by the equipment grips. Special care was given to the specimens' fixation in the grips by adjusting the clamping torque to limit the specimen pre-tension and slippage during testing. Pre-tests on dummy specimens without inserting the rubber band resulted in failure frequently occurring in the fixation area due to the grip-pinching effect.

Tensile tests were carried out at a feed rate of 0.1 mm/min and an acquisition rate of 10 Hz. The feed speed was 0.1 mm/min so that specimen failure would occur after 0.5 to 2 min from the test start. Data acquisition was performed using TW Essential software (version 4.3.1.375) [34]. The modulus of rupture, strain at break, and modulus of elasticity (MOE) were calculated using load–displacement data. The MOE was calculated based on 5% elongation following the default software procedures. The MOE calculus based on manually fitting the elastic interval showed minor deviations from the software output and, therefore, was disregarded.

2.2.2. Statistical Analysis

The data distribution was tested using a one-sample Kolgomorov-Smirnov test using the kstest Matlab2023a function, which resulted in the rejection of the null hypothesis. The Kolgomorov–Smirnov test is a non-parametric test that was used due to its sensitivity to differences between expected and observed distributions and sample size independence [35]. Statistical analysis was conducted using a one-way analysis of variance (ANOVA) to measure the statistical significance of the sliced lamellae thickness's impact on the sliced lamellae mechanical properties, namely the strain at rupture, modulus of rupture (MOR), and MOE. Following the experimental design, grouping was conducted based on the thickness, with four resultant categories, i.e., 1.5, 2.5, 3.5, and 4.5 mm. A one-way ANOVA was performed using anova1 function in Matlab2023a. Post hoc tests were performed to compare the variance between individual groups using a contrast analysis. The contrast analysis was executed using the multcompare Matlab2023a function using Tukey's Honestly Significant Difference (HSD) procedure.

2.2.3. Digital Image Correlation Investigation

The application of the DIC method involves ensuring that the whole surface has unique, random, and high-contrast identifiers to allow the correspondence problem to be solved [36]. A common practice is that the sample's surface must be coated with a white background and black speckled paint layer, or use other synthetic identifiers such as markers [37].

The specimens were prepared for the DIC investigation via manually spraying acrylic paint (Corro Protect) with two hours of drying between the white background layer and the black-coloured speckles. The spraying distance and intensity were adjusted to obtain a black-coloured speckle size of approximately 5 pixels. The specimens were tested 19 h after the final coat application and stored at 60% RH and 23 °C to reach approximately 12% EMC.

The DIC acquisition setup (Figure 1C) was based on a single DMK 33GX183 camera (2D-DIC analysis) equipped with a Fujinon CF25ZA-1S 23MP 25 mm focal length lens at f/1.8. The illumination source was a single-ring SVL R80 light. The image resolution was 26.5 pixels/mm. The image acquisition was performed at two-second intervals until failure detection.

The two-dimensional augmented Lagrangian digital image correlation (2D-ALDIC) algorithm was used for the DIC analysis [37]. The algorithm is based on local and global subset displacement convergence, resulting in a low cost and high precision [38,39]. The deformation gradients and strains were computed using the finite difference method for the Eulerian strain type [39]. For the DIC parameters, the subset size was 20 pixels and the subset step was 10 pixels.

3. Results and Discussion

3.1. Mechanical Properties

The results of the MOE, MOR, and strain at break are presented in Figure 2. The MOR and strain at break generally increase with an increase in the slicing thickness, demonstrated by the increasing median value (red line), the position of the interquartile range (the box) (Figure 2), and the mean group values (Table 1). The MOR was significantly lower for the 1.5 mm thick sliced lamellae, with a mean of 0.8 N mm^{-2}, compared to 1.16–1.43 N mm^{-2} (mean value) for the 2.5 to 4.5 mm thick sliced lamellae, with no significant differences between the 2.5–4.5 mm groups (Table 1). The positive correlation between the slicing thickness and MOR is thought to be due to the increased thickness of the unchecked wood with increasing sliced lamellae thickness, and aligns with the findings on rotary-cut birch veneers [23].

The failure of the sliced lamellae that were subject to tension in the perpendicular-to-grain direction occurred mainly in the radial direction. The sliced lamellae's reduced strength, i.e., several times lower than alternative species studied [6,23], were probably due to the oak species' loose-cut slicing check characteristics and the material's low strength features, such as large earlywood vessels [40].

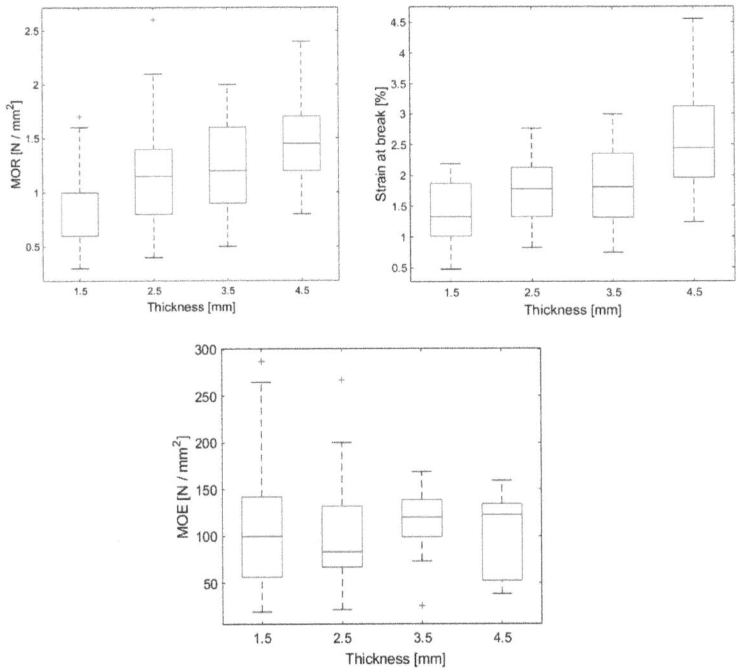

Figure 2. Effect of sliced lamellae thickness on the MOR (**upper left**), strain at break (**upper right**), and MOE (**lower**).

Table 1. MOR, MOE, and strain at break descriptive statistics and ANOVA results.

Thickness (mm)	No. Samples	Mean	Coeff. of Variation (%)	Standard Error	C.I. 95% of the Mean		2-Way ANOVA *p*-Value	*p*-Value ANOVA Contrasts Thickness (mm)			
								1.5	2.5	3.5	4.5
MOR (N mm^{-2})											
1.5		0.80	47.88	0.07	0.66	0.95		NA	8.6×10^{-3} *	3.4×10^{-4} *	3.9×10^{-7} *
2.5	30	1.16	43.33	0.09	0.98	1.35	8.4×10^{-7} *	8.6×10^{-3} *	NA	0.8	0.058
3.5		1.26	33.49	0.08	1.11	1.42		3.4×10^{-4} *	0.8	NA	0.35
4.5		1.43	28.33	0.08	1.28	1.59		3.9×10^{-7} *	0.058	0.35	NA
Strain at break (%)											
1.5		1.37	36.33	0.09	1.18	1.56		NA	0.054	0.035 *	5.4×10^{-11} *
2.5	30	1.79	32.57	0.11	1.57	2.01	1.6×10^{-10} *	0.054	NA	0.99	1.17×10^{-5} *
3.5		1.82	34.41	0.11	1.58	2.05		0.03 *	0.99	NA	2.4×10^{-5}
4.5		2.64	29.50	0.14	2.34	2.93		5.4×10^{-11} *	1.17×10^{-5} *	2.4×10^{-5} *	NA
MOE (N mm^{-2})											
1.5		110.7	62.03	12.5	85.0	136.3		NA	0.99	0.99	0.55
2.5	30	100.5	55.06	10.1	79.8	121.1	0.56	0.99	NA	0.99	0.47
3.5		118.1	26.38	5.7	106.5	129.7		0.99	0.99	NA	0.61
4.5		104.5	39.38	7.6	88.8	120.1		0.55	0.47	0.61	NA

*—statistically significant at $p < 0.05$, C.I.—confidence interval.

The results indicate that the strain at break increased with the increasing sliced lamellae thickness from 1.37% for 1.5 mm to 2.64% for the 4.5 mm thick sliced lamellae (Table 1). The obtained results for the strain at break were statistically significant at a 95% confidence interval for the 1.5 compared to the 3.5 and 4.5 mm and 3.5 to 4.5 mm thick sliced lamellae group comparisons. The increase in the strain at break with the increasing sliced lamellae thickness could be explained by a crack opening displacement at the crack initiation point [41]. Therefore, sliced lamellae consisting of large slicing checks experience a significant opening phase under low stress at the beginning of loading. Hence, ascertaining the effect of clamping devices and specimen geometry should be further evaluated to validate the measurements' accuracy. Other potentially significant variations in the sliced lamellae properties impacting the results may be due to various factors such as raw material properties, drying, and slicing parameters, which were beyond the current study's experimental design limitations.

The impact of the slicing thickness on the MOE perpendicular to the grain illustrates a somewhat independent variation (Figure 2 and Table 1). The obtained MOE of 110.7 N mm^{-2} for the 1.5 mm thick sliced lamellae was about six-fold lower than the reported 0.5 mm thick beech sliced lamellae results [6], and about eight times lower than oak solid wood reported values [42]. These aspects reiterate plain-sliced oak lamellae's significantly diminished stiffness due to slicing check damage.

Overall, plain-sliced oak lamellae's mechanical response behaviour resembles a quasi-brittle material, similar to solid wood. The most commonly identified stress–strain curves are shown in Figure 3. The stress–strain curves can be generally characterised by a nearly linear increase over the test course, with some specimens exhibiting a plastic strain behaviour during loading, followed by a continuous linear stress–strain increase (Figure 3 right). A possible reason for this behaviour may lie in the crack opening phase, followed by a fibre-bridging effect [1].

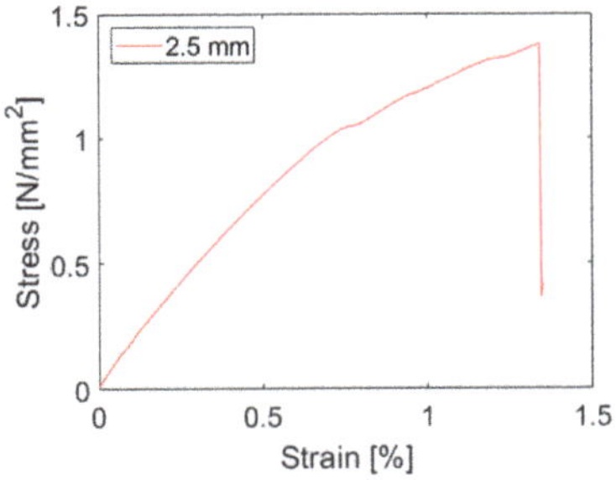

Figure 3. Examples of most common stress–strain curves observed on oak sliced lamellae; elastic followed by plastic deformation (**left**) and elastic with plastic strain mid-deformation (**right**).

3.2. Slicing Check Characteristics

The mean check depth ratio per studied thickness group and examples of the slicing checks' appearance are presented in Figure 4. The check depth ratio decreased with the increasing sliced lamellae thickness, i.e., 68% for the 1.5 mm thick sliced lamellae to 50% for the 4.5 mm thick sliced lamellae. The results indicate a negative correlation between check depth ratio and the strength properties, MOR, and strain at break (Figure 4).

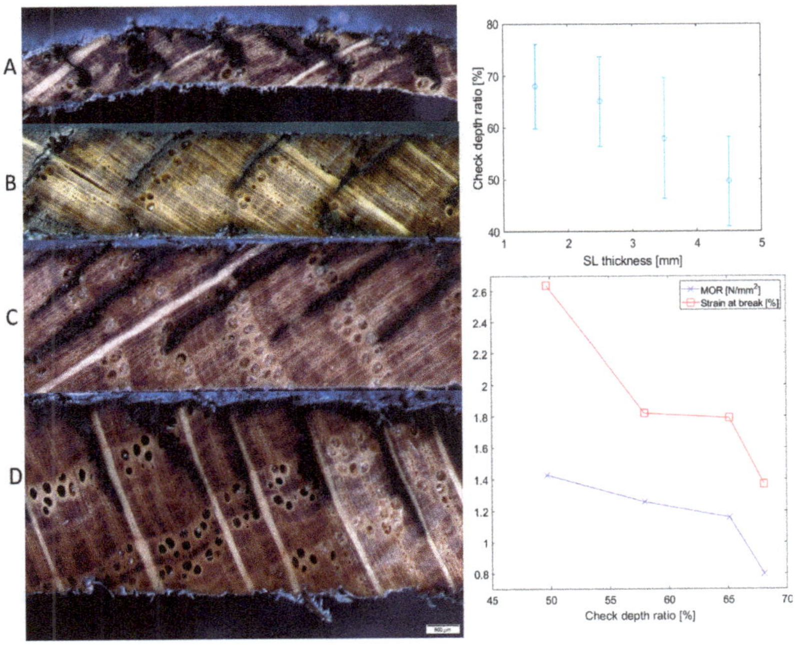

Figure 4. Example of slicing checks highlighted by a stain for (**A**) 1.5 mm, (**B**) 2.5 mm, (**C**) 3.5 mm, and (**D**) 4.5 mm thick oak sliced lamellae (left); mean check depth ratio for the evaluated SL thicknesses (upper right) and mean MOR and strain at break as functions of the check depth ratio (lower right).

Extensive slicing checks when slicing oak species may be mainly attributed to the presence of multiseriate rays that constitute stress–concentration areas during slicing [4,14]. Additionally, large earlywood vessels constituting regions of low mechanical properties facilitate crack propagation [40,43]. Overall inefficiencies in oak species wood plasticisation due to anatomical characteristics, e.g., tyloses [15], may also contribute to extensive slicing checks. If plain-sliced oak lamellae are used for structural applications, the middle range of the selected thickness groups could be beneficial, i.e., 2.5–3.5 mm. Lamellae slicing at 4.5 mm and above would provide only marginal benefits in terms of yield and productivity when compared to sawn solid wood while the decreasing material mechanical properties significantly.

Hence, the use of oak sliced lamellae for structural applications is rather unlike; therefore, the current study's results could benefit a more common use for decorative products such as flooring and furniture panel production. Understanding a raw material's most critical mechanical properties can facilitate further testing method development and product optimisation. Further studies should ascertain the impact of sliced lamellae thickness on surface checking in sliced-lamellae-based finished products. A possible impact of sliced lamellae thickness on surface checking could enable the use of mechanical properties, e.g., MOR and strain at break, as predictors for the defect occurrence. The benefit of using mechanical properties as a quality control method is due to the potential correlation between static and dynamic methods and the use of non-destructive means for the latter.

3.3. DIC Analysis

The DIC analysis revealed more insight into the fracture behaviour of the tested sliced lamellae. Figure 5a–d illustrate a common fracture mechanism depicting stress concentration around the most extensive slicing checks. A fracture occurred in the radial–longitudinal plane. The strain maps in Figure 5(Ia,IIa) show the detection of significant strain corresponding to several slicing checks at the beginning of loading. By the end of the loading, as observed in Figure 5(Ib,IIb), the failure would occur following extensive slicing checks.

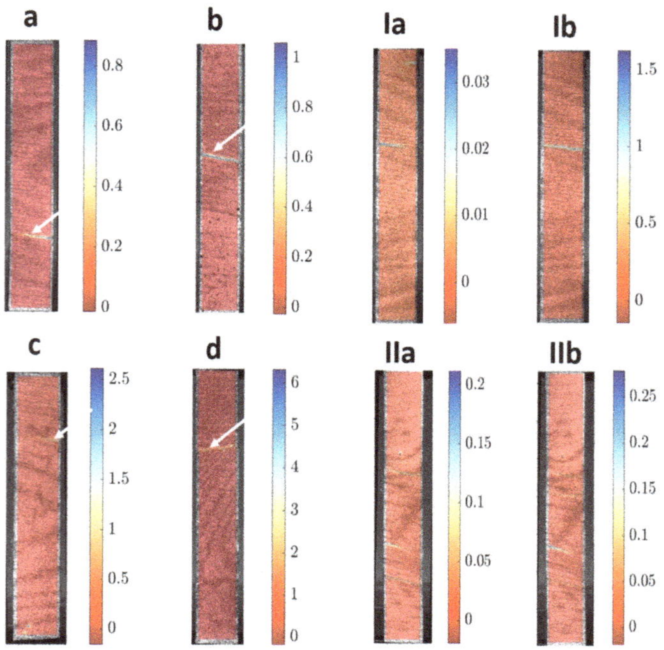

Figure 5. Principal maximal strain for (**a**) 1.5 mm, (**b**) 2.5 mm, (**c**) 3.5 mm, and (**d**) 4.5 mm thick sliced lamellae (arrows indicate fracture area); (**Ia**,**IIa**) strain concentration around areas of slicing checks at the beginning of loading and (**Ib**,**IIb**) at the breaking point.

Generally, principal maximal strain tends to increase with increasing sliced lamellae thickness (Figure 5), similar to the strain rate from experimental data. The method should be further explored to enable robust and interchangeable results between the two methods.

4. Conclusions

Oak sliced lamellae are a quasi-brittle material, affected significantly by slicing check characteristics. The novelty of this study is highlighted by the correlation between the check depth ratio, which is influenced by the lamella thickness, and the sliced lamellae strength and strain at failure. In the elastic domain, the stiffness was not significantly impacted by slicing check characteristic variations.

The obtained mechanical properties for the oak sliced lamellae indicate inferior mechanical performance compared to other reported species and corresponding solid wood species. The DIC method has shown that slicing checks are sliced lamellae's stress concentration and ultimate failure areas. The DIC strain data facilitated the interpretation of failure mechanisms in sliced lamellae.

Author Contributions: Conceptualisation, V.G. and J.J.; methodology, V.G.; software, V.G.; validation, V.G.; formal analysis, V.G.; investigation, V.G.; resources, J.J.; data curation, V.G.; writing—original draft preparation, V.G. and J.J.; writing—review and editing, V.G. and J.J.; visualisation, V.G.; supervision, J.J.; project administration, J.J.; funding acquisition, J.J. All authors have read and agreed to the published version of the manuscript.

Funding: This research was funded by the Knowledge Foundation.

Data Availability Statement: The data presented in this study are available on request from the corresponding author. The data are not publicly available due to data quality assurance.

Conflicts of Interest: The authors declare no conflict of interest.

References

1. Niemz, P.; Teischinger, A.; Sandberg, D. *Springer Handbook of Wood Science and Technology*; Springer International Publishing AG: Cham, Switzerland, 2023.
2. Vasileva, R. LVL is Unlocking the 'Timber Age'. 2019. Available online: https://www.raute.com/blog/industry-insights/lvl-is-unlocking-the-timber-age/ (accessed on 21 September 2023).
3. FEP. *FEP 2020 Statistics*; FEP General Assembly: Trenton, NJ, USA, 2021.
4. Lutz, J.F. Wood veneer: Log selection, cutting, and drying. In *Technical Bulletin No.1577*; USDA Forest Products Laboratory: Madison, WI, USA, 1977.
5. Marchal, R.; Mothe, F.; Denaud, L.E.; Thibaut, B.; Bleron, L. Cutting forces in wood machining—Basics and applications in industrial processes. A review COST Action E35 2004-2008: Wood machining—Micromechanics and fracture. *Holzforschung* **2009**, *63*, 157–167.
6. Pfriem, A.; Buchelt, B. Influence of the slicing technique on mechanical properties of the produced veneer. *Eur. J. Wood Prod.* **2011**, *69*, 93–99. [CrossRef]
7. Rohumaa, A.; Hunt, C.G.; Hughes, M.; Frihart, C.R.; Logren, J. The influence of lathe check depth and orientation on the bond quality of phenol-formaldehyde—Bonded birch plywood. *Holzforschung* **2013**, *67*, 779–786. [CrossRef]
8. DeVallance, D.B.; Funck, J.W.; Reeb, J.E. Douglas-fir plywood gluebond quality as influenced by veneer roughness, lathe checks, and annual ring characteristics. *For. Prod. J.* **2007**, *57*, 21–28.
9. Purba, C.Y.C.; Pot, G.; Viguier, J.; Ruelle, J.; Denaud, L. The influence of veneer thickness and knot proportion on the mechanical properties of laminated veneer lumber (LVL) made from secondary quality hardwood. *Eur. J. Wood Prod.* **2019**, *77*, 393–404. [CrossRef]
10. Pałubicki, B.; Marchal, R.; Butaud, J.-C.; Denaud, L.-E.; Bléron, L.; Collet, R.; Kowaluk, G. A Method of Lathe Checks Measurement; SMOF device and its software. *Eur. J. Wood Prod.* **2010**, *68*, 151–159. [CrossRef]
11. Denaud, L.; Bléron, L.; Ratle, A.; Marchal, R. Online control of wood peeling process: Acoustical and vibratory measurements of lathe checks frequency. *Ann. For. Sci.* **2007**, *64*, 569–575. [CrossRef]
12. Antikainen, T.; Eskelinen, J.; Rohumaa, A.; Vainio, T.; Hughes, M. Simultaneous measurement of lathe check depth and the grain angle of birch (Betula pendula Roth) veneers using laser trans-illumination imaging. *Wood Sci. Technol.* **2015**, *49*, 591–605. [CrossRef]
13. Dupleix, A.; Denaud, L.-E.; Bleron, L.; Marchal, R.; Hughes, M. The effect of log heating temperature on the peeling process and veneer quality: Beech, birch, and spruce case studies. *Eur. J. Wood Prod.* **2013**, *71*, 163–171. [CrossRef]

14. Rohumaa, A.; Viguier, J.; Girardon, S.; Krebs, M.; Denaud, L. Lathe check development and properties: Effect of log soaking temperature, compression rate, cutting radius and cutting speed during peeling process of European beech (*Fagus sylvatica* L.) veneer. *Eur. J. Wood Prod.* **2018**, *76*, 1653–1661. [CrossRef]
15. Schramm, A. *A Complete Guide to Hardwood Plywood and Face Veneer*; Purdue University Press: West Lafayette, IN, USA, 2003.
16. Cassens, D.; Leng, Y.L.; McCabe, G. Face check development in veneered furniture panels. *For. Prod. J.* **2003**, *53*, 79–86.
17. Buchelt, B.; Wagenführ, A.; Dietzel, A.; Raßbach, H. Quantification of cracks and cross-section weakening in sliced veneers. *Eur. J. Wood Prod.* **2018**, *76*, 381–384. [CrossRef]
18. Grubii, V.; Johansson, J. An image processing algorithm for detection and analysis of slicing checks. *Eur. J. Wood Wood Prod.* **2021**, *79*, 999–1006. [CrossRef]
19. Leavengood, S.; Funck, J.W.; Reeb, J.E. A note on face veneer checking in maple plywood. *Int. Wood Prod. J.* **2011**, *2*, 120–123. [CrossRef]
20. Burnard, M.; Leavengood, S.; Muszyński, L.; Ganio, L. Investigating face veneer check development in decorative plywood panels: The impact of four common manufacturing factors. *Eur. J. Wood Prod.* **2019**, *77*, 961–979. [CrossRef]
21. Christiansen, A.W.; Knaebe, M. *Diagnostic Guide for Evaluating Surface Distortions in Veneered Furniture and Cabinetry*; General Technical Report FPL-GTR-143; US Forest Service, Forest Products Laboratory: Madison, WI, USA, 2004.
22. Grubîi, V.; Johansson, J. Performance of multi-layered wood flooring elements produced with sliced and sawn lamellas. *Pro Ligno* **2019**, *15*, 166–172.
23. Pramreiter, M.; Stadlmann, A.; Huber, C.; Konnerth, J.; Halbauer, P.; Baumann, G.; Müller, U. The Influence of Thickness on the Tensile Strength of Finnish Birch Veneers under Varying Load Angles. *Forests* **2021**, *12*, 87. [CrossRef]
24. Kang, H.-Y.; Muszynski; Milota, M.R.; Kang, C.-W.; Matsumura, J. Preliminary tests for optically measuring drying strains and check formation in wood. *J. Fac. Agric. Kyushu Univ.* **2011**, *56*, 313–316.
25. van Blokland, J.; Olsson, A.; Oscarsson, J.; Daniel, G.; Adamopoulos, S. Crack formation, strain distribution and fracture surfaces around knots in thermally modified timber loaded in static bending. *Wood Sci. Technol.* **2020**, *54*, 1001–1028. [CrossRef]
26. Ljungdahl, J.; Berglund, L.A.; Burman, M. Transverse anisotropy of compressive failure in European oak—A digital speckle photography study. *Holzforschung* **2006**, *60*, 190–195. [CrossRef]
27. Fernández Zúñiga, E.; García Esteban, L.; García Fernández, F.; de Palacios, P.; García-Iruela, A.; García, I.; Lozano, M.; Fernández-Canteli, A. Applying Digital Image Correlation (DIC) Techniques to Characterise Plywood According to UNE-EN 314 Standards: Bonding Quality Tests. *Forests* **2022**, *13*, 2135. [CrossRef]
28. Peng, M.; Ho, Y.-C.; Wang, W.-C.; Chui, Y.H.; Gong, M. Measurement of wood shrinkage in jack pine using three dimensional digital image correlation (DIC). *Holzforschung* **2012**, *66*, 639–643. [CrossRef]
29. Burnard, M.D.; Muszyński, L.; Leavengood, S.; Ganio, L. An optical method for rapid examination of check development in decorative plywood panels. *Eur. J. Wood Prod.* **2018**, *76*, 1367–1377. [CrossRef]
30. Grubîi, V.; Johansson, J.; Dagbro, O. Measurement of surface-checking in sliced lamellae-based engineered wood flooring using digital image correlation. *Eur. J. Wood Wood Prod.* **2023**, *81*, 1427–1436. [CrossRef]
31. *DIN 52188*; Bestimmung der Zugfestigkeit Parallel zur Faser. Dtsch. Inst. für Norm.: Berlin, Germany, 1979.
32. Okamoto, S.; Akiyama, N.; Nakamura, N.; Aoki, K.; Inayama, M. Estimation of the perpendicular-to-the-grain tensile strength of Scots pine glued laminated timber via three-point bending tests. *J. Wood Sci.* **2021**, *67*, 33. [CrossRef]
33. MTS Systems Corporation. Detailed Technical Specifications—Comparative. 2023. Available online: https://www.mts.com/en/products/materials/static-materials-test-systems/exceed-electromechanical (accessed on 12 September 2023).
34. MTS Systems Corporation. *MTS TestSuiteTM TW Software*; Version 4.3.1.375; MTS Systems Corporation: Technology Drive: Eden Prairie, MN, USA, 2023.
35. Witte, R.S.; Witte, J.S. *Statistics*; J. Wiley & Sons Hoboken, NJ: Hoboken, NJ, USA, 2010.
36. Sutton, M.A.; Schreier, H.; Orteu, J.-J. *Image Correlation for Shape, Motion and Deformation Measurements: Basic Concepts, Theory and Applications*; Springer: New York, NY, USA, 2009.
37. International Digital Image Correlation Society. *A Good Practices Guide for Digital Image Correlation*; Jones, E.M.C., Iadicola, M.A., Eds.; International Digital Image Correlation Society: Washington, DC, USA, 2018.
38. Yang, J.; Bhattacharya, K. Augmented Lagrangian Digital Image Correlation. *Exp. Mech.* **2019**, *59*, 187–205. [CrossRef]
39. Yang, J. *Augmented Lagrangian Digital Image Correlation (2D_ALDIC)*; CaltechDATA: London, UK, 2020.
40. Savill, P.S. Anatomical characters in the wood of oak (Quercus-robur L and Q-Petraea Liebl) which predispose trees to shake. *Commonw. For. Rev.* **1986**, *65*, 109–116.
41. Boatright, S.W.J.; Garrett, G.G. The Effect of Microstructure and Stress State on the Fracture-Behavior of Wood. *J. Mater. Sci.* **1983**, *18*, 2181–2199. [CrossRef]
42. Blumer, S.; Niemz, P.; Serrano, E.; Gustafsson, P.J. Moisture induced stresses and deformations in parquet floors: An experimental and numerical study. *Wood Res.* **2009**, *54*, 89–101.
43. Wang, H.H.; Youngs, R.L. Drying stress and check development in the wood of two oaks. *Iawa J.* **1996**, *17*, 15–30. [CrossRef]

applied sciences

MDPI

Article

Wood Utilization in Windmill Mechanisms on Sikinos Island (Greece)

Evangelia Agnantopoulou [1], Ioannis Barboutis [2] and Vasiliki Kamperidou [2,*]

[1] Directorate of Forests of Cyclades, Ministry of the Environment and Energy, 84100 Ermoupoli, Cyclades, Greece; e.agnantopoulou@eedpaig.ypen.gr
[2] Laboratory of Wood Technology, Department of Forestry and Natural Environment, Aristotle University of Thessaloniki, 54635 Thessaloniki, Greece; jbarb@for.auth.gr
* Correspondence: vkamperi@for.auth.gr

Abstract: Windmills constitute a valuable part of cultural heritage, especially in Greece, as unique structures of popular architecture and know-how of the pre-industrial era. Their wooden mechanisms were of exceptional constructional ingenuity, with the Mediterranean ones bearing a vertical wing and a rotating roof ("trula"), with a manual torsion mechanism that allowed for operation in all wind directions. Sikinos is a small Aegean island characterized by rich landscapes, which do not have wood-producing forests, but only sparse *Juniperus phoenicea* shrub land mixed with evergreen hardwood species and the presence of numerous sclerophyllus vegetation species (maquis). Three abandoned windmills are still preserved on the island today, but only two of them appear to have their wooden mechanisms left. In the one windmill mechanism examined, it was found that different wood species were used to manufacture its individual parts. Oak wood was used in most of the large elements, with the exception of the sprattle beam ("zigos"), which is made of fir wood. The local juniper was not used in any crucial part of the mechanism, though it was used only as a structural material on the first floor ("anogio") flooring, along with large-diameter olive trunks. The findings of this study highlighted the thorough knowledge of the properties of various wood species and the effective use of simple tools toward the construction of functional and effective windmill mechanisms.

Keywords: architecture; building; construction; culture; heritage; island; properties; structure; timber; windmill; wood

Citation: Agnantopoulou, E.; Barboutis, I.; Kamperidou, V. Wood Utilization in Windmill Mechanisms on Sikinos Island (Greece). *Appl. Sci.* **2023**, *13*, 9216. https://doi.org/10.3390/app13169216

Academic Editors: Lidia Gurau, Mihaela Câmpean and Emilia-Adela Salca

Received: 19 July 2023
Revised: 8 August 2023
Accepted: 11 August 2023
Published: 14 August 2023

1. Introduction

Wood is undoubtedly an important element of cultural heritage, as it has been the raw material of a large number of products (the Greek word used for the "material" comes from the ancient word "ύλη", which means both "forest" and "wood") since prehistoric times, essentially contributing to human survival and the development of civilization. As a material, it has offered mankind a variety of possibilities, including the manufacture of furniture, tools, constructions, musical instruments, various objects of daily use, etc. This multivariate use transformed wood from a natural material to a cultural product and led to the development of a particularly interesting culture, the "wood culture" [1]. In recent years, there has been an international effort to promote and highlight this culture, whereas in 2007, the International Wood Culture Society (IWCS), a non-profit, non-governmental international network dedicated to the research, education and promotion of wood culture was founded and supported to work in this direction.

The windmills were impressively cylindrical, stone-built buildings resembling to towers [2], adapted to the local construction practices of each island [3], which were built usually in a prominent [2], windy location, near paths [3], closer to settlements than fields [4]. Despite the fact that the time of their appearance is not yet totally clear, during the 16th century, they were already widespread in the Aegean area, grinding mainly grain for the production of flour and other seeds for the production of animal feed. At the beginning of the 20th century, they gradually began to be abandoned, first on big islands and gradually on smaller and isolated ones. The main motives of depreciation were the supply of flour from other places, development of bakeries providing bread, sale of barley for beer production, etc. [3], until they finally fell into disuse due to the appearance of diesel engines in the technological era [2].

Windmills constitute a remarkable, complex example of a mechanism of the pre-industrial era [2,5], with the Mediterranean ones usually including vertical windmill sail frames (windmill wings) and a rotating roof ("trula") [3], with a manual turning mechanism, which enabled operation with all wind directions [2]. Depending on the number of floors, the windmills were classified into three categories:

- Single-floor buildings (ground floor and "anogio");
- Two-floor ones (including the ground floor, mezzanine floor and "anogio");
- Three-floor ones (including the ground floor and two floors).

The "anogio" was the most important space, since that was the location of the drive mechanism and millstones for grinding, as well as the torsion mechanism of the rotating roof ("trula") [3]. Access to the "anogio" was implemented through a narrow, stone, spiral staircase that was always built to the left of the entrance [2], while the rest of the internal structures were, on most islands, wooden, at least for single- and two-floor windmills [3], with the flooring of the upper floor consisting of wooden beams of large diameters so that they could withstand the weight of the millstone [6].

The construction of a windmill, in combination with the necessary large wooden mechanism construction (analyzed in the following section), was considered an extremely expensive investment, many times higher than other buildings (e.g., churches), carried out mainly through the prominent ones of these islands since it required years to build and import of appropriate materials (e.g., wood for parts of the mechanism) [3], while its presence during those years was a sign of economic prosperity [7].

Windmills constitute a valuable part of cultural heritage, presenting a dual character, as they are architectural monuments, examples of folk architecture and knowledge, but also examples of wooden mechanisms of special manufacturing intelligence in relation to their mode of operation [6] that smartly combined simple engineering principles [8,9].

Windmills in some areas have already been studied, such as those in the province of ancient Halicarnassus in Turkey [8], Iran [10], Castile of Spain [11] and Scotland [12]. In general, differences have been detected between those in the East and the West, with the East windmills mainly bearing horizontal windmill wing frames ("fteroti") which rotate horizontally on the top of the building, and the West ones bearing vertical rigid windmill wing frames, rotating vertically on the side of the building. In windmills in the Mediterranean area, there was an interesting fusion of Eastern and Western features, with a dominant form that included a vertical flexible windmill sail frame and rotating roof, a construction that probably evolved and became standardized over the centuries [8]. In Greece, windmills were located mainly on islands, since they required a constant supply of wind energy [3,13], even though the rest of Greece was quite familiar with these conditions as well [14]. Although a few horizontal windmills have also been found (Karpathos), vertical ones with a rotating roof dominated, while fixed ones that operated with a single wind direction were rare and limited mainly to Crete and Karpathos [15].

The moving part of the windmill (mechanism) was traditionally made of wood [16] and was the most important part. In general, it consisted of a horizontal beam (Figure 1), which carried the system for converting the force of the wind into motion on the outside and a large wooden wheel on the inside, the rotary motion of which set the millstone system in motion. The turning mechanism allowed to simply and manually change the position of the above system, so that the mill worked with all wind directions [2]. In general, the mechanism construction was based on the raw materials available in the area [14], as well as on those that could be imported. This especially concerned places where the local timber amount was not adequate, mainly for the construction of the windshaft, wooden ring located at the top of the windmill on which the roof rotated and the wallower, where wood imported from Mount Athos or Asia Minor were used [3], or wherever else it was available. For example, in Karpathos, "Tsikoudia" (*Pistacia terebinthus*) was imported from the island of Tilos in the Dodecanese area [15].

Figure 1. Main wooden parts of the windmill mechanism (a. "katoplakes", b. "panarika", c. windshaft, d. wheel, e. "zigos", f. "bratsolia", g. "antixono").

The wooden mechanism of the windmill, as well as the wooden mechanisms of water and oil mills, are complex examples of mechanisms of the pre-industrial era [2,5], and therefore monuments of technological cultural heritage [17] that gradually disappear or have already disappeared, both due to abandonment and looting for other uses [14]. Wood mechanism was replaced with mechanisms made of other materials in the case of water mills [18] and oil mills [2,19].

In the current study, a thorough examination of the Sikinos island windmills and their mechanisms is carried out, aiming to highlight the way of thinking and action of windmill manufacturers of that pre-industrial era (and therefore, of other wooden structures of that time), as well as the use of tools, manufacturers know-how and potential deep knowledge of different wood species properties and behavior.

2. Sikinos Island

The island of Sikinos is located at the south part of the Cyclades island complex (a group of varying-sized islands scattered over the deep blue waters of the Aegean Sea), between the Folegandros island (west) and the Ios island (east), covering an area of approximately 41 km^2. Despite the rich landscape, the island does not have wood-producing forests, as along with the rest of the Cycladic islands. These areas include only sparse and low-height juniper-*Juniperus phoenicea* shrubs that dominate, with evergreen hardwood species (mainly *Pistacia lentiscus*), pistacia also growing in areas where juniper is not spread. A characteristic of the above-mentioned areas is the presence of numerous sclerophyllus (fryganic) vegetation species. Moreover, the Sikinos island did not have a natural port, and a ferry boat approached the island for the first time in the year 1988. Consequently, due to the absence of a port, there was an isolation of the island and difficulties in transporting materials, which contributed to the preservation of a large part of traditional constructions, nowadays offered for a thorough study.

Therefore, the Sikinos island was chosen to be the study area of this research, since it was it was considered of even greater interest to record and study windmills constructed during the pre-industrial era on an island that had neither wood-producing forests, nor a port (at that time) and as a result, making the difficulties of such windmill construction even greater.

3. Windmill Mechanisms on the Sikinos Island

In the "Kastro" district, at the steep northern edge of Sikinos Chora (main town), three abandoned windmills (Figure 2) are preserved today, one of which, in the northeast, is partially demolished. According to Vaos and Nomikos [3], there were six windmills in Chora (Figure 3).

Figure 2. The "Kastro" district of Sikinos Chora and the two of three abandoned windmill buildings, pointed out with arrows in a photograph taken in the year of 2019.

Figure 3. Windmills in Sikinos Chora at the end of the 19th c. [20].

According to the testimonies of residents, the rest of the windmill buildings were destroyed at least 80 years ago, which is also confirmed by a photo taken before 1935, which includes only the three remaining windmills at that time (Figure 4). Overall, taking into account all the relevant findings (with testimonies and old photographs, etc.), it can be concluded that before 1900, the Sikinos Chora had ten windmills: six in the "Kastro" district and four in the "Chorio" district, arranged in five groups of two. All these windmills were privately owned by wealthy residents of the island, as continued to this day. The remaining windmills are being gradually destroyed because of their abandonment, even though they constitute cultural heritage monuments with a dual meaning; on the one hand, as architectural structures, and on the other hand, as complex examples of wooden mechanisms of the past. It is worth noting that none of them have been classified as historical monuments to be preserved, while no effort has been implemented by the state and the local government to protect and highlight their existence and value. The danger can be detected both in the buildings and their wooden mechanisms, which are susceptible to biological attacks, but also to illegal removal for other uses.

Figure 4. Windmill buildings in Sikinos Chora in a photo taken before 1935 [21].

The mechanisms are being preserved today only in the two most western-located windmills (Figure 5), as the wooden part has been completely removed and destroyed in the one located further east. Both windmills that retained the wooden mechanism nowadays are of the vertical single-floor type, with a rotating roof ("trula"), which allows for the main mechanism to be repositioned so that the mill can operate with all wind directions, although they no longer include the external conversion frame of wind power in motion through sails. One of them, in the north–northeast side of Chora, was in operation until 1977 and is considered to be in a relatively good condition, retaining part of the conical roof, while it remains locked by its owners.

Figure 5. The two windmills on Sikinos in which the wooden mechanisms are preserved.

The other windmill, located in the northwest of Chora and the subject of a thorough study, stopped its operation possibly in the 1960s, while nowadays, it does not have a door and is completely abandoned. Its mechanism includes the following main parts of the large dimensions (Figure 1):

- The manual roof rotation–orientation system (Figures 1 and 6), which includes a wide fixed wooden ring (probably nailed to "hidden" vertical wooden parts located inside the masonry), consisting of individual pieces of wood ("katoplakes", Figures 1a and 6a), on which rotation occurs in a special groove, a second wooden ring (Figure 1b), also consisting of thoroughly sanded pieces of wood ("panarika"), which have a special projection tongue (Figure 6b) in the center, entraining the entire upper part of the mechanism in the rotational movement;
- The long horizontal beam (windshaft/"axoni") (Figure 1c), which formerly externally carried the system of converting the force of the wind into motion by means of the sails, and internally, the vertical break wheel, the rotary motion of which set the upper millstone in motion through one more wooden component and a vertical iron shaft ("royal iron");
- The horizontal beam, called "zigos" (yoke), known also as the "sprattle beam" (Figure 1e), is placed lower than the windshaft, perpendicular to it and parallel to the wheel (from the side of the cogs, the teeth of the break wheel), through the center of which the "royal iron" passes, while the ends of the "zigos" terminate at two other beams ("bratsolia", sprattle beam supporting arms) for better weight distribution (Figure 1f), which have probably been cut off and reattached using metal parts;
- The "antixono" (anti-axle) is another horizontal beam on which the windshaft rests on its rear side to limit the pressure exerted on the movable base of the torsion system (Figure 1g).

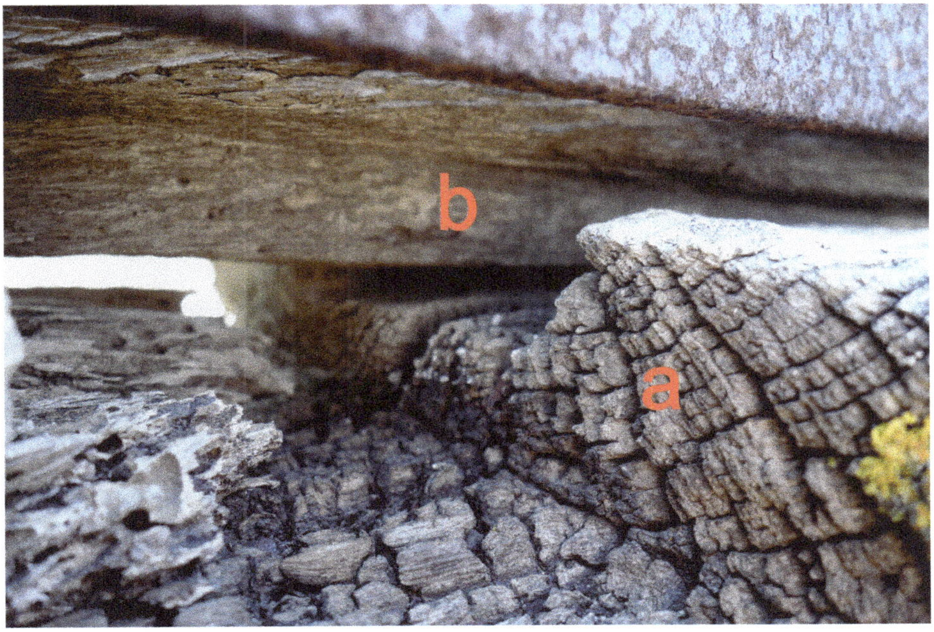

Figure 6. The manual roof rotation–orientation system (a. "katoplakes", b. "panarika").

The break wheel consists of "stavros" (the clasp arms of the break wheel) and the rim (Figure 7). The "stavros" consists of four continuous or interrupted double rectangular wooden parts, fastened together with bolts, while a square gap is formed in the center of the "stavros" for anchoring the wheel to the windshaft. Also, the rim is composed of four curved pieces of wood joined together with a lap joint, and includes 48 gaps that carry corresponding cogs (Figure 5), which are held by nails at the back of their protrusion. The rotary motion of the wheel sets the upper millstone in motion by means of a large-diameter wooden cylinder called the "anemi" (Figure 8b), located at the bottom of the wheel, which is intended to accommodate smaller cylinders, "pentaravda" (rod system, Figure 8a), that create the necessary vacancies for the cogs of the wheel to enter, thus causing the rotational movement of the wallower, but also of the "royal iron" passing vertically through both the interior of the "anemi" and the millstone system. The "anemi", in combination with "pentaravda", creates the wallower, something that resembles a cage-type gear. This type of wallower gear can be constructed or repaired using very simple tools, since the teeth are not integrated into the central body of the gear, and for its construction, special cuttings are not required. In this way, the vertical rotational movement of the break wheel is transformed into a horizontal rotational movement of the wallower and correspondingly of the upper millstone, with a simultaneous multiplication of rotations (speed), due to the difference in the number of cogs of the wheel and the number of recesses of the wallower (48:8 = 6, which represents the "speed ratio"), which means that one complete rotation of the wheel corresponds to six rotations of the wallower and the respective millstone.

Figure 7. The break wheel of the windmill (a. rim, b. "stavros", c. windshaft, d. cogs).

Figure 8. Wallower gear (a. "pentaravda", b. "anemi", c. teeth of the brake wheel).

Additionally, the mechanism includes several other parts of smaller dimensions, some of which are referred in the following:

- On the ground floor of the windmill building, the main part of the auxiliary system was used for adjusting the millstones;
- The flooring of the "first floor" of windmills is made of slates fixed to unprocessed, or almost unprocessed, logs of various cross-sections;
- Also, the windmill mechanism no longer carries the outer part of converting the energy of the wind into motion, and therefore, there is no current evidence of the timbers of the aerial system used to hang the sails, and the way they were connected to the shaft;
- The usual number of "antennas" (supporting sail spars) was 10 [2,3] or 12 [2,3,15], although there are a few cases with 8 or 14 "antennas";
- The wooden strips of curved cross-section sides ("violas") together form a hollow cylinder found at the rotation points of the windshaft in order to reduce its wear, and could be quite easily replaced with other new pieces of wood when the need arises (Figure 9).

Figure 9. "Violas" (a), found at the windshaft (b), and antixono (c).

4. Wood Species Used in Windmills (Sikinos)

In the windmill mechanism on Sikinos, various wood species were used for the construction of its individual parts (Table 1 and Figure 10). In particular, in most large-sized parts (windshaft, "panarika", "katoplakes", "antixono", "bratsolia"), deciduous oak, red (*Q. cerris* and *Q. Trijana*) or white (all the other *Quercus* sp.) (according to the wood identification keys of European timber species [22] (p. 582), was used, with the exception of the "zigos", which is made of fir (*Abies* sp.) wood. Additionally, since the wheel is made of many different individual parts, it was usually constructed employing a variety of wood species. Thus, two of the four parts of the wheel rim were made of plane (*Platanus* sp.) wood species and one part was made of elm (*Ulmus* sp.) wood. Unfortunately, from the fourth part, it was not feasible to obtain a sample to be analyzed. Also, chestnut (*Castanea sativa*) was used for the "stavros" construction and evergreen oak wood (kermes: *Quercus coccifera;* or holm oak: *Quercus ilex*) for the cogs. For the part of "anemi", ash (*Fraxinus* sp.) wood was used, while for the smaller parts, the use of both olive (*Olea europaea*) and evergreen oak (*Quercus* sp.) was frequently applied. The local juniper (*Juniperus* sp.) was not used in any key part of the mechanism, probably because of the various wood structure defects it presents [23]. Nevertheless, it was used as a structural element on the flooring of the "anogio" together with olive logs of a large diameter, but also in the three holes that start from the "anogio" and end approximately in the middle of the ground floor, carrying two small logs, horizontally placed, used for the tying of the cord, in order to stop the rotation of the "sail system".

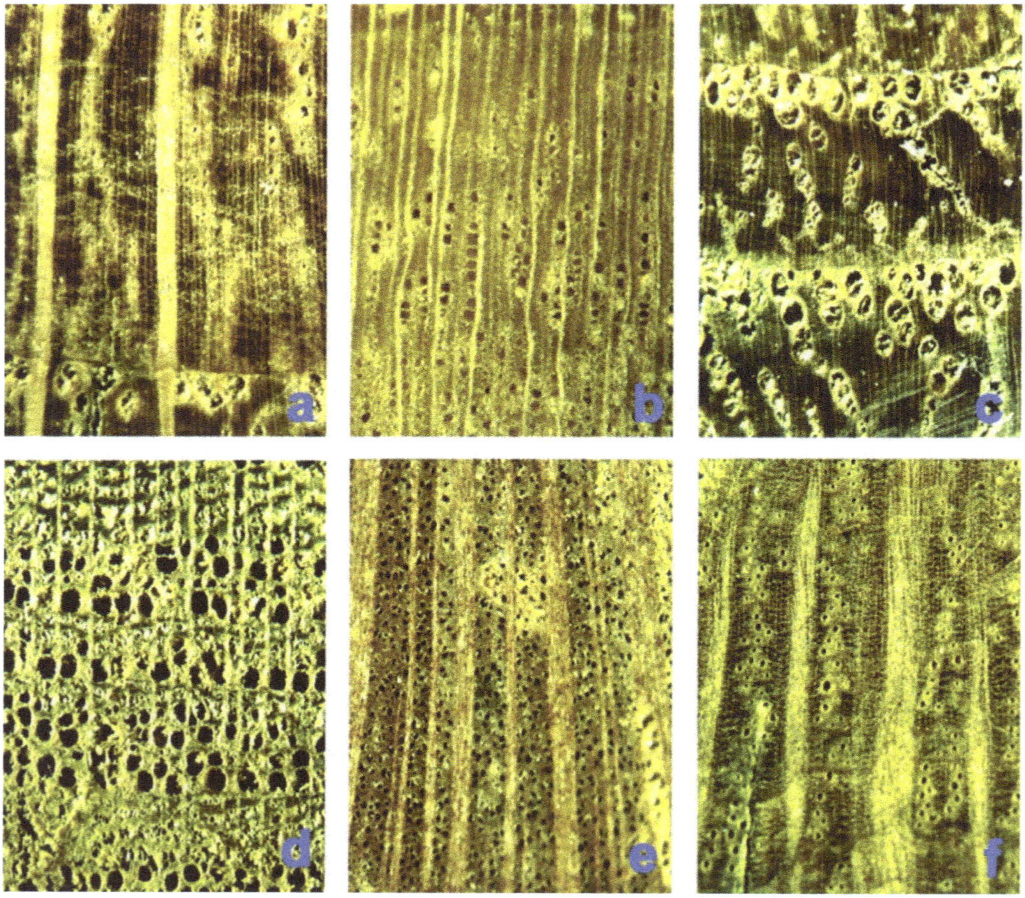

Figure 10. Wood species used in the windmill mechanism on Sikinos ((**a**). red oak (*Quercus* sp.), (**b**). olive (*Olea Eurepaea*), (**c**). chestnut (*Castanea sativa*), (**d**). elm (*Ulmus* sp.), (**e**). plane (*Platanus* sp.), (**f**). holm oak (*Quercus ilex*)) (stereomicroscope 16×, Nikon SMZ800).

Table 1. Sikinos windmill mechanism wood species scientifically recognized: comparison with those reported in the literature (empirically recognized [2,3,15]).

Mechanism Component	Wood Species in the Sikinos Mechanism	Reference Wood Species (Empirically Identified)		
		Cyclades [3]	Ios of Cyclades [15]	Kasos of Dodekanese [2]
Windmill Shaft	Deciduous oak	Oak, pine (Pinus sp), cypress (*Cupressus* sp.), fir	Pitch pine (*Pinus rigida*)	Cypress, fir
"Panarika" and "Katoplakes"	Deciduous oak (*Quercus* sp.)	Plane, wild olive, olive, fir, cedar, pine	Elm	Kermes oak, wild olive, mulberry (*Morus* sp.)
"Antixono"	Deciduous oak	Oak, pine, cypress, fir	Cypress	
"Bratsolia"			Elm	
Wheel cogs and "Pentaravda"	Evergreen oak (Kermes oak or holm oak)	Kermes oak, wild olive	-	Kermes oak, wild olive, mulberry
"Zigos"	Fir	Fir, oak, pine, cypress	Elm	-
"Stavros"	Chestnut	Plane, wild olive, olive, fir, cedar (*Cedrus* sp.), pine	-	Kermes oak, wild olive, mulberry
Wheel rim	Plane, elm	Plane, wild olive, olive, fir, cedar, pine	-	Kermes oak, wild olive, mulberry
"Anemi"	Ash	Plane, wild olive, olive, fig (*Ficus carica*), fir, cedar, pine	-	Kermes oak, wild olive, mulberry

Several wood species coincide with those reported in the literature relating Cycladic windmills [3], which are empirically recognized (Table 1) in the use of:

- Oak (*Quercus* sp.) wood for the windshaft, "antixono" and "bratsolia", as well as kermes oak (*Quercus coccifera*) for the wheel cogs and "pentaravda";
- Fir (*Abies* sp.) wood for the "zigos";
- Plane (*Platanus* sp.) wood for the rim of the break wheel,

While there are no similarities with the windmill studied on the island of Ios (Cyclades island), which provides evidence of for the overwhelming use of elm wood [15].

Furthermore, there are no common wood species with the windmills reported on the Kasos (Dodecanese), with the exception of the use of kermes oak wood for the cogs of the wheel and the "pentaravda", although the available literature is very limited [2], covering only a few individual parts of the mechanism.

It is concluded that for the construction of the studied windmill mechanism on Sikinos, hardwood species were used, with the exception of one beam that comes from a softwood species (fir). A clear predominance of oak can even be detected, which is found in almost all very long (windshaft) or relatively long parts ("panarika", "katoplakes", "antixono", "bratsolia"), though also in smaller parts (cogs, "pentaravda"). In general, oak is a hardwood species of high mechanical strength, with the deciduous oak showing moderate-to-very high density (0.45–0.95 g/cm^3) and evergreen oak being very heavy (0.80–1.10 g/cm^3) [22]. The latter (kermes oak or holm oak) was the most suitable species for the construction of the cogs and "pentaravda", since these parts were under a lot of mechanical stress because of the continuous rotational movement. Of course, the choice of wood species was also determined by the availability and possibility of their supply, since finding oaks was impossible on most of the Cyclades islands, with a potential exception of the low-height kermes oak species. Thus, the large shaft, which is made of oak wood, probably red, i.e., from the species of turkey oak (*Quercus cerris*) or Macedonian oak (*Quercus trojana*), may come either from Northern Greece or Asia Minor, where the main distribution areas of the above species are located. Corresponding information is available from the study of Vaos and Nomikos [3], who reported timber imports from Mount Athos or Asia Minor mainly for the windshaft, "katoplakes" and "anemi", respectively. A heavy weight (0.82–1.02 g/cm^3) and high hardness are also evident in olive wood [22], which was used for the construction of individual components of small dimensions and could have come from the Sikinos island due to the high availability. In fact, the choice of olive wood species for the "violas" construction, i.e., at the pivot points of the shaft, facilitated its rotational movement, preventing damage, attributed to the natural oiliness of the specific wood. Furthermore, the ash wood used for the construction of "anemi" is a species of moderate-to-heavy weight (0.52–0.87 g/cm^3) [22] and hard, though of a high modulus of elasticity, with great mechanical strength and particular resistance to the impact of bending forces [24], properties that are both of great significance because of its continuous rotation, as well as in cases where, because of an inappropriate adjustment of the wheel position, its cogs may collide with the body of the "anemi".

On the contrary, for the construction of the wheel, with the exception of the cogs, the wood of plane, elm and chestnut were used [22], with the plane and elm being mainly of medium weight (0.60–0.72 g/cm^3 and 0.46–0.70 g/cm^3, respectively) and hardness, while the chestnut shows a low-to-medium weight (0.37–0.70 g/cm^3) and low-to-medium hardness [22].

The use of these wood species is probably related to the desire to decrease the weight of the structure for an easier rotation due to the large size of the wheel, which produced a higher pressure on the cogs. That is the reason why they were made of a hard and very heavy wood species (evergreen oak). Also, the use of chestnut in the "stavros" construction is justified by the high dimensional stability of this species. In this way, the risk of developing defects in the connection of the individual parts of the wheel is eliminated.

Additionally, for the construction of the wheel rim, some species with an already curved trunk or characterized by the ease of the curvature had to be used, since the rim of the wheel was composed of four curved pieces. Plane and elm species are, among others, some of the species that bend comparatively easier [24]. The use of elm in constructions with curved parts, such as the wheels of carriages, has been established since ancient times [25]. The curved wooden part, in some cases, seems to be achieved by applying a combination of curvature and a suitable cutting.

On the contrary, in the rotation and orientation system of the "fteroti" ("panarika", "katoplakes"), many individual pieces were used that created an almost polygonal structure in places, thus limiting the necessity of using already curved elements. The choice of hardwood species for specific components, attributed to their great ease of bending, is also reinforced by the fact that softwood species that are difficult to be bent were not used, despite their high strength in combination with their relatively low weight (like fir wood presenting a density of 0.44 g/cm^3) [22] (p. 111), although the reduction in weight of the overall mechanism was obviously desired, aiming to make it easier to operate. The same conclusion can be drawn from the use of elm species in the aforementioned individual parts of the windmill mechanism studied on the island of Ios [15]. Of course, the availability of various wood species seems to have been a decisive factor that forced the use of the same species in several individual sections, where possible, based on the properties that each application required, depending on the use–location of the windmill mechanism. This also proves the thorough empirical knowledge of each wood species property. Additionally, it should be noted that some of these components might have been replaced over the windmills' long service lifespan using any wood species available at that period.

Regarding the biological resistance of wood species that have been applied, most demonstrate a long natural durability (oak, olive, chestnut) [22], although the mechanism was intended to work in a sheltered area. Therefore, wood biological durability may not be the primary selection criterion, with the exception of the unsheltered "fteroti" mechanism that protruded and was exposed to both biotic and abiotic actions. However, due to abandonment and subsequent destruction/removal of the roof, the wooden mechanism has been directly exposed to abiotic factors and biological attacks for many years, as a result of which it appears to bear disintegration signs, corrosion and cracks, with the greatest extent of attack being found in the wood species of ash used in the part of the "anemi", which is susceptible to the catastrophic action of wood-destroying microorganisms, such as fungi (durability class 4 to 5 according to EN 350:2016) [22], followed by the species of fir, used in the "zigos". Extensive bio-deterioration also appears in one of the four parts that make up the wheel rim. However, due to the extensive biological attack in relation to the rest of the region, it is assumed that this wood species is different from the other two species used (plane, elm), possibly suggesting a later addition or at least another part of the trunk, as the natural strength varies not only among different species, but also among different parts of the same species or the same log/trunk (heartwood and sapwood) [24,26].

As it appears from the limited oral testimonies of elderly residents, no means were used in the past to protect these wooden parts that have survived biological attacks, although in the literature, there is a reference to the initial protection of the shaft via a burning–smoking process, which was later replaced by the use of pine tar [3]. Of course, parts of the studied windmill mechanism (mainly "violas", cogs, "pentaravda" and "panarika") had been regularly smeared with grease or soapy water, which, however, did not aim to protect against biological attacks, but facilitate the rotational movement and the subsequent damage reduction.

5. Conclusions

As a unique example of the popular pre-industrial-era architecture and technological advancements, windmills constitute an important component of cultural heritage. The Sikinos island windmills, as well as the Mediterranean windmills, typically included a vertical sail frame and a rotating roof ("trula"), with a manual torsion mechanism that permitted for operation in all wind directions. Their wooden mechanisms were of exceptional technical ingenuity. Sikinos is a small Aegean island with intense landscapes and scant *Juniperus phoenicea* vegetation, combined with evergreen hardwood species and an abundance of sclerophyllus vegetation species, without wood-producing forests. For this reason, it was of even greater interest to record and study the windmills of this particular island. Currently, only two of the island's three abandoned windmills still have their wooden mechanisms.

It was discovered that several wood species were utilized to manufacture each element of the two windmill mechanisms that were studied. With the exception of the "zigos" part, which was constructed with fir wood, most substantial pieces were constructed using oak wood. The indigenous juniper, together with large-diameter olive trunks, was employed only for the structural elements of the "anogio" flooring, without being used in any essential part of the mechanism. The present study highlights the level of know-how in the construction of windmills, and the thorough knowledge of the characteristics and properties of the different available wood species that has been established since ancient times. Evidence of the utility, durability and simplicity of these windmill mechanisms, the techniques employed in windmill construction and the proper application of different wood species to serve ideally at the appropriate point in the construction, in accordance with their properties and characteristics, were recorded in order not to lose this existing knowledge. Undoubtedly, the goal of future research should be to accurately date the specific windmills on Sikinos, as well as examine other Greek islands' corresponding wooden monuments of cultural heritage. The current study, as well as future studies that are going to be implemented, aims to raise awareness in the local community and motivate the stakeholders toward the achievement of governmental support for the protection, preservation and projection of these significant cultural monuments.

Author Contributions: Conceptualization, I.B.; methodology, E.A. and I.B.; software, E.A. and I.B.; validation, E.A.; formal analysis, E.A. and I.B.; investigation, E.A.; resources, E.A.; data curation, E.A., I.B. and V.K.; writing—original draft preparation, E.A. and I.B.; writing—review and editing, V.K.; visualization, I.B.; supervision, I.B.; project administration, E.A. and I.B. All authors have read and agreed to the published version of the manuscript.

Funding: This research received no external funding.

Institutional Review Board Statement: Not applicable.

Informed Consent Statement: Not applicable.

Data Availability Statement: No new data were created.

Conflicts of Interest: The authors declare no conflict of interest.

Abbreviations

Terminology used for Cycladic windmills

"anemi"	The root cylinder of the wallower gear. "Anemi", with the "pentaravda", constitutes constitutes the wallover, a horizontal cage-type gear, driven by the brake wheel.
"anogio"	First floor.
"antenes"	Spar support sails.
"antixono"	Anti-axle, a horizontal beam on which the windshaft rests on its rear side (also known as the tail beam).
"axoni"	Windshaft.
"bratsolia"	Supporting arms of the "zigos", also referred to as cross beams in previous studies.
"fteroti"	Windmill sail structural system outside the windmill building (windmill wings).
"katogio"	Basement.
"katoplakes"	A wide, fixed wooden ring with a special groove, placed on top of the windmill.
"panarika"	Wooden ring with a special tongue in the center, moving over "κatoplakes" and entraining the entire upper part of the mechanism in rotational movement.
"patari"	Mezzanine floor.
"pentaravda"	Cylindrical rods, the teeth of the wallower.
"roda"	The brake wheel.
"stavros"	Cross, clasp arms of the break wheel.
"trula"	The rotating roof.
"violes"	The curved cross-sectional strips at the pivot points of the windshaft.
"zigos"	The sprattle beam, a horizontal beam which supports the windshaft (known also as the yoke).

References

1. Hatzitoulousis, S. Wood as Archaeological Material in Prehistory: The Case Study of the Neolithic Lake Side Settlement at Dispilio Kastoria. Ph.D. Thesis, Department of History and Archaeology, Aristotle University of Thessaloniki, Thessaloniki, Greece, 2006; p. 493. [CrossRef]
2. Skoulias, M. *The Windmills and the Plowing of Kasos*; Ministry of the Aegean and Municipality of Kasos: Athens, Greece, 2001; 178p.
3. Vaos, Z.; Nomikos, S. *The Windmill in the Cyclades*; Dodoni: Athens, Greece, 1993; 406p.
4. Radford, A.; Clark, C. *Cyclades-Study of an Anonymous Architecture*; Doumani, O.B., Oliver, P., Eds.; Architectural Issues; Shelter in Greece: Athens, Greece, 1974; pp. 64–82.
5. Sarantakou, E. Architecture of the Rural Residence in the Aegean. Cultural Gateway of the Aegean Archipelago—Greater Hellenism Foundation. 2005. Available online: http://www2.egeonet.gr/Forms/fLemmaBodyExtended.aspx?lemmaID=6979 (accessed on 13 January 2018).
6. Tapanli-Dania, M.; Malakos, G.; Diagouma, A. Restoration of a windmill in Perama, Lesvos. In Proceedings of the 2nd National Conference "Mild Interventions for the Protection of Historic Structures", Thessaloniki, Greece, 14–16 October 2004; Thessaloniki, TEE and Ephorate of Newer Monuments of Central Macedonia. Volume 2, pp. 91–101.
7. Kotsakis, A. The Formation of the Cultural Landscape of the Island of Naxos from 1204 to the Foundation of the Greek State. Ph.D. Thesis, University of Peloponnese, Kalamata, Greece, 2013; p. 566.
8. Medioli, A.; Özkan, S.; Plunz, R. The Grain Cycle and a Windmill at a Village on the Aegean. Middle East Technical University. *J. Fac. Archit.* **1977**, *3*, 39–62.
9. Barboutis, I.; Koutis, P.; Kamperidou, V.; Chryssoulaki, S. Knowledge and quality of woodworking in ancient Greece; revealed from a simple finding. In Proceedings of the 12th International Conference "Wood Science and Engineering in the Third Millennium", Brașov, Romania, 7–9 November 2019; Transilvania University of Brașov: Brașov, Romania, 2019; pp. 233–240.
10. Saeidian, A. Windmills in Iran. *Elixir Hist. Preserv.* **2012**, *42*, 6507–6515.
11. Pérez-Martín, E.; Herrero-Tejedor, T.R.; Gómez-Elvira, M.Á.; Rojas-Sola, J.I.; Conejo-Martin, M.Á. Graphic study and geovisualization of the old windmills of La Mancha (Spain). *Appl. Geogr.* **2011**, *31*, 941–949. [CrossRef]
12. McLaren, T. Old Windmills in Scotland, with Special Reference to the Windmill Tower at Dunbarney, Perthshire. *Proc. Soc. Antiqu. Scotl.* **1945**, *79*, 6–14. [CrossRef]

13. Siaxampani-Stefanou, X. Byzantine and Post-Byzantine Watermills in the Vicinity of Thessaloniki. Ph.D. Thesis, Polytechnic School of Aristotle University of Thessaloniki, Thessaloniki, Greece, 2014; p. 360.
14. Koumanoudis, I. *Windmills: Kythira-Mani-Argolis-Arcadia-Messinia-Akarnania & Three Opposite Islands (Kalamos-Kastos-Meganisi)*; Technical Chronicles of the Technical Chamber of Greece: Athens, Greece, 2004; Issue 6; 29p.
15. Leimona-Trembela, E. *Aegean Windmills*; (In Greek: "Αιγαιοπελαγίτικοι ανεμόμυλοι"); Technical Chronicles of the Technical Chamber of Greece: Athens, Greece, 1974; pp. 319–338.
16. Hunt, D. Properties of wood in the conservation of historical wooden artifacts. *J. Cult. Herit.* **2012**, *13S*, S10–S15. [CrossRef]
17. Axiotis, E. Watermills of Lesvos. Ph.D. Thesis, Department of Cultural Technology and Communication, University of the Aegean, Mytilini, Greece, 2008; 237p.
18. Valavanidou, A. The Watermills of Macedonia. Ph.D. Thesis, Polytechnic School of Aristotle University of Thessaloniki, Thessaloniki, Greece, 2008. Volume A. 889p.
19. Dimitropoulos, D. Olive mills, mills, ovens, churches in the island area in the 17th century—An approach to the issue of co-ownership based on the example of Mykonos. *Memory* **1994**, *16*, 37–70.
20. Byzantine and Christian Museum. *Journeys of Georgios Lambakis*—Γεωργίου Λαμπάκη περιηγήσεις—*Calendar*; Lapourtas, A., Ed.; Ministry of Culture—Byzantine and Christian Museum: Athens, Greece, 2000; 60p.
21. Perilla, F. (Ed.) *Les îles de la Grèce*; Athenes aprox.1935, 130 pages, 4o, Greece; 87p. (In French)
22. Tsoumis, G. *Structure, Properties and Utilization of Wood*, 1st ed.; Aristotle University of Thessaloniki Publications Service: Thessaloniki, Greece, 1983; 351p.
23. Kamperidou, V.; Aidinidis, E.; Barboutis, I. Impact of Structural Defects on the Surface Quality of Hardwood Species Sliced Veneers. *Appl. Sci.* **2020**, *10*, 6265. [CrossRef]
24. Voulgaridis, H. *Quality and Uses of Wood*; Association of Greek Academic Libraries: Athens, Greece, 2015; Available online: http://hdl.handle.net/11419/5260 (accessed on 29 April 2018).
25. Kakaras, I. Applications of Wood in Antiquity. In *Technology of Wooden Structural Constructions*; Kakaras, I., Ed.; Ion: Athens, Greece, 2013; pp. 101–172.
26. Kamperidou, V. The Biological Durability of Thermally-and Chemically-Modified Black Pine and Poplar Wood Against Basidiomycetes and Mold Action. *Forests* **2019**, *10*, 1111. [CrossRef]

applied sciences

MDPI

Article

System Connector "Double Dovetail Tenon"—A Wood–Wood Based Alternative for Connecting CLT-Elements

Andreas Pomaroli and Anton Kraler *

Unit for Timber Engineering, University of Innsbruck, Technikerstraße 13, A-6020 Innsbruck, Austria; andreas.pomaroli@uibk.ac.at
* Correspondence: anton.kraler@uibk.ac.at

Featured Application: Ecological alternative for wall–ceiling connections with cross-laminated timber elements. No additional metal fasteners are required for connecting and suspending the ceiling elements.

Abstract: In timber construction, there are still no standardized wall–ceiling system connectors made of wood for cross laminated timber construction (CLT). Based on experimental investigations of the bar-shaped dovetail connections, a wood–wood system connection was developed at the University of Innsbruck. First, the traditional dovetail connection for bar-shaped beam connections was investigated. The findings showed the high potential of force transmission in the flank area, especially for beam connections of the same height, where the height of the pre-timber is too small. To date, the load-bearing capacity of the flanks has not been taken into account in the calculation methods. To increase the force transmission paths in the base of the tenon, it was shaped in steps. This led to a load redistribution and thus delayed the failure of the connection. The knowledge gained formed the basis for an independent connecting element that resulted in the so-called "double dovetail". The connector is a three-dimensional, statically effective double dovetail element. The wood system connector is made of block-glued laminated veneer lumber (LVL) and is manufactured in a CNC-milling process. The results of the tests showed the high performance potential of the LVL system connector, which can play an important role in future timber construction.

Keywords: wood–wood connection; dovetail connections; shear force bearing capacity; flank participation; ductile behavior; wall–ceiling connections; cross laminated timber

Citation: Pomaroli, A.; Kraler, A. System Connector "Double Dovetail Tenon"—A Wood–Wood Based Alternative for Connecting CLT-Elements. *Appl. Sci.* **2023**, *13*, 11727. https://doi.org/10.3390/app132111727

Academic Editor: Laurent Daudeville

Received: 14 September 2023
Revised: 23 October 2023
Accepted: 24 October 2023
Published: 26 October 2023

1. Introduction

Wood–wood connections have the potential to make an important contribution to the timber construction industry and at the same time meet all code requirements in terms of load-bearing capacity and dimensional stability. Through the specific application of wood materials, they can also be used as an independent wood–wood fastener. Therefore, they offer an alternative to metal system connectors and contribute to climate protection by reducing CO_2 emissions. The use of such connectors requires more detailed scientific studies and the revision or review of valid calculation approaches. The inclusion of modern wood materials as well as multidimensional geometries, which are precisely manufactured by CNC-milling, must also be considered. The system connector made of wood is an alternative to the common metal system connectors. The wood–wood fastener can be produced in-house or industrially. This adds value to the wood construction industry. The aim is to apply the potential of the findings presented on bar-shaped connections by means of the coupling element "double dovetail pin" to surface wall–ceiling connections made of CLT.

The structural design of a dovetail connection is based on calculation formulas that distinguish between the failure of a transverse connection of the main girder (mode 1)

and a notching failure of the secondary girder (mode 2). The current version of the building authority approval (ABZ) Z-9.1-649 [1] already takes into account the use of glued laminated veneer lumber (LVL) as a material for bar-shaped connections. However, the wood connection is milled into the beams and thus its position depends on the connection placement. In new Eurocode 5 formulas [2], the rounded dovetail connection within bar-shaped connections will be introduced. The calculation formula is given for mode 2—notching failure of the secondary girder—but not for mode 1—transverse tensile failure of the main girder—which can occur even earlier with same-level beam connections. In his work, Tannert [3] dealt in depth with the dovetail connection and its beneficial form and also recognized the potential of alternative load transfer paths and multi-step load application. In a topic related publication, Tannert et.al. also described experimental and numerical investigations on the rounded dovetail joint (RDJ) in bar-shaped connections [4]. Furthermore, reinforcement measures with self-tapping screws by Tannert and Lam [5] and an adhesive layer on beam connections by Tannert [6] were investigated. Schickhofer et.al. [7] also investigated experimental geometries of dovetail connections in tests and FE-simulations of bar combinations. Blaß and Enders-Comberg [8] experimented with measures to increase friction within the tenon contact by special shaping of the flanks of RDJ and they also showed the possibilities of multi-step load transfer paths within a stepped connection for truss nodes of timber constructions. Also, the experimental investigations described in Braun et.al. [9] show the potential of CNC-manufactured wood–wood joints with experimental geometries and multi-step load transfer paths. All geometry parameters of the main beam and the secondary beam of a rounded dovetail joint (RDJ) are illustrated in Figure 1.

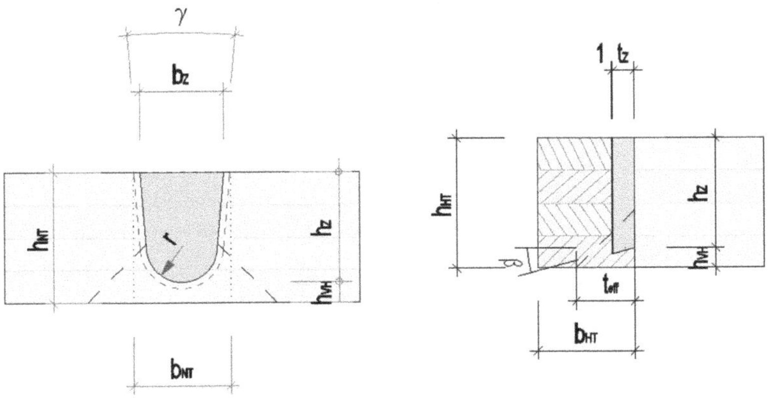

(a) (b)

Figure 1. Geometry parameters of the main beam (HT) and secondary beam (NT) according to [1] (**a**) View with assumed load dispersion (**b**) Section showing connection depth t_{eff}.

The description of the connection failure is based on equivalent formulas of ÖNORM EN 1995-1-1 [10], introduced by Ehlbeck and Görlacher for the transverse connection [11] and Gustafsson for the notching [12]. An important parameter of the cross connection (mode 1) is the effective connection depth t_{eff}. Particularly in the case of small pre-timber heights h_{VH} below the base of the mortise—as occurs with beam combinations of the same height h_{HT} and h_{NT}—this mode of failure and thus the parameter t_{eff} is decisive. The greater the connection depth t_{eff}, the higher the load-bearing capacity and, at the same time, the higher the ductility of the connection. In order to improve this failure mode without reinforcing mechanical fasteners (e.g., with screws), adjustments are required with regard to permissible and calculable shapes and forms. That means, connections with small pre-timber heights h_{VH} result in very low effective connection depths t_{eff}, which is caused by a low resisting transverse tension volume. In this case, the connection failure happens in a rather brittle way and leads to large covariances (COV) and thus lowers the characteristic calculation value. The investigations presented below show that premature failure, due to flaws or non-existent redistribution possibilities (e.g., small pre-timber height), can be prevented within the connection by a multi-step shape of the tenon base. This redistribution possibility is created by the application of force to several higher levels, which leads to a delayed, more ductile load-bearing connector behavior. Increasing the flank load-bearing capacity by widening the notch taper angle γ could also ensure optimal load-bearing behavior for beam combinations of the same height. In this case, the redistribution forces are directed toward the flank surfaces and thus the effective connection depth t_{eff} is also increased. This increase also has a favorable effect on the potential residual load-bearing capacity and robustness of the construction.

With the two variants mentioned, the stepped shape of the tenon base and widening of the notch taper angle, means improved force redirection possibilities are created and thus a larger resisting transverse tension volume can be achieved. The decisive factor here is ultimately a bigger effective connection depth t_{eff}. The weak points mentioned are not or only inadequately taken into account in current calculation formulas. This publication intends to provide additional input for improved techniques.

Based on the knowledge gained from experiments with bar-shaped dovetail tenon joints, and taking into account assembly and material-related properties, a wood–wood system joint in the form of a "double dovetail" for cross laminated timber (CLT) elements was developed. Due to the cross-layer arrangement of the board layers, CLT offers new possibilities for wood–wood connections. This means that the surface components enable wider notch taper shapes, and the crosswise arrangement allows a locking effect of the board layers, which helps to ensure the force transmission paths via the flanks. Therefore, not only the stability but also the long-term serviceability of the structure is given. The separation of the wood–wood connection from the load-bearing element (CLT) as an independent coupling element not only provides an economical solution, but also a variety of possible shapes based on applied mechanics.

The decisive factor for the need to design a "double dovetail" system connector was that there had been no standardized and economical wood–wood connection for CLT that could absorb and transmit the loads that occur. The current main application of the wood–wood connector made of LVL-C according to LVL Merkblatt [13] is for the wall–ceiling connection in CLT element construction as shown in Figure 2.

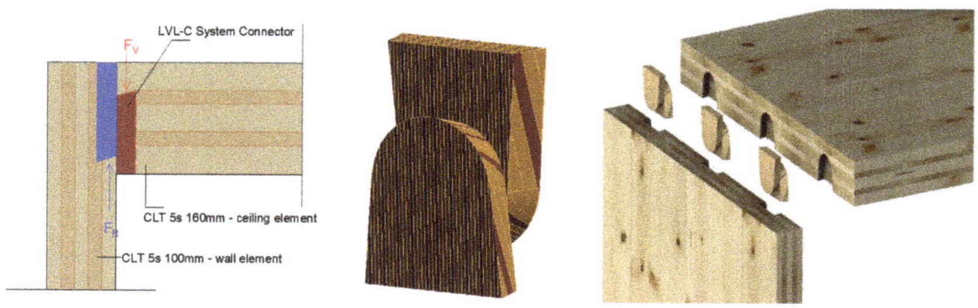

Figure 2. Wood–wood system connection for CLT wall–ceiling joints.

2. Materials and Methods

This section describes the materials and methods used in the investigations carried out. In order to keep the scope of the investigation within feasible limits, the research results on moisture behavior are not completed yet and therefore cannot be presented in this paper. All wood materials used in the tests were conditioned in a reference climate (relative humidity 65% at 20 °C) for more than 4 weeks before the laboratory tests were carried out. Wood moisture contents ω and the respective wood densities ρ of all specimens were examined after the tests according to EN 13183-1 [14]. They are listed together with the ultimate loads F_v and initial stiffnesses K_{ser} for all test series.

2.1. Bar-Shaped Investigations on Load Bearing Optimization of RDJ

The aim of the tests was to demonstrate the performance of the dovetail joint as a new option for wood–wood joints. In particular, the properties of the individual geometry components can be demonstrated on equal-height beam connections, which are realized by means of glulam beam combinations. The test setup shown in Figure 3 was used for all laboratory tests carried out according to EN 26891 [15] on the bar-shaped beam combinations of wood grade GL24h classified in EN 14080 [16].

(a)

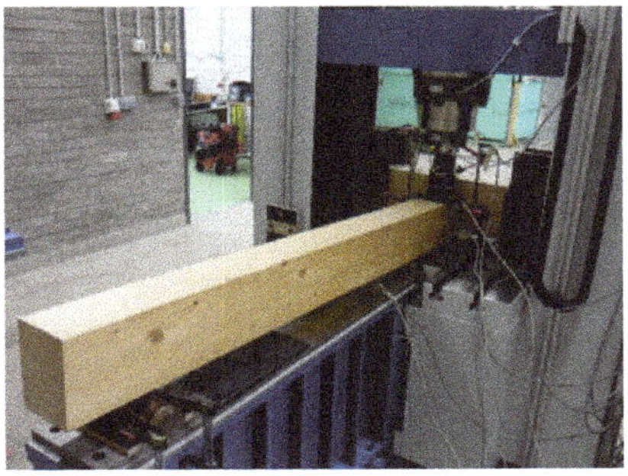

(b)

Figure 3. Test setup for bar-shaped connections (**a**) Front view (**b**) Side View.

In order to eliminate the effects of notching and to provoke purely main girder-side failure, shear tests were carried out. For the tests, a secondary girder with L = 1600 mm was connected in the middle of the main girder L = 800 mm by means of a traditional milled dovetail joint. The joints were milled with the CNC-milling machine Hundegger Robot Drive© 5-axis. The load induction takes place close to the connection via a steel plate 100 × 100 mm. The support points of the beam are also underlaid with steel plates of the same size to avoid local press-in. To absorb torsion due to eccentric load application, a fork support with steel angle structures is attached. The vertical deformations at the connection as well as the torsion rigid support of the main girder are recorded by means of displacement transducers. For each series, five tests were carried out at the Technical Testing and Research Institute (TVFA) Innsbruck, Austria. All test specimens were stored in a reference climate (relative humidity 65% at 20 °C) until equilibrium moisture content was reached.

FEM—Model Data Values

All test results of the bar-shaped connections were compared with FEM simulations created with Ansys© Mechanical 2022 R1. The rounded dovetail joint (RDJ) was modelled with a frictional contact with a coefficient of friction of $\mu = 0.2$. Quadratic *Solid 187* elements for the connection area and quadratic *Solid 186* elements for the beam sections outside the connection area were used for the mesh.

Stiffness values for solid spruce wood C24 according to EN 338 [17] and yield stresses according to Sandhaas and van de Kuilen [18] are applied for the bilinear material model with Hill yield criterion and isotropic hardening in the course of the simulations. The following values, listed in Table 1, were used within all simulations of the bar-shaped connections.

2.1.1. Load-Bearing Capacity of the Tenon Base and Flanks

For this purpose, glulam beam combinations with cross sections of the main and secondary beam of 120 × 160 mm (series REF 01, ZGR 01 and FLA 01) and with cross sections of 160 × 160 mm (series REF 02, ZGR 02 and FLA 02)—all with strength grade GL24h—were investigated for their load-bearing capacity and post-cracking behavior.

Figure 4 features series ZGR 02 and FLA 02 which represent the load bearing components of series REF 02, which is shown in Figure 5.

Table 1. Material values in FEM—simulations of bar-shaped connections.

Material Property	Value	Description
[-]	[MPa]	[-]
E_0 (axial)	11,000	Modulus of elasticity (axial)
E_{90} (perp.)	370	Modulus of elasticity (perp.)
G	690	Shear Modulus
G_{roll}	65	Rolling Shear Modulus
$f_{y,0}$ (axial)	36	Yield stress (axial)
$f_{y,90}$ (perp.)	4.3	Yield stress (perp.)
f_v	6.9	Yield stress (shear)
$f_{v,roll}$	1.5	Yield stress (rolling shear)
T	1	Tangent Modulus

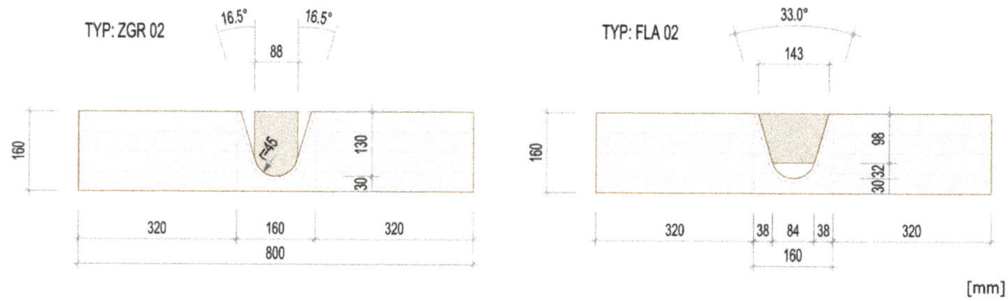

Figure 4. Tenon base shape ZGR 02 without flanks; flank shape FLA 02 without tenon base.

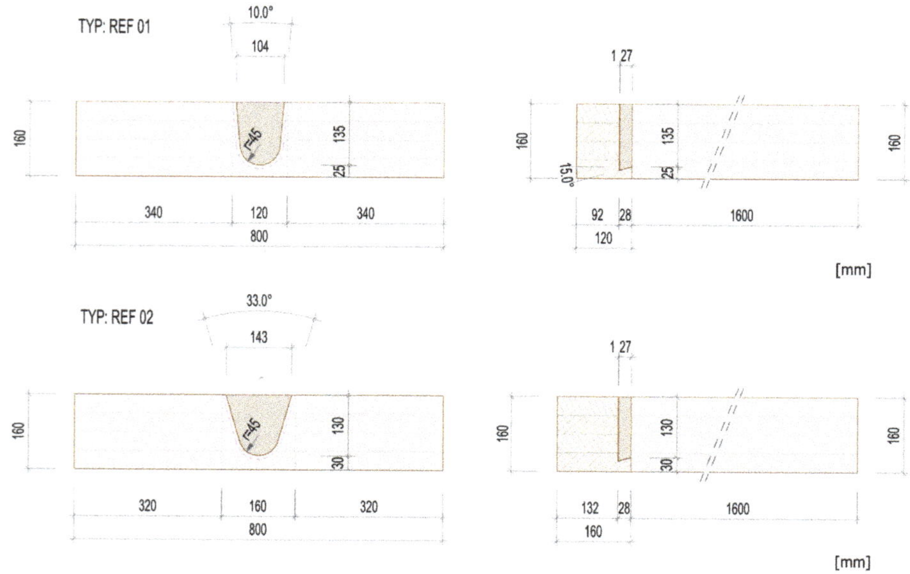

Figure 5. REF 01 and REF 02 reference series—investigation of load-bearing components.

Series REF 01 (see Figure 5) is computed based on calculation formulas in Z-9.1-649 [1] for transverse connection failure (mode 1). This includes an open tenon geometry with a 10° notch taper angle γ, which is still permissible in the design approach based purely on the load-bearing capacity of the tenon base. A statically optimized dovetail connection based on [1] is obtained for equal-height, 160 mm high girder combinations by equal ultimate load values of the main and secondary girder-side failure modes with a pre-timber height of 25 mm and a tenon height in the shear joint of h_Z = 135 mm. The tenon width and the opening angle of the notch taper are determined by using the full beam width, as recommended in [1]. Series REF 02 cannot be calculated on the basis of existing calculation methods because with a notch taper angle of 33°, the permissible opening dimension of the tenon geometry is exceeded. On the safe side, however, the calculated value of series REF 01 can be taken into account, as there is an almost identical tenon base area with the same fillet radius. The two reference geometries, series REF 01 and series REF 02, each represent a combination of two bearing components: that of the tenon base and that of the flanks. They are then divided into their bearing components by examining only the tenon base and only the flanks, respectively. Since the bearing loads of the reference shapes are divided according to the stiffnesses on the respective transmission paths, a corresponding division ratio of the force transmission on the reference geometries can be defined.

The laboratory tests include the determination of load capacities, joint stiffness and post-failure behavior. The results of the reference forms were compared with the calculation according to Z-9.1-649 [1]. In principle, the basic tenon form series ZGR 01 is to be used for the calculation design purpose, because only this reflects the (time-independent) load-bearing capacity of the dovetail geometry. If the pre-timber height of the main girder is low, e.g., the girder combinations are the same height, the calculated value must be questioned. In these cases, depending on the respective ratio of the stiffnesses, redistribution processes towards the flanks take place. This is the only way to achieve the calculated value. When the bottom of the tenon begins to fail, the residual load-bearing capacity is finally ensured by the flanks, which are then in a full friction state.

2.1.2. Stepped Tenon Base

An equal-height glulam beam combination GL24h with a cross-section of the main and secondary beam of 120 × 160 mm—all with strength grade GL24h—and the geometrically smallest possible pre-timber height h_{VH} series REF 03 illustrated in Figure 6 was also tested. It can be assumed that the bottom of the spigot already has such low stiffness that the loads are transferred, to a large extent, via the upper area of the fillet. The flanks also contribute to the load transfer and provide a redistribution option in the event of failure. Due to the lack of alternative transfer paths, a classic tenon with orthogonal dimensions would be critical here. However, the dovetail tenon shows a considerable load capacity in series REF 03. Its geometry will be compared in the following, with the developed stepped tenon base series STZ (see Figure 6), in order to explore the transition to the reference geometry series REF 01. The multi-step tenon base is a further alternative for improving the properties of the dovetail tenon joint. All geometries of this investigation comply with the specifications regarding the notch taper angle according to Z-9.1-649 [1].

2.2. *Coupling Element for CLT Wall–Ceiling Connections—Double Dovetail*

The aim of our considerations was to detach the connection from the supporting element and enable the creation of a separate connector made of laminated veneer lumber (LVL), which can be integrated into the surface components. The coupling element "Double dovetail" is produced from panel elements of the wood-based material Kerto©Q (Metsä, Finland) approved in Z-9.1-847 [19] by means of CNC manufacturing. The connector consists of two dovetail tenon shapes, which are arranged in opposite directions at their overlapping surface. It can be machined in Edgewise (HK) and Flatwise (FK) arrangements of the veneers. The fiber orientations of the longitudinal and transverse layers are arranged Edgewise to ensure eccentric force application in combination with high stiffness in the insertion direction. Flatwise, the coupling element is additionally reinforced with 12 × ϕ10 beech dowels to prevent shear failure.

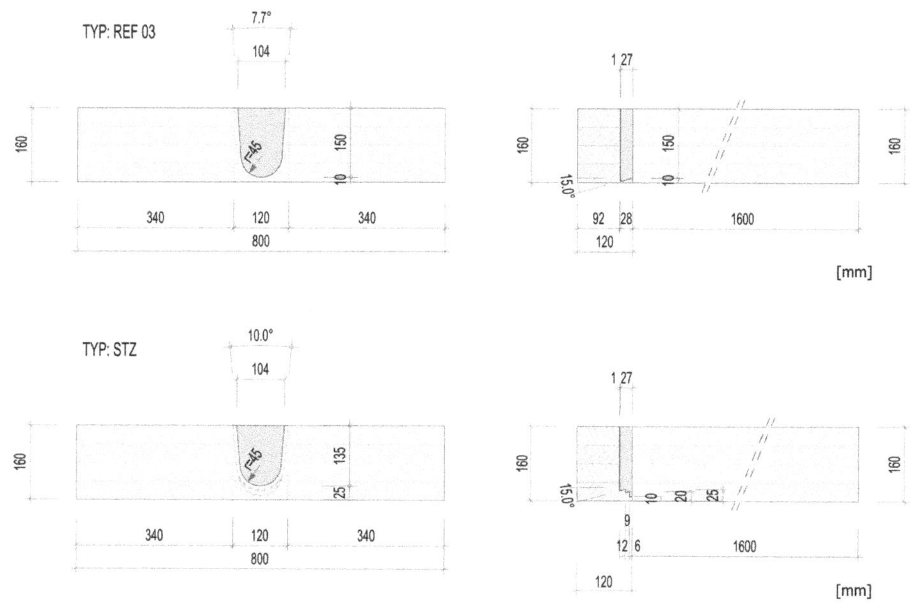

Figure 6. Test series REF 03 and STZ of the study on the stepped tenon base.

Cross laminated timber (CLT) offers optimum conditions for geometries with wide notch taper angles due to the dimensional stability provided by the locked board lamellas. On the wall side, the load is usually transferred from the connection via the top layer end grain of the wall element. The wall element can theoretically only fail in transverse tension as a result of large support rotations. This could be proven by pure rotation tests on cantilevers. However, for single span girder systems up to l = 5.0 m, lower support rotations occur than the permissible limit torsion, so that this failure mode does not happen in practice. The slab-side connection combination is relevant for the design. This consists of ceiling-side tenons as well as the CLT ceiling element, which is comparable to an equal-height connection of a beam connection.

For geometric optimization of the ceiling–side tenon connection, load-bearing tests are carried out on CLT combinations with wall elements CLT 5 s 100 mm (5 × 20 mm) and ceiling elements CLT 5 s 160 mm (40/20/40/20/40 mm). The board lamellas of all CLT elements are of C24 quality classified in EN 338 [17]. From the test specimens, the effective bending stiffness and shear stiffness are first determined by global Young's modulus measurement and shear field measurement according to EN 408 [20]. As a test setup, 3-point bending tests (test setup shown in Figure 7) according to EN 26891 [15] are performed on both side tested 1.5 m long CLT elements with an effective span during testing of 1.0 m. The load is applied in the center of the field via a steel plate. The CLT elements for the tests are 400 mm wide. The selected width represents the assumption of a connector spacing value and at the same time complies with the edge and center distances according to the specifications of valid calculation models for bar-shaped connections [1]. The connection in the tests is made by a floor element suspended on one side in a wall element. On the opposite side, the CLT element is supported on a sliding roller bearing. The wall element is mounted torsion-rigid in the test by means of a fork bearing. Both the vertical deformation at the connection and the torsion of the ceiling element at the connection are recorded during the tests by means of displacement transducers. For vertical displacement measurements, transducers are placed on both sides of the ceiling in positions of the load induction and the system connection. Horizontal displacement measurements on both sides of the top and the bottom of the connection evaluate the rotational behavior of the system connection. In addition, an optical measurement on the surface of the ceiling in the area of the system connection is performed in all tests.

(a) (b)

Figure 7. Test setup of the 3-point bending test for wall–ceiling connections made of CLT (**a**) Front view (**b**) Side view.

The "double dovetail" system connector for wall–ceiling connections made of CLT can be designed in three variants, shown in Figure 8.

(a) (b) (c)

Figure 8. (**a**) Variant V1 (**b**) Variant V2 (**c**) Variant V3 of the wood–wood system connector.

The main fiber direction is always oriented in the insertion direction for both veneer arrangements. The alignment of the LVL can be seen in the system connector variants in Figure 8.

Variant V1: The system connector is based on the requirements set out in Z-9.1-649 [1] and includes flanks inclined up to 6° on both sides.

Variant V2: By stepping the tenon base, a noticeable increase in load-bearing capacity is achieved, especially with thin floor slabs. V1 and V2 are manufactured in Edgewise veneer arrangement and from one piece.

Variant V3: Manufactured from one piece or from two parts glued together. The veneers are oriented in a flatwise arrangement. The shear surface is reinforced by means of glued-in beech dowels across the coupling element. This veneer arrangement is particularly favorable with regard to the dimensional stability of the system connector. Furthermore, the individual production of the dovetail tenons allows a material-saving combination of two different tenon geometries.

The development of the reference geometry of the timber system connection is based on REF 01 of the bar combinations 120 × 160 mm tests and is initially determined for the wall-side tenon WS of the system connector. The height of the ceiling-side tenon of the system connector must be adjusted in compliance with a notch ratio of $\alpha = 0.5$ accordingly [1,10] and an undamaged cover lamella. However, the notch taper should be designed with a similar opening angle γ as for the wall-side tenon. The timber system connector Variant 01 (V1) is defined according to these criteria. Spruce LVL-32C [13] with the veneer arrangement in Edgewise design is used as the material for the system connector. In the shear area, the design-relevant ceiling-side tenon is 120 mm wide and 112 mm high. Optimization potential is attainable by increasing the tenon width as well as by reducing the tenon height.

In the following, two further variants will be investigated. Variant 02 (V2) with a stepped tenon base on the ceiling side (veneers in Edgewise design) and Variant 03 (V3) with an extended notch taper angle. By extending the mortise and tenon angle in Variant 03, the direction of the funneling in the wood system connection must be rotated by 90° due to the increased load transfer via the flank sides. This means the arrangement of the veneers is Flatwise. Furthermore, longitudinal dowelling of the shear surface of the two dovetail segments by means of 12 beech dowels is required. Several test series (see Figure 9) with four tests each are carried out for the three variants. The laboratory tests were performed at the Technical Testing and Research Institute (TVFA) of the University of Innsbruck.

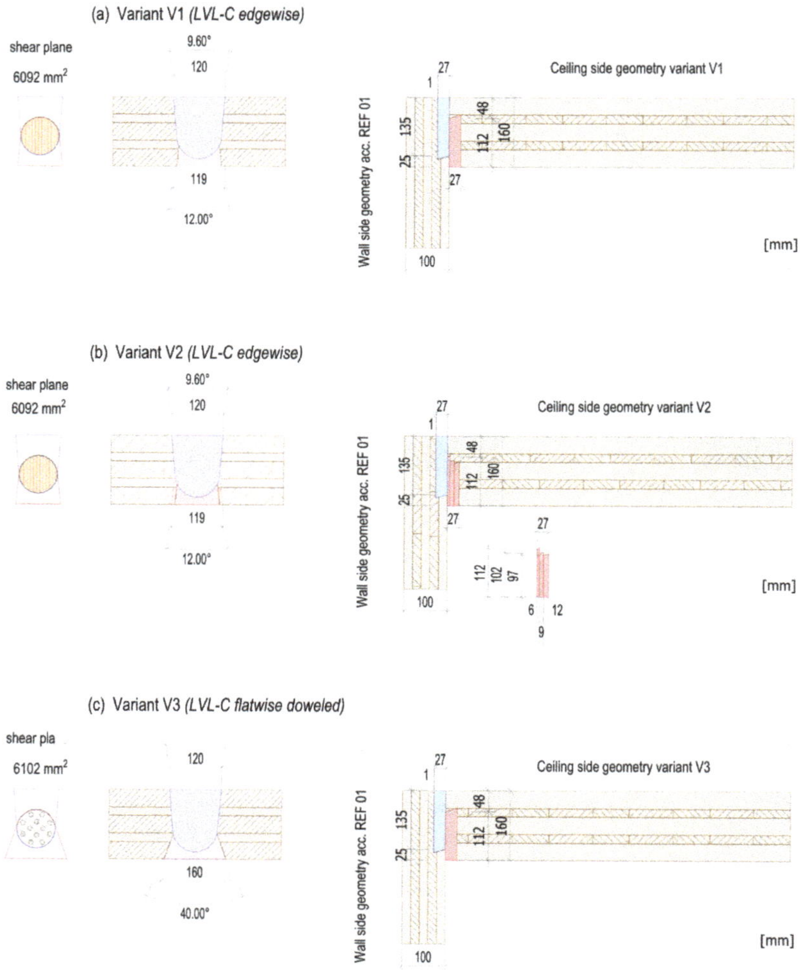

Figure 9. Test series (**a**) Variant V1 (**b**) Variant V2 (**c**) Variant V3 of the wood–wood system connector for wall–ceiling connections made of CLT.

3. Results and Discussion

3.1. Bar-Shaped Investigations for Load Bearing Optimization of RDJ

3.1.1. Bar-Shaped Investigations—Load-Bearing Part of the Tenon Base and Flanks

Equal-height GLT beam combination of 120 × 160 mm with a 10° opening angle

When evaluating the GLT girder combinations series REF 01, ZGR 01, FLA 01, it can be seen that the failure modes of the reference form REF 01 as well as the basic tenon form ZGR 01 are almost identical. In both cases, a transverse tension failure of the main girder occurs, see Figure 10. The effective depth t_{eff} in the test runs is approximately $0.75 \cdot b_{HT}$. This value is significantly lower than the uniform calculation value of t_{eff} = 100 mm for transverse connection failure (mode 1) for single-sided dovetail connections specified in Z-9.1-649 [1]. Since t_{eff} is a significant factor in the calculation formula for the load-bearing capacity of the transverse connection, it can be determined that the calculated characteristic load-bearing capacity for mode 1 according to [1] is not achieved in the test runs of series ZGR 01, which takes the shrinkage of the flanks into account, if the pre-timber height h_{VH} = 0.15 h_{HT} is given. The REF 01 series reaches the calculated value during the test, but only due to the existing flank load components. These are not taken into account in the calculation formula of the Z-9.1-649 [1] approval. Lower pre-timber heights h_{VH} = 0.05 h_{HT} (REF 03 series under 3.1.2) lead to lower connection depths t_{eff} and thus to larger deviations of the bearing capacity of the tapping base from the calculated value for mode 1 according to [1].

Figure 10. Damage patterns and cracks for series with 120 mm × 160 mm beam combinations—REF 01 (**top**), ZGR 01 (**middle**) and FLA 01 (**bottom**).

To determine the separately investigated load-bearing components of the tapping base and the flanks, their stiffness is determined by means of series ZGR 01 and FLA 01. These are then related to the total stiffness of the reference form series REF 01. In the course of the investigation of REF 01, it is found that the stiff tenon base is initially the main force transmission path of the joint. With increasing changes in the stiffness ratios, due to incipient damage in the tenon base, rearrangement processes occur towards the flanks. The residual load-bearing capacity of the wood connection, which is approximately 75% of the ultimate load, is ensured by the flanks. Since the force transmission via the flanks has little stiffness, noticeable deformations occur after failure of the bearing capacity of the tenon base. This effect is clearly visible in series FLA 01, since the trunnion does not rest on the bottom of the tenon hole. That means, the bottom of the main beam remains undamaged, see Figure 10, and the stiffness of the connection is represented only by that of the flanks. For series FLA 01 the stiffness is therefore very low. All load-deformation mean value curves for 120 mm × 160 mm beam combinations are shown in Figure 11. Their loads and stiffnesses are listed in Table 2.

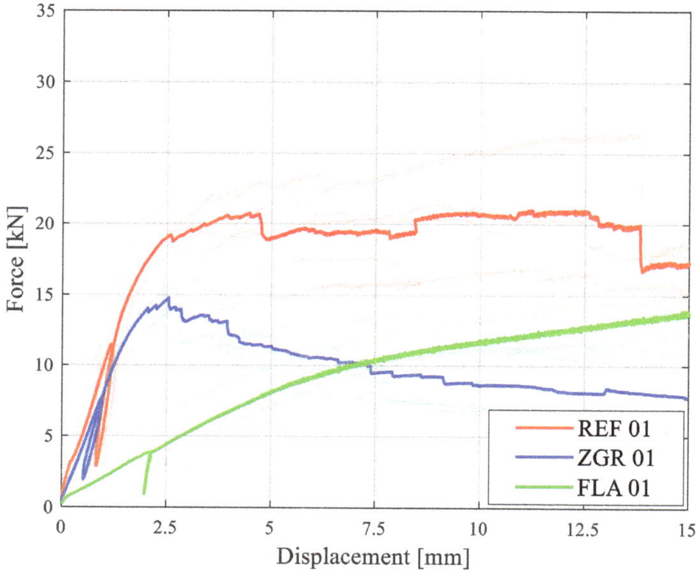

Figure 11. Load-deformation mean value curves (thick colored lines, listed in legend) of the related single tests (thin colored lines) for 120 mm × 160 mm beam combinations.

Table 2. Evaluation of the tests—loads and stiffnesses according to EN 14358 [21].

Series [-]	Tests [-]	ω_{mean} [%]	ρ_{mean} [kg/m^3]	$F_{v,mean}$ [kN]	$F_{v,k}$ [kN]	COV [%]	$K_{ser,mean}$ [N/mm]	$K_{ser,k}$ [N/mm]	COV [%]
REF 01	5	13.1	446	22.1	16.5	12.1	9246	8884	11.8
ZGR 01	5	12.2	426	15.2	13.0	6.2	8031	7897	5.0
FLA 01	5	12.7	435	13.8	11.3	7.8	1419	1357	11.8

The investigations of the stiffness on GLT girder combinations with cross sections of 120 × 160 mm show that the participation of the load transfer over the flanks, which contributes to series REF 01, amounts to 15% (see Figure 12). As already mentioned, in common calculation models the flank load transfer part is not considered. However, at low pre-timber heights, redistribution processes from the bottom of the journal to the flanks already occur at small load steps. The flanks guarantee a residual load-bearing capacity of almost 75%, which is important for the robustness of the connection and the structure as a whole. The characteristic load of the reference geometry REF 01 achieved in the test, corresponds exactly to the calculated value according to [1]. However, the value of the pure tenon base form ZGR 01 is lower. In the case of beam combinations of the same height, the calculated value according to [1] is only achieved in the present test runs if the flank load-bearing component is taken into account.

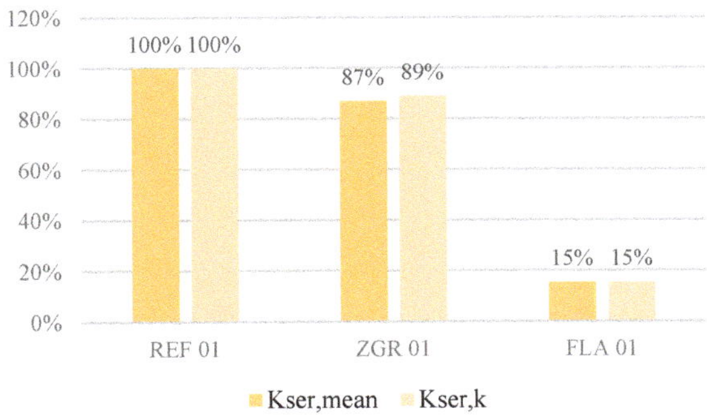

Figure 12. Weighted stiffnesses of Series 01 (graphs).

Equal-height GLT girder combination of 160 × 160 mm with an opening angle of 33°

In order to demonstrate the optimization potential for shapes with a larger notch taper angle γ, GLT beam combinations series REF 02, ZGR 02, FLA 02 were investigated in the same way as in the previous series. The failure mode of the reference geometry REF 02 now differs from the corresponding basic tenon shape ZGR 02 by a multi-step force application due to the load-bearing components of the flanks. This leads to a higher total resultant force transmission, which is shown by a greater effective depth t_{eff} of the reference shape compared to the basic tenon shape. The FLA 02 series flank form is again tested without the tenon base. The load achieved with the FLA 02 series is higher than with the FLA 01 due to the wider flank angle. This is also visible in the damage patterns of the main beam (see Figure 13).

Figure 13. Damage patterns and cracks for series with 160 mm × 160 mm beam combinations—REF 02 (**top**), ZGR 02 (**middle**) and FLA 02 (**bottom**).

The basic tenon form, ZGR 02, shows exactly the same values as the corresponding basic tenon form, ZGR 01. Its calculated load-bearing capacity can also be estimated on the safe side by using the calculation formula for mode 1 in [1]. The significant increase in the load-carrying capacity of the reference form REF 02 compared to REF 01, therefore results exclusively from the higher flank load-bearing component because of the wider notch taper angle. The ductile behavior of the connection and the robustness of the structure as a whole can also be guaranteed by widening the notch taper angle γ. The increase in load-bearing capacity before failure and the residual load-bearing capacity at failure of the connection depend exclusively on the flank load-bearing component and thus on the inclination of the flanks. The potential of this load increase for the REF 02 series is 66%, compared to the ultimate load of the basic tenon form ZGR 02, which is relevant for the design. After failure of the tenon base, approximately 90% of the ultimate load at an opening angle of 33° is available as residual load-bearing capacity through the flanks in the case of large deformations. Taking the flank load-bearing components into account in the calculation formulas would lead to a significant increase in the ultimate load potential of the timber connection at large opening angles. Due to the early onset of redistribution effects with beam combinations of the same height, these could also be included in the ultimate load calculation—assuming appropriate milling accuracy and limitation to NKL 1. The flank load-bearing components certainly ensure the robustness of the structure as a result of their considerable residual load-bearing capacity. All load-deformation mean value curves for 160 mm × 160 mm beam combinations are shown in Figure 14. Their loads and stiffnesses are listed in Table 3.

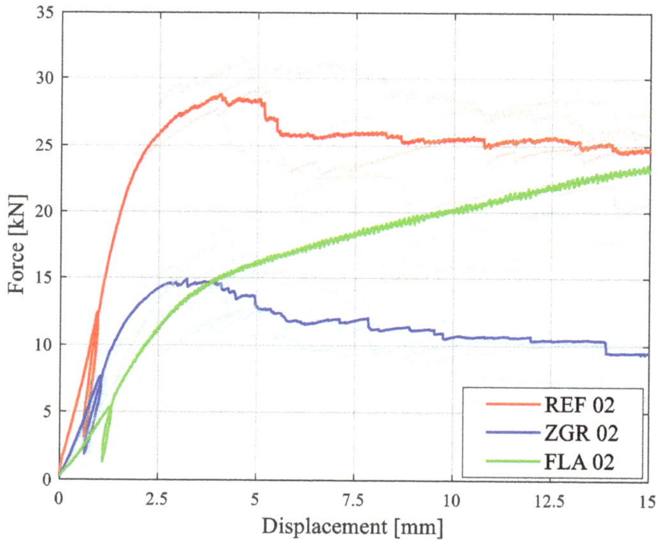

Figure 14. Load–deformation mean value curves (thick colored lines, listed in legend) of the related single tests (thin colored lines) for 160 mm × 160 mm beam combinations.

Table 3. Evaluation of the tests—loads and stiffnesses according to EN 14358 [21].

Series [-]	Tests [-]	ω_{mean} [%]	ρ_{mean} [kg/m^3]	$F_{v,mean}$ [kN]	$F_{v,k}$ [kN]	COV [%]	$K_{ser,mean}$ [N/mm]	$K_{ser,k}$ [N/mm]	COV [%]
REF 02	5	13.0	452	29.9	26.1	5.0	12,688	12,368	6.7
ZGR 02	5	12.6	460	15.9	12.7	9.0	7905	7715	7.2
FLA 02	5	12.3	456	25.0	15.1	18.9	5107	4802	18.0

The contribution of the flanks to the force transmission can be illustrated by comparing the stiffness of the results on GLT girder combinations with cross sections of 160 × 160 mm of the REF 02, ZGR 02 and FLA 02 series. This results in a load participation of the flanks of 40%, which contributes to the reference form REF 02 (see Figure 15).

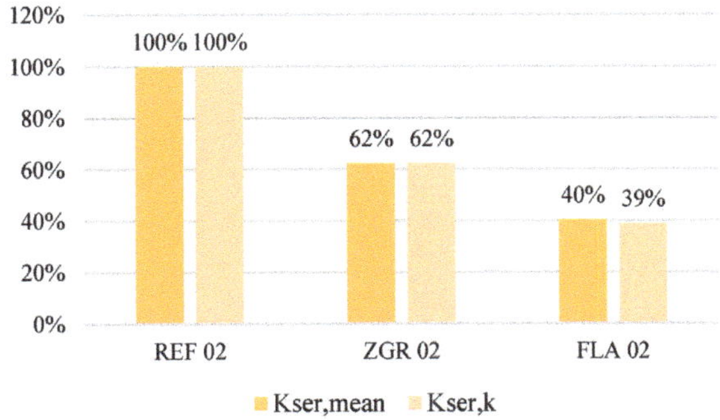

Figure 15. Weighted stiffnesses of Series 02 (graphs).

FEM—Validation

Stiffness values for solid spruce wood C24 according to EN 338 [17] and yield stresses according to Sandhaas and van de Kuilen [18] are applied for the bilinear material model with Hill yield criterion and isotropic hardening in the course of the simulations (Table 1). All results of the test runs of the reference geometries REF 01 and REF 02 can thus be validated in numbers (see Figure 16).

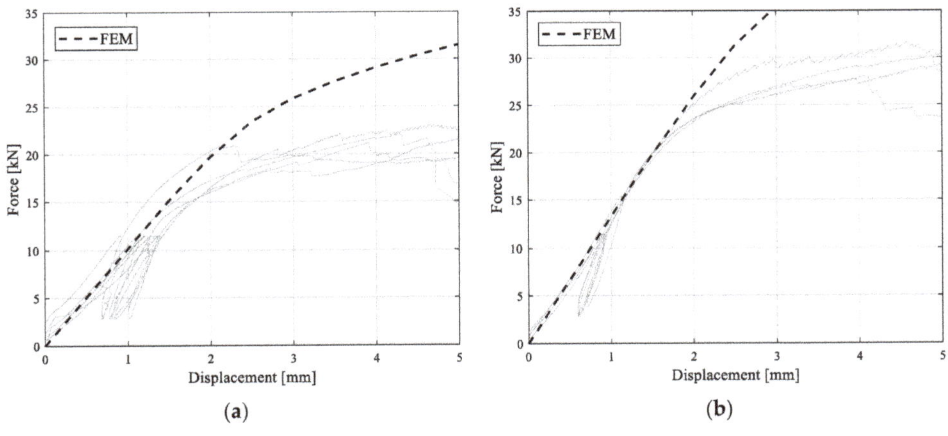

Figure 16. FEM—Validation of the test runs (thin solid lines) (**a**) REF 01 (**b**) REF 02.

The flank load components can be confirmed in FE simulations of the two reference geometries REF 01 and REF 02 (Table 4). The same applies to the fact that both tenon bases show identical load-bearing behavior in the FE simulations (cf. series ZGR 01 and ZGR 02). This confirms that improved load bearing and stiffness properties in the tests result exclusively from increased properties of the flanks (see Figure 17).

Table 4. FE—Comparison of stiffnesses with test series.

Series	$K_{ser,mean}$		$K_{ser,FEM}$	
[-]	[N/mm]	[%]	[N/mm]	[%]
REF 01	9246	100%	10,145	100%
ZGR 01	8031	87%	8506	84%
FLA 01	1419	15%	1639	16%
REF 02	12,688	100%	13,257	100%
ZGR 02	7905	62%	8081	61%
FLA 02	5107	40%	5223	39%

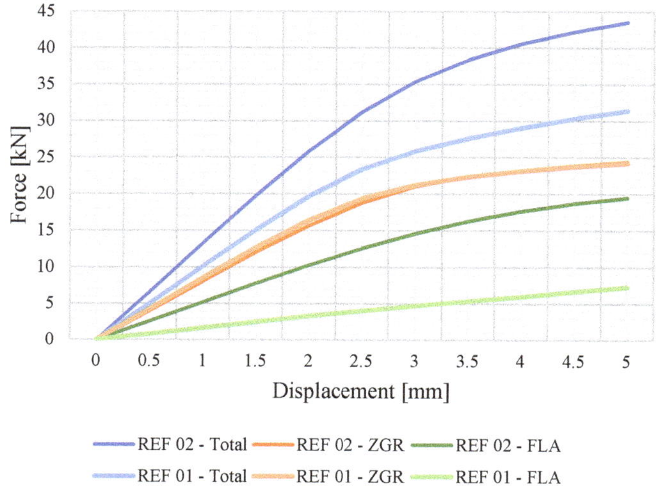

Figure 17. FEM—Comparison of the load-bearing behavior of reference shapes REF 01 and REF 02.

3.1.2. Bar-Shaped Investigations—Stepped Tenon Base

Girder combinations with low pre-timber heights h_{VH} at the main girder fail by exceeding the transverse tension strength. The main reason is that there is too little transverse tension volume available due to insufficient connection depth t_{eff} below the fillet of the main girder. Moreover, the combination of a low transverse tension strength of the wood and load introductions with a low pre-timber height, lead to premature transverse tension failure resulting from defects in the wood.

The total dimensions of the pre-timber of the main beam is composed of:

$$h_{VH} = (h_{HT} - h_{NT}) + (h_{NT} - h_Z) \tag{1}$$

As already mentioned, the unfavorable case of load distribution occurs with beam combinations of the same height, as frequently applied in practice. Z-9.1-649 [1] does not limit this dimension parameter and allows t_{eff} = 100 mm in any case in the transverse tension check for mode 1 failure (transverse connection—failure of the main girder side).

In order to investigate the bearing behavior in such a case, a test series of 120 × 160 mm GLT-girder combinations with minimum pre-timber heights and permissible geometry parameters was developed according to [1]. The REF 03 series was thus tested in five tests and on the same test setup as in the previous investigations with respect to mode 1 failure according to [1]. Furthermore, a series is developed following the measure of a "double dovetail tenon" set by Tannert in [3]. Step-shaping the depth of the tenon forms a three-row shape of the stepped tenon base, starting from the fillet of the tenon. The individual levels of the steps are optimized in terms of their widths and heights so that the uppermost step has the maximum width and can thus take the most load. The aim here is to distribute the load application over several levels. This results in a more favorable distribution of the transverse tension stresses in the area of the fillet of the mortise of the load-bearing side and a delayed failure of the main girder. By introducing part of the forces at the higher level (uppermost step), a greater connection depth t_{eff} is achieved and thus a greater transverse tension volume is attained. As a result, even imperfections in the wood have hardly any influence on the load transfer.

Due to the stepping of the tenon base, the connection depth t_{eff} of series STZ increases by approximately 50% (see Figure 18). The loads of the mean value curve of series STZ can almost reach those of the reference geometry REF 01 at 15 mm deformation (see Figure 19). Therefore, REF 01 is taken into account in all comparisons.

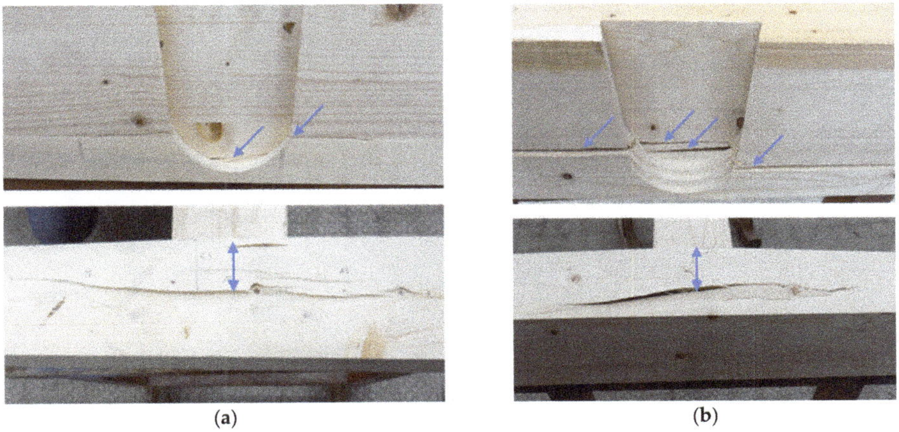

Figure 18. Damage patterns and cracks for (**a**) REF 03 series (**b**) STZ series.

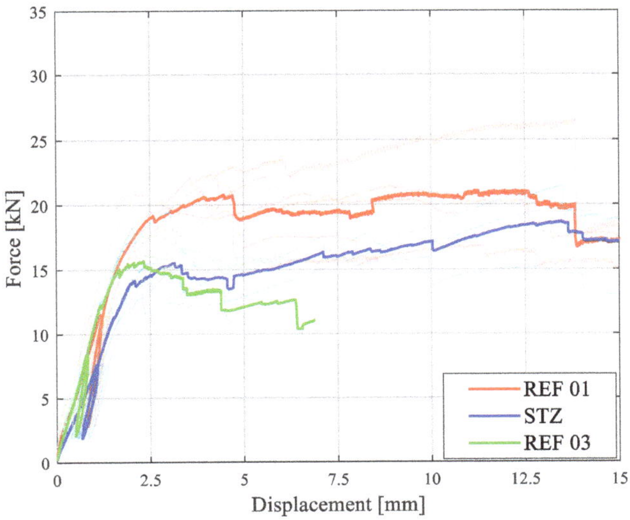

Figure 19. Load-deformation mean value curves (thick colored lines, listed in legend) of the related single tests (thin colored lines) for 120 mm × 160 mm beam combinations.

The high spread of the ultimate loads in the REF 03 series with minimum pre-timber height is striking, which is due to the effects of local defects and thus premature failure. The tests of this reference geometry spread with respect to the ultimate loads $F_{v,k}$ at a covariance COV of 18% (see Table 5). The minimum initiation height from the beginning of the fillet of the mortise to the lower edge of the girder is only capable of ensuring a connection depth t_{eff} of 50 mm. The widely spreading values are due to local defects, which lead to different transverse tension crack propagation. This results in premature failure below the tenon base of the load-bearing main girder without redistribution possibilities. In contrast, the stepped tenon base geometry of the STZ series exhibits a high dispersion in terms of its stiffness. This is due to the form closure on several levels and the associated tolerances, i.e., until the closing of the multi-stage overall system. However, the load transfer to the multiple levels results in a significant increase in the ultimate load, which has a positive effect on the characteristic calculation value $F_{v,k}$ in particular (see Figure 20). The initial stiffness values K_{ser} spread with a covariance (COV) of up to 32% (see Table 5). However, if deformations of up to 15 mm are allowed, the connection force-locks at all step levels. The mean load curve of the STZ series thus approaches that of the reference form REF 01 at 15 mm deformation.

Table 5. Evaluation of the tests—loads and stiffnesses according to EN 14358 [21].

Series [-]	Tests [-]	ω_{mean} [%]	ρ_{mean} [kg/m^3]	$F_{v,mean}$ [kN]	$F_{v,k}$ [kN]	COV [%]	$K_{ser,mean}$ [N/mm]	$K_{ser,k}$ [N/mm]	COV [%]
REF 01	5	13.1	446	22.1	16.5	12.0	9246	8884	11.8
STZ	5	12.3	439	18.7	13.8	12.0	7102	6338	32.4
REF 03	5	12.7	450	17.1	10.8	17.7	9562	8927	20.0

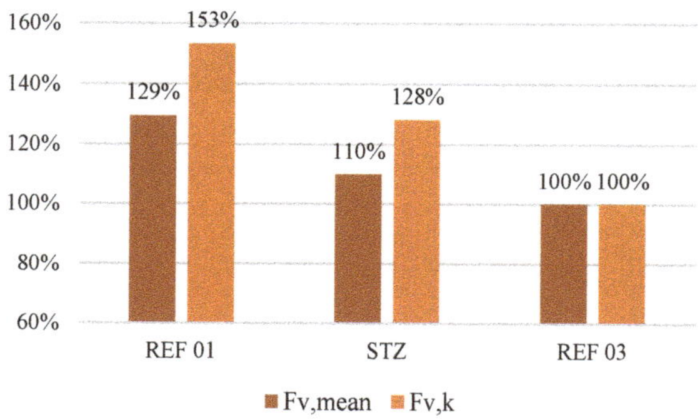

Figure 20. Weighted strengths of Series 03 (graphs).

By stepping the tenon base, a height distribution of the load application can be achieved, which is also reflected in the fracture patterns. Failure is thus not determined by a defect in the transverse tension volume. Ultimately, all test specimens reach a similar ultimate load level after deforming to a higher or lesser degree. The individual levels of the stepped tenon base communicate with each other, so that in the event of an individual level yielding, the load can be redistributed to other, intact levels.

In the reference form REF 03, the entire force is applied via the single "step" located far below, with a low connection depth t_{eff}. As a result, only a small transverse tension volume of the main girder is activated in the transverse connection. If the load introduction is distributed over several levels according to the geometry of the STZ series, an overall higher force introduction is achieved. In this way, approximately 75% of the width of the 120 × 160 mm main beam b_{HT} can be activated as the resisting connection depth t_{eff}. The stepped tenon base is designed so that the step with the lowest connection depth also introduces the lowest load component. The two higher steps introduce higher load components with greater effective connection depths. This means that the widest step at the top transfers more than half of the resultant force to the load-bearing main beam.

FEM—Validation

The positive effect of the stepping can also be seen in the course of numerical simulations (see Figure 21).

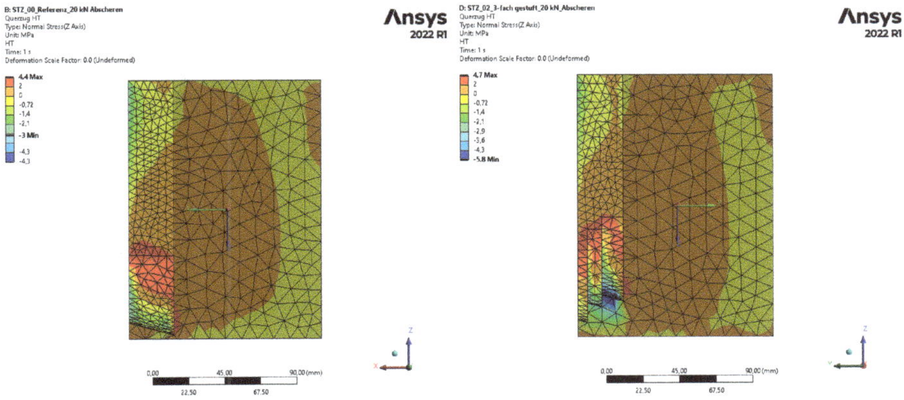

(a) (b)

Figure 21. FEM—Transverse tension and compression stresses (z-direction) in the hole of the main girder (**a**) Reference REF 03 section (**b**) Step shape STZ section.

Stiffness values for solid spruce wood C24 according to EN 338 [17] and yield stresses according to Sandhaas and van de Kuilen [18] are applied for the bilinear material model with Hill yield criterion and isotropic hardening in the course of the simulations (Table 1). All results of the test runs of the reference geometry REF 03, as well as the stepped tenon base geometry STZ, can thus be validated in numbers (see Figure 22).

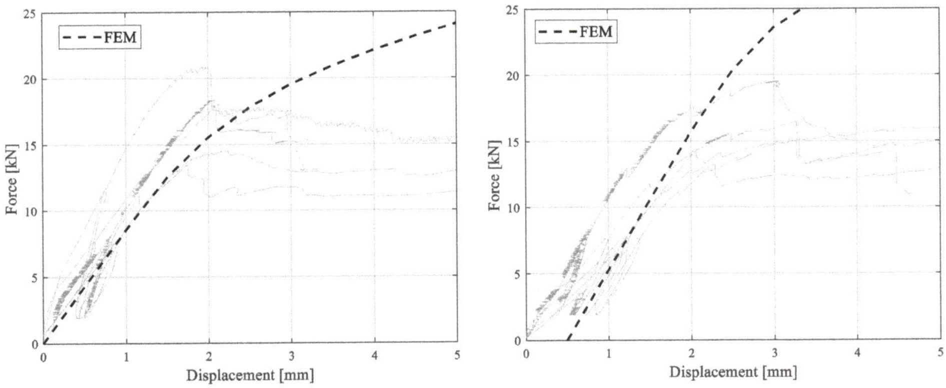

(a) (b)

Figure 22. FEM- Validation of the test runs (thin solid lines) (**a**) REF 03 (**b**) STZ.

Due to the delayed closure of the entire system in the tests with the stepped tenon base shape STZ, the FE curve had to be shifted by 0.5 mm.

The mean value curves of the test results of the REF 01 and STZ series show equally high ultimate loads (see Figure 19), although those of the stepped tenon base form are only reached at deformations of 15 mm. In the FE model, these series show the same initial stiffnesses K_{ser} (see Figure 23). The initial stiffness K_{ser} is overestimated in the FE model for the STZ series, since the problem of closing the individual steps resulting from a lack of fitting accuracy is not covered by the model.

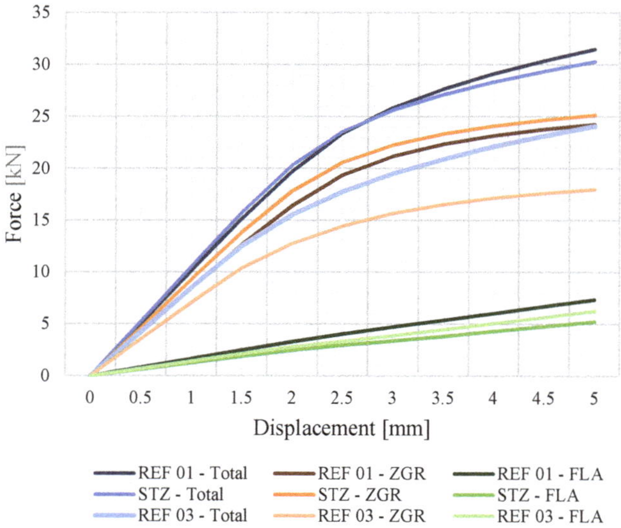

Figure 23. FEM—Comparison of the load-bearing behavior of reference shapes REF 01 and REF 03 with stepped tenon base shape STZ.

Reference form REF 03 shows the lowest initial stiffness, *K*ser, in the FE model (see Figure 23). The ultimate loads in the tests of the REF 03 series, shown in Figure 19, are all reached within deformations of 5 mm. Furthermore tests of REF 03 show a higher initial stiffness, $K_{ser,mean}$, as in the FE model because of a positive "outlier" in the tests (see Table 6).

Table 6. FEM—comparison of stiffnesses with test series.

Series	$K_{ser,mean}$		$K_{ser,FEM}$	
[-]	[N/mm]	[%]	[N/mm]	[%]
REF 01	9246	100%	10,145	100%
STZ	7102	77%	10,563	104%
REF 03	9562	103%	8526	84%

The individual steps of the tenon base attract different load proportions (see Figure 24). The widest, step 3 at the top, absorbs more than half of the load share of the entire tenon base. In general, the FE model for the STZ geometry shows a distribution of the load-bearing components between the tenon base and the flanks of approximately 88% to 12% (total value $K_{ser,FEM}$ of STZ in Table 6). This distribution thus, corresponds to that of the reference shape REF 01 (see Figure 23).

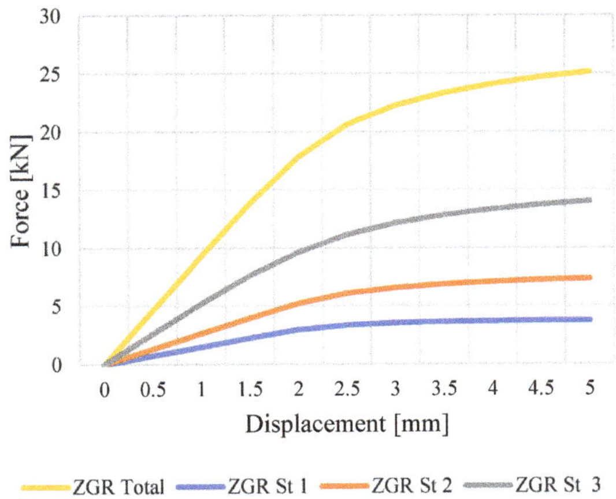

(**a**)

Part	$K_{ser,FEM}$	Distribution
[-]	[N/mm]	[%]
ZGR Total	9306	100%
ZGR Step 1	1481	16%
ZGR Step 2	2644	28%
ZGR Step 3	5181	56%

(**b**)

Figure 24. (**a**) Load-bearing components of the steps (graphs) (**b**) Tabular listing.

3.2. Coupling Element for CLT Wall–ceiling Connections—Double Dovetail

In principle, the failure modes shown in Figure 25 are possible:

- Mode 1: Failure of the cover lamella of the CLT wall element (transverse tension, or transverse compression).
- Mode 2: Failure of the LVL system connector (combined shear, bending tension failure)
- Mode 3: Failure of the cover lamella of the CLT ceiling element (transverse tension with rolling shear).

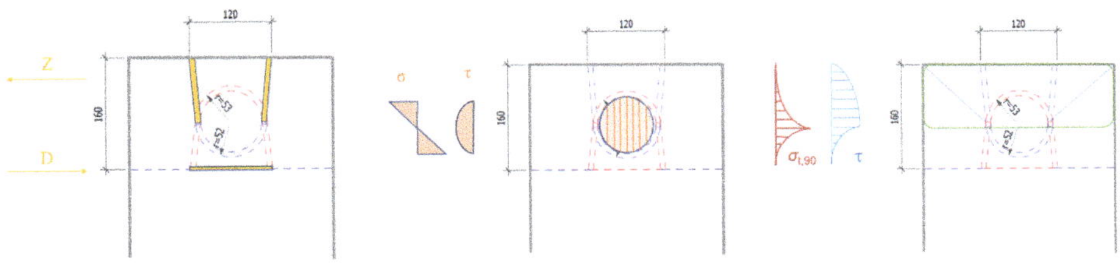

(**a**) (**b**) (**c**)

Figure 25. Failure modes of the system connection. (**a**) Cover lamella wall side (**b**) System connector (**c**) Cover lamella ceiling side.

In the present studies of the variants of the wall–ceiling connection, the failure is due to the notching of the cover lamella from the CLT (mode 3) or shearing of the system connector (mode 2). Calculation models for notched CLT-elements are described in Serrano et al. [22], Serrano and Danielsson [23] and Malagic et al. [24]. Values of tests conducted with the "double dovetail tenon" on CLT 5 s 160 mm elements, indicate much higher load-bearing capacities of the rounded dovetail joint geometry as for the fully notched CLT-elements with width b_Z and height h_Z. Test values for fully notched bar-shaped CLT 5 s 160 mm cross sections can be found in Friberg [25]. The load bearing capacities of mode 3 of the system connector can therefore be best compared with fully notched elements with comparable heights (h_Z-r). The reasoning is that the load transmission starts from the beginning of the rounding of the rounded dovetail joint down to the tenon base contrary to notched elements, as all the load transmission has been transferred to the same level support. Shear strength values for edgewise and flatwise veneer arrangements of Kerto© LVL-Q (see [19]) can be used for the calculation of mode 2 load bearing capacities. The reason being that nearly the same strength value was reached in shear tests of the connector with flatwise and edgewise arranged dowels. Failure on the wall side (mode 1) is not decisive for any of the variants due to the low support torsion that occurs. The eccentricity limit center is also maintained by means of a stiff bearing with regard to torsion of the wall element.

In the course of optimization measures, which have the objective of delaying the failure on the ceiling side, a significant load increase is achieved for V2 and V3 compared to V1. The connection is optimized in such a way that the failure of the cover lamella of the CLT ceiling element is delayed in V2 by stepping the tenon base on the ceiling side, or it even occurs on the side of the system connector in V3. V2 and V3 fail at the same load level, which corresponds to an optimization of these two failure modes.

The variants investigated are characterized by a ductile failure of the connection, whereby variant V2 and V3 do not fail up to a 10 mm displacement (see Figure 26). In variant V2, this is due to a deeper force transmission introduced over several levels, which leads to a delay in failure of the CLT on the ceiling side as compared to variant V1. In variant V3, the connector fails on shear, whereby the doweling by means of 12 ϕ10 grooved beech dowels ensures the necessary ductility with a flatwise arrangement of the veneers. Damage patterns of Variant V2 and Variant V3 are shown in Figure 27.

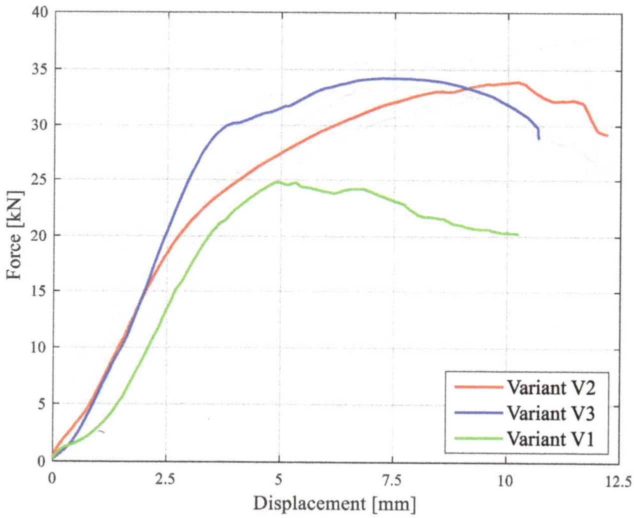

Figure 26. Load-deformation mean value curves (thick colored lines, listed in legend) of the related single tests (thin colored lines) of wall–ceiling connections with wood–wood system connectors Variant V1, Variant V2 and Variant V3.

The V2 and V3 variants are optimized geometries with the same load-bearing capacity level at different failure modes for the present combination of the same-height connection of a CLT 160 5 s ceiling element to a CLT 100 5 s wall element by means of a pure wood–wood connection, which is ensured by system connectors of the wood-based material Kerto©Q.

In the tests, variants V2 and V3 show a load-bearing capacity that is 50% higher than that of variant V1. Their initial stiffness values K_{ser} are also 4% and 15% higher than those of variant V1. All values are listed in Table 7. The ultimate loads and stiffnesses achieved allow CLT floor elements b = 1.25 m to be connected to CLT wall elements by means of two or three connectors and a center distance of the coupling elements of 600 mm or 400 mm, depending on the live loads to be applied, in purely static terms up to a span of 5.0 m in building construction at service class 1 (NKL1).

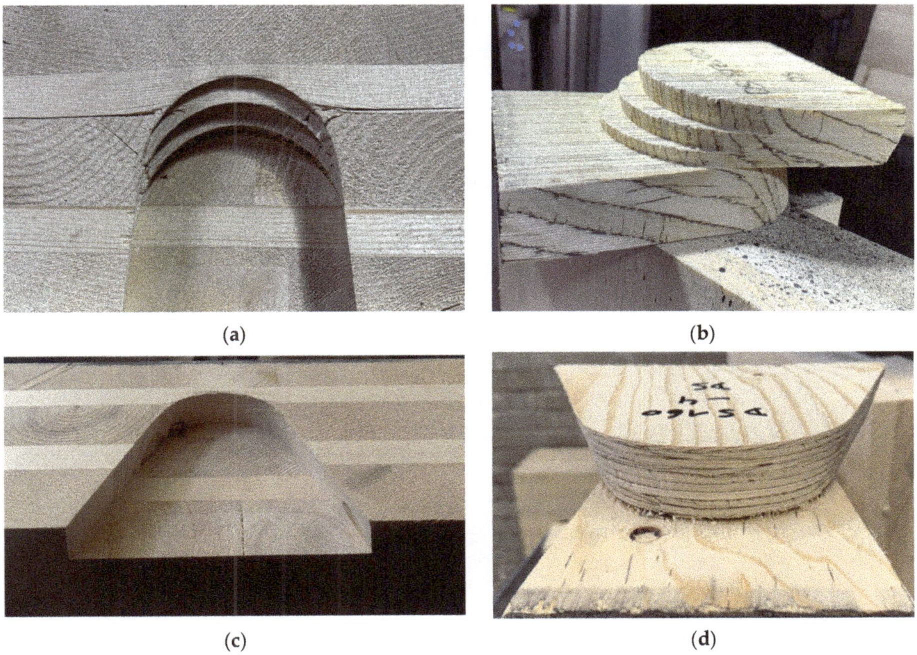

Figure 27. Damage patterns for (**a**) Variant 2—ceiling hole decisive (**b**) Variant 2—connector (**c**) Variant 3—slab hole (**d**) Variant 3—connector decisive.

Table 7. Load capacities and initial stiffnesses of the three variants of the wood system connection.

Series [-]		Tests	F_{mean} [kN]	F_{Rk} [kN]	COV [%]	K_{ser} [N/mm]
Variant V1	EDGEWISE	4	26.0	19.1	11.0%	8060
Variant V2	EDGEWISE	4	35.3	29.4	7.0%	8383
Variant V3	FLATWISE (doweled)	4	34.4	30.0	5.0%	9300

4. Conclusions

In the case of equal-height connections of dovetail joints with a low pre-timber height, the widening of the notch taper angle leads to significantly improved load-bearing behavior. This applies to both bar-type connections and the new wood–wood system connection for CLT wall–ceiling connections. The values of tests conducted with the "double dovetail tenon" on CLT-elements also indicate much higher load-bearing capacities of the rounded dovetail joint geometry as for the fully notched CLT-elements.

The available investigations have shown that a robust and secure wood–wood connection is ensured by the dovetail geometry together with its inherent flank inclination. The more the flanks are inclined, the higher is their load-bearing capacity before failure. That also applies to the residual load-bearing capacity of the connection after failure. This always remains the same for any material combination, irrespective of time. Similar properties are produced by stepping the tenon base of a dovetail joint. This activates a larger volume of the load-bearing component during force application. It reduces the influence of local flaws and thus delays the failure of the connection. The abovementioned optimization measures have a very great influence on component connections at the same height, which also makes them suitable for floor elements that are connected to wall elements by means of the system connector. The load-bearing capacity of wall–ceiling connections made of

CLT could be increased by 50% by such geometric optimization. The wood–wood system connection consisting purely of wood-based materials is thus a structural alternative to mechanical system connectors for ceiling loads and spans relevant in building construction. It meets the specifications of load-bearing structures with regard to ductility and robustness, even in the connection area.

Preliminary numerical moisture simulations of the system connection also indicate promising results. With Edgewise veneer arrangement, the coupling element swells and shrinks mainly transverse to the insertion direction. With Flatwise veneer arrangement, dimensional stability is provided by wood fiber directions in all three axes: transverse plies transverse to the direction of insertion, beech dowels in the thickness direction, and the main fiber direction in the direction of insertion. System connectors with this veneer arrangement remain dimensionally stable along the flanks. Further investigations on this topic will be part of future research work [26].

5. Patents

The "double dovetail" system connector, with all the variants shown, has also been protected by a patent since July 2023.

Author Contributions: Conceptualization, A.P. and A.K.; methodology, A.P. and A.K.; writing—original draft preparation, A.P.; writing—review and editing, A.K; supervision, A.K.; project administration, A.K.; funding acquisition, A.K. All authors have read and agreed to the published version of the manuscript.

Funding: The work on the wood–wood connections has been funded by the innovation promotion of the country Tirol, Austria (Land Tirol) within the research projects "CNC-manufactured wood–wood connections", Reference WA-540-02-00606/01-0010 and "wood–wood system connection for multi-storey solid wood construction", Reference F.42814/9-2022 which were both processed together with the commercial partners Holzbau Lengauer-Stockner (Schwoich, Austria), Holzbau Saurer (Höfen, Austria), and FS1, Fiedler-Stöffler (Innsbruck, Austria).

Institutional Review Board Statement: Not applicable.

Informed Consent Statement: Not applicable.

Data Availability Statement: The data presented in this study are available on request from the corresponding authors. The data is not publicly available due to ongoing research.

Acknowledgments: The authors would like to thank the Technical Testing and Research Institute (TVFA) Innsbruck for carrying out the experimental campaign. Special thanks are extended to the partners of the research projects "CNC-manufactured wood–wood connections" and "wood–wood system connection for multi-storey solid wood construction"—Lengauer-Stockner, Saurer and FS1—and the innovation promotion of the country Tirol. The computational results presented here have been achieved (in part) using the LEO HPC infrastructure of the University of Innsbruck.

Conflicts of Interest: The authors declare no conflict of interest.

References

1. *Z-9.1-649;* Schwalbenschwanz-Verbindung in Bauteilen. Deutsches Institut für Bautechnik: Berlin, Germany, 2022.
2. *CEN/TC 250/SC 5;* prEN 1995-1-1 for Formal Enquiry—Eurocode 5: Design of Timber Structures—Part 1-1: General Rules—General Rules and Rules for Buildings. CEN (European Committee for Standardization) CEN: Brussels, Belgium, *unpublished.*
3. Tannert, T. Structural Performance of Rounded Dovetail Connections. Ph.D. Thesis, University of British Columbia, Vancouver, BC, Canada, 2008.
4. Tannert, T.; Lam, F.; Vallée, T. Structural performance of rounded dovetail connections: Experimental and numerical investigations. *Eur. J. Wood Wood Prod.* **2011**, *69*, 471–482. [CrossRef]
5. Tannert, T.; Lam, F. Self-tapping screws as reinforcement for rounded dovetail connections. *Struct. Control. Health Monit. Off. J. Int. Assoc. Struct. Control. Monit. Eur. Assoc. Control. Struct.* **2009**, *16*, 374–384. [CrossRef]
6. Tannert, T. Improved performance of reinforced rounded dovetail joints. *Constr. Build. Mater.* **2016**, *118*, 262–267. [CrossRef]
7. Schickhofer, G. *Prüfung und Optimierung von CNC-Gefertigten Schwalbenschwanzzapfverbindungen;* Forschungsbericht Holz-Holz-Verbindungen: Graz, Austria, 2014.

8. Blaß, H.J.; Enders-Comberg, M. Fachwerkträger für den industriellen Holzbau. In *Karlsruher Berichte zum Ingenieurholzbau No. 22*; KIT Scientific Publishing: Karlsruhe, Germany, 2012.
9. Braun, M.; Pantscharowitsch, M.; Kromoser, B. Experimental investigations on the load-bearing behaviour of traditional and newly developed step joints for timber structures. *Constr. Build. Mater.* **2022**, *323*, 126557. [CrossRef]
10. *ÖNORM EN 1995-1-1*; Eurocode 5: Design of Timber Structures—Part 1-1: General—Common Rules and Rules for Buildings (Consolidated Version). CEN (European Committee for Standardization) CEN: Brussels, Belgium, 2019.
11. Ehlbeck, J.; Görlacher, R. *Tension Perpendicular to the Grain in Joints, STEP 1—C2*; Centrum hout: Almere, The Netherlands, 1995.
12. Gustafsson, P.J. *Notched Beams and Holes in Glulam Beams, STEP 1—B5*; Centrum hout: Almere, The Netherlands, 1995.
13. IngenieurHolzbau. *de. Finnish Woodworking Industries: Furnierschichtholz-Merkblatt (LVL)—2. Auflage*; Studiengemeinschaft Holzleimbau e.V.: Wuppertal, Germany, 2020.
14. *EN 13183-1*; Moisture Content of a Piece of Sawn Timber—Part 1: Determination by Oven Dry Method. CEN (European Committee for Standardization) CEN: Brussels, Belgium, 2004.
15. *EN 26891*; Timber structures—Joints Made with Mechanical Fasteners—General Principles for the Determination of Strength Aand Deformation Characteristics. CEN (European Committee for Standardization) CEN: Brussels, Belgium, 1991.
16. *EN 14080*; Timber Structures—Glued Laminated Timber—Requirements. CEN (European Committee for Standardization) CEN: Brussels, Belgium, 2013.
17. *EN 338*; Structural Timber—Strength Classes. CEN (European Committee for Standardization) CEN: Brussels, Belgium, 2016.
18. Sandhaas, C.; Van de Kuilen, J.W. Material model for wood. *Heron* **2013**, *58*, 171–191.
19. *Z-9.1-847*; Bauarten Mit Furnierschichtholz "Kerto-S", "Kerto-Q" und "Kerto-Qp". Deutsches Institut für Bautechnik: Berlin, Germany, 2019.
20. *EN 408*; Timber Structures—Structural Timber and Glued Laminated Timber—Determination of Some Physical And Mechanical Properties. CEN (European Committee for Standardization) CEN: Brussels, Belgium, 2012.
21. *EN 14358*; Timber Structures—Calculation and Verification of Characteristic Values. CEN (European Committee for Standardization) CEN: Brussels, Belgium, 2016.
22. Serrano, E.; Gustafsson, P.J.; Danielsson, H. Prediction of load-bearing capacity of notched cross laminated timber plates. In *International Network on Timber Engineering Research-Meeting 52*; Timber Scientific Publishing: Karlsruhe, Germany, 2019.
23. Serrano, E.; Danielsson, H. Fracture Mechanics Based Design of CLT-plates—Notches at Supports and Half-and-Half Joints. In *International Network on Timber Engineering Research-Meeting 53*; Timber Scientific Publishing: Karlsruhe, Germany, 2020.
24. Malagic, A.; Augustin, M.; Silly, G.; Thiel, A.; Schickhofer, G. Load-bearing Capacity and Fracture Behaviour of Notched Cross Laminated Timber Plates. In Proceedings of the International Network on Timber Engineering Research (INTER) Meeting, Online, 19 August 2021; Paper INTER/54-12-5. Timber Scientific Publishing: Karlsruhe, Germany, 2021; pp. 16–19.
25. Friberg, A. Bärförmåga för KL-trä Med Urtag—Provning och Beräkningsmetoder [Load-Bearing Capacity of CLT with Notches—Testing and Calculation Methods]. Report THID-5526. Bachelor Thesis, Division of Structural Mechanics, Lund University, Lund, Sweden, 2017. (In Swedish).
26. Pomaroli, A. Holz-Holz Systemverbindung für Wand-Deckenanschlüsse aus Brettsperrholzelementen. Ph.D. Thesis, Universität Innsbruck, Innsbruck, Austria. *manuscript under preparation*.

applied sciences

MDPI

Article

Drying Behaviour of 50 mm Thick Turkey Oak Lumber

Iulia-Teodora Deaconu, Sergiu-Valeriu Georgescu and Mihaela Campean *

Department of Wood Processing and Wood Products, Faculty of Furniture Design and Wood Engineering, Transilvania University of Brasov, 500036 Brasov, Romania; iulia.deaconu@unitbv.ro (I.-T.D.); sergiu.georgescu@unitbv.ro (S.-V.G.)
* Correspondence: campean@unitbv.ro; Tel.: +40-726686725

Abstract: Turkey oak lumber is difficult to dry, being prone to cracks even at low temperatures. The main objective of the research was to test different drying conditions in order to establish whether air-drying or kiln-drying from a green state was more suitable; if the lumber width had an influence on the drying quality; if prior steaming had a positive effect upon the kiln-drying time and quality; if end cracks could be avoided by protecting the board ends; and what the most appropriate kiln-drying conditions were, with conventional drying. The research results recommend kiln-drying as a better alternative than air-drying in terms of both time and quality, but very mild conditions must be applied. Wide lumber parts cracked both when air-dried and when kiln-dried. All defect-free lumber parts dried accurately, with no deterioration. Steaming had a positive effect upon the drying quality; no steamed part cracked. The end protection of the boards was efficient only in the case of kiln-drying.

Keywords: Turkey oak; air-drying; kiln-drying; drying rate; drying quality

Citation: Deaconu, I.-T.; Georgescu, S.-V.; Campean, M. Drying Behaviour of 50 mm Thick Turkey Oak Lumber. *Appl. Sci.* **2023**, *13*, 10676. https://doi.org/10.3390/app131910676

Academic Editor: Giuseppe Lazzara

Received: 21 August 2023
Revised: 20 September 2023
Accepted: 23 September 2023
Published: 26 September 2023

1. Introduction

The drying of *Quercus* wood species represents a challenge for the industry due to the high density, the heterogenous structure, the sensitivity to temperatures higher than 60–70 °C, and the tendency to crack, especially along the rays. The reference literature [1–6] recommends for European oak wood (*Quercus robur* and *Quercus petraea*) drying schedules with low temperatures (30–max.40 °C) and high relative air humidity (85–95%) during the first stage of actual drying (free water removal) and low drying gradients (1.5–2.0) during the second stage (drying below the fiber saturation point) in order to prevent cracking, but these conditions lead to very long drying times. Therefore, it is tempting to apply air-drying at least until the fiber saturation point is reached, in order to save energy. However, this approach is not always an appropriate one, because the lumber is exposed to biological attack and case-hardening [4] during this period. This often affects the wood quality, and sometimes it slows down the subsequent kiln-drying.

Turkey oak (*Quercus cerris*) belongs to the group of red oaks [7]. It has a high proportion of sapwood (50%) and significantly higher density (ρ_0 = 728 kg/m^3) than sessile oak (ρ_0 = 667 kg/m^3), and its fiber saturation moisture content is lower (23–24%), according to determinations performed by the authors, compared with sessile oak (26–27%). These properties make it more difficult to dry than European oak lumber. The difficult drying, combined with the scarcity of the reference sources, results in the reticence of the industry to valorize it [8].

Steaming was found as a solution to reduce wood hygroscopicity and also to attenuate the color differences within lumber boards that contain both sapwood and heartwood [9,10]. According to [9], the best results in terms of the equilibrium moisture content (EMC) reduction and color uniformization were obtained at 120 °C and 18 h. It is also known that steaming has a favorable effect upon the drying of wood, especially in the case of hardwoods. This is because it temporarily plastifies wood and thus allows a more rapid water removal, with fewer cracks. However, the steaming of Turkey oak wood is a difficult task. Poorly chosen steaming parameters (temperature and duration) may cause internal surface cracks, collapse, or warp.

A combined steaming–air-drying–thermo-vacuum treatment was applied by [11] to Turkey oak wood, in order to obtain a wooden material with improved processing properties. The freshly harvested logs were steamed at 110 °C and 140 kPa for 24 h, then cooled for 5 days and then cut into boards. The boards were air-seasoned for 5 months, until they reached a moisture content of 20–25%, then half of them were introduced in a press-vacuum plant (PVP) [12] and the other half in a thermo-vacuum cylinder (TVP) (both kilns being manufactured by WDE Maspell Italy). The PVP treatment avoided the deformation of the boards. The TVP treatment produced a more severe darkening. The incipient steaming had a decisive effect upon the EMC reduction. It was also observed that the defect-free boards suffered no qualitative damage, but the boards containing knots, for example, displayed severe cracks along a great part of the board.

According to the authors' knowledge, in Romania, there is only one plant where vacuum-press drying is applied to oak lumber. Conventional drying is by far the most common method applied. According to a survey conducted in Romanian wood-processing companies [8], 30% of them perform kiln-drying from a green state, while 60% of the companies prefer pre-drying by air-seasoning followed by conventional kiln-drying. Most of the respondents (70%) indicated as a necessary optimization direction the reduction in the drying time and of the cracking tendency, especially with thicker grades (>40 mm).

Therefore, the main objective of the present research was to test different drying conditions using the common drying methods (air-drying and conventional kiln-drying), in order to establish an optimum combination of temperature and air humidity for the drying of 50 mm thick Turkey oak lumber, so as to avoid cracking but also to optimize the drying time.

2. Materials and Methods

The wooden material used within this research consisted of 0.6 m long, 15–30 cm wide, and 50 mm thick lumber parts, cut from 14 logs existing in the log yard of a wood processing company, having average diameters of 45–50 cm and an age of 90–95 years. They were cut so that some of the boards included only heartwood, and some included both heartwood and sapwood.

Half of the selected boards, originating from the same plot, were stacked under a shed, tied together (Figure 1a), and exposed for 7 months (from 30 November 2022 to 3 July 2023) to air-drying conditions. The other half were stacked and kiln-dried in a FEUTRON KPK200 climate chamber (Figure 1b), with controlled air parameters.

(**a**) (**b**)

Figure 1. Mini-stacks of 50 mm thick Turkey oak lumber prepared for air-drying (**a**) and for kiln-drying in a KPK200 climate chamber (**b**).

Half of the kiln-dried lumber boards were previously steamed at 100 °C for 48 h in an industrial steaming chamber.

In both drying tests, half of the lumber pieces placed within the same stack were end-protected. For the air-dried boards, a paraffin-based sealer was used, and for the kiln-dried boards, a mixture made with PVA adhesive and sawdust was used. Also, in both drying situations, the boards were dried under pressed conditions, in order to avoid deformations.

Before stacking, specimens were sampled (Figure 2) in order to assess the inital state of the boards, namely, the average initial moisture content, the moisture gradient across the lumber thickness, and the eventual inner stresses. The visible cracks were marked with white chalk.

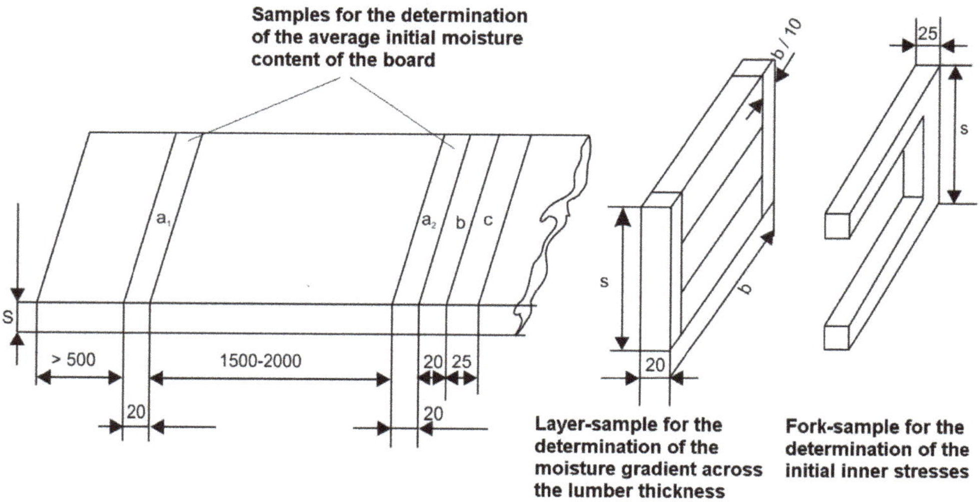

Figure 2. Samples for the evaluation of the initial state of the lumber boards: a_1 and a_2—samples for determination of the average initial moisture content of the board; b—layer sample; c—fork sample.

The climatic conditions recorded during the air-drying exposure period are presented in Figure 3. The moisture content of the air-dried boards was monitored monthly by means of a resistive-type moisture meter type LG9NG from Holzmeister (Italy). Both the core moisture content (at 20 mm depth) and that in the surface layer (at 10 mm depth) were recorded, in order to assess not only the drying rate but also the evolution of the moisture gradient across the lumber thickness.

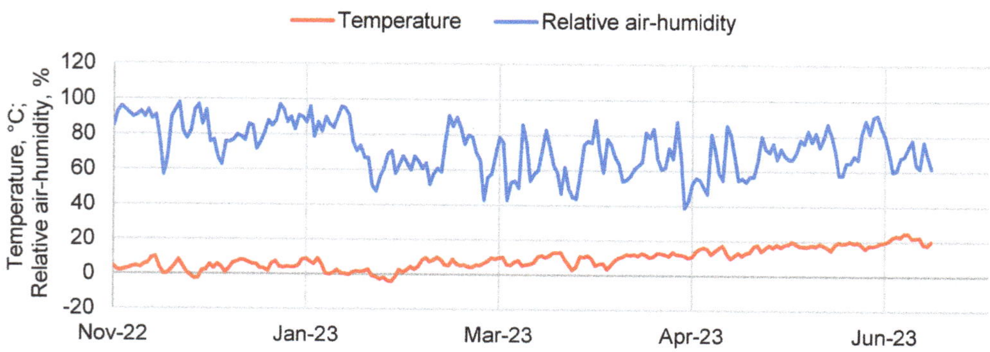

Figure 3. Climatic conditions during the air-drying.

The drying schedule applied within the kiln-drying test is presented in Table 1. It was established based on the reference literature [6,13–16] and on the authors' experience with oak lumber drying. In this trial, the moisture content was continuously monitored by means of V2A sensors with insulated rods from Gann (Germany) (Figure 4a). The 25 mm long electrodes were used to assess the core moisture content, and short (15 mm) electrodes were placed nearby in the board (Figure 4b) to assess the moisture content in the surface layer.

Table 1. Drying schedule applied for the kiln-drying of 50 mm thick Turkey oak wood.

Phase	Wood Moisture Content, %	Air Temperature, °C	Equilibrium Moisture Content, %
Initial heating	$MC_{initial}$	30	18
Actual drying	$MC_{initial} \ldots MC_{FSP}$ *	30	18 … 15.5
	MC_{FSP} * … MC_{final}	30 … 60	15.5 … 6.5
Cooling	MC_{final}	30	-

* The fiber saturation moisture content of Turkey oak wood was estimated at 23–24% [9].

(**a**) (**b**)

Figure 4. Monitorization of the moisture content of wood during kiln-drying by two pairs of V2A sensors, of different lengths (**a**), in order to measure separately the core and surface MC of the board (**b**).

The drying diagram was drawn for each load, and the drying rate (w) was calculated according to Equation (1), as a function of the moisture decrease interval (ΔMC) and the time interval (Δt) during which this moisture decrease occurred:

$$w = \frac{\Delta MC}{\Delta t}, \ \%/\text{day} \tag{1}$$

The final quality control included the visual examination of each board and registration of the established defects within a drying record sheet, where also observations regarding the initial state of the examined board were noted. Additionally, a case-hardening test [17] for randomly chosen boards was performed (Figure 5).

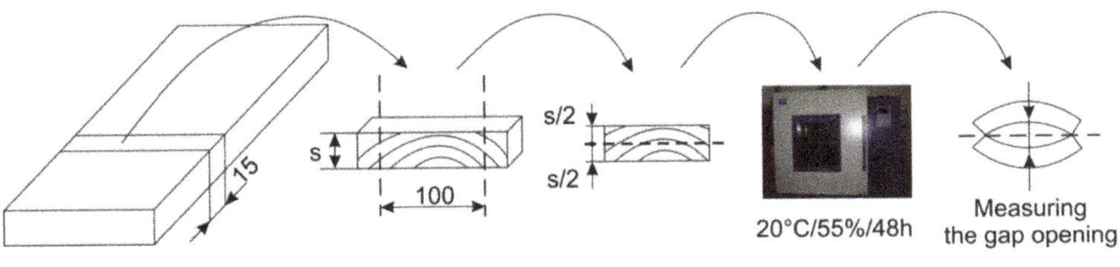

Figure 5. Method for establishing the case-hardening degree after drying.

Through the adopted conditions, this reserach envisaged establishing the following:

- Which drying method is more suitable for Turkey oak lumber (air-drying vs. conventional kiln-drying from a green state);

- If the lumber width has an influence upon the drying quality and if the presence of sapwood and heartwood within the same piece of lumber favor (or not) the fissuring tendency;
- If sealing the lumber ends is an effective method to prevent end cracking in the case of 50 mm thick Turkey oak lumber and also, which sealer provides the better results (paraffin vs. sawdust mixed with PVA adhesive);
- If steaming is an efficient method to reduce the drying rate and/or to improve the drying quality in the case of Turkey oak wood.

Finally, some recommendations for the industry concerning the drying conditions to be applied in the case of 50 mm thick Turkey oak lumber were formulated.

3. Results

3.1. Initial Evaluation of the Lumber Boards

The evaluation of eight aleatory selected boards revealed an average initial moisture content of 55.5 ± 1.13%, an average moisture gradient across the lumber thickness of 12 ± 1%, and the absence of inner stresses (according to the fork samples, which all maintained absolutely straight teeth after cutting) (Figure 6).

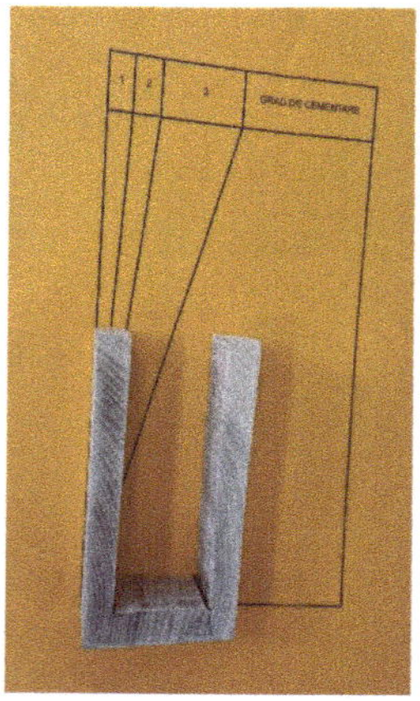

Figure 6. Assessing the initial inner stresses in Turkey oak lumber.

The steamed boards displayed numerous cracks (Figure 7).

Figure 7. Surface cracks after the steaming of Turkey oak boards.

3.2. Air-Drying

The evolution of the moisture content of the monitored boards throughout the air-drying period is presented in Figure 8.

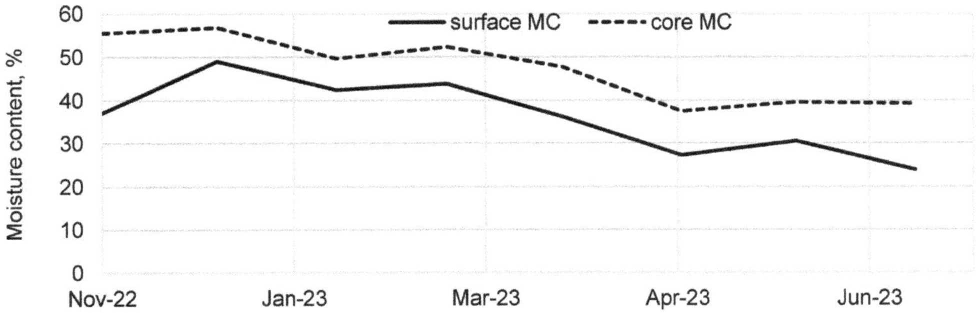

Figure 8. Drying diagram for air-dried Turkey oak lumber.

It can be noticed that after 7 months of exposure, the Turkey oak lumber eliminated only free water, 12% of its moisture content (from an average of 47% to an average of 35%), thus remaining at a moisture content above the fiber saturation point. The highest drying rate (w = 0.13%/day) was attained in June, but the average over the whole period was only 0.06%/day.

The moisture gradient across the lumber thickness in Turkey oak decreased from the initial 18% to 5–6% during the winter months, but it increased again a little toward the end of February, when the relative air humidity dropped from 80–90% to 60–70% (see Figure 3).

According to the obtained results, it can be stated that the air-drying of 50 mm thick Turkey oak lumber allowed the removal of ca. 2% of free water per month from November to April and ca. 4% of free water per month from May to October, in this geographic area (with a continental temperate climate). Under these circumstances, depending also on the initial moisture content, it might take 9–12 months for the 50 mm thick Turkey oak lumber to reach the fiber saturation point.

This result is in good correlation with data provided by an industrial survey [10], which indicated a duration of 9 months for the moisture content reduction from 45% to 35% in the case of the same species. According to the measured data, the air-drying time could be reduced if the lumber was stacked in March. The drying rate was increased during the spring and summer months, but increases in the moisture gradient across the lumber thickness and case-hardening hindered drying during the same period.

The obtained results showed that the air-drying of Turkey oak lumber required more than that of European oak or red oak. According to the reference literature, 50 mm thick European oak (*Quercus robur*) dries in 100 days from 50% to 25% when stacked in February and in 70 days when stacked in May [3], and 4/4 red oak lumber needs 2.5 months to dry from 80% to 21% when stacked in March [5].

The present research did not allow assessing the time needed to air-dry this species below the fiber saturation point.

The results of the final quality control, when all test boards were visually examined, revealed that 37% of the evaluated boards displayed surface and end cracks. The cracks occurred only on the wide boards (with b = 30 cm). None of the narrow boards (b = 15 cm) displayed any type of cracks, neither in the heartwood nor in the sapwood or in the transition area between the sapwood and heartwood. The end-sealing with a paraffin-based sealer was ineffective: 75% of protected pieces displayed end cracks (Figure 9). The case-hardening test confirmed the accumulation of moderate internal stresses through drying) as all evaluated boards recorded a gap opening of 2.5 ± 0.5 mm (medium case-hardening) (Figure 10). Crosscutting the test boards revealed that no internal cracks occurred.

Figure 9. End cracks in air-dried 50 mm thick Turkey oak lumber despite paraffin-based end protection.

Figure 10. Case-hardening sample of air-dried 50 mm thick Turkey oak lumber (gap opening = 2.5 mm).

3.3. Kiln-Drying

The drying diagram recorded for the kiln-dried batch is presented in Figure 11.

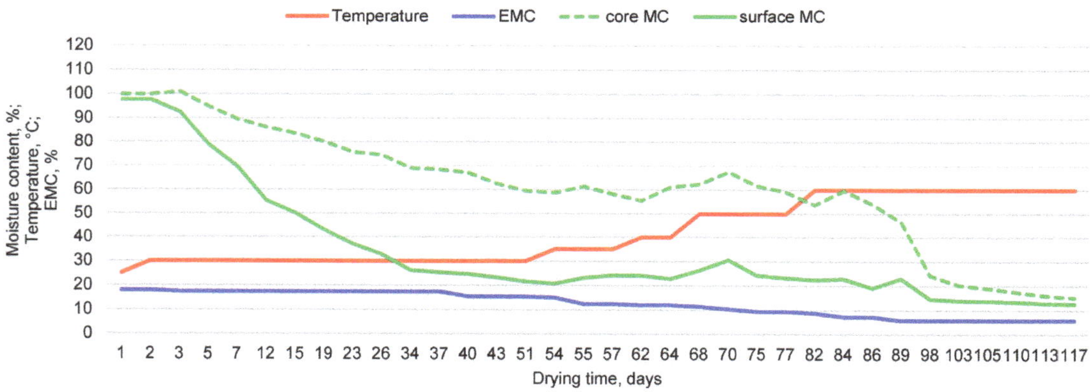

Figure 11. Drying diagram for kiln-dried Turkey oak lumber.

The kiln-drying allowed the drying from an average initial moisture content of 68% (determined by means of the oven-dry method, considering that in this stage the sensors provided no precise reading) to an average of 15% within 117 days, in the absence of forced air circulation. This corresponds to a drying rate of 0.45%/day. The free water was removed within 35 days (which corresponded to a drying rate of 1.3%/day). The average drying rate for the bound water removal (from 24% to 15%) was 0.8%/day. The same as with air-drying, the kiln-drying times were higher than indicated by the reference literature for European oak [3,13,14], but this result could not be considered concludent, because the kiln-drying was performed with no air circulation. Kiln-drying in an industrial dryer was envisaged as up-following stage of research, in order to obtain a concludent result concerning the kiln-drying time of Turkey oak.

The moisture gradient across the lumber thickness increased during the first stage of drying up to 40% until the surface layer reached 24%. At this point, the moisture content in the surface layer remained almost constant for 5 weeks, and the constant drying rate period began, when the core moisture content started decreasing, reducing the moisture gradient across the lumber thickness as well. Toward the end of the process, the difference between the core and surface MC reached 3%, even without conditioning, which was a very good result. This correlated well with the acceptable case-hardening values (<3 mm) that were obtained for all tested pieces.

The final quality control revealed that 88% of the test pieces remained intact and displayed no type of cracks. None of the steamed pieces cracked, and furthermore, the cracks caused by steaming closed. Only one piece of lumber displayed several surface cracks and had internal cracks as well. This was a narrow part containing exclusively heartwood, with unprotected ends. No wide board and no end-protected board displayed cracks.

The positive effect of steaming was also demonstrated by the results of the case-hardening test performed after drying and cooling, which revealed gap openings of 2.2 $\pm$ 0.2 mm for the steamed pieces (Figure 12a) and higher (2.5 $\pm$ 0.5 mm) for the unsteamed ones (Figure 12b).

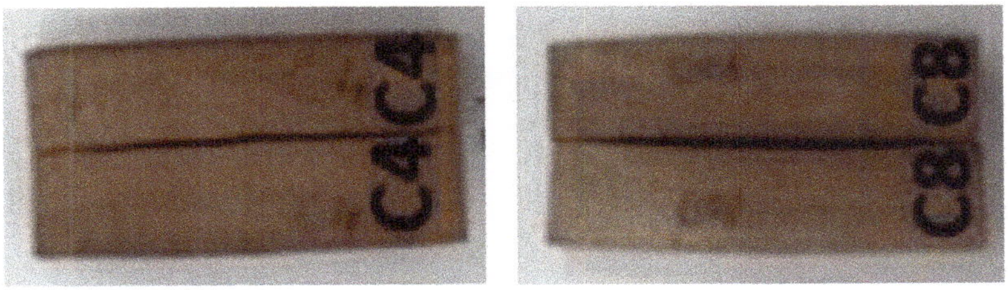

(**a**) (**b**)

Figure 12. Case-hardening samples of kiln-dried 50 mm thick Turkey oak lumber: (**a**) steamed board (gap opening = 2 mm) and (**b**) unsteamed board (gap opening = 3 mm).

4. Conclusions

The drying tests revealed that Turkey oak lumber is indeed diffcult to dry conventionally but not impossible. The following conclusions can be formulated:

1. The drying quality very much depended on the wood quality and the growth defects. The presence of knots, eccentric pith, and inclined grain induced incipient inner stresses that were difficult to handle. Most pieces with such defects cracked and/or warped during drying.
2. Kiln-drying from a green state provided better results than air-drying, in terms of both quality and time, due to the possibility of rigorous air humidity control at high values during the first stage of drying.
3. The most important issue with the 50 mm thick lumber was the moisture gradient across the lumber thickness, which was high (10–18%) from the beginning of the process. Turkey oak is very sensitive to temperature, so as soon as the temperature rises (even under natural conditions, when shifting from winter to spring), if the air humidity is not high enough, the lumber will crack. By rigorously controlling the air humidity, kiln-drying created better conditions to keep the inner stresses under control. Maintaining the temperature at 30 °C and the air humidity at 80–85% until the surface layer (measurement at 8–10 mm depth inside wood) reached 24% helped to avoid cracks in all defect-free boards that were tested.
4. The gradual increase in the temperature up to 60 °C (maximum), by maintaining the drying gradient at 1.5–2.0, allowed good drying quality, with no cracks, of the defect-free boards, at an average drying rate of 0.45%/day.
5. Steaming prior to kiln-drying had a positive effect upon the drying quality. No influence upon the drying time was observed.

6. The lumber width influenced the drying quality: narrow boards (b = 15 cm), even if they contained both sapwood and heartwood, dried better than wide boards (b = 30 cm).
7. End protection by means of paraffin-based sealer during air-drying did not help preventing cracks. Protection with a PVA adhesive + sawdust mixture applied on the kiln-dried boards provided much better results. No protected board cracked.

Author Contributions: Conceptualization, I.-T.D. and M.C.; methodology, I.-T.D. and M.C.; graphics, S.-V.G.; validation, I.-T.D., S.-V.G. and M.C.; formal analysis, M.C.; investigation, I.-T.D.; resources, I.-T.D.; data curation, I.-T.D.; writing—original draft preparation, I.-T.D.; writing—review and editing, I.-T.D.; visualization, I.-T.D.; supervision, M.C. All authors have read and agreed to the published version of the manuscript.

Funding: This research was funded as part of the research programme of the Transilvania University of Brasov.

Institutional Review Board Statement: Not applicable.

Informed Consent Statement: Not applicable.

Data Availability Statement: The data presented in this study are available on request.

Conflicts of Interest: The authors declare no conflict of interest.

References

1. Janik, W. *Handbuch der Holztrocknung*; VEB Fachbuchverlag Leipzig: Leipzig, Germany, 1965.
2. Kollmann, F.F.P.; Cote, W.A. *Principles of Wood Science and Technology. I. Solid Wood*; Springer Verlag Berlin: Berlin/Heidelberg, Germany, 1968.
3. Marinescu, I. *Wood Drying*; Technical Publishing House: Bucharest, Romania, 1979. (In Romanian)
4. Lamb, F.M.; Wengert, E.M. *Techniques and Procedures for the Quality Drying of Oak Lumber*; Department of Wood Science and Forest Product, Virginia Tech.: Blacksburg, VA, USA, 1990.
5. Denig, J.; Wengert, E.M.; Simpson, W.T. *Drying Hardwood Lumber*; USDA Forest Service, Forest Products Laboratory: Madison, WI, USA, 2000.
6. Truebswetter, T. *Holztrocknung: Verfahren zur Trocknung von Schnittholz-Planung von Trocknungsanlagen*, 2nd ed.; Fachbuchverlag Leipzig im Carl Hanser Verlag: Leipzig, Germany, 2006.
7. Richter, H.G.; Dallwitz, M.J. *Commercial Timbers: Descriptions, Illustrations, Identification, and Information Retrieval*; DELTA Database: 2000. Available online: https://delta-intkey.com/wood/en/index.htm (accessed on 15 October 2022).
8. Deaconu, I.; Campean, M. Survey concerning the challenges of industrial drying of Quercus lumber. *Pro Ligno* **2022**, *18*, 36–44.
9. Todaro, L.; Zuccaro, L.; Marra, M.; Basso, B.; Scopa, A. Steaming effects on selected wood properties of Turkey oak by spectral analysis. *Wood Sci. Technol.* **2012**, *46*, 89–100. [CrossRef]
10. Tolvaj, L.; Molnar, S. Color homogenization of hardwood species by steaming. *Acta Silv. Lignaria Hung.* **2006**, *2*, 105–112.
11. Ferrari, S.; Allegretti, O.; Cuccui, I.; Moretti, N.; Marra, M.; Todaro, L. A revaluation of Turkey oak wood (*Quercus cerris* L.) through combined steaming and thermo-vacuum treatments. *BioResources* **2013**, *8*, 5051–5166. [CrossRef]
12. Allegretti, O.; Cividini, R.; Valenti, L.; Negri, M. Vacuum press drying process for beech. Laboratory tests on the influences of drying parameters on quality. In Proceedings of the IUFRO Wood Drying Conference, Brasov, Romania, 24–29 August 2003.
13. Hildebrand, R. *Die Schnittholztrocknung*; Robert Hildebrand Maschinenbau: Oberboihingen, Germany, 1979.
14. Cividini, R. *Conventional Kiln-Drying of Lumber*; NARDI S.p.A.: Chiampo, Italy, 2001.
15. McMillen, J.M. Drying stresses in red oak. *For. Prod. J.* **1955**, *5*, 71–76.
16. Wengert, E.M. *Drying Oak Lumber*; Dept. of Forestry, University of Wisconsin-Madison: Madison, WI, USA, 1990.
17. *ENV 14464:2002*; Sawn Timber. Method for Assessment of Case-Hardening. Available online: https://standards.iteh.ai/catalog/standards/cen/016624f6-9158-4a8e-83bb-3ca54dc4659d/env-14464-2002 (accessed on 22 August 2023).

applied sciences

MDPI

Article

Drying Behaviour of Western Hemlock with Schedules Developed for Norway Spruce and Scots Pine

Ebrahim Hajian [1,*], José Couceiro [1], Lars Hansson [1,2] and Dick Sandberg [1]

[1] Wood Science and Engineering, Department of Engineering Sciences and Mathematics, Luleå University of Technology, Forskargatan 1, 93187 Skellefteå, Sweden; jose.couceiro@ltu.se (J.C.); laha@ntnu.no (L.H.); dick.sandberg@ltu.se (D.S.)

[2] Department of Ocean Operations and Civil Engineering, Faculty of Engineering, Norwegian University of Science and Technology, 6025 Ålesund, Norway

* Correspondence: ebrahim.hajiyan@ltu.se

Abstract: Determining moisture content (MC) distribution during the drying of porous materials such as wood is crucial for developing drying schedules and assessing their suitability to achieve optimised processes. This study aimed to determine the causes of the unique drying behaviour and the well-known unusual longer drying time of western hemlock compared to other similar softwoods. In situ X-ray computed tomography (CT) was used to study the evolution of MC in timber during the drying process. The drying behaviour of western hemlock (*Tsuga heterophylla* (Raf.) Sarg.) was compared with Norway spruce (*Picea abies* (L.) Karst.) and Scots pine (*Pinus sylvestris* L.) from green to oven-dried condition with industry-proposed drying schedules used for steering a custom-made experimental kiln combined with a CT scanner. CT scanning was performed at 30 min intervals during the complete drying period of 30 h, and the CT images were processed to calculate the MC evolution within the specimen. Western hemlock showed a considerably slower capillary-phase drying and did not go into the transition and diffusion phases when a schedule adapted to pine and spruce drying was applied for its drying. CT images and MC gradient calculations showed a lower drying rate and severe non-uniformity in MC distribution, which could be due to the effect of higher green MC and the presence of wet pockets. Furthermore, the evaporation front at the first 5 h of drying receded faster into the hemlock specimen, and as drying proceeded, it slowed down compared to other specimens.

Keywords: CT scan; experimental research; image processing; moisture content; porous material; timber drying; wet pockets

Citation: Hajian, E.; Couceiro, J.; Hansson, L.; Sandberg, D. Drying Behaviour of Western Hemlock with Schedules Developed for Norway Spruce and Scots Pine. *Appl. Sci.* **2023**, *13*, 11083. https://doi.org/10.3390/app131911083

Academic Editors: Lidia Gurau, Mihaela Câmpean and Emilia-Adela Salca

Received: 17 September 2023
Revised: 7 October 2023
Accepted: 7 October 2023
Published: 9 October 2023

1. Introduction

The reduction in the moisture content (MC) of green sawn timber is inevitable due to the hygroscopic nature of wood, which will absorb or desorb water depending on the surrounding conditions. It is also crucial in order to reach the required mechanical properties of wooden products, lower weight, etc. [1,2]. Desorption or adsorption of water in wood under a certain MC level, known as the fibre saturation point (FSP), is associated with shrinkage and swelling, respectively, which causes stresses in the wood material that may result in distortion and cracking of the sawn timber. Drying of sawn timber must thus be performed in a controlled way to reduce the undesired consequences of water removal. Kiln drying is the most common industrial method to dry softwood timbers, and it is performed on large volumes of green sawn timber by blowing air through the timber packages placed in the kiln. The timber is stacked with a defined distance between each layer of timber, and the air condition in the kiln is regulated by adjusting temperature and relative humidity (RH) over time. Psychrometric measurements are achieved for monitoring the process, which is set through the so-called drying schedule, which regulates these parameters with the aim of reaching a final target MC of the timber without causing distortion, cracking, or other quality lowering changes of the sawn timber.

Green wood contains liquid water in the cell lumen (free water) and water chemically bonded to the cell wall (bound water). At the initial stages of the drying process, the so-called capillary regime, liquid water is transported from the inner parts of the wood to the surface, where it evaporates. As the moisture level is lowered and the FSP is reached, the moisture transport transitions into being diffusion-driven when the bound water is removed [3].

The anatomical structure of a tree can be divided into the sapwood, which is the outermost part that transports water from the roots to the leaves, and into heartwood, which is the central part of the stem that does not transport free water. Nevertheless, the structure and properties of wood vary largely between species and within a single tree, making it complicated to extrapolate drying methods, as different woods may absorb or desorb water differently before reaching the equilibrium moisture content (EMC). Western hemlock, growing in coastal regions of North America [4], and Norway spruce and Scots pine, growing in Scandinavia, are common softwood species used in sawmilling, with very extended use in timber engineering. However, western hemlock has been considered more difficult to dry than the Scandinavian-grown Norway spruce and Scots pine, and the reasons for that have been suggested to be a higher initial MC in green conditions, a higher extractive content, and a lower permeability [5,6]. A motive for this study is that when drying technologies from Scandinavia were transferred to North America, the drying times for hemlock were reported to be approximately twice those for spruce and pine. This unexpected behaviour was surprising, given that these three species are softwoods with a close range of density values. Literature records show that hemlock is prone to a phenomenon known as wet pockets (also known as wet wood or sinker wood), formed mainly due to bacterial infection in the living tree, and occurs mostly in the heartwood and in the transition zone between heartwood and sapwood [5,7]. Wet pockets manifest during drying as localised regions within the timber with significantly higher MC, often above FSP; meanwhile, other regions are well below it. This could be one of the causes for the drying of hemlock to be considerably more time-consuming than spruce and pine.

In practice, wet pockets are assumed to be the reason for increased drying time, higher energy consumption, and non-uniform final MC of the timber, which may cause quality-reducing defects in drying. Kozlic et al. [8] found extractive subtraction by dissolving in chemical composition to be more effective in achieving uniformed final MC than raising the temperature. Ward et al. [9] confirmed Kozlic's results and observed a reduced presence of wet pockets in young western hemlock trees and proposed sorting the timber prior to drying according to MC, preferably determined by using electrical resistance MC meters. Modern tools such as machine learning and stochastic models are applied prior to drying to sort western hemlock timbers based on the initial MC with the goal of achieving homogeneity in the final MC [10–12]. Implementing post-drying equalisation of the MC may be effective in preventing non-uniformity in the MC levels by decreasing dry bulb temperature, which leads to reducing over-drying, while it prolongs drying time [13]. Berberović et al. [14] found an inverse correlation between basic density and heartwood percentage of hemlock on drying rate, while initial MC was positively correlated. Chaffe et al. [15] also reported pit aspiration causing a reduction in diffusion coefficient as a possible reason for the slow drying rate.

X-ray computed tomography (X-ray CT) is one of the most accurate non-destructive methods utilised in industry for measuring the inner properties of logs. X-ray CT detects the density of a material, which can be used to evaluate density distribution, and on a lab-scale, can also be used to determine the MC distribution in the entire timber volume. Different methods for determining MC precisely from CT data have been investigated [16]. Analysing CT data would also lead to enhancing our comprehension of the effect of important parameters in drying behaviour, such as temperature and density, by improving the numerical modelling of moisture gradients and flow speed, especially in transition, diffusion, and conditioning regimes [17].

The objective of this study was to determine the causes of the well-known longer drying time and drying behaviour (wet pockets) for western hemlock compared to Norway spruce and Scots pine using X-ray computed tomography (CT), a technology that allows for internal imaging of the wood structure due to density variations, combined with timber drying in a custom-made experimental drying kiln. Developing the knowledge of the dominant wood-drying regime by utilising CT image processing can lead to providing an optimised drying schedule that minimises energy consumption. The drying behaviour of heartwood and sapwood of the three species was studied for pine and spruce industrial drying schedules. Furthermore, a comparative assessment was conducted to determine if high green MC and wet pockets of western hemlock were visible and if they affected the drying rate and regimes.

2. Materials and Methods

A total of three specimens of western hemlock, two specimens of Norway spruce, and one specimen of Scots pine with the size of 500 × 80 × 36 mm were selected based on their sapwood and heartwood content, aiming to obtain specimens with almost more than 90% volume of the content of either heartwood or sapwood. During the process of sampling, storage in the freezer, and handling, there were measures taken to prevent any undesired loss of moisture prior to the experiments. The wood specimens were wrapped in plastic and stored in a freezer. The cross-sectional ends of the specimens were covered in heat-resistant silicone to prevent drying in the longitudinal direction during the experiments, thereby avoiding the "end-effects". As indicated in Figure 1, the green specimens were thawed in water just before the drying experiment started to maintain surface MC during the start-up of the dryer.

Figure 1. Specimens in the drying chamber. The white layer on the surfaces is ice due to the water thawing.

2.1. Gravimetric MC Measurements

The green MC of all the specimens was determined with the gravimetric method [18]:

$$MC_G = \frac{MC_G - MC_{OD}}{MC_{OD}} \quad (1)$$

where M_G and M_{OD} are the green and the oven-dried weight, i.e., the weight at 0% MC, respectively. These values are tabulated in Table 1, and they are in the range of values mentioned in the Forest Products Laboratory's Wood Handbook [18].

Table 1. Specimens' green moisture content at the start of the drying experiments (MC_0).

Drying Run	Specimens	MC_0 (%)
I	Pine sapwood	107
	Hemlock sapwood	176
II	Spruce heartwood	35
	Hemlock heartwood	51
III	Spruce sapwood	95
	Hemlock sapwood	146

2.2. Kiln Drying and CT Scanning

This study was carried out by using a custom-made drying kiln coupled with a medical CT scanner installed at Luleå University of Technology (LTU) in Skellefteå, as shown in Figure 2.

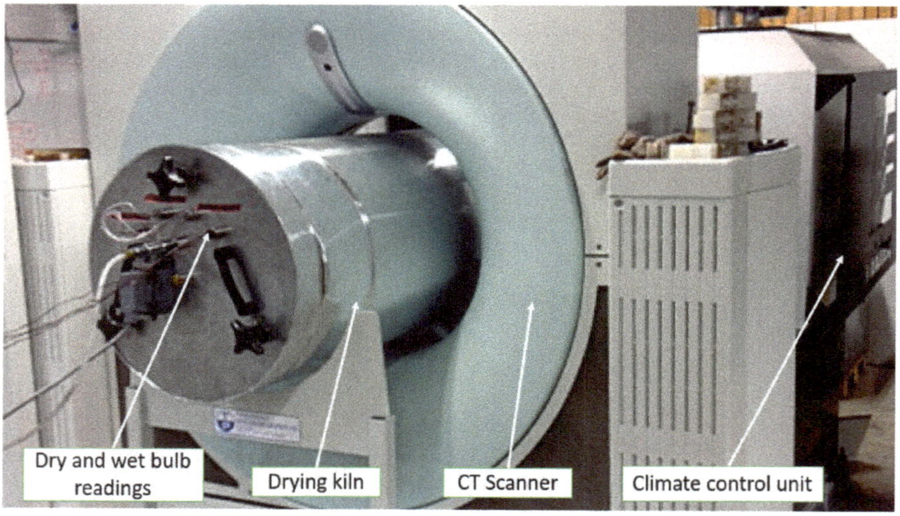

Figure 2. Kiln drying and simultaneously CT scanning at Luleå University of Technology.

The drying kiln is made of aluminium to allow for X-rays passing the envelope. The volume of the tube is approx. 0.3 m^3. The heating coils have a capacity of 3 kW; the airflow is provided by a 5 kW electric motor-driven fan that can be adjusted by variable frequency drive between 1 to 8 m/s. The temperature range of the kiln is from room temperature to 220 °C, and dry-bulb and wet-bulb temperature readings monitor the relative humidity. Moist air is extracted from the kiln through venting. The software controlling the drying parameters is Valmatics 4.0 [Valutec AB, Skellefteå, Sweden].

A medical Siemens Somatom Emotion CT scanner [Siemens, Munich, Germany] was used in combination with the drying kiln. Two-dimensional CT images of the timber cross section were taken at 0.5 h time intervals with a pixel resolution of 1.7 pixels per mm.

2.3. Drying Schedule

Figure 3 presents the drying schedules implemented in the industry (Valutec AB) for spruce and pine and applied in this experiment to reach a target MC of 18% in all drying runs.

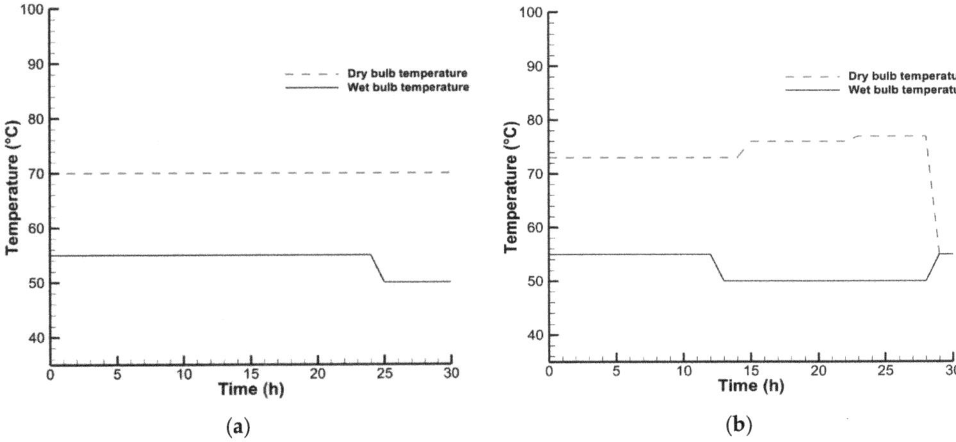

Figure 3. Drying schedules used in the experiments for drying runs: (**a**) hemlock/pine sapwood (I), and (**b**) hemlock/spruce heartwood and hemlock/spruce sapwood (II and III).

2.4. MC Calculations from the CT Images

The CT images of specimen cross sections were converted into density images by calibrating them with water and air phantoms included in the scanning zone, and the *MC* was calculated in a Python algorithm for the advanced image processing, which transformed the shrinkage deformation/distortion of the specimens due to shrinkage occurring below FSP to the shape in green state (registration transformation) [19,20]. This transformation is necessary for the comparison of a specific region within a specimen at two different MCs, where the drying has resulted in a *MC* change that, in turn, has caused shrinkage and distortion of the specimen. These calculated *MC* values from the image processing of *CT* data are compared with the *MC* measured according to the gravimetric method:

$$Error = \frac{MC_{OD} - MC_{CT}}{MC_{OD}} \tag{2}$$

where MC_{OD} is the gravimetrically obtained *MC*, and MC_{CT} is the *MC* obtained through the processing of the *CT* images.

2.5. Drying Rate

The drying rate is here defined as the proportion of evaporable water during each time step of the drying process [8]:

$$E = \frac{MC - EMC_1}{MC_0 - EMC_1} \tag{3}$$

where *E* is the fraction of evaporable water left in the wood; *MC* is the current moisture content of the specimen at any given time t during the drying; EMC_1 and MC_0 are the equilibrium moisture content under the specific drying conditions (climate) and the moisture content at the beginning of the drying of the wood, respectively.

2.6. Moisture Gradient

The *MC* gradient is a parameter capable of providing a more comprehensive explanation of the influence of wet pockets on drying behaviour, specifically within the heartwood of hemlock and spruce specimens. For this aim, the Esping method [21] is implemented by dividing specimens' cross sections into sections, which calculate the MC gradient (MC_{grad}) by taking the maximum and average MC of each section, and can be formulated as follows:

$$MC_{grad} = \frac{MC_{\max} - MC_{avg}}{MC_{avg}} \quad (4)$$

where $MC_{\max}$ and MC_{avg} state the maximum and average values of MC in each section of the specimen.

3. Results and Discussions

3.1. Average MC Loss

The average MC loss during the drying runs I, II, and III is shown in Figure 4. As expected, Scots pine sapwood dries faster than hemlock to the target MC (Figure 4a), while for hemlock, it did not reach even FSP, which was reported to be around 17% [22]. It can be observed that below the FSP, the steepness of the MC curve for pine is reduced, showing the lower drying rate of the diffusion regime. Figure 4b indicates that the slope of the MC loss curves for hemlock and spruce heartwood is similar, while spruce heartwood reached the target MC faster than the hemlock heartwood. The specimen of hemlock sapwood cracked significantly during the drying test, which meant that the moisture content could not be calculated reliably based on the CT measurements. Therefore, this specimen has been excluded from the results analysis, as can be seen in Figure 4c. There is, however, an evident difference in drying behaviour between spruce heartwood and spruce sapwood, where the sapwood drying occurred initially under the effect of heat transfer with the evaporation of free liquid water, while in heartwood, most of the time, it is a diffusion regime where there is no free water (Figure 4b,c).

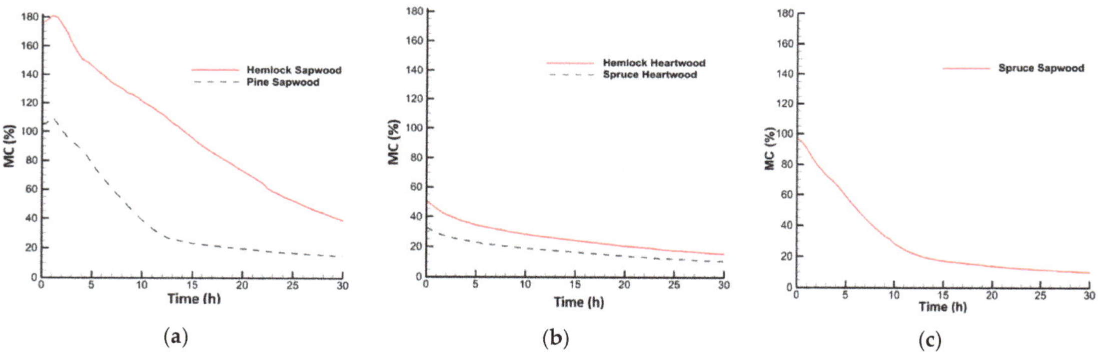

Figure 4. Average MC distribution for (**a**) hemlock/pine sapwood, (**b**) hemlock/spruce heartwood, and (**c**) spruce sapwood.

Table 2 shows the comparison of green MC measurements obtained through the gravimetric method and CT method for the three drying runs according to Equation (2), which shows the differences between the two methods. Error values are below 4.5% between the gravimetrically measured MC calculation and the average MC from CT image processing. As already mentioned, substantial cracks appeared during the drying of hemlock sapwood in drying run III, which makes it not feasible to analyse the MC changes with good precision, and CT images could not be processed for bringing into the discussions.

Table 2. The errors of the MC were obtained from the CT-processed images as a percentage of the gravimetrically measured MC.

Drying Run	Specimens	Error (%)
I	Pine sapwood	2.03
	Hemlock sapwood	0.24
II	Spruce heartwood	4.53
	Hemlock heartwood	1.51
III	Spruce sapwood	2.71

3.2. MC Distribution

Figure 5 shows the pixel-wise MC calculated from the density-based CT images for three drying runs for each 5 h of drying up to 30 h, and as explained, the last drying run only included spruce sapwood.

In all drying runs, pine and spruce reached the desired target MC and capillary, transition, and diffusion regimes occurred. The moisture reduction starts from the sawn-timber surfaces, creating a drier "shell". Liquid water migrates from the inner regions of the specimens toward their surface (the capillary phase). The water undergoes rapid evaporation on the specimen's surface, faster than new water is added to this region, creating the "shell". The dominant heat-transfer regime is convection through air, which is needed to transport the evaporated water away from the specimen. As drying continued and liquid free water reduced, the diffusion phase with conduction-dominated heat transfer was governed, which, because of the high initial MC of hemlock sapwood, after 30 h liquid water observed in the images, proved that it was still in the capillary phase.

Figure 5a shows that once the surface has dried out in both pine and hemlock specimens, the evaporation front recedes deeper towards the centre of the timber, resulting in the formation of a dry layer on the surface and a still moist core. This recession is faster for pine compared with hemlock except for the first 5 h of the drying process, so it can be inferred that two different drying behaviours occurred in hemlock compared to pine. When the moisture content moved from above FSP to below FSP for pine, the drying process was slowed down. This can be observed in pine specimens for around 10 h and 15 h, which showed free water discontinuity in the pine and implies a transition to diffusion known as irreducible saturation, where shrinkage begins on the outer region of the timber. However, the hemlock specimen remained at the capillary regime. As can be seen, it might be said that the MC distribution and changes in hemlock sapwood are uniform during the drying, and there is no abnormal local high MC, which can be inferred as the absence of wet pockets. At the same time, a high initial MC impacts on prolonging the drying time.

Figure 5b indicated MC loss of spruce and hemlock heartwood, where spruce reaches the desired target MC while hemlock has abnormally high MC regions (wet pockets), which makes it have high MC resistance to drying in some regions, even when other regions are below FSP or target MC, as shown in 20 h of drying. This specimen faces overdrying and high local MC issues, which enhance drying stress and defects. Permeability is a critical factor in wood drying, especially during the initial stages when water migrates to the surface of the wood through capillary flow. As the drying process progresses, the permeability diminishes due to pit aspiration [23]. Apart from the inherent higher initial MC and wet pockets, lower permeability in hemlock sapwood and heartwood compared with spruce in all directions could be another reason for having prolonged drying time [24,25]. The behaviour of the evaporation front is not like the first run, and it can be said that in all drying times, it recedes slower into hemlock heartwood compared to spruce heartwood.

Figure 5c shows that the MC change for the spruce sapwood specimen started from high green MC, which achieved the desired target MC uniformly with the proposed drying schedule. Compared with drying spruce heartwood specimens, it can be seen that free water in sapwood is the reason for having a more extended capillary regime than heartwood species, which mostly is bound water diffusion-controlled.

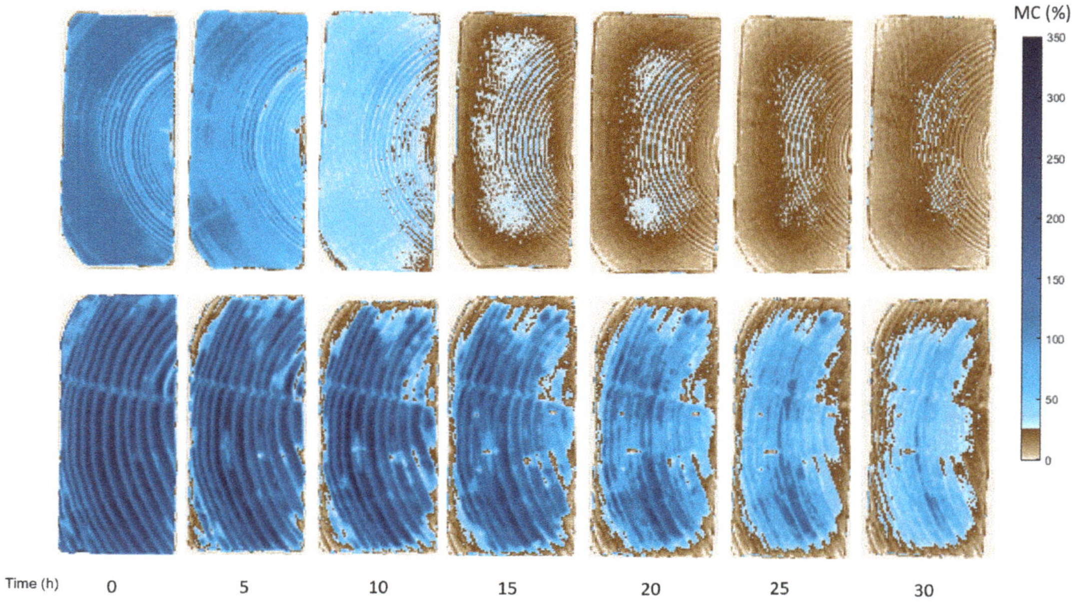

(**a**)

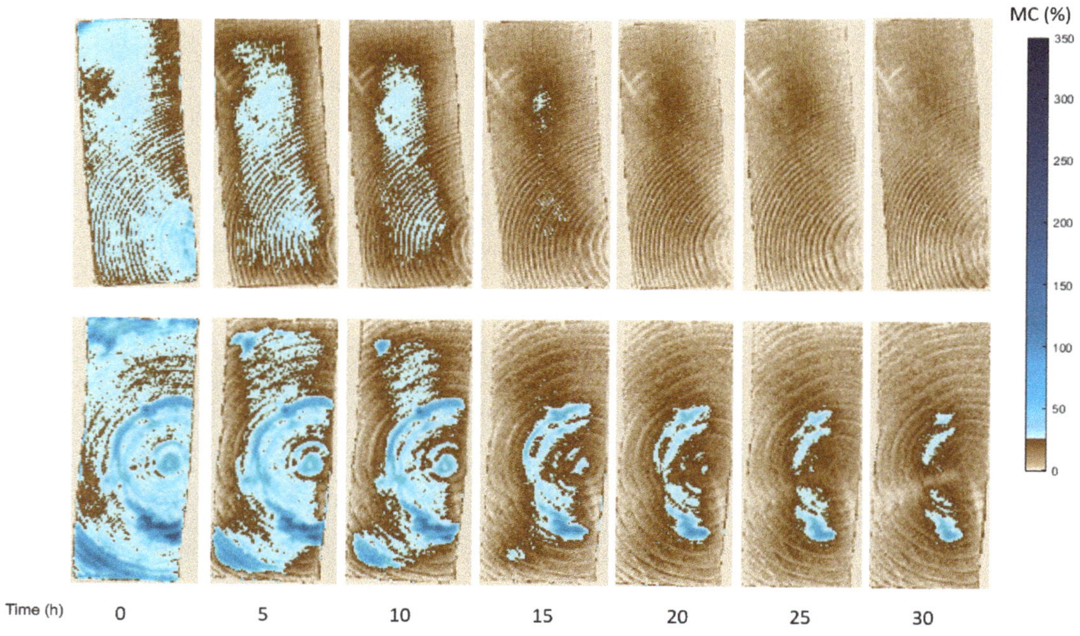

(**b**)

Figure 5. *Cont.*

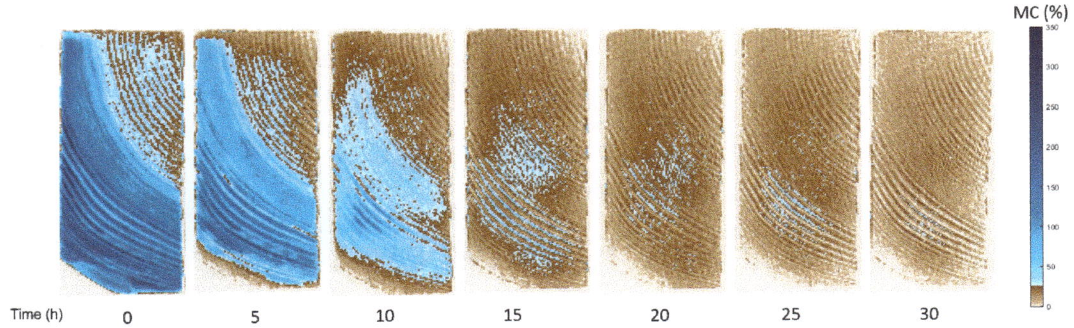

(c)

Figure 5. CT images of MC change during drying runs of (**a**) pine (**above**)/hemlock (**below**) sapwood, (**b**) spruce (**above**)/hemlock (**below**) heartwood, and (**c**) spruce sapwood.

3.3. Drying Rate

To obtain a more comprehensive assessment of the differences in drying behaviour among the specimens, calculating the drying rate (E) according to Equation (3) can provide a more insightful perspective.

It can be observed in Figure 6a that during the initial hour of drying, there is a minor rise in E. This increase can be because the wood absorbs moisture from the surrounding environment. At the beginning of the drying process, the wet-bulb temperature was lower than the temperature on the timber surface, but after reaching equilibrium, it began to decrease, and the evaporation front started to recede into the wood [26]. In the first drying run, approximately 1.5 h was required for beginning the MC decrement and warming up the equipment; however, for the second drying run (Figure 6b), this time was not needed since the kiln was already settled and warmed up from the first run. Since the last hour of the drying schedule of spruce heartwood was assigned for conditioning and a higher value of EMC, to have a unified MC distribution in the specimens, drying did not occur, and E went to zero, which does not imply that there was no evaporable water left in the specimens. Figure 6a shows that as the drying process progressed, the residual evaporable water within the specimens was close. After approximately 5 h, the difference in E value between the specimens increased dramatically. Pine sapwood showed faster water evaporation than hemlock, which also corresponds with what was observed in CT images where in the first 5 h of drying, the evaporation front receded faster into the hemlock sapwood. Afterwards, it slowed down compared with pine. This phenomenon can be explained by the lower permeability and porosity that obstructed void spaces within the hemlock and restricted the movement of liquid water towards the surface. This behaviour repeated for the second run, and the value of E was lower for spruce heartwood in all drying times, while there is not much difference in the steepness of the graphs, and CT images showed regions with abnormally high MC for hemlock that the E value can't represent, and MC gradient investigation is needed.

3.4. MC Gradient

As demonstrated in Figure 7, in the CT measurements, each section of the specimen covers three pixels with the size of 0.5892 mm in the x direction and the whole of the y direction.

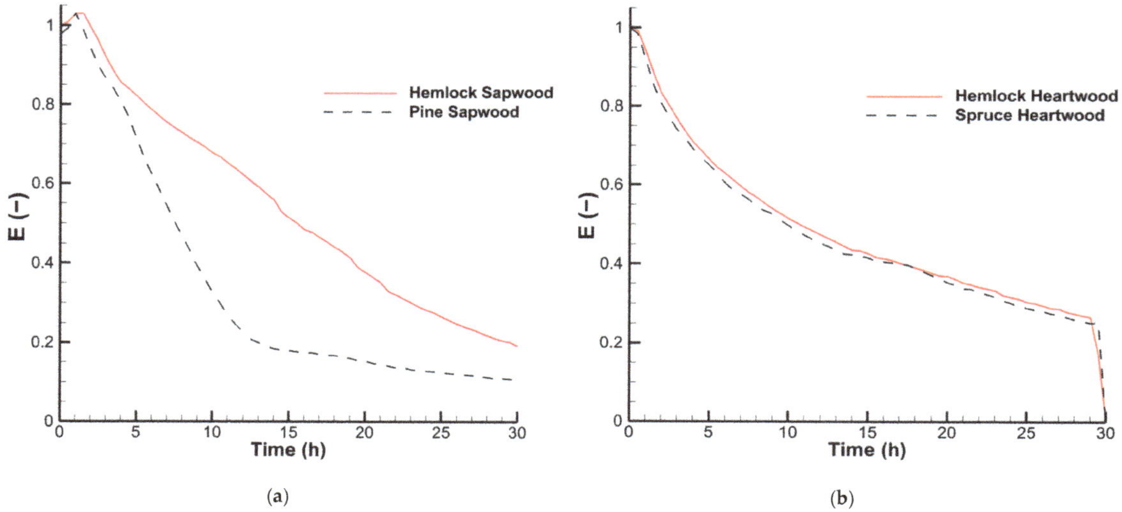

Figure 6. Evaporable water left in the specimens, i.e., drying rate (E): (**a**) hemlock/pine sapwood, and (**b**) hemlock/spruce heartwood.

Figure 7. Sections for the calculation of MC gradients.

By evaluating E trends, since the curve steepness between the species is close, the MC gradient approach provides a more comprehensive elucidation regarding the presence of wet pockets, which most likely occurred in the heartwood. Figure 8a demonstrates the MC gradients calculated by Equation (4) for spruce and hemlock heartwood in four drying times and demonstrates that values of MC gradients in most of the drying times and sections are more intense in hemlock than the spruce, up to even six times higher in some sections. By comparing the MC gradient of spruce (Figure 8a), it can be inferred that as the drying process progressed, MC distribution tended towards greater uniformity, and 5 h MC gradient in most of the sections was exceeded over the 20 h when the specimens were dried. This behaviour cannot be observed in the hemlock specimen (Figure 8b) since, in distinct regions, the transition occurred from the capillary into diffusion-dominated regimes, while the rest is still in a capillary-controlled state, and the MC gradient becomes more intense in 20 h even compared with 5 h, which showed unusual resistance to drying that may indicate wet pocket presence.

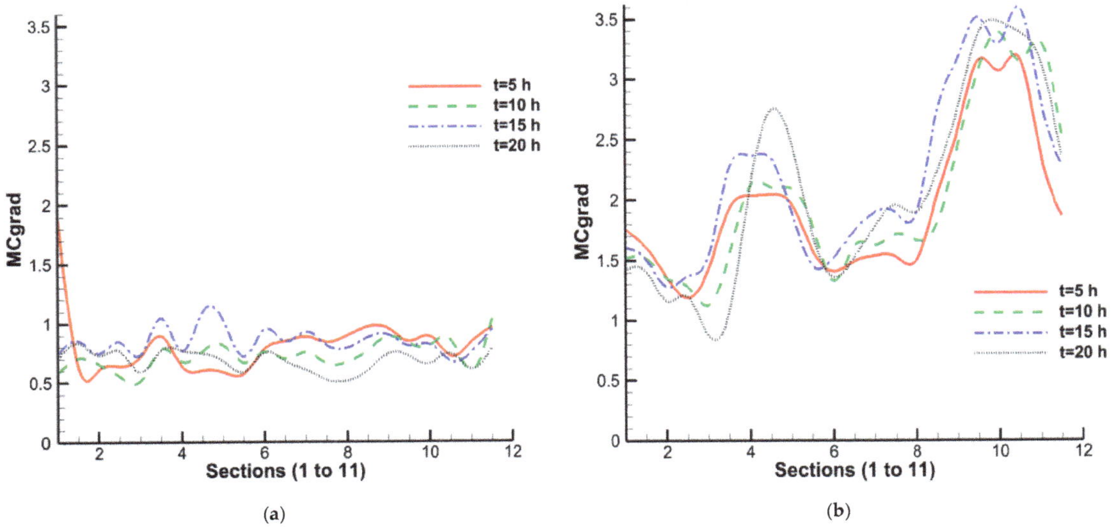

Figure 8. MC gradients in cross section of (**a**) spruce heartwood and (**b**) hemlock heartwood.

4. Conclusions

The purpose of this study was to determine the causes of the unique drying behaviour and the well-known longer drying time of western hemlock compared to Norway spruce and Scots pine. Due to a higher initial moisture content and the presence of wet pockets, particularly in the heartwood, hemlock fails to achieve the target moisture content when subjected to the same drying schedule as Norway spruce and Scots pine. Different drying behaviours were observed for hemlock sapwood, i.e., the evaporation front receded faster into the specimen compared with pine sapwood in the first 5 h, and after that, this receding slowed down. Analysing MC gradients of hemlock heartwood indicated regions with abnormally high MC, known as wet pockets. Average MC did not provide comprehensive information on drying behaviour and necessitated the utilisation of moisture content gradient analysis.

A drying schedule adapted for timber prone to create wet pockets may result in over-drying for species not prone to the presence of wet pockets, if mixed timber species are dried together in the same kiln. It is also recommended to sort hemlock timbers before drying into high-content sapwood and high-content heartwood, respectively, and adapt the drying schedule to the type of wood. This approach aims to mitigate the challenge of non-uniformity in MC. Additionally, for heartwood stacks, suitable conditioning measures need to be implemented.

Further research can be implemented by providing a drying schedule suited for hemlock heartwood and sapwood on a lab scale and applying the related schedules with MC gradient measurements on an industrial scale, which was the limitation of this study due to the lack of hemlock specimens.

Author Contributions: Conceptualisation, E.H. and J.C.; methodology, E.H., J.C., L.H. and D.S.; software, L.H. and E.H.; validation, E.H., J.C., L.H. and D.S.; formal analysis, E.H.; investigation, E.H.; writing—original draft preparation, E.H.; writing—review and editing, E.H., J.C., L.H. and D.S.; visualisation, E.H.; supervision, D.S. All authors have read and agreed to the published version of the manuscript.

Funding: This research received no external funding.

Institutional Review Board Statement: Not applicable.

Informed Consent Statement: Not applicable.

Data Availability Statement: Not applicable.

Conflicts of Interest: The authors declare no conflict of interest.

References

1. Sehlstedt-Persson, M. Impact of Drying and Heat Treatment on Physical Properties and Durability of Solid Wood. Ph.D. Thesis, Luleå University of Technology, Wood Science and Engineering, Skellefteå, Sweden, 2008.
2. Yin, Q.; Liu, H.-H. Drying stress and strain of wood: A review. *Appl. Sci.* **2021**, *11*, 5023. [CrossRef]
3. Niemz, P.; Teischinger, A.; Sandberg, D. *Springer Handbook of Wood Science and Technology*; Springer Nature: Cham, Switzerland, 2023.
4. Jozsa, L.; Munro, B.; Gordon, J.R. *Basic Wood Properties of Second-Growth Western Hemlock*; Special Publication SP-38; Forintek Canada Corp.: Fredericton, NB, Canada, 1998.
5. Shahverdi, M.; Oliveira, L.; Avramidis, S. Kiln-drying optimisation for quality pacific coast hemlock timber. *Dry. Technol.* **2017**, *35*, 1867–1873. [CrossRef]
6. Moya, R.; Muñoz, F.; Jeremic, D.; Berrocal, A. Visual identification, physical properties, ash composition, and water diffusion of wetwood in Gmelina arborea. *Can. J. For. Res.* **2009**, *39*, 537–545. [CrossRef]
7. Ward, J.C.; Pong, W.Y. *Wetwood in Trees: A Timber Resource Problem*; General Technical Report PNW-112; US Department of Agriculture, Forest Service, Pacific Northwest Forest and Range Experiment Station: Portland, OR, USA, 1980; p. 60.
8. Kozlik, C.J. Problems of drying western hemlock heartwood to a uniform final moisture content. In *Proceedings of the 21st Annual Meeting of Western Dry Kiln Clubs*; Washington-Idaho-Montana Seasoning Club: Missoula, MT, USA, 1970; pp. 55–61.
9. Ward, J.C.; Kozlik, C.J. Kiln drying sinker heartwood from young-growth western hemlock: Preliminary evaluation. In *Proceedings of the 26th Annual Meeting of Western Dry Kiln Clubs*; Oregon State University: Corvallis, OR, USA, 1975; pp. 44–63.
10. Rahimi, S.; Avramidis, S.; Lazarescu, C. Estimating moisture content variation in kiln dried Pacific coast hemlock. *Holzforschung* **2022**, *76*, 26–36. [CrossRef]
11. Rahimi, S.; Nasir, V.; Avramidis, S.; Sassani, F. Benchmarking moisture prediction in kiln-dried Pacific Coast hemlock wood. *Int. Wood Prod. J.* **2022**, *13*, 219–226. [CrossRef]
12. Elustondo, D.; Avramidis, S. Stochastic numerical model for conventional kiln drying of timbers. *J. Wood Sci.* **2003**, *49*, 485–491. [CrossRef]
13. Rohrbach, K.; Oliveira, L.; Avramidis, S. Drying schedule structure and subsequent post-drying equalisation effect on hemlock timber quality. *Int. Wood Prod. J.* **2014**, *5*, 55–64. [CrossRef]
14. Berberović, A.; Milota, M.R. Impact of wood variability on the drying rate at different moisture content levels. *For. Prod. J.* **2011**, *61*, 435–442. [CrossRef]
15. Chafe, S.A. Relationship between equilibrium moisture content and specific gravity in wood. *J. Inst. Wood Sci.* **1991**, *12*, 119–122.
16. Hansson, L.; Couceiro, J.; Fjellner, B.-A. Estimation of shrinkage coefficients in radial and tangential directions from CT images. *Wood Mater. Sci. Eng.* **2017**, *12*, 251–256. [CrossRef]
17. Florisson, S.; Hansson, L.; Couceiro, J.; Sandberg, D. Macroscopic X-ray computed tomography aided numerical modelling of moisture flow in sawn timber. *Eur. J. Wood Wood Prod.* **2022**, *80*, 1351–1365. [CrossRef]
18. Ross, R.J. *Wood Handbook: Wood as an Engineering Material*; General Technical Report FPL-GTR-282; Department of Agriculture, Forest Service: Madison, WI, USA, 2021.
19. Hansson, L.; Cherepanova, E. Determination of wood moisture properties using a CT-scanner in a controlled low-temperature environment. *Wood Mater. Sci. Eng.* **2012**, *7*, 87–92. [CrossRef]
20. Poupet, B.; Couceiro, J.; Florisson, S.; Hansson, L.; Sandberg, D. Estimation of moisture distribution in sawn timber using computed tomography. In Proceedings of the WCTE 2023—2023 World Conference on Timber Engineering, Oslo, Norway, 19–22 June 2023; 7p.
21. Esping, B. *Trätorkning 2: Torkningsfel-åtgärder. [Wood Drying 2: Drying Failure and Actions]*; Swedish Institute of Wood Technology Research (Trätek): Stockholm, Sweden, 1988. (In Swedish)
22. Koumoutsakos, A.; Avramidis, S. Mass transfer characteristics of western hemlock and western red cedar. *Holzforschung* **2002**, *56*, 185–190. [CrossRef]
23. Hansmann, C.; Wimmer, R.; Teischinger, A. Permeability of wood—A review. *Drev. Vysk.* **2002**, *47*, 1–16.
24. Rayirath, P.; Avramidis, S. Some aspects of western hemlock air permeability. Maderas. *Cienc. Y Tecnol.* **2008**, *10*, 185–193.
25. Moore, J. *Wood Properties and Uses of Sitka Spruce in Britain*; Research Report; Forestry Commission: Edinburgh, Scotland, 2011; 60p.
26. Pang, S.; Wiberg, P. Model predicted and CT scanned moisture distribution in a *Pinus radiata* board during drying. *Holz Als Roh-Und Werkst.* **1998**, *56*, 9–14. [CrossRef]

applied sciences

MDPI

Article

Chemical Composition and Color of Short-Rotation Teak Wood Thermally Modified in Closed and Open Systems

Felipe da Silva Gomes [1], Miquéias de Souza Reis [1], Anna Clara Oliveira Rupf [1], Paulo Henrique dos Santos Silvares [1], Jessica Sabrina da Silva Ferreira [1], Luciano da Costa Dias [1], João Paulo Sardo Madi [2], Pedro Henrique Gonzalez de Cademartori [2], Humberto Fantuzzi Neto [1], Juarez Benigno Paes [1] and Djeison Cesar Batista [1,*]

[1] Department of Forest and Wood Sciences, Federal University of Espírito Santo, Jerônimo Monteiro 29550-000, Brazil
[2] Graduate Program in Forest Engineering, Federal University of Paraná, Curitiba 80210-170, Brazil
* Correspondence: djeison.batista@ufes.br

Abstract: Although the effect of thermal modification (TM) on teak wood color is well documented, few studies have been carried out on closed-system processes, and it remains unclear what the effect is of different processes on the same material. This work aimed to verify the effect of closed- and open-system processes of TM on the color of short-rotation teak wood. Thermally modified wood (TMW) was evaluated in a closed system at 160 °C (CS160) and in an open system at 185 °C and 210 °C (OS185 and OS210). We measured the moisture content (initial and final) of the wood and the corrected mass loss (CML). The chemical analyses encompassed the contents of alpha-cellulose, hemicelluloses, lignin, and extractives (total, in acetone and dichloromethane). Wood color was measured before and after TM according to the CIEL*a*b* color space. It was possible to achieve the same color using different processes of thermal modification (CS160 and OS210). TM reduced wood lightness (L*), red–green chromaticity coordinate (a*), and yellow–blue chromaticity coordinate (b*). L* and a* had the biggest and smallest variations, respectively. TMW color was significantly changed, even at the mildest condition tested (OS185, 0.33% CML).

Keywords: *Tectona grandis*; thermal modification; chemical analyses; CIELab

Citation: Gomes, F.d.S.; Reis, M.d.S.; Rupf, A.C.O.; Silvares, P.H.d.S.; Ferreira, J.S.d.S.; Dias, L.d.C.; Madi, J.P.S.; Cademartori, P.H.G.d.; Neto, H.F.; Paes, J.B.; et al. Chemical Composition and Color of Short-Rotation Teak Wood Thermally Modified in Closed and Open Systems. *Appl. Sci.* **2023**, *13*, 13050. https://doi.org/10.3390/app132413050

Academic Editors: Emilia-Adela Salca, Lidia Gurau and Mihaela Câmpean

Received: 22 September 2023
Revised: 10 October 2023
Accepted: 11 October 2023
Published: 7 December 2023

1. Introduction

Thermal modification (TM) is a suitable industrial process for improving the properties of wood from planted forests, which is composed basically of lower-quality juvenile wood [1]. The main industrial processes of wood modification are in operation in Europe, where thermal modification dominates the market with a production volume of 1.1 million m^3 [2].

The high temperatures used in thermal modification (140 °C to 230 °C) degrade the wood's fundamental chemical constituents, mainly hemicelluloses, which are among the most labile to heat. As a product, thermally modified wood (TMW) has properties that are different from the original wood (greater dimensional stability and greater durability against microorganisms), which can be useful in diversifying wood applications [1–4].

ThermoWood® (Helsinki, Finland) is the leading commercial processor of TMW, with a production of around 240,00 m^3 in 2020. Thermal modification is carried out in kilns at atmospheric pressures (ca. 0.101 MPa) and temperatures ranging from 185 °C to 212 °C [5]; therefore, regarding the pressure inside the equipment, it is classified as an open-system process [1,6]. On the other hand, processes carried out at super-atmospheric pressure are classified as closed systems [1,6]. Batista and Bernadi [7] compared four closed-system industrial processes and reported pressures from 0.3 MPa to 2.0 MPa and temperatures from 140 °C to 190 °C. Wentzel et al. [6,8] pointed out that even at lower temperatures and cycles, closed-system processes can have higher effects on wood properties than open-system ones.

Regarding the process, in addition to the pressure, the combination of temperature and duration will play key roles in different levels of alteration in the chemical composition of TMW [1]. In turn, these chemical changes [1,2,4,6,8] will be responsible for the altered properties of TMW, where the most important are enhanced dimensional stability and biological durability, as well as control of color changes [1,2].

When revising the literature, comparing the results of different TM processes found in different studies is difficult due to the interpretation of the interaction of many variables altogether. Another factor that makes comparisons difficult is the heterogeneity of the wood material itself, even among the same species. This is because complex external factors (climate, soil, planted vs. natural forests) will produce wood with different qualities [9,10] resulting in an effect greater than that of the process itself. To overcome this difficulty, the corrected mass loss (CML) has been suggested as a quality control marker to compare the effect of different processes on TMW properties [6,8,11,12].

The difference between mass loss and CML is that the latter discounts the extractive content in TMW, which is particularly suitable when comparing open- and closed-system processes. That is because in closed systems, condensable (volatiles and water) and thermally degraded products are accumulated in the TMW, whereas in open systems, these products are removed from the atmosphere [1,6,11].

Teak (*Tectona grandis* L.f.) is one of the most valued woods in the international market because of its excellent characteristics, such as durability, dimensional stability, and color (heartwood). This has led many countries on all continents to plant teak forests, resulting in around 7 million hectares, mainly in India (37%), Indonesia (21%), and Thailand (12%) [13]. However, Silva et al. [10] highlighted that teak wood from fast-growing forests under the age of 10 years old can be unsuitable for solid products.

Regarding teak wood color, it was found that young trees have a larger proportion of sapwood [10], which is yellowish, clearly demarked from the heartwood, and has a lower market value. This has led some researchers to study the suitability of TM to darken and homogenize the color of teak wood from short-rotation plantations as a means of its valorization [14,15]. Since color is one of the most important wood properties for the wood market, other authors also studied the effect of TM on the teak wood color [16–21]. However, the majority of the studies used open-system processes [15–20], whereas few studies used different processes, such as thermo-vacuum [14] and pressurized closed systems [21].

As mentioned before, it is difficult to compare the effects of different TM processes from different studies because of the different materials tested, even from the same species. To resolve this difficulty, we designed this work, which aimed to verify the effect of closed- and open-system processes of TM on the color of short-rotation teak wood.

2. Materials and Methods

2.1. Wood Sampling and Treatments

We tested juvenile teak wood (*Tectona grandis* L.f.) from the first thinning (six years old) of a planted forest in Cáceres, Mato Grosso, Brazil. The material was supplied by the company TWBrazil, Ponta Grossa, Paraná, Brazil, where the industrial process of thermal modification is carried out. We sampled 15 tangential battens (25 × 90 × 2250 mm) from the company's stock, composed of heartwood and without pith.

Each batten was planed and then trimmed into three samples of 500 mm (longitudinal), which were equally distributed among three groups (Figure 1) corresponding to the thermal modification treatments that were assessed (Table 1). The objective of this sampling scheme was to minimize the effect of the material on the results. Specimens were produced for moisture content (MC) determination by the oven-drying method (Figure 1), and the mean initial MC was 12.8% (±0.77%), which is suitable for thermal modification.

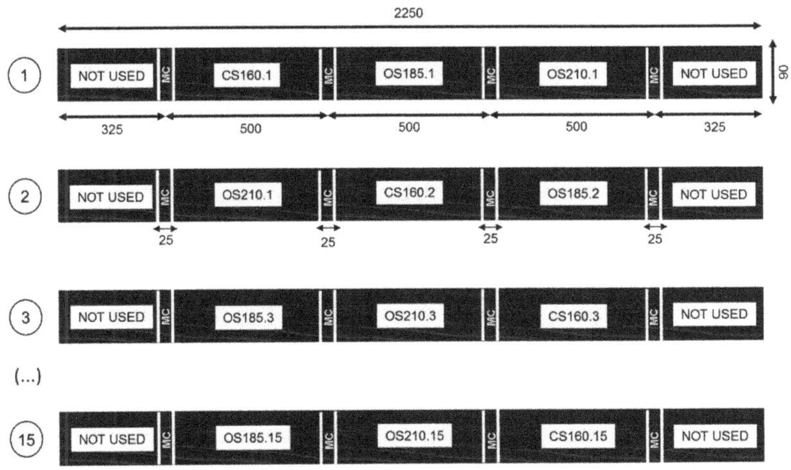

Figure 1. Sampling scheme. MC: moisture content specimens; units in millimeters.

Table 1. Treatments' description.

Treatment	Process	Temperature	Observation
CS160	Closed system, pressurized	160 °C	Temperature adopted by the company where the process was carried out
OS185	Open system, non-pressurized	185 °C	Temperature similar to the ThermoWood process for hardwoods: Thermo-S product (185 °C) and Thermo-D (210 °C) [9]
OS210	Open system, non-pressurized	210 °C	

2.2. Thermal Modification

TMW of the CS160 treatment was produced in an industrial autoclave with a capacity of 6 m^3 and the samples of this treatment were loaded along with the company's process wood. The process lasted around 720 min and was carried out according to a schedule developed by the company, as follows: (i) initial heating up to 110 °C at 1.30 °C min^{-1}; (ii) holding at 110 °C for 30 min; (iii) heating up to 140 °C at 0.50 °C min^{-1}; (iv) holding at 140 °C for 30 min; (v) heating up to 160 °C at 0.60 °C min^{-1}; (vi) holding at 160 °C for 90 min, when the process reached the highest pressure (0.60 MPa); (vii) cooling down to 60 °C at 0.25 °C min^{-1}, when the process is finished.

TMW of the treatments OS185 and OS210 was produced in a 250 L electrical oven (Linn Elektro Therm, Bad Frankenhausen, Germany) and we adopted a similar schedule to the one used by Lopes et al. [15,18], who worked with the same wood species and used equipment of the same model. The schedule was: (i) initial heating up to 100 °C at 0.50 °C min^{-1}; (ii) heating up to the target temperature (185 °C or 210 °C) at 2.60 °C min^{-1}; (iii) holding the target temperature for 180 min; (iv) switching off the heating system and consequent cooling down to 60 °C. The complete process took 362 min and 372 min, respectively, for OS185 and OS210.

For the quality control of thermal modification, we measured the initial MC (before the process), the final MC (after the process), and the corrected mass loss (CML, %). These analyses were performed according to the method adopted by Wentzel et al. [6], with all 15 samples of each treatment described as follows.

Moisture content was measured with the oven-drying method, and the initial moisture content (MCi, %) of each sample was estimated based on the mean MC between specimens A and B (Figure 2). Specimen C was used for chemical analyses. Then, sample D, for thermal modification (Figure 2), was weighed to the nearest 0.01 g, resulting in the initial mass (Mi, g). Based on Mi and MCi, it was possible to calculate the estimated initial oven-dried mass ($M_{0\%}$) of sample D, as in Equation (1).

$$M_{0\%} = \frac{(100 \cdot Mi)}{(100 + MCi)} \quad (1)$$

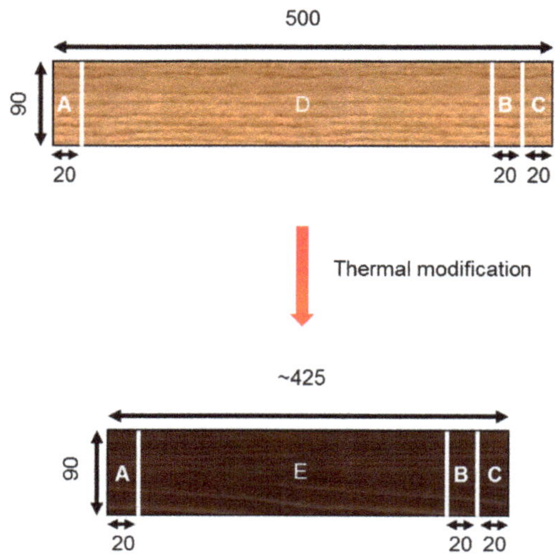

Figure 2. Representation of an individual sample before (**upper**) and after (**lower**) thermal modification; units in millimeters. A and B: moisture content specimens; C: chemical analyses specimens; D: sample for thermal modification; E: corresponding D sample after thermal modification.

After thermal modification, the samples were immediately weighed (Mf, g) and the other, A and B, specimens (Figure 2) were taken for calculation of the final moisture content (MCf, %). Based on Mf and MCf, it was possible to calculate the estimated final oven-dried mass ($Mf_{0\%}$), as in Equation (2).

$$Mf_{0\%} = \frac{(100 \cdot Mf)}{(100 + MCf)} \quad (2)$$

The mass loss (ML, %) caused by the thermal modification process was calculated with Equation (3).

$$ML = \frac{(M_{0\%} - Mf_{0\%})}{M_{0\%}} \,.\, 100 \quad (3)$$

2.3. Chemical Analyses

For chemical analyses (Table 2), we used specimen C (Figure 2). These specimens were transformed into chips in a hammer mill and ground into sawdust in a Wiley mill. The sawdust passed through two sieves (40 and 60 mesh), and the latter portion was conditioned (25 ± 2 °C and 65 ± 5% relative humidity—RH) for two weeks before analyses. The chemical analyses were performed in triplicate, except for extractive analyses, which were performed in five replicates.

The corrected mass loss (CML, %) was calculated with Equations (4)–(6).

$$CML = \frac{(CM_{0\%} - CM_{f0\%})}{CM_{0\%}} \cdot 100 \quad (4)$$

Table 2. Summary of chemical analyses.

Chemical Characteristic Evaluated	Observation	Reference
Total extractives	Use of toluene instead of benzene	T 264 cm-97 [22]
Extractives	Extraction with acetone and dichloromethane	T 204 cm-97 [23]
Insoluble Lignin	Insoluble Klason lignin	Gomide and Demuner [24]
Soluble Lignin	Acid soluble lignin	Goldschmid [25]
Holocellulose	Reaction with sodium chlorite and acetic acid	Wise, Murphy, and D'Addieco [26]
Alpha-cellulose	Reaction of the holocellulose with 17.5% sodium hydroxide	Adaptation of T 203 cm-99 [27]
Hemicellulose	Obtained by the difference between holocellulose and alpha-cellulose	-

In which CML is corrected mass loss (%); $CM_{0\%}$ is corrected estimated initial oven-dried mass (g); $CM_{f0\%}$ is corrected estimated final oven-dried mass (g).

$$CM_{0\%} = M_{0\%} - (M_{0\%} \cdot E_u) \tag{5}$$

In which $CM_{0\%}$ is corrected estimated initial oven-dried mass (g); $M_{0\%}$ is estimated initial oven-dried mass (g); E_u is mean total extractive content of untreated wood (%).

$$CMf_{0\%} = Mf_{0\%} - (Mf_{0\%} \cdot E_{tmi}) \tag{6}$$

In which $CMf_{0\%}$ is corrected estimated final oven-dried mass (g); $Mf_{0\%}$ is estimated final oven-dried mass (g); E_{tmi} is mean total extractive content of thermally modified wood of treatment *i* (%).

2.4. Color Measurements

The evaluation of the colorimetric parameters was performed according to the CIEL*a*b* color space. The colorimetric parameters L* (lightness), a* (red–green chromaticity coordinate), and b* (yellow–blue chromaticity coordinate) were measured using a Konica-Minolta (Tokyo, Japan) (model CM2500D) portable spectrophotometer with aperture diameter of 3 mm, standard illuminant D65, observation angle of 10 °, and specular light included. Before measurements, the equipment was calibrated with the black and white standard plates that accompanied the equipment. The chroma (C* coordinate or saturation) and the hue angle (h°) were calculated, respectively, with Equations (7) and (8) [28]. The five colorimetric parameters were used to classify the wood color according to the cluster classification proposed by Camargo and Gonçalez [28].

$$C^* = \left[\left(a^*\right)^2 + \left(b^*\right)^2\right]^{0.5} \tag{7}$$

$$h^\circ = \tan^{-1}\left(\frac{b^*}{a^*}\right) \tag{8}$$

The samples were measured before (D, Figure 2) and after (E, Figure 2) thermal modification, where the former measurements represented the untreated treatment. The readings were taken at three points of each sample (Figure 3) on the tangential surface closer to the bark.

The variation in the colorimetric parameters (ΔL*, Δa*, and Δb*) was calculated based on the deduction of the readings before and after the thermal modification, and the total color difference (ΔE*) was calculated with Equation (9), modified from [29]. ΔE* was qualitatively classified according to Toyoda et al. [29], representing levels of visual perception (Table 3).

$$\Delta E^* = \sqrt{\left(\Delta L^*\right)^2 + \left(\Delta a^*\right)^2 + \left(\Delta b^*\right)^2} \tag{9}$$

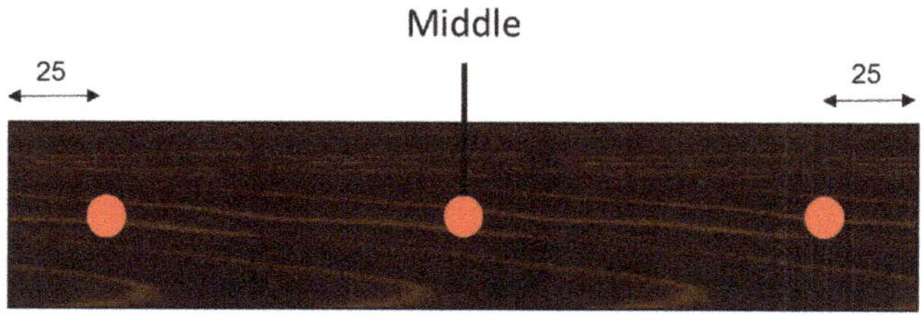

Figure 3. Positioning scheme (red circles) for colorimetric readings; units in millimeters.

Table 3. Perceived difference after thermal modification (ΔE*) [29].

Total Color Variation (ΔE*)	Classification
0.0–0.5	Trace
0.6–1.5	Slight
1.6–3.0	Noticeable
3.1–6.0	Appreciable
6.1–12	Much

2.5. Statistical Analyses

Statistical analyses were performed in a completely randomized design, and the significance level adopted was up to 5% probability for all tests. Initial and final moisture contents and corrected mass loss results were compared and discussed based on the mean and standard deviation of the treatments.

The analysis of variance (ANOVA) was adopted to verify the effect of treatments on the results of chemical analyses. For the validation of the ANOVA, the tests of Bartlett and Shapiro–Wilk were previously applied to verify, respectively, the homogeneity of the variances and the normal distribution. In cases of rejection of the null hypothesis (F-test; $p < 0.05$), the Tukey test was applied to differentiate the means.

For cases in which there was no homogeneity of variances and data normality ($p < 0.05$), the H-test of Kruskal–Wallis was applied, which provides a non-parametric method for ANOVA. In cases of rejection of the null hypothesis, the Bonferroni test was applied to differentiate the mean scores.

The linear coefficient of correlation (r) was calculated between the corrected mass loss and the results of the chemical analyses. The significance of r was tested with the Student *t*-test.

3. Results and Discussion

3.1. Moisture Content (MC) and Corrected Mass Loss (CML)

The mean initial moisture content ranged from 12.51% (CS160) to 11.40% (OS210), resulting in a mean amplitude of only 1.11 percentage points (p.p.) (Figure 4). This indicates that the initial MC of the samples under the different treatments was homogeneous and had no impact on further processing. For all treatments, the final MC was lower than the initial MC, indicating the effect of the drying steps during thermal modification in both closed and open systems.

The most remarkable result in Figure 4 is that different processes caused different effects on the final MC. CS160 had the highest final MC, reaching 9.09%, which represents a drying of only 3.42 p.p. compared to the initial MC. In contrast, the open-system treatments resulted in a final MC close to 0%. This discrepancy is mainly due to the types of heating and equipment used. In the CS160, the system was closed, which prevented the elimination of gases generated in the thermal modification process. In addition, heating was performed by injecting steam into the autoclave, which increased the availability of water in the system (increased relative humidity). On the other hand, in OS185 and OS210, the gases generated during the thermal modification were continuously eliminated. Wentzel et al. [11] also reported this same pattern for thermally modified Eucalyptus nitens in closed and open systems.

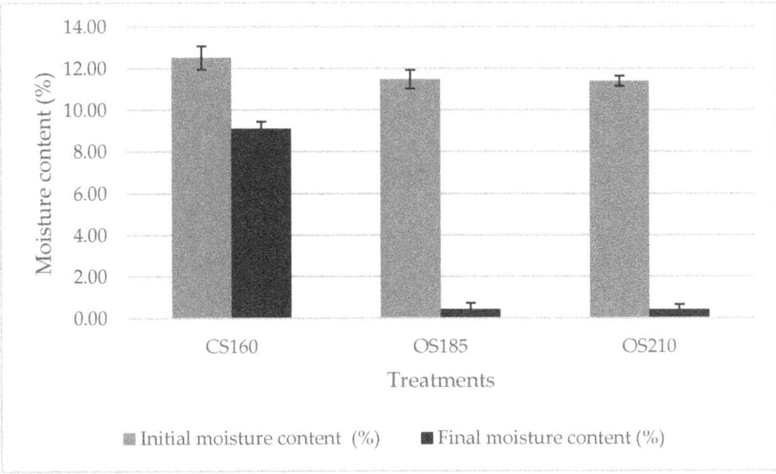

Figure 4. Initial and final moisture content of wood by treatment.

It is also noted that, for the open-system treatments, the final MCs were similar, regardless of the final temperature (185 °C and 210 °C). This indicates that the 25 °C temperature difference had no significant impact on the final MC, and the equipment worked similarly to a dry kiln.

For Boonstra et al. [30], the lower final MC is considered a positive effect for the open-system process because the wood can have a higher mechanical resistance, which can compensate, in part, for the reduction caused by the mass loss imparted by the process. However, the same effect will result in wood that is more difficult to machine process.

The results of CML per treatment are shown in Figure 5. During thermal modification, cell wall constituents are degraded and can be accounted for as extractives during Soxhlet extraction. Furthermore, depending on the type of process, thermal modification can cause the volatilization of extractives and other molecules resulting from cell wall degradation. Thus, CML is a useful tool for neutralizing the effect of wood extractives on mass loss results.

It is noted in Figure 5 that the different processes of thermal modification influenced the CML results. The CML was close to zero (0.33%) for OS185, indicating that this treatment slightly affected the wood constituents, even the most thermally labile ones. This suggests that the OS185 treatment was ineffective, basically resulting in a drying process of the wood.

OS210 had a slightly higher CML (2.44%) than OS185, showing the effect of increasing temperature on mass loss for the same kind of process (open system). Open-system processes carried out at atmospheric pressure depend mainly on heat to induce mass loss. Despite the high temperatures (185 °C and 210 °C), the low mass losses may be related to the overall short cycles of thermal modification in laboratory-scale equipment, including the duration of maintaining the maximum temperature. These CML results raise doubts about the effectiveness of the adopted schedules since the main benefits of the process are associated with mass loss, such as reduced hygroscopicity and increased resistance to biodeterioration by fungi [11].

In contrast, CS160 had the highest CML, reaching 18.47%. In addition to heat, the closed-system process has pressure and a saturated atmosphere as mass loss inducers, which explains the higher CML, even at a lower temperature. This result was also observed by Wentzel et al. [6] on the thermal modification of eucalyptus wood in closed and open systems.

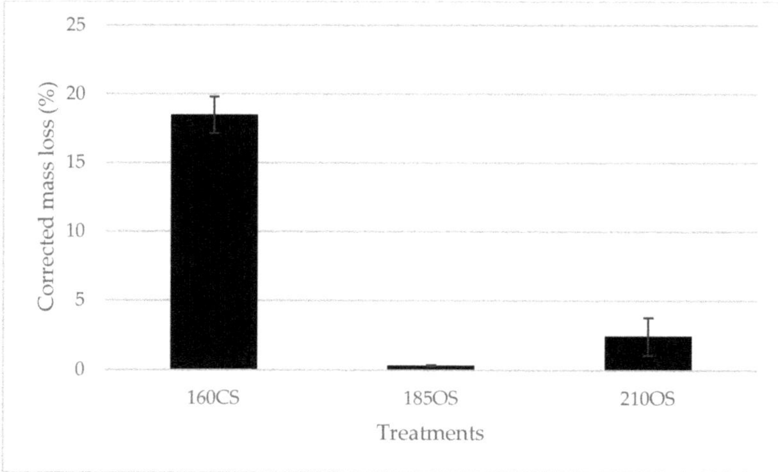

Figure 5. Corrected mass loss (%) per treatment.

Studies argue that the behavior of wood degradation varies according to the heating medium used, which is due to the surrounding atmosphere [31]. The closed system (such as CS160) allows for the presence of steam, facilitating wood degradation compared to using temperature alone. In this process, hemicelluloses and cellulose are depolymerized by hydrolysis in the presence of water, which is the main cause of the loss of resistance to degradation of these polymers. Both Modes et al. [31] and Severo and Tomaselli [32] stated that wood exposed to high relative humidity degrades predominantly by acid hydrolysis, with a higher degradation rate than thermal or pyrolytic degradation. Water is responsible for the breakdown of acetyl groups through the formation of acetic acid, which is responsible for further wood degradation.

According to the CML results, a greater effect on chemical composition and wood color is expected for CS160, followed by OS210. On the other hand, OS185 should show similar results to the untreated wood.

3.2. Chemical Composition

The results of the chemical analyses of the structural components of the cell wall are shown in Table 4. The results of the alpha-cellulose content revealed that CS160 had a higher absolute mean than untreated wood, while for the open-system treatments (OS185 and OS210), it was lower than the untreated wood. Despite this, none of the means differed significantly from untreated wood, which suggests that the thermal modification did not have a significant effect on the alpha-cellulose content. This result reinforces the high thermal stability of this cell wall constituent [33].

The thermal modification had varying effects on hemicellulose content: while CS160 had the lowest mean (8.90%), OS185 did not differ statistically from untreated wood, and OS210 was 3 p.p. lower than untreated but with statistical significance. The coefficient of linear correlation between CML and hemicellulose content was very high, negative, and significant ($r = -0.996$ *), indicating that the higher the CML, the lower the hemicellulose content. Indeed, the correlation between CML and the structural components of the cell wall was only significant ($p < 0.05$) for the hemicellulose content.

Hemicelluloses are the most fragile components of the cell wall due to their thermal sensitivity [3,20–22]. The CS160 treatment showed a higher CML (Figure 5) due to the greater degradation of hemicelluloses (Table 4). In contrast, OS185 had a CML close to zero, indicating low thermal degradation, with a non-significant difference in the hemicellulose content compared to the control. These results confirmed what had been forecasted in the previous discussion of the results of CML.

Table 4. Mean results of structural components of wood per treatment.

Treatment	Alpha-Cellulose (%)	Hemicelluloses (%)	Insoluble Lignin (%)	Soluble Lignin (%)	Total Lignin (%)
Untreated	48.21 AB	24.24 a	32.34 A	1.89 a	34.23 A
	(0.74)	(2.10)	(2.48)	(0.45)	(2.34)
CS160	53.01 A	8.89 c	33.36 A	1.61 b	34.97 A
	(4.15)	(6.36)	(9.76)	(5.40)	(9.57)
OS185	46.04 AB	24.55 a	32.29 A	1.88 a	34.17 A
	(0.27)	(1.57)	(0.52)	(2.32)	(0.42)
OS210	45.72 B	21.20 b	33.554A	1.72 b	35.27 A
	(0.59)	(4.93)	(1.26)	(3.32)	(1.32)
Shapiro–Wilk test	0.0065 *	0.41 ns	0.0356 *	0.8384 ns	0.0359 *
Bartlett test	0.0034 *	0.59 ns	0.0058 *	0.0642 ns	0.0046 *
F-test	-	239.3 *	-	14.94 *	-
H-test	9.84	-	3.20	-	3.21
r	0.909 ns	−0.996 *	0.580 ns	−0.880 ns	0.487 ns

Means followed by the same lowercase or uppercase letter in the columns do not differ ($p < 0.05$) according to the Tukey and Bonferroni tests, respectively. ns: not significant ($p > 0.05$). *: significant ($p < 0.05$). Results in parentheses are the coefficient of variation (%).

Teak wood contains several substances in its vessels, which are solubilized during chemical analyses. In thermal modification, especially, the most volatile extractives tend to disappear or degrade, as pointed out by Lengowski et al. [21]. However, during the degradation of the structural components of the cell wall, new compounds are formed and can be extracted from the wood. The application of high temperatures degrades the hemicelluloses, resulting in the formation of acetic acid, which acts as a catalyst for depolymerization and increases polymer breakdown. This leads to the formation of formaldehyde, furfural, and other aldehydes. In addition, part of the organic material can also be degraded and solubilized, justifying the reduction in the hemicellulose content after thermal modification at 160CS and 210OS.

There was no statistical difference among the treatments for total lignin and insoluble lignin contents, and, along with the results of alpha-cellulose, this indicates that the mass loss was mainly due to the degradation of hemicelluloses. Accordingly, the values (Table 4) of r for CML and total and insoluble lignin were low and non-significant.

Some studies [21,34] mentioned an apparent increase in the lignin content, meaning that there is no formation of this polymer during the treatment, but a condensation of the lignin structure due to the degradation of other wood components, mainly hemicelluloses. Temperatures above 200 °C result in greater availability of hemicellulose degradation products, which can contribute to the lignin condensation reaction [34]. Another explanation for this apparent increase in lignin content is a change in the balance of the cell wall mass once these analyses are gravimetric. However, this difference had no significant effect on the results of total lignin content between the tested treatments.

On the other hand, there was a significant effect of thermal modification of CS160 on soluble lignin when compared to untreated wood. The CML was high enough to cause a significant reduction in soluble lignin, and, in general, the higher the CML, the lower the content of it (r = −0,880 ns). However, the content of soluble lignin was too low (ranging from 1.61% to 1.89%) to significantly impact the total amount of lignin.

The results of the extractive content are shown in Table 5.

Table 5. Means of extractive content according to the solvent.

Treatments	Total Extractives (%)	Dichloromethane Extractives (%)	Acetone Extractives (%)
Untreated	8.50 c (2.11)	4.25 b (5.07)	2.15 d (1.52)
CS160	15.17 a (0.55)	8.50 a (2.84)	7.79 a (1.21)
OS185	7.71 d (0.79)	1.76 c (2.83)	2.72 c (0.83)
OS210	8.59 b (1.09)	4.61 b (3.11)	4.39 b (3.91)
Shapiro–Wilk test	0.88 ns	0.57 ns	0.14 ns
Bartlett test	0.54 ns	0.33 ns	0.08 ns
F test	2806.7 *	728.02 *	4020.1 *
r	0.994 *	0.911 ns	0.968 *

Means followed by the same lowercase letter in a column do not differ ($p < 0.05$) according to Tukey's test. ns: not significant ($p > 0.05$). *: significant ($p < 0.05$). Results in parentheses are the coefficient of variation (%).

CS160 had the highest total extractive content, while the open-system treatments (OS185 and OS210) had means closer to the untreated wood, although all means differed significantly from each other. The extraction performed with the sequence ethanol:toluene, ethanol, and hot water, accounted for as total extractives, showed the highest absolute means compared to the other solvents. This sequential extraction uses a mixture of solvents, allowing the removal of several classes of extractives with different molecular weights [23]. Indeed, CS160 had the highest means of extractive content of all tested solvents. There was a strong, positive, and significant correlation between CML and total extractive content (r = 0.994 *) indicating an increase in extractives with mass loss. The higher mass loss, associated with the impossibility of the removal of gases from the equipment, resulted in a higher amount of extractives in the closed-system than in the open-system processes.

The open-system treatments (OS185 and OS210) showed different trends compared to untreated wood in terms of extractives in acetone and dichloromethane. For extractives in acetone, OS185 had a mean similar to the untreated, while the mean of OS210 was twofold that of the untreated. For extractives in dichloromethane, OS210 had a mean similar to that of the untreated, while OS185 had a mean of less than half of it.

In the acetone extraction process, it is suggested that 185 °C was not sufficient to volatilize the polar extractives, while 210 °C increased the extractive content due to mass loss. On the other hand, in dichloromethane extraction, 185 °C was sufficient to volatilize the non-polar extractives, while 210 °C produced new substances that compensated for the volatilized fraction.

The trend observed in the extraction with dichloromethane is in line with the literature for open-system processes [3,35]. The extractive content of thermally modified wood can vary compared to untreated wood, increasing or reducing depending on the process type and schedule (time, temperature), which affect the chemical reactions during thermal modification, forming different types of substances.

Comparing acetone (polar) and dichloromethane (nonpolar) solvents, the latter showed, in general, greater extracting capacity. Gomide et al. [36] and Lopes [37] stated that dichloromethane is more advantageous when you want to extract lipophilic compounds from wood, such as fats, waxes, terpenes, sterols, resins, and non-volatile hydrocarbons [23,38].

3.3. Color Measurements

Initially, the means of the colorimetric parameters L*, a*, b*, C*, and h° of teak wood before thermal modification were 53.87, 11.52, 23.41, 26.09, and 63.80°, respectively. They were calculated based on data from all 45 samples (three groups of 15, Figure 1) before thermal modification. According to the classification of Camargos and Gonçalez [28], the color of untreated teak wood was classified as "light brown". This was the same color as reported by Gasparik et al. [17] for teak wood from Myanmar. Although the authors did not provide extra information about the material (e.g., tree age, planted x natural forests), it is likely to be heartwood because of the lightness result (47.0).

Chávez-Salgado et al. [39] and Qiu et al. [40] stated that secondary metabolites influence the color and smell of teak wood. Extractives contain chromatic compounds, such as phenolic hydroxyl groups, carbonyl groups, and double bonds, which impart color to the wood. In the specific case of teak, the unique color is promoted by the presence of 2-methyl-anthraquinone.

The results of the mean colorimetric parameters L*, a*, and b* are plotted in Figure 6, where it is possible to see that the wood before the thermal modification is grouped close to each other, indicating that they had a similar color. This was expected according to the design of the initial sampling (Figure 1). However, the color of untreated wood for each treatment had a different classification [28], as can be seen in Figure 7. Although this difference is not too perceptible to human vision, the method adopted identified three different colors, which indicates how sensitive it is.

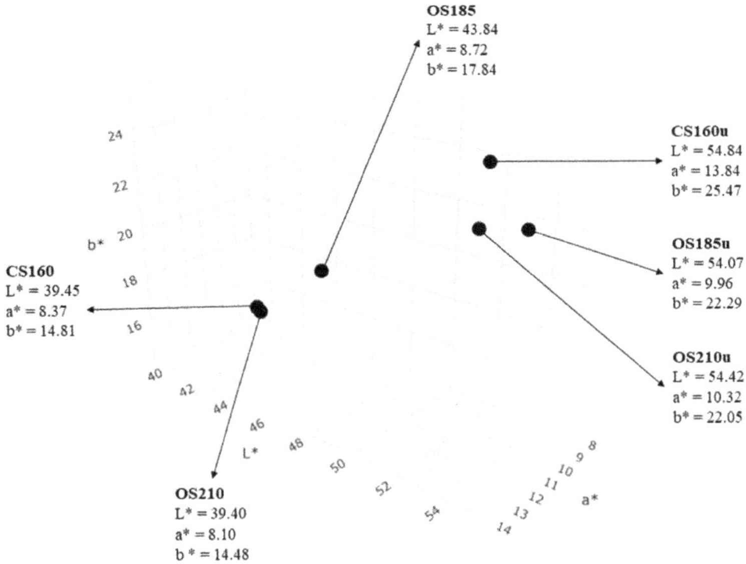

Figure 6. Three-dimensional graphic positioning for each treatment. Treatments marked with "u" mean wood before thermal modification (untreated).

TMW shifted downward to different positions in Figure 6 from its respective untreated wood, indicating the effect of the process on the original wood color. According to the classification of ΔE*, the change in wood color after thermal modification was "much" (from 6.1–12.0, Table 3) for all treatments. In Figure 8, it is possible to see that the means of ΔE* were 19.74, 11.65, and 15.37, respectively, for 160CS, 185OS, and 210OS. Indeed, the means of ΔE* of 160CS and 210OS did not fit the scale presented in Table 1, which suggests that, for thermally modified wood, the scale should be enlarged. We highlight that the classification was not designed for analyzing the color change of TMW but rather for weathered wood [29]. Although 160CS had a higher ΔE*, it was lower than that reported (24.11) by Lengowsky et al. [21], who studied the same closed-system process and temperature.

Treatment	Before thermal modification	After thermal modification	ΔE* classification [1]
160CS	"Brownish yellow [2]"	"Dark brown"	"Much"
185OS	"Olive brown"	"Olive"	"Much"
210OS	"Light brown"	"Dark brown"	"Much"

Figure 7. Classifications of color and total color variation (ΔE*). [1] Toyoda et al. [29]. [2] Camargos and Gonçalez [28].

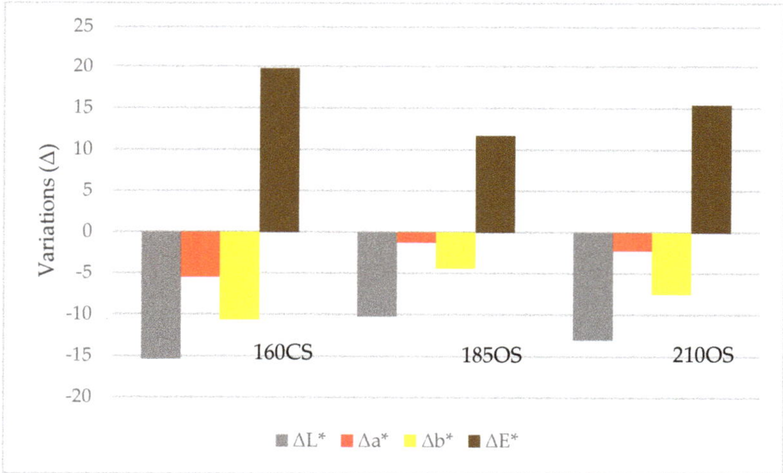

Figure 8. Variations in colorimetric parameters by treatment.

Anyway, TMW became darker compared to the untreated wood, which is represented by the reduction in lightness (ΔL*) for all treatments (Figure 8) and can be seen in the color chart presented in Figure 7. According to the results of other studies [14,18,19,21], the reduction in lightness is the most relevant color change related to thermally modified teak wood, which was also verified in our work. All structural cell wall components, as well as extractives, can contribute to changing the wood color. The darker color can be attributed to the formation of hemicellulose degradation products (except in the case of OS185), changes in the extractives, and the formation of oxidation products, such as quinones, as well as the percolation of extractives to the samples' surface [17,41].

Variations in L*, a*, and b* coordinates followed the same trend as ΔE*, with greater absolute variation (in modulus) in CS160 and lesser variation in the OS185 treatment. This highlights the smaller effect of thermal modification on the red–green (a*) pigmentation of teak wood when compared to the yellow–blue (b*) pigmentation, which was also verified in other studies [18,19,21].

Despite the absolute means and the different colors before thermal modification (Figure 7), the TMW of CS160 and OS210 achieved the same color ("dark brown"). Notice how close these treatments are in Figure 6 when compared to OS185. This result was unexpected, considering that CS160 and OS210 were carried out with different temperatures and processes, which resulted in different corrected mass losses (Figure 5), as well as different extractive contents (Table 5). However, since ΔE* was higher for CS160 (19.74) than for OS210 (15.37), if the wood color of both treatments was the same before thermal modification, TMW CS160 would likely be darker. But this hypothesis needs further investigation.

Based on the results of CS160 and OS210, it seems that the CML is not related to the color of TMW from different processes. On the other hand, it makes sense when comparing TMW from the same treatment, as OS210 had higher CML and ΔE* than OS185. This pattern was verified by other authors in open-system thermal modification, i.e., the increase in temperature resulted in increased ΔE* [17–19]. Willems et al. [11] provided a literature review and discussed that there is a good dependency between mass loss, chemical composition, fungal durability, and hygroscopicity. However, these relations cannot be directly transferred to other properties, such as color, and further investigation on this topic is necessary.

Even the mildest condition of thermal modification (OS185) was able to change the color of TMW (ΔE* = 11.65) compared to untreated wood. This result was similar to those obtained in open-system processes at 180 °C by other authors [17,18], which were, respectively, 9.67 and 11.97. There was no significant difference in the chemical composition of the cell wall components between untreated wood and OS185 (Table 4). On the other hand, there was a significant difference in the extractive content between untreated wood and OS185 (Table 5), where the most relevant (reduction in 2.49 p.p.) was in the dichloromethane content. Thus, it seems that the combined effect of the volatilization of apolar extractives and the drying of hygroscopic water (from 11.48% to 0.43%, Figure 4) was responsible for this color change. Notice that this drying effect also happened in treatments CS160 and OS210.

4. Conclusions

It was possible to achieve the same color using different processes of thermal modification at 160 °C in a closed system and 210 °C in an open system.

Thermal modification reduced wood lightness (L*), red–green chromaticity coordinate (a*), and yellow–blue chromaticity coordinate (b*). L* and a* had the biggest and smallest variations, respectively. The color of thermally modified wood was significantly changed, even at the mildest condition tested: 185 °C, open system, and 0.33% corrected mass loss.

Even at a lower temperature (160 °C), the corrected mass loss was higher (18.47%) in the closed system than in the open system (185 °C: 0.33% and 210 °C: 2.44%), indicating that the type of process had more influence on mass loss than the temperature. This resulted in different effects on the chemical composition, with a higher effect at 160 °C in the closed system on the hemicellulose content (reduced) and the extractive content (increased for all solvents). The hemicelluloses were the most susceptible components of the cell wall to thermal modification.

There was no difference in the chemical composition of untreated wood and thermally modified wood in an open system at 185 °C, except for the extractive content in dichloromethane, which was reduced in the latter.

We suggest that more studies are needed to clarify the effect of extractive content (in different solvents) on the color of thermally modified wood, mainly in cases where different processes are used. These studies could include the identification of the different extractives using chromatographic techniques.

Author Contributions: Conceptualization, D.C.B., P.H.G.d.C., H.F.N. and J.B.P.; Methodology, D.C.B., F.d.S.G., M.d.S.R., A.C.O.R. and L.d.C.D.; Software, P.H.d.S.S., A.C.O.R. and J.P.S.M.; Validation, D.C.B., P.H.G.d.C., H.F.N. and J.B.P.; Formal analysis, F.d.S.G., J.S.d.S.F. and P.H.d.S.S.; Investigation, F.d.S.G., M.d.S.R., A.C.O.R. and L.d.C.D.; Data curation, D.C.B., F.d.S.G. and P.H.d.S.S.; Writing—original draft preparation, F.d.S.G., P.H.d.S.S. and J.S.d.S.F.; Writing—review and editing, D.C.B., J.S.d.S.F., P.H.d.S.S., P.H.G.d.C., H.F.N. and J.B.P.; Supervision, D.C.B., P.H.G.d.C., H.F.N. and J.B.P.; Project administration, D.C.B.; Funding acquisition, D.C.B. All authors have read and agreed to the published version of the manuscript.

Funding: This research was financed in part by the Coordenação de Aperfeiçoamento de Pessoal de Nível Superior, Brasil (CAPES), Finance Code 001, scholarship to F.d. S.G.; the National Council for Scientific and Technological Development, CNPq, grant number 422637/2018-1; and CAPES/Fundação de Amparo à Pesquisa e Inovação do Espírito Santo (FAPES), through the "Programa de Desenvolvimento da Pós-graduação" (PDPG), Consolidação dos Programas de Pós-graduação da Área de Ciências Agrárias do Estado do Espírito Santo, Ciências Florestais, call FAPES/CNPq N° 23/2018—PRONEM, process number 2021-FDGS5 and grant agreement 131/2021.

Institutional Review Board Statement: Not applicable.

Informed Consent Statement: Not applicable.

Data Availability Statement: The data presented in this study are available on request from the corresponding author. The data are not publicly available because they are part of a project funded by a Brazilian Government funding agency. Therefore, before sharing the data, the authors would need prior authorization.

Acknowledgments: The authors express their gratitude to the company TWBrazil for donating the material, thermally modifying it in closed system, and shipping it to the execution institution.

Conflicts of Interest: The authors declare no conflict of interest.

References

1. Hill, C. *Wood Modification: Chemical, Thermal and Other Processes*; John Wiley & Sons: West Sussex, UK, 2006; p. 264. [CrossRef]
2. Jones, D.; Sandberg, D. A review of wood modification globally–updated findings from COST FP1407. In *Interdisciplinary Perspectives on the Built Environment*; IPBE: Izola, Slovenia, 2020; Volume 1. [CrossRef]
3. Esteves, B.M.; Pereira, H.M. Wood modification by heat treatment: A Review. *BioResources* **2009**, *4*, 370–404. [CrossRef]
4. Minkah, M.A.; Afrifah, K.A.; Batista, D.C.; Militz, H. Chemical and Mechanical Characterization of Thermally Modified *Gmelina arborea* Wood. *Les/Wood* **2021**, *70*, 31–44. [CrossRef]
5. International Thermowood Association. *ThermoWood®Handbook*; International ThermoWood Association: Helsinki, Finland, 2021; p. 56.
6. Wentzel, M.; Fleckenstein, M.; Hofmann, T.; Militz, H. Relation of chemical and mechanical properties of *Eucalyptus nitens* wood thermally modified in open and closed systems. *Wood Mater. Sci. Eng.* **2018**, *14*, 165–173. [CrossRef]
7. Bernardi, L.P.; Batista, D.C. TanWood®: The Brazilian Process of Thermal Modification of Wood. In *Book of Abstracts of 9th European Conference on Wood Modification (ECWM9)*; SHR: Arnhem, The Netherlands, 2018; p. 138.
8. Wentzel, M.; Rolleri, A.; Presenti, H.; Militz, H. Chemical analysis and cellulose crystallinity of thermally modified Eucalyptus nitens wood from open and closed reactor systems using FTIR and X-ray crystallography. *Eur. J. Wood Wood Prod.* **2019**, *77*, 517–525. [CrossRef]
9. Tsoumis, G. *Science and Technology of Wood: Structure, Properties, Utilization*; Verlag Kessel: St. Paul, MN, USA, 1991; p. 494.
10. Silva, J.G.M.; Oliveira, R.F.; Rocha, S.M.G.; Vidaurre, G.B.; Latorraca, J.V.F.; Carmo, F.H.D.J. Qualidade da madeira da teca. In *Teca (Tectona grandis Lf.) No Brasil*, 1st ed.; Reis, C.A.F., Oliveira, E.B., Santos, A.M., Eds.; Embrapa: Brasília, Brazil, 2023; p. 734.
11. Willems, W.; Altgen, M.; Militz, H. Comparison of EMC and durability of heat treated wood from high versus low water vapour pressure reactor systems. *Int. Wood Prod. J.* **2015**, *6*, 21–26. [CrossRef]
12. Batista, D.C.; Silvares, P.H.S.; Wentzel, M.; Brischke, C.; Militz, H. Influence of closed and open systems on the durability of short-rotation *Eucalyptus grandis* wood against rot-fungi. In *11th European TMT Workshop*; IHD: Dresden, Germany, 2023; pp. 44–50.
13. Tomaselli, I.; Hirakuri, S.R. Mercado mundial da madeira de teca. In *Teca (Tectona grandis Lf.) No Brasil*, 1st ed.; Reis, C.A.F., Oliveira, E.B., Santos, A.M., Eds.; Embrapa: Brasília, Brazil, 2023; p. 734.
14. Cuccui, I.; Negro, F.; Zanuttini, R.; Espinoza, M.; Allegretti, O. Thermo-vacuum modification of teak wood from fast-growth plantation. *BioResources* **2017**, *12*, 1903–1915. [CrossRef]
15. Lopes, J.O.; Garcia, R.A.; Nascimento, A.M.; Latorraca, J.V.F. Color uniformization of the young teak wood by heat treatment. *Rev. Árvore* **2014**, *38*, 561–568. [CrossRef]

16. Garcia, R.A.; Lopes, J.O.; Nascimento, A.M.; Latorraca, J.V.F. Color stability of weathered heat-treated teak wood. *Maderas. Cienc. Y Tecnol.* **2014**, *16*, 453–462. [CrossRef]
17. Gašparík, M.; Gaff, M.; Kačík, F.; Sikora, A. Color and chemical changes in teak (*Tectona grandis* L. f.) and meranti (*Shorea* spp.) wood after thermal treatment. *BioResources* **2019**, *14*, 2667–2683. [CrossRef]
18. Lopes, J.O.; Garcia, R.A.; Latorraca, J.V.d.F.; Nascimento, A.M.D. Color change of teak wood by heat treatment. *Floresta E Ambiente* **2014**, *21*, 521–534. [CrossRef]
19. Méndez-Mejías, L.D.; Moya, R. Effects on density, shrinking, color changing and chemical surface analysis through FTIR of *Tectona grandis* thermo-treated. *Sci. For.* **2016**, *44*, 811–820. [CrossRef]
20. Méndez-Mejías, L.D.; Moya, R. Effect of Thermo-Treatment on the Physical and Mechanical, Color, Fungal Durability of Wood of Tectona Grandis and Gmelina Arborea from Forest Plantations. *Mater. Sci.* **2018**, *24*, 59–68. [CrossRef]
21. Lengowski, E.C.; Bonfatti Júnior, E.A.; Nisgoski, S.; Muñiz, G.I.B.d.; Klock, U. Properties of thermally modified teakwood. *Maderas Cienc. Tecnol.* **2021**, *23*, 1–26. [CrossRef]
22. Technical Association of Pulp and Paper Industry. *T264 cm: Preparation of Wood for Chemical Analysis*; TAPPI: Norcross, GA, USA, 1997.
23. Technical Association of Pulp and Paper Industry. *T204 cm: Solvent Extractives of Wood and Pulp*; TAPPI: Norcross, GA, USA, 1997.
24. Gomide, J.L.; Demuner, B.J. Determinação do teor de lignina em material lenhoso: Método Klason modificado. *O Pap.* **1986**, *47*, 36–38.
25. Goldschimid, O. Ultraviolet Spectra. In *Lignins: Occurrence, Formation, Structure and Reactions*; John Wiley & Sons: New York, NY, USA, 1971; p. 916.
26. Wise, L.E.; Murphy, M.; D'Addieco, A.A. Chlorite holocellulose, its fractionation and bearing on summative wood analysis and studies on the hemicelluloses. *Pap. Trade* **1946**, *122*, 35–43.
27. Technical Association of Pulp and Paper Industry. *T203 cm-99: Alpha-, Beta- and Gamma-Cellulose in Pulp*; TAPPI: Norcross, GA, USA, 1999.
28. Camargos, J.A.A.; Gonçalez, J.C. Applied colorimetry as instrument in the elaboration of a timber color chart. *Bras. Florest.* **2001**, *71*, 30–41.
29. Toyoda, T.; Azuma, M.; Hikita, Y. Weather Testing of Timber—Discoloration. In Proceedings of the 32nd Annual Meeting of The International Research Group on Wood Preservation (IRG/WP, 01-20221), Nara, Japan, 20–25 May 2001; p. 7.
30. Boonstra, M.J.; van Acker, J.; Tjeerdsma, B.F.; Kegel, E.V. Strength properties of thermally modified softwoods and its relation to polymeric structural wood constituents. *Ann. For. Sci.* **2007**, *64*, 679–690. [CrossRef]
31. Modes, K.S.; Santini, E.J.; Vivian, M.A.; Haselein, C.R. Effect of heat treatment on mechanical properties of *Pinus taeda* and *Eucalyptus grandis* woods. *Ciênc. Florest.* **2017**, *27*, 291–302. [CrossRef]
32. Severo, E.T.D.; Tomaselli, I. Effect of vaporization treatment in logs and saw timber of *Eucalyptus dunnii* on the static bending. *Floresta* **2003**, *29*, 37–51. [CrossRef]
33. Sundqvist, B. Colour Changes and Acid Formation in Wood during Heating. Ph.D. Thesis, Luleå Tekniska Universitet, Luleå, Sweden, 2004.
34. Pratiwi, L.A.; Darmawan, W.; Priadi, T.; George, B.; Merlin, A.; Gérardin, C.; Gérardin, P. Characterization of thermally modified short and long rotation teaks and the effects on coatings performance. *Maderas. Cienc. Tecnol.* **2019**, *21*, 209–222. [CrossRef]
35. Esteves, B.M.; Graça, J.; Pereira, H.M. Extractive composition and summative chemical analysis of thermally treated eucalypt wood. *Holzforschung* **2008**, *62*, 344–351. [CrossRef]
36. Gomide, J.L.; Colodette, J.L.; Oliveira, R.C.D.; Silva, C.M. Technological characterization of the new generation of Eucalyptus clones in Brazil for kraft pulp production. *Rev. Árvore* **2005**, *1*, 129–137. [CrossRef]
37. Lopes, O.R. The Influence of Extractives and Heartwood/Sapwood Ratio on Near-Infrared Evaluation of Wood for Pulp Production. Master's Thesis, Universidade Federal de Viçosa, Viçosa, Minas Gerais, Brazil, 2008.
38. Baptista, I.; Miranda, I.; Quilhó, T.; Gominho, J.; Pereira, H. Characterisation and fractioning of *Tectona grandis* bark in view of its valorisation as a biorefinery raw-material. *Ind. Crops Prod.* **2013**, *50*, 166–175. [CrossRef]
39. Chávez-Salgado, L.P.; Vandenbossche, V.; Vilarem, G. *Tectona grandis* Linn. f. secondary metabolites and their bioactive potential: A review. *Iforest Biogeosci. For.* **2022**, *15*, 112–120. [CrossRef]
40. Qiu, H.; Liu, R.; Long, L. Analysis of chemical composition of extractives by acetone and the chromatic aberration of teak (*Tectona grandis* L. f.) from China. *Molecules* **2019**, *24*, 1989. [CrossRef]
41. Bekhta, P.; Niemz, P. Effect of high temperature on the change in color, dimensional stability and mechanical properties of spruce wood. *Holzforschung* **2003**, *57*, 539–546. [CrossRef]

applied sciences

MDPI

Article

Effects of Solvents Used for Conservation/Restoration Treatments on Damaged Linden Panels of Cultural Heritage Assets

Anamaria Moraru-Avram [1,*], Constantin Ștefan Ionescu [2,*] and Aurel Lunguleasa [1]

1 Wood Processing and Design of Wooden Product Department, Transilvania University of Brasov, 500038 Brasov, Romania; lunga@unitbv.com
2 Laboratory of Restoration and Research "Restaurare Ionescu Constantin", 550234 Sibiu, Romania
* Correspondence: anamaria.avram@unitbv.ro (A.M.-A.); ionescu.constantin.stefan@unitbv.ro (C.Ș.I.); Tel.: +40-76-665-1672 (A.M.-A.); +40-74-148-5550 (C.Ș.I.)

Abstract: Heritage objects with wooden supports can degrade in inappropriate storage conditions or when microclimate factors contribute to the development of biological attacks. Another issue regarding the deterioration of artifacts is the lack of a full understanding of material properties and their behavior during restoration treatments. In this paper, we note the strengthening treatments of artifacts with severely damaged wood and the various treatments against bio-pests. The influence of solvent on dimensional changes was observed for water, acetone, and white spirit. Acetone was found to cause the greatest swelling and deformation of the treated panels. The present work highlights the importance of choosing not only the correct types of solvents for the solubilization of synthetic resins, as well as those used in conservation-restoration treatments, but also the effects they have on polychrome wood panels that have been degraded by xylophagous insects.

Keywords: heritage artefacts; wooden support; damaged linden; dimensional modifications; treatments; solvents; conservation; restoration

Citation: Moraru-Avram, A.; Ionescu, C.Ș.; Lunguleasa, A. Effects of Solvents Used for Conservation/Restoration Treatments on Damaged Linden Panels of Cultural Heritage Assets. *Appl. Sci.* **2023**, *13*, 11148. https://doi.org/10.3390/app132011148

Academic Editor: Giuseppe Lazzara

Received: 7 September 2023
Revised: 7 October 2023
Accepted: 8 October 2023
Published: 10 October 2023

1. Introduction

The degradation of cultural heritage objects with wooden supports is a result of the passage of historical time and the fact that wood is a hygroscopic, vulnerable material that responds negatively to environmental conditions in inappropriate storage or exhibition spaces and is prone to physical, chemical, and especially biological deterioration.

Cultural heritage, as defined in Romanian legislation, represents the totality of assets that reflect cultural values and require protection through scientific, legal, and technical measures for conservation and restoration [1]. Romania is part of numerous international organizations (e.g., UNESCO), which work to protect and promote cultural heritage worldwide.

The conservation and restoration of works of art with wooden supports are distinct disciplines that are focused on maintaining the integrity of the original materials and their cultural value. Conservators/restorers must deeply understand the materials used and apply appropriate methods and techniques to conserve works of art [2,3]. Solving practical problems involves research to extend scientific discoveries with applicability in the field of heritage conservation; heritage conservation specialists "need scientific examinations of collections" [4].

It is important that a cultural heritage object is understood and interpreted both as a whole and as a composite element. It is made of different adjacent or overlapping materials (Figure 1), which have different behaviors, and can be composed as follows: the wooden support, successive layers of glue in different concentrations, the preparation layer (the primer), over which the pictorial layer (the artistic creation) is placed, and the protective varnish. Thus, the wooden support and the complex preparations (the primer) have strengthening, balancing, equalizing, and finishing roles, over which structure the artistic creation is developed (with an aesthetic and/or thematic role). This composite ensemble, in its historical course, accumulates and transmits education, culture, and civilization.

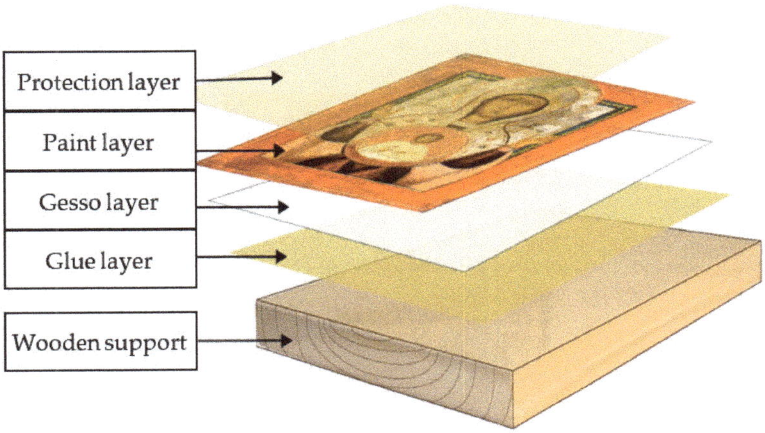

Figure 1. The successive arrangement of the layers that form the art object.

Both the wooden support and the primer layer, together with the painting film, have elastic properties but exhibit different degrees of elasticity. The cracking of the primer layer (gesso) is often responsible for cracks appearing in paintings. The anisotropy of the wood support, manifesting in the dimensional expansion induced by humidity, inevitably causes deformations of the panels [5]. The relationship between these deformations and the risk of physical damage to the paint layers must be established via modeling and experimental research [5].

We reiterate the importance of knowing and having access to the control mechanisms of dimensional stability, indicating not only the ways in which cultural assets are (or can be) affected but also the processes that are applied to improve dimensional stability. "Stability is defined as: values at which a piece of wood contracts or swells in response to a change in its moisture content" [6].

Wood swelling, as an external dimensional change, is influenced by density and tree species [7,8] and occurs in water vapor, liquid water, or organic solvents. Severe changes in its microclimate, compared to the considered equilibrium humidity parameters of RH 55 ± 5% and a temperature of 22 ± 2 °C, can cause swelling or the dimensional contraction of a panel, producing internal cracks, creating swelling forces in the wood, and initiating splits and fractures in the support or between the support and the painting's layers [9]. Dimensional changes are different in the three sectioning directions (anisotropy), the most significant being produced in the tangential direction [10,11], with values of between 6 and 12% in the tangential direction, 3–6% in the radial direction, and 0.1–0.3% longitudinally [12–14], all these changes occurring in the range of the FSP (fiber saturation point). Rowell (2005) [15] states that for wood to swell, "water or another swelling agent must penetrate the cell wall", and that moisture does not "concentrate in one place, but spreads throughout the cell wall structure". In a fiber saturation point (FPS) ranging up to 30%, where water localization in wood is initiated at the cellular level [13], volumetric swelling or contraction occurs, which is also known as "wood movement" [16]. The importance of these changes—as seen through the lens of degradation—can be understood as a bonding problem or the result of deformations and internal stresses [17]. Once the wood swells, internal stresses are produced, called swelling pressures [18,19], which cause deformations. In this context, the moisture content in the wood cells determines a torsional behavior [12]. The time taken to reach the maximum swelling pressure is shorter for thin wooden elements [20].

In addition to dimensional instability, high microclimate conditions favor the development of bio-pests. When wood is wet, it is susceptible to biodeterioration by fungi that decompose it [21], or by xylophagous insects, either independently or symbiotically. The moisture content (MC) of wood affects its performance due to the moisture's influence on its physical and mechanical properties [22,23]. The mechanical properties, the dimensional changes, the tendency to develop a certain drying defect, and the necessary conditions for the installation and development of biological deterioration are all influenced by wood's moisture content [24].

During restoration interventions on art objects, it is crucial to consider the behavior of their materials, as they can influence the support and the painting layers. An improvement in the performance of wood should be achieved without additional non-beneficial effects [25]. Solvents are frequently used, either for dissolving polymers, for strengthening purposes, or for antiseptic treatments. The controlled impregnation of wood with monomers can improve its resistance to damage [26]. Evaluation of the impregnation degree [27] must be a ratio between the maximum level of treatment and its efficiency and can be evaluated by weight gain (WPG) [26] and hardness determinations [27]. The consolidation treatments applied to art objects are closely related to the degree of damage: the support can be degraded and exhibit a slightly reduced surface and volume, but situations can frequently be encountered when the surface and volume, in particular, are severely damaged, resulting in significant losses and a reduction in wood density of between 20% and 50% compared to the healthy constituent/reference wood. Such situations will require multiple and complex treatments [28].

Solvents can temporarily swell the wood, which returns to its original dimensions as the solvent evaporates [29]. Due to the hygroscopic nature of wood, it is susceptible to swelling when in contact with liquids [30], depending on the type of liquid being absorbed. Polar liquids with high binding capacity can cause greater swelling, while non-polar or weakly polar organic solvents lead to lesser swelling [31]. The choice of solvents and the composition of the wood strongly influence its swelling. For example, toluene causes minimal swelling, while acetone can lead to swelling that is 3–4.7 times greater than that with toluene [32,33].

The swelling process of wood is complex and is affected by the moisture content and pre-treatment of the wood. Swelling in ethanol–water mixtures can significantly affect the wood structure by partially dissolving lignin [34]. In addition, solvents of low polarity can effectively reduce the stresses in stressed wood, as indicated by the longitudinal shrinkage pattern [31].

A non-exhaustive analysis of the specialized literature shows that most studies focus on new and healthy wood, which does not support polychromy or artistic painting.

These studies examine the sensitivity of wood to the absorption of liquids, such as water [15,32,35,36] or organic solvents [15,33], and focus on wood swelling as a result of these processes.

Different from the specialized literature, our study, and its novelty, refer not only to the dimensional changes produced by swelling but especially to deformations: curvatures and torsions that appear during partial immersion when the wood is in contact with selected solvents. Our study also brings to the forefront and evaluates the behavior of historical wood that has been severely damaged by wood-boring insects, as is found in cultural heritage objects, where a polychrome layer, composed of protein-primer-pigment glue, is present on one of the surfaces (the tangential surface).

Unlike other research, where the samples are conditioned in the oven before the experiments are carried out, in the present case, the samples have a special character where the experiments simulate the real treatments conducted during restoration. Art objects cannot be placed into ovens and conditioned according to the requisite standards, so the dimensional changes can exhibit different and increased variations. In this regard, Mantanis (1995) argues that the presence of water in many organic solvents significantly influences the swelling of wood.

The purpose of the research is to observe the effect of solvents used in conservation and restoration treatments (treatment with liquid solutions—solvents for consolidation or treatment with biocides) on the historical wood (*Tilia cordata*) found in cultural heritage assets that have been severely and progressively damaged by xylophagous insects. All this is necessary to extend and save the life of heritage objects.

2. Materials and Methods

For this study, lime wood (*Tilia cordata*), which had been damaged by xylophagous insects and recovered from historical ensembles, and which required replacement due to static considerations, was used. In choosing the wooden material, it was taken into account that polychromy, without aesthetic importance, is present on one of the faces, composed of primer (gesso), a film of a unitary red shade of paint, and a protein glue binder.

The recovered wood was previously thermally and hygroscopically balanced in a controlled environment for 6 months at 20 ÷ 25 °C and RH of 55 ÷ 60%.

Initially, the linden planks had varied sizes and humidity of 8.4 ÷ 9.6%. The historical wood was not processed beforehand with a machine and it was not necessary to prepare it in the same way as a new standardized wood plank.

It was decided to cut and form 15 equal samples, with dimensions of L = 90 × l = 130 × h = 25 ÷ 27 mm; 9 of these were sectioned manually in the longitudinal direction, thus obtaining 18 samples with dimensions of L = 90 × l = 130 × h = 12 mm (±2 mm)—of which 9 samples had a surface with polychromy (P) and 9 samples were without polychromy (FP), respectively, on the back of the panel. The remaining 6 samples retained a thickness of 27 mm.

In the dimensional formation of the samples, it was decided to keep a ratio of 1:5 in relation to the painted panels, which was taken as a reference, and a minimum number of 10–12 annual rings was taken into account for the samples with a thickness of 27 mm. For the sections of 10–12 mm in thickness, there should be a minimum of 5–7 rings (Figure 2).

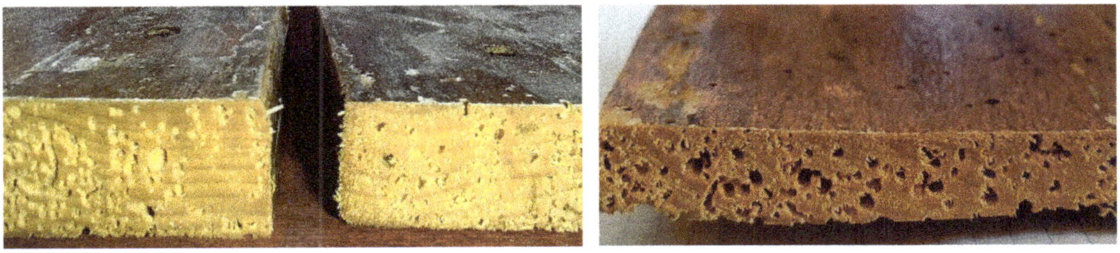

Figure 2. Identification of annual rings in the two types of specimens.

Three common liquids were selected for the art conservation treatments: water, acetone, and white spirit D40. These substances are generally used for the transfer of active biocide substances against bio-pests, such as fungi and xylophagous insects, to the wooden support. They are also used to solubilize reinforcing resins.

It was necessary to observe the changes not only on samples with a thickness of 27 mm but also on those with a thickness of 10–12 mm.

The 6 samples with a thickness of 27 mm were immersed, in groups of 2, in each type of selected solvent. The sectioned specimens (12 mm) were also immersed in liquids, with 6 for each solvent, respectively, employing 3 specimens with polychromy and 3 without polychromy.

Partial immersion in the solvents was employed to the level of half the thickness of the samples (Figure 3). This is because, in current practice, art objects with polychromy cannot be totally immersed in solvents or treatment liquids.

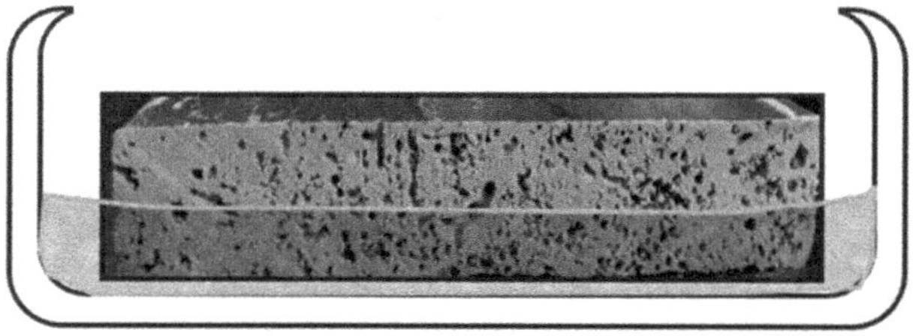

Figure 3. Representation of the technique for immersing the samples.

The samples with a thickness of 12 mm were placed two-by-two in the containers, taking into account the fact that absorption was carried out from the sectioned (cut) face. In the case of the 27-mm-thick specimens, they were placed with the polychrome face outward.

For the samples with a thickness of 27 mm, a quantity of 300 mL of solvent per container was used, and for the samples with a thickness of 12 mm, 150 mL of solvent was used per container. In order to be able to determine the absorption of the solvent, both the mass and the volume of liquid used/set of samples were taken into account Table 1.

Table 1. Solvents used for the experiment and their parameters.

Cleaners	Density (g/cm^3)	Quantity Used/Container/Specimen Set (g)	Quantity Used/Container/Specimen Set (mL)
Water	1.00	150/300	150/300
Acetone	0.7899	118.49/237.97	150/300
White spirit (D40)	0.789	118.35/236.7	150/300

The damaged volume of the samples was calculated, and a quantity of solvent was taken into account, which, in terms of volume, should not exceed approx. 10% of the total volume of the immersed sample. The immersion time was also taken into account and limited to 15 min—dictated by simulating a treatment stage and in agreement with Rowell (2005), who mentions that swelling does not occur constantly and uniformly; the most important swelling occurs in the first 15 min after immersion, namely, 5.3%; at 30 min, the swelling ratio is 7%, while at 60 min, it is 8.1%.

Apparatus. The preparation and cutting of the samples were performed with a Sheppach 250 electric circular saw, and the sectioning in half, along the longitudinal direction, was performed with a manual saw.

Wood moisture was measured with a digital hygrometer with pins—a digital moisture meter with ±4% deviation and 0.1 accuracy. To measure the mass, the analytical scale Kern EHA 500-1 was used, with a precision of 0.01 g, and the dimensions were measured using the digital caliper.

The variations in deformation measurements were determined with an HBM Machines dial gauge, and the depth of curvature was measured with the Parkside PTM 2 A1 digital apparatus. The devices can be seen in Figure 4.

To determine the density, which can be defined as the ratio between mass and volume, the ISO 3131-1975 [37] standard was used for the samples with moisture content at the time of testing. Also based on the same standard, the conventional humidity was determined as a ratio between the mass in the dry state and the volume of the sample with a high moisture content—after extraction from (partial) immersion. Thus, Equations (1a) and (1b) were used:

$$\rho_W = \frac{m_w}{v_w} \left[g/\,cm^3\right] \quad (1a)$$

where

ρ_w = moisture density at the time of testing;
m_w = mass of the sample at the set humidity at the time of testing, in g;

v_w = volume of the sample at the set humidity at the time of testing, in cm^3.

$$\rho_y = \frac{m_i}{v_u} \left[g/\,cm^3 \right]. \quad (1b)$$

Here,

ρ_y = conventional density;
m_i = mass of the dry/conditioned specimen, initially, in g;
v_u = the volume of the sample at the set humidity, at the moment of extraction from the immersion in cm^3.

Figure 4. The measuring devices used to determine deformation.

In order to determine the moisture content, the principle found in the EN 13183-1/2002 [38] standard was used; however, due to the fact that historical wooden elements recovered from cultural assets cannot be fully subjected to the rigors of the standards, the calculation formulas and application principles for the field of cultural property restoration were adopted. Since the historical wood pieces cannot be brought to 0% humidity, especially in real conditions when there is polychromy, it was decided to condition the recovered samples in a controlled environment. The recorded humidity had values of between 8.4 and 9.6%—considered the initial humidity (in relation to the standard, u0). Thus, to obtain the moisture content after partial immersion, Formula (2) will be used:

$$MC = \frac{m_u - m_i}{m_i} \cdot 100[\%] \quad (2)$$

where

MC = moisture content, as a percentage;
m_u = wet mass of the sample extracted from immersion, in g;
m_i = initial mass, in g.

Linear swelling was calculated according to the ISO 4859-1982 [39] standard, using Formula (3):

$$\alpha_{TG} = \frac{l_{max} - l_{min}}{l_{min}} \cdot 100[\%] \quad (3)$$

where

α_{TG} = swelling on the tangential section (TG), expressed as a percentage;
l_{max} = size of the wet samples when extracted from immersion solvents, in mm;
l_{min} = is the size of the dry samples, before immersion, in mm.

The amount of absorbed solvent (Abs.), measured in grams, was calculated with Formula (4a), while Formula (4b) was used to determine the absorbed volume:

$$M_{ABS} = m_u - m_i[\text{g}] \tag{4a}$$

where

M_{ABS} = amount absorbed, expressed in g;
m_u = wet mass, after extraction from the solvent, in g;
m_i = initial mass, before treatment, in g.

$$V_{ABS} = \frac{M_{ABS}}{\rho_{solvent}}\left[\text{cm}^3\right] \tag{4b}$$

Here,

V_{ABS} = absorbed volume, expressed in cm^3;
M_{ABS} = amount absorbed, expressed in g;
$\rho_{solvent}$ = solvent density, expressed in g.

Dimensional variation was determined with Formula (5). The readings were obtained with the dial gauge and the digital device. Measurement points were taken on the surface of the specimen, at the same points, before and after immersion.

$$\Delta = l_u - l_i\ [\text{mm}] \tag{5}$$

Here,

Δ = variation, in mm;
l_u = value of the measured point on the measuring axis after immersion, in mm;
l_i = value of the measured point on the initial measurement axis (before immersion), in mm.

Conventional measurement points on the horizontal axes, D-YD-C, A-YA-B, and XA-XB, were considered, in accordance with Figure 5a. To counteract and balance both the initial deformations and those resulting from extraction from the liquid, compensation wedges with a length of 60 mm and an inclination of from 1 mm to 70 mm were used, as shown in Figure 5b.

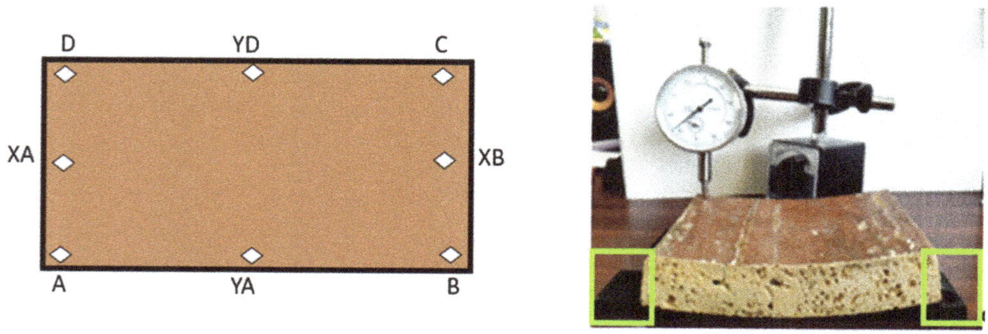

(a) (b)

Figure 5. (**a**) Marking the measurement points on the samples and the coordinates on the axes; (**b**) measuring the deflection with the dial gauge, highlighting the compensation wedges.

3. Results

3.1. Fluid Absorption and Swelling Dimensional Changes

The absorption of moisture implicitly leads to the volume increase of the wood.

The absorption of the amount of solvent was determined, both in grams and in cm^3, and it was found that the highest absorbed volume was of acetone, followed by water, and then by white spirit (according to Table 2). It is found that the absorption in grams is higher in the case of water, compared to acetone and white spirit, and the humidity of the samples extracted from water had the highest values (Figure 6).

Table 2. Means obtained from the determinations for the 12-millimeter-thick sectioned samples.

Determinations	Sample Code	Solvent Used		
		Water	White Spirit	Acetone
The mass before the immersion [g]	I.FP	56.93	56.55	57.33
	II.P	63.14	62.88	62.33
Initial volume [cm^3]	I.FP	137.95	140.15	136.04
	II.P	138.01	143.5	135.31
The mass after immersion [g]	I.FP	96.06	87.69	92.97
	II.P	111	95.52	100.46
Volume after immersion [cm^3]	I.FP	151.1	145.23	200.72
	II.P	156.99	144.87	228.21
Solvent absorption [g]	I.FP	39.13	31.14	35
	II.P	47.86	32.64	38.13
Solvent absorption [cm^3]	I.FP	39.13	39.92	44.30
	II.P	47.86	41.85	48.27
Absorption from the total used (150 mL) [%]	I.FP	26.09	26.62	29.54
	II.P	31.91	27.90	32.18
Solvent content/test tube, relative to the initial volume and the volume of solvent absorbed [%]	I.FP	28.365	28.486	32.567
	II.P	34.679	29.161	35.671
Initial density [g/cm^3]	I.FP	0.413	0.403	0.426
	II.P	0.458	0.438	0.461
Density after immersion [g/cm^3]	I.FP	0.6357	0.6038	0.4632
	II.P	0.7071	0.6593	0.4402
Apparent density [g/cm^3]	I.FP	0.3768	0.3894	0.2856
	II.P	0.4022	0.4340	0.2731
The growth in volume [%]	I.FP	9.532	3.625	47.545
	II.P	13.753	0.955	68.657

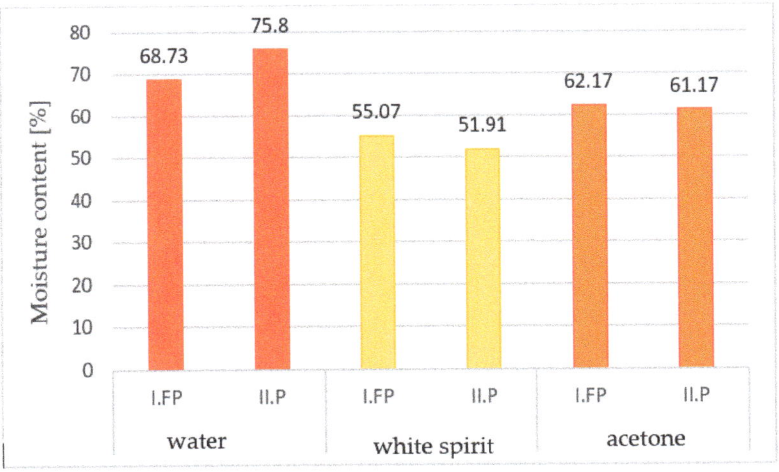

Figure 6. The moisture contents of the samples as a percentage, after immersion in solvents for 15 min.

The moisture content, obtained mathematically, can be attributed to the fact that the samples experienced an intense xylophagous attack that caused galleries in the wood. The wooden substance was lost and was replaced by the liquid absorbed by the wood.

The results regarding the determinations of mass, volume, solvent absorption, and density, both before and after immersion of the thin specimens, can be seen in Table 2.

The densities reached by the samples after immersion can be seen in Figure 7.

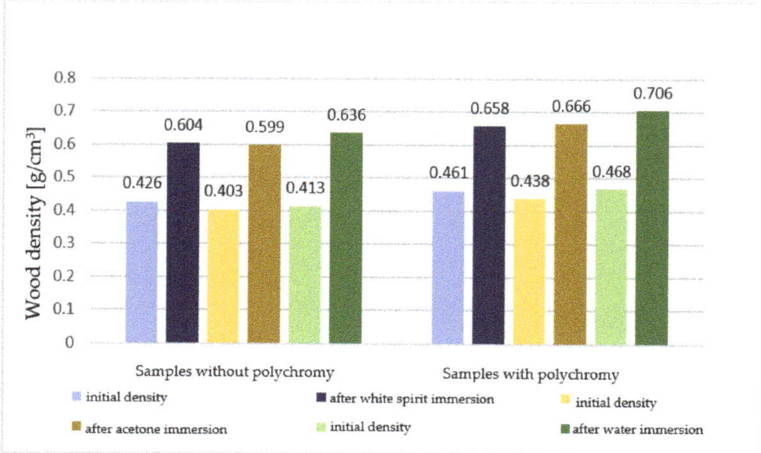

Figure 7. Densities (g/cm^3) of the sectioned samples before treatment and after extraction from the solvents.

The same methodology was applied to the samples with a thickness of 27 mm, and the results thus obtained can be seen in Table 3. Observing the results obtained in Table 3 and comparing them with those in Table 2, it can be seen that the behavior of the wood is different depending on its thickness, the type of liquid, and the length of time during which the liquid acts on the wood.

Comparing the two types of samples with different thicknesses (27 mm and 12 mm), the difference in volume growth can be seen in Figure 8.

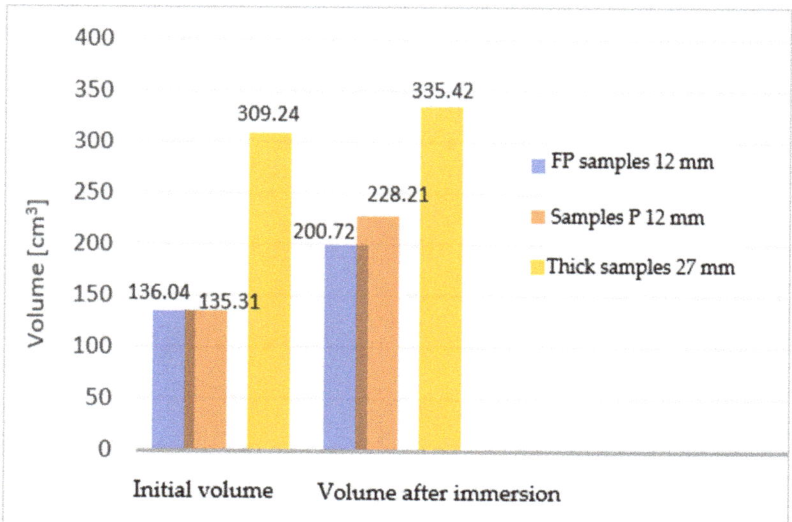

Figure 8. The difference in volume growth for the two types of samples.

Table 3. Means obtained from the determinations for the samples with a thickness of 27 mm.

Determinations	Solvent Used		
	Water	White Spirit	Acetone
Sample code	1P	3P	5P
Mass before the immersion [g]	125.32	134.11	130.48
Initial volume [cm^3]	284.87	308.74	309.24
Mass after immersion [g]	176.84	210.245	224.7
Volume after immersion [cm^3]	301.72	312.33	335.42
Solvent absorption [g]	51.52	76.135	94.22
Solvent absorption [cm^3]	51.52	97.61	119.27
Absorption from total used (300 mL) [%]	17.17	32.54	39.76
Solvent content/test tube, relative to initial volume and volume of solvent absorbed [%]	18.085	31.615	38.567
Initial density [g/cm^3]	0.439	0.434	0.421
Density after immersion [g/cm^3]	0.586	0.673	0.669
Apparent density [g/cm^3]	0.415	0.429	0.389
Increase in volume [%]	5.91	1.16	8.46

For this experiment, we wanted to report the changes regarding swelling in the tangential direction because this section produces the greatest change, according to the specialized literature [12,13]. These changes are noted in Tables 4 and 5. Moreover, the research refers to and applies to the specific case of painted panels, representing art objects that require impregnation treatments, and these panels were usually developed on the tangential section.

Table 4. Dimensional changes in tangential average values with swelling.

Solvent	Sample Code	Swelling Difference between Initial Measurements and after Partial Immersion (15 min)			
		Initial Tangent Dimensions (mm)	Tangential Dimensions after Immersion (mm)	Swelling (mm)	Dimensional Variation ($\Delta\alpha$) (%)
Water	I.FP **	130.88 (0.97) *	136.26 (0.83) *	5.38	4.11
	II.P ***	130.83 (0.72) *	137.54 (0.97) *	6.71	5.13
White spirit	I.FP **	132.06 (0.67) *	132.46 (0.67) *	0.4	0.30
	II.P ***	132.84 (0.81) *	132.89 (0.24) *	0.05	0.04
Acetone	I.FP **	130.04 (0.86) *	138.95 (2.60) *	8.91	6.85
	II.P ***	130.04 (0.86) *	140.26 (0.89) *	10.22	7.86

* Standard deviation; ** FP wood without polychromy; *** P wood with polychromy.

Table 5. Tangential swellings in the case of specimens with a thickness of 27 mm.

Solvent Sample Code		Swelling Difference between Initial Measurements and after Partial Immersion (15 min)			
		Initial Tangent Dimensions (mm)	Tangential Dimensions after Immersion (mm)	Swelling (mm)	Dimensional Variation (Δα) (%)
Water	1P (A1)	130.39	133.03	2.64	2.02
	2P (A2)	129.14	131.54	2.40	1.86
White spirit	3P (A3)	128.93	129.64	0.71	0.55
	4P (A4)	129.82	130.20	0.38	0.29
Acetone	5P (A5)	129.60	134.70	5.10	3.94
	6P (A6)	129.45	134.62	5.17	3.99

Acetone was observed to cause the most significant swelling of damaged wood.

For example, in the case of samples without polychromy, the swelling was 65.61% greater in the case of immersion in acetone than in water.

Regarding the polychromy samples, the swelling variation was 52.31% greater in the case of immersion in acetone compared to water.

Comparing the white spirit solvent with acetone, it was observed that the swelling in acetone was 2127% higher for the samples without polychromy and 20,340% higher for the polychromy samples.

The swelling ratio in the tangential direction was also determined for the thicker specimens. It was found that in these cases, acetone also produced the greatest swelling, but this was smaller compared to the sectioned pieces (12 mm thick).

It can be concluded that, depending on the thickness of the wood and the time it remains under the action of the solvent, the swelling ratio is higher or lower.

3.2. Dimensional Changes—Wood Deformation Produced by Solvents

As a result of the experiment, it is evident that there are changes in the flatness of the samples. The deformation could be observed visually and was also determined using a dial gauge and digital depth gauge measurements. The values obtained for the three solvents used in the experiment, for both thin and thick samples, can be seen in Tables 6–8.

Table 6. Deformations produced in the samples that were partially immersed in water (in mm).

Code / Sample		Before Treatment			After Immersion			The Difference between Points on the Horizontal Axis		
Solvent	Axis of Determination	Points on the Axis						$\Delta\mu = l_{\text{after immersion}} - l_{\text{before immersion}}$		
Water		D	YD	C	D1	YD1	C1			
SECTIONAL SAMPLES 12 mm										
I P	D-YD-C	5	7.88	8.43	7.35	5.59	10.15	2.35	−2.29	1.72
	XA-XB	5		7.55	7.2		10.31	2.2	0	2.76
	A-YA-B	4.3	5.94	5.64	7.44	5.38	10.71	3.14	−0.56	5.07
I FP	D-YD-C	5	5.18	6.27	7.78	5.15	10.4	2.78	−0.03	4.13
	XA-XB	6.43		6.2	9.67		10.48	3.24	0	4.28
	A-YA-B	7.47	5.8	5	10.7	6.61	10.56	3.23	0.81	5.56
II P	D-YD-C	4	5.1	6.75	4.23	5.4	6.8	0.23	0.3	0.05
	XA-XB	3.45		5.27	4		6	0.55	0	0.73
	A-YA-B	3.19	3.45	3.88	3.7	4.21	4.84	0.51	0.76	0.96
II FP	D-YD-C	5.7	5.65	5.55	8.39	5.58	8.35	2.69	−0.07	2.8
	XA-XB	6.66		5.63	7.65		7.73	0.99	0	2.1
	A-YA-B	7.27	5.78	5.68	9.15	5.18	6.85	1.88	−0.6	1.17

Table 6. *Cont.*

Code / Sample / Solvent: Water	Axis of Determination	D	YD	C	D1	YD1	C1	The Difference between Points on the Horizontal Axis $\Delta\mu = l_{after\ immersion} - l_{before\ immersion}$		
		Before Treatment			After Immersion					
		SECTIONAL SAMPLES 12 mm								
III P	D-YD-C	7.51	7.35	5.77	10.68	8.44	7.09	3.17	1.09	1.32
	XA-XB	8.82		6.49	10.51		7.78	1.69	0	1.29
	A-YA-B	10.13	9.66	7.31	10.51	8.48	7.96	0.38	−1.18	0.65
III FP	D-YD-C	4.79	5.1	5.65	5.62	4.35	5.92	0.83	−0.75	0.27
	XA-XB	5.44		5.17	6.1		5.93	0.66	0	0.76
	A-YA-B	5.77	5.12	4.64	6.64	4.77	5.74	0.87	−0.35	1.1
		SECTIONAL TESTS 27 mm								
A1 (1P)	D-YD-C	5.26	6.74	6.18	7.8	6.97	7.47	2.54	0.23	1.29
	XA-XB	6.1		5.75	7.76		7.14	1.66	0	1.39
	A-YA-B	6.37	6.86	5.79	7.76	6.95	7.13	1.39	0.09	1.34
A2 (2P)	D-YD-C	7.64	8.45	8.62	8.88	9.31	11.73	1.24	0.86	3.11
	XA-XB	8.38		8.36	8.36		11.52	−0.02	0	3.16
	A-YA-B	8.99	8.7	8.1	8.38	8.28	11.42	−0.61	−0.42	3.32

Table 7. Deformations produced in the samples that were partially immersed in white spirit.

Code / Sample / Solvent: White Spirit	Axis of Determination	D	YD	C	D1	YD1	C1	The Difference between Points on the Horizontal Axis $\Delta\mu = l_{after\ immersion} - l_{before\ immersion}$		
		Before Treatment			After Immersion					
		SECTIONAL SAMPLES 12 mm								
I P	D-YD-C	5.69	7.94	8.55	5.9	7.63	8.86	0.21	−0.31	0.31
	XA-XB	5.38		7.32	5.4		7.7	0.02	0	0.38
	A-YA-B	4.6	6.08	5.9	4.58	5.65	6.09	−0.02	−0.43	0.19
I FP	D-YD-C	5.08	4.95	7.47	5.78	5.33	8.74	0.7	0.38	1.27
	XA-XB	6.47		7.94	6.91		9.2	0.44	0	1.26
	A-YA-B	7.68	6.17	8.08	8.1	6.6	9.09	0.42	0.43	1.01
II P	D-YD-C	4.72	7.29	5.54	4.75	7.34	6.06	0.03	0.05	0.52
	XA-XB	5.71		4.66	5.36		5.06	−0.35	0	0.4
	A-YA-B	6.53	6.6	4.04	6.28	6.49	4.17	−0.25	−0.11	0.13
II FP	D-YD-C	6.13	6.29	8.5	6.31	6.42	8.8	0.18	0.13	0.3
	XA-XB	5		7.62	6.04		8.09	1.04	0	0.47
	A-YA-B	5.54	5.06	6.38	5.81	5.16	6.66	0.27	0.1	0.28
III P	D-YD-C	7.8	8.31	6.32	8.1	8.33	6.55	0.3	0.02	0.23
	XA-XB	9.76		6.96	9.67		7.22	−0.09	0	0.26
	A-YA-B	10.41	11.82	7.79	10.56	10.75	8.11	0.15	−1.07	0.32
III FP	D-YD-C	5.25	5.86	7.23	5.46	5.75	6.85	0.21	−0.11	−0.38
	XA-XB	5.29		6.39	5.83		6.25	0.54	0	−0.14
	A-YA-B	5.2	4.95	5.26	5.98	5.32	5.74	0.78	0.37	0.48
		SECTIONAL SAMPLES 27 mm								
A3 (3P)	D-YD-C	8.23	9.43	10.02	8.35	9.45	10.12	0.12	0.02	0.1
	XA-XB	8.24		9.46	8.3		9.62	0.06	0	0.16
	A-YA-B	7.92	8.69	8.56	8.03	8.75	8.71	0.11	0.06	0.15
A4 (4P)	D-YD-C	8.34	8.64	8.97	8.32	8.53	8.94	−0.02	−0.11	−0.03
	XA-XB	8.13		8.75	8.25		8.79	0.12	0	0.04
	A-YA-B	8.3	8.64	8.29	8.3	8.64	8.33	0	0	0.04

Table 8. Deformations produced in the samples that were partially immersed in acetone.

Code Sample / Solvent: Acetone	Axis of Determination	Before Treatment D	Before Treatment YD	Before Treatment C	After Immersion D1	After Immersion YD1	After Immersion C1	The Difference between Points on the Horizontal Axis $\Delta\mu = l_{after\ immersion} - l_{before\ immersion}$		
					SECTIONAL SAMPLES 12 mm					
I P	D-YD-C	6.75	7.41	7.31	9.23	7.1	7.2	2.48	−0.31	−0.11
	XA-XB	7.39		7.24	8.88		8.84	1.49	0	1.6
	A-YA-B	7.98	7.75	7.08	8.86	6.45	7.98	0.88	−1.3	0.9
I FP	D-YD-C	6.06	6.21	9.6	6.29	5.98	9.27	0.23	−0.23	−0.33
	XA-XB	7.59		9.74	7.73		9.73	0.14	0	−0.01
	A-YA-B	8.9	7.38	9.91	8.91	7.55	10.1	0.01	0.17	0.19
II P	D-YD-C	5.2	7.79	6.7	6.74	5.92	8.87	1.54	−1.87	2.17
	XA-XB	5.87		5.95	6.48		8.62	0.61	0	2.67
	A-YA-B	6.65	6.7	4.85	6.31	6.18	8.82	−0.34	−0.52	3.97
II FP	D-YD-C	7.51	7.61	8.21	8.27	6.76	9.31	0.76	−0.85	1.1
	XA-XB	7.92		7.66	8.55		8.51	0.63	0	0.85
	A-YA-B	8.18	7.02	6.99	9.89	5.75	7.55	1.71	−1.27	0.56
III P	D-YD-C	8.56	8.86	6.9	11.73	9.22	8.82	3.17	0.36	1.92
	XA-XB	10.04		7.69	11.73		8.83	1.69	0	1.14
	A-YA-B	11.2	11.39	8.49	11.74	9.14	8.61	0.54	−2.25	0.12
III FP	D-YD-C	6.38	5.51	6.22	6.54	5.15	6.5	0.16	−0.36	0.28
	XA-XB	7.52		6.41	7.42		6.56	−0.1	0	0.15
	A-YA-B	8.52	6.52	6.4	8.03	6.15	6.62	−0.49	−0.37	0.22
					SECTIONAL SAMPLES 27 mm					
A5 (5P)	D-YD-C	8.36	8.45	8.92	8.92	9.17	9.46	0.56	0.72	0.54
	XA-XB	8.36		8.84	8.95		9.7	0.59	0	0.86
	A-YA-B	8.36	8.56	8.76	9.03	9.27	9.61	0.67	0.71	0.85
A6 (6P)	D-YD-C	8.6	8.44	8.68	9.09	9.76	10.05	0.49	1.32	1.37
	XA-XB	8.48		8.77	9.15		10.02	0.67	0	1.25
	A-YA-B	8.4	8.49	8.92	9.14	9.78	9.98	0.74	1.29	1.06

3.2.1. Results for Water Samples

The water-immersed specimens, both the thin (12 mm) and the thick (27 mm) samples, curved either concavely or convexly. Where the values are negative as a result of the immersion, a change is produced that is in a direction lower than the initial value, resulting in deformation in the form of either bends or twists.

Figure 9 identifies the changes that occurred after immersion in water in the samples with a section of 27 mm, as shown on the three horizontal axes. Pronounced deformation can be observed, compared to the original shape of the specimen.

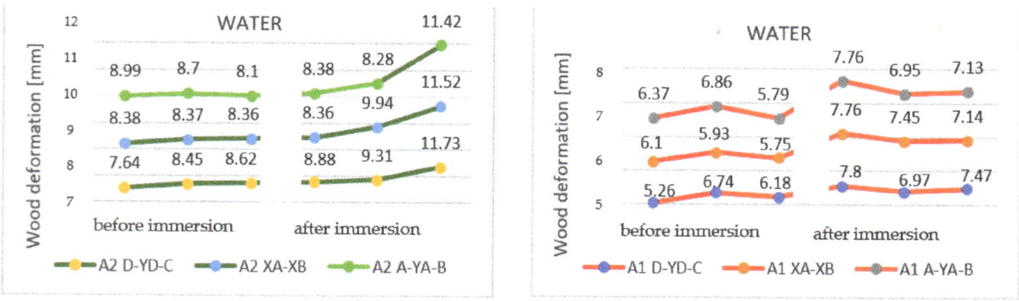

(a) (b)

Figure 9. The deformations on the horizontal axis, shown on the three measuring lines, for the specimens with a thickness of 27 mm: (**a**) sample A2 (2P); (**b**) sample A1 (1P).

The different deformation variations between the points were measured initially and after, as shown on the axis D-DY-C; D1-YD1-C1 is also highlighted in Figure 10a,b.

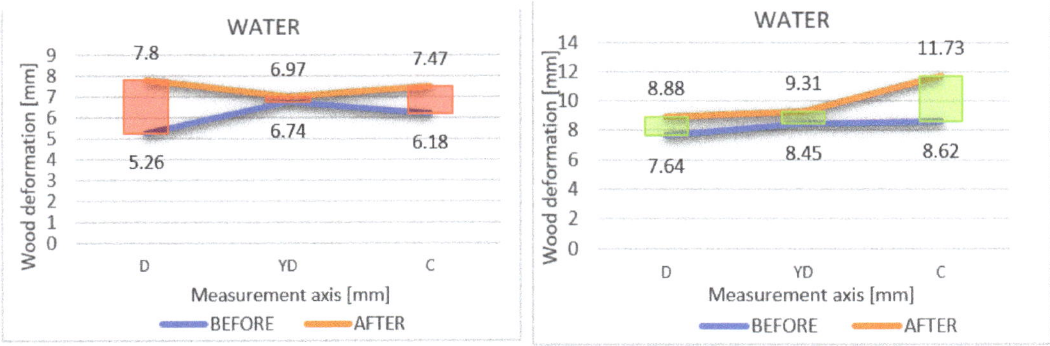

(**a**) (**b**)

Figure 10. (**a**,**b**) The different variations in deformation between the points, measured initially and after immersion, on the axis D-DY-C and D1-YD1-C1, respectively.

The thin samples changed after being extracted from the water and strong curvatures were found, with a curvature radius of between 4 and 10 mm (Figures 11 and 12a) In their case, unlike with the thick samples, torsions (Figure 12b) were present with values of between 4 and 7 mm.

Figure 11. Strong curvatures are visible after extracting the sectioned samples from the water.

(**a**) (**b**)

Figure 12. (**a**) Curvature of 10 mm after extraction from water; (**b**) torsion after extraction from water.

Figure 13 shows the changes that occurred after immersion in water for the tested thin samples.

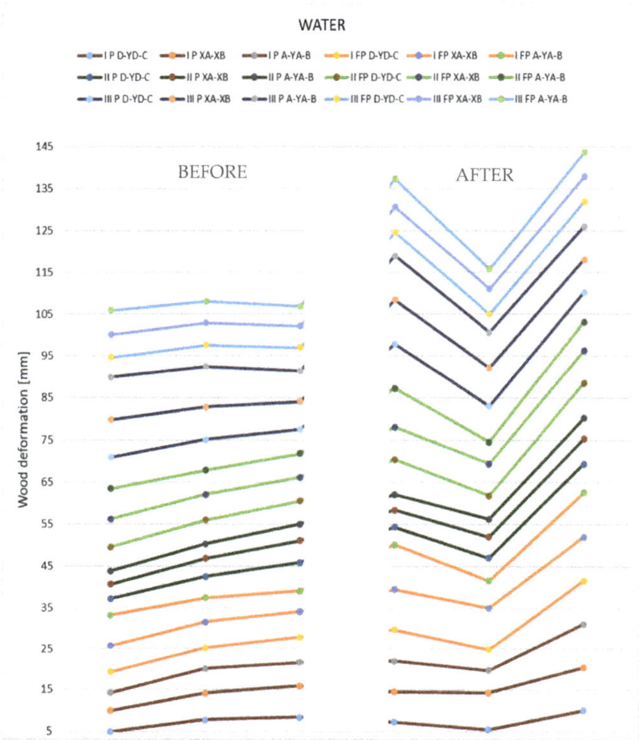

Figure 13. Representation of the deformation values obtained from measurements of thin 12 mm specimens.

3.2.2. Results Obtained with White Spirit

In the case of samples immersed in white spirit, it is evident that the wood absorbed a significant amount of solvent (31.89 g and 40.88 cm^3). However, it is clear that the tested samples showed fewer twists and bends compared to the other two solvents. The results obtained from the experiment can be seen in Table 7.

In the case of specimens with a thickness of 27 mm, it can be observed that the deformations identified in Figure 14 were uniform and were greatly reduced throughout the samples.

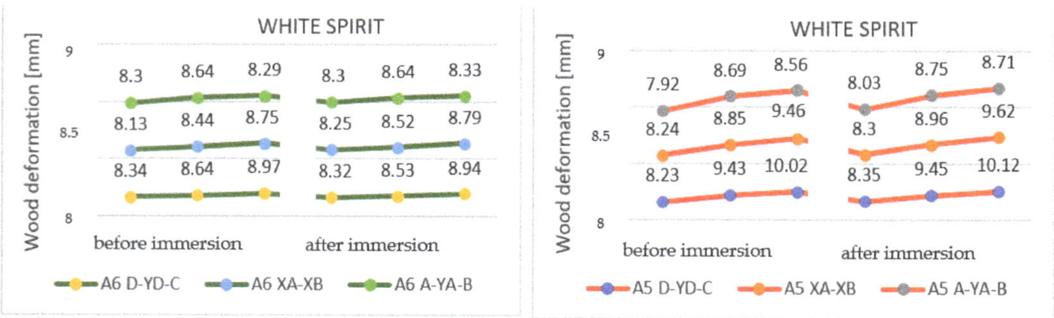

(**a**) (**b**)

Figure 14. Dimensional variations after immersion in white spirit: (**a**) sample A4 (4P); (**b**) sample A3 (3P).

The resulting deformations in the thin samples (12 mm) were more obvious, and, to exemplify them on the surface, they were marked on the measuring axes in Figure 15.

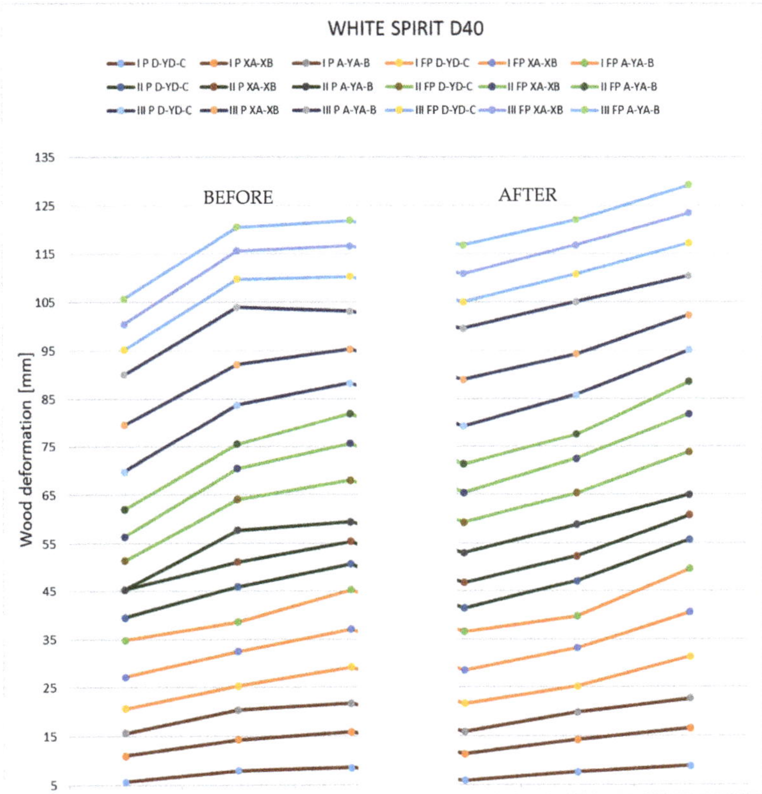

Figure 15. Deformation after immersion in white spirit.

3.2.3. Results Obtained for Acetone

After immersing the 27-millimeter samples in acetone, it was evident that two types of deformations occurred: pronounced curvatures and torsions with reduced values. These can be seen in Figure 16a,b.

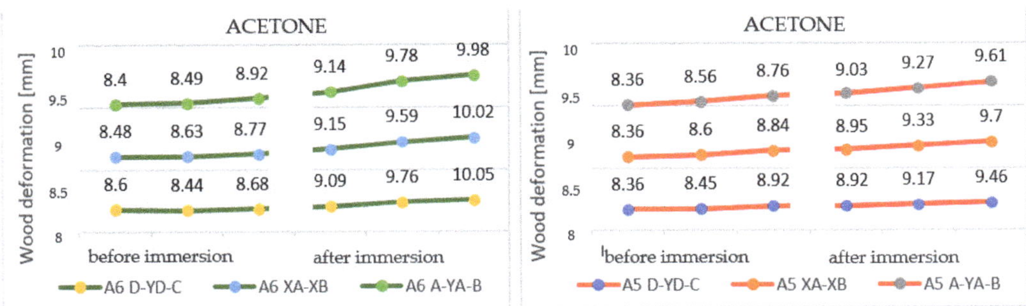

(**a**) (**b**)

Figure 16. Changes following the immersion in acetone of thick samples: (**a**) sample A6 (6P); (**b**) sample A5 (5P).

The 12-millimeter sectioned samples show different and non-uniform curvatures and torsions; these deformations are noticeable for both types of samples (with polychromy and without polychromy).

For example, for the thin sample without painting (1FP), deformations of the convex curvature type, with approximately 3–4 mm of curvature (arrow) and torsion of 6 mm are evident (Figure 17).

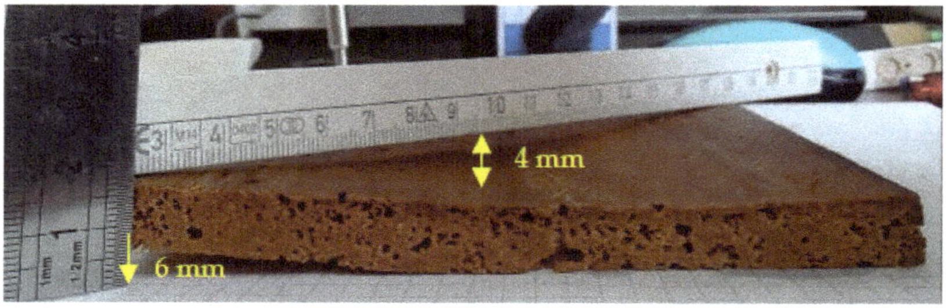

Figure 17. Thin specimen without painting, with strong twisting.

In the case of the polychromy specimen (3P), there are torsional deformations of approximately 8–12 mm, and a curvature of 4 mm (Figure 18), while the specimen without painting, 1FP, does not show significant deformations, with only a slight torsion of about 1–1.5 mm (Figure 19).

Figure 18. Sample 3P shows significant twists and bends after extraction from the solvent.

Figure 19. Sample 1FP: after extraction from the acetone, a slight torsion is observed, compared to the other samples extracted from the same solvent.

In the following graph (Figure 20), it is possible to see the deformations on the three measuring axes for the sectioned samples (three with polychromy and three without polychromy).

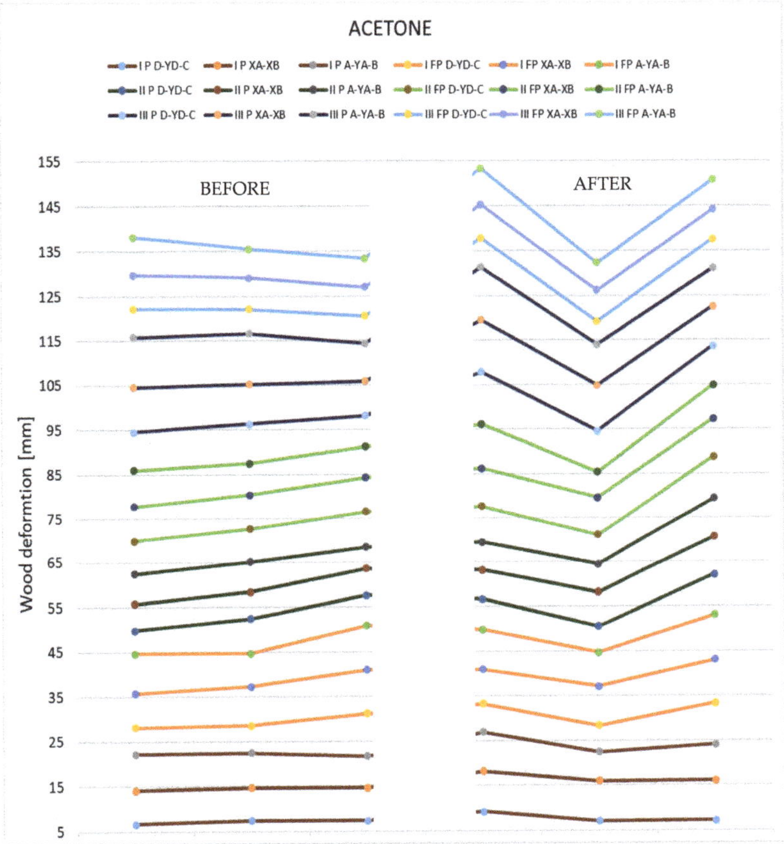

Figure 20. Deformations produced by acetone in the thin specimens.

4. Discussion

Before applying a solvent treatment to a damaged wood object, it is necessary and important to first determine its volume, in order to be able to determine the amount of solvent to be introduced. In the case of art objects that have been severely damaged by xylophagous insects, where the damage is not uniform, the use of an excessive amount of solvent for a treatment application is difficult to control, and consequences may occur for the support but especially for the polychromy layer. The migration of the solvent to the surface with the painting, in relation to the heavily damaged wood, can happen in a very short time, thereby destabilizing it. The choice of a fast-evaporating solvent will also cause dimensional changes to the wooden support, as well as cracks and fractures. In the case of solvents with a slow evaporation rate, there is both the advantage of a deep treatment and the disadvantage of increasing the carrier mass, with the evaporation time being long.

For this experiment, the amount of solvent was limited because total immersion of the samples was not desired since polychromic art objects cannot be fully immersed. Restorative treatments, whether preventive or curative, that are applied to prevent or stop bio-pests, or for the purposes of consolidation, will generally require multiple, repeated sessions. In this context, wood undergoes dimensional changes and deformations through repeated swelling and contraction. This stress will inevitably transfer to the layers of polychromy.

In restoration laboratory practice, during strengthening treatments with synthetic resins, deformations of geometry and dimensions appeared on the panels showing severe damage caused by wood-eating insects. The treatment was performed with Paraloid B72, at 10%, in acetone. The changes observed were mainly in the geometry of the panel, when, after introducing the consolidant, the panel curves according to its constraints and the positioning of the work surface (Figure 21). The curvature radius was 7.93 mm for a width of 630 mm [28].

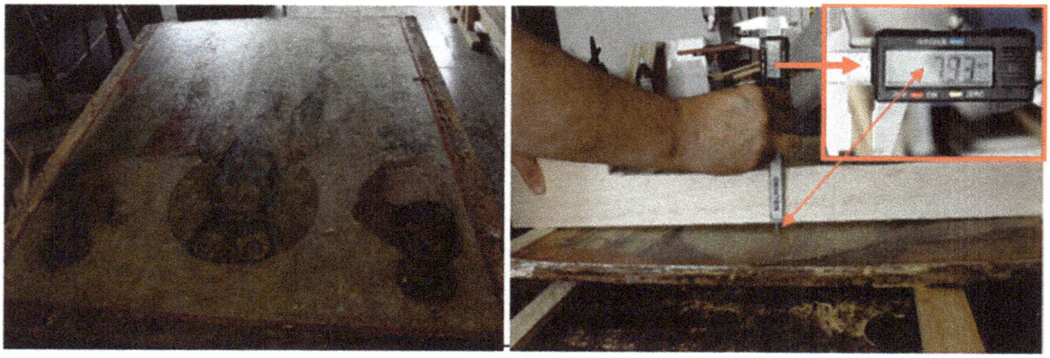

(a) (b)

Figure 21. (**a**) Panel flatness before treatment; (**b**) changes produced during treatment.

From this observation, it is clear that the panel changed not only its geometry but also its dimensions during the treatment. The present study was developed to ascertain the causes of these deformations, with a return to the original shape after the completion of the treatment and the evaporation of the solvent that solubilized the consolidant. This warping and the changes in geometry and dimension effects, although only 1% to 3%, actually cause stress to the transfer media on the paint layer, resulting in cracks, delamination, and gaps.

4.1. Fluid Absorption and Swelling Dimensional Changes

Regarding the swelling of historical wood damaged by xylophagous insects, it can be seen that depending on the level of damage [10,28], the thickness of the wood, and the time in which it is in contact with a solvent, it will dimensionally change differently (Figure 22).

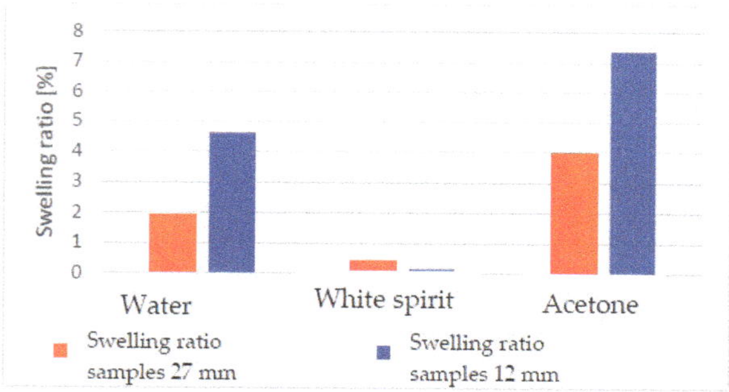

Figure 22. The swelling ratio according to the thickness of the specimens and the immersion time (15 min for both types).

In the case of less polar solvents, the migration inside the samples was very fast, at less than 2 min for acetone and up to 5 min for white spirit, compared to water, where, due to the increased polarity, the absorption occurred more slowly; the absorption/migration time increased to 10 min (Figure 23).

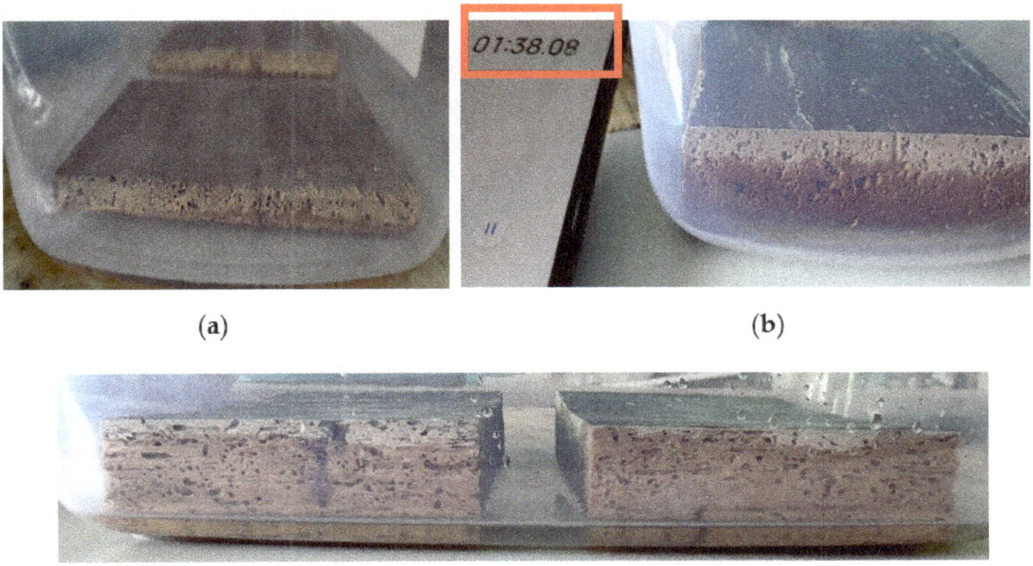

(a) (b) (c)

Figure 23. Migration of the solvent into the wood after the first 2 min: (**a**) water; (**b**) white spirit; (**c**) acetone.

The solvents were absorbed in different amounts and volumes (g and cm^3). Along with this absorption, different degrees of swelling occurred, as exemplified in Tables 5 and 6, which show that the largest volume is represented by the swelling in acetone, at approximately 68.65%. If we refer to the initial mass of the samples and the maximum volume obtained after extraction, it can be concluded that the mass and volume after immersion do not increase equally or in the same proportions with any of the solvents used.

During the experiments observing the changes in the wood that were produced by the solvents, it was found that the greatest absorption occurred in the samples that were partially immersed in water; however, the greatest level of swelling occurred with the acetone solvent.

There are different mass increases in all three solvents. In the case of water and white spirit, evaporation is slower and remains longer in the volume of wood, while acetone, being very volatile, evaporates much faster.

During the research, it was found that in the case of acetone, once it evaporated from the wood, the initial mass of the sample decreased from 63 g to 58 g. This phenomenon can be explained by the fact that the rapid evaporation of acetone also involves water from the initial moisture content of the wood in the evaporation process, thereby reducing the density of the wood.

In the case of the white spirit solvent, after immersion, the mass of the sample increased to 212 g, resulting in a quantity of 77 g of remaining solvent in the wood. After 24 h of slow evaporation, the mass reached 182 g, including 47 g of solvent, an additional 30 g having evaporated. It was found that after 24 h, approx. 38.96% of the absorbed amount had evaporated.

Analyzing the samples with polychromy, it was found that white spirit, even if it migrates to the painted surface, does not then destabilize the polychrome layer, does not wet it, and does not produce additional cracking, as seen in the case of acetone or water.

Regarding the samples immersed in acetone, it was observed that micro-cracks appeared on the wooden surface and on the polychrome layer, where the already existing micro-cracks became more pronounced, making the layer unstable and brittle (Figures 24 and 25).

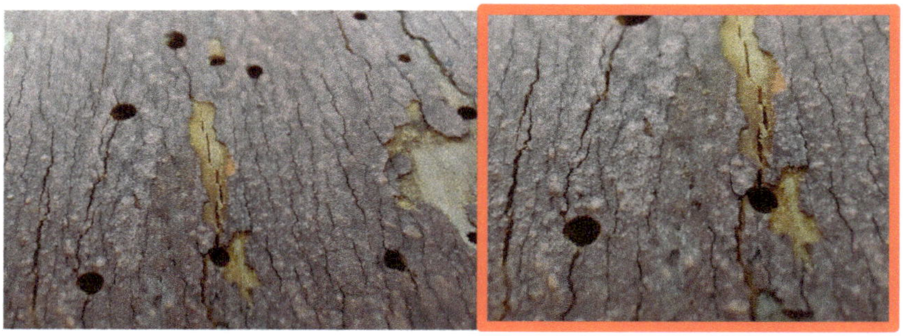

Figure 24. Cracks produced in the wood under the polychrome layer, with accentuation of the originally existing cracks.

Figure 25. Magnified view of the cracks and swelling after immersion in acetone.

On the polychrome samples that were in contact with water, it was found that after migrating to the polychrome surface, the water caused wetting and destabilized the layer. On the reverse side, the initial deep cracks stood out, an accentuated cracking of approximately 0.8 to 1 mm occurred, and this continued throughout the thickness of the specimen (Figure 26).

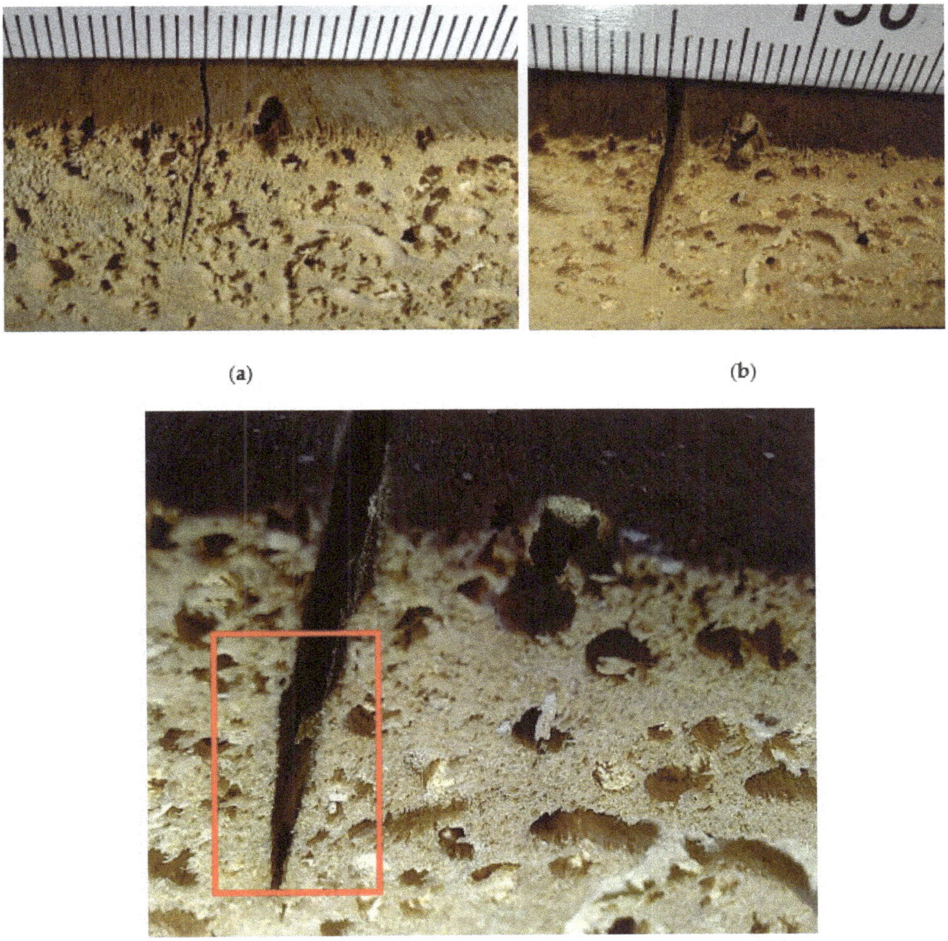

(a) (b) (c)

Figure 26. Cracking of wood in water: (**a**) wood crack before immersion; (**b**) crack spacing after immersion; (**c**) the freshly detached woody fraction can be seen.

4.2. Dimensional Changes—Wood Deformations Produced by Solvents

Deformations following immersion in water. Once the liquid is absorbed, the wood increases in volume and deforms differently, depending on the solvent.

Concave or convex curvatures, as well as twists, can be seen to occur. It is clear that the deformations are not always influenced by the annual rings, and the curvature does not occur only in the opposite direction to the annual rings.

In Figure 27, we can see an example of a situation with concave curvature toward the face with polychromy, producing an agglomeration of the painted surface, detachments in the "roof" (exfoliated painted layer), and, eventually, gaps in the surface. When the curvature is convex to the polychromy layer, it will stretch with the curvature of the wood, but since it does not have the same elasticity, cracks and gaps will occur.

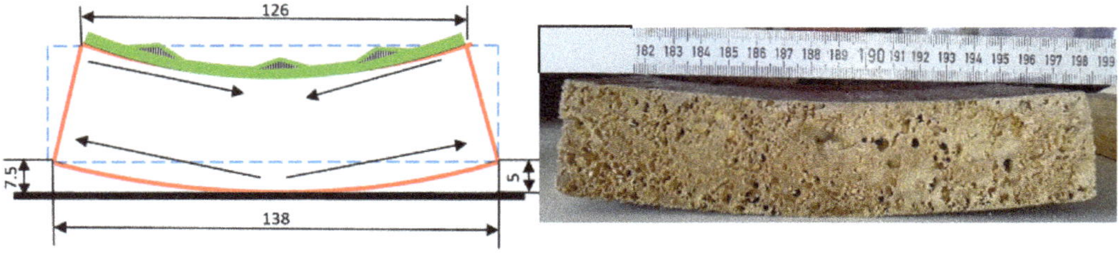

Figure 27. Graphic representation of concave deformation relative to the surface with polychromy, producing detachments in the roof.

In the case of samples with a thickness of 27 mm, extracted from the water and left to dry in the open air for 24 h, on the polychromy face, deformations of the concave curvature type are present, sample 1P having a curvature radius of approximately 6–7 mm (Figure 27). It can be seen that the deformation continues; the curvature on the convex side increases to 140 mm, and the bend increases from 7.5 mm (Figures 27 and 28) to 10 mm.

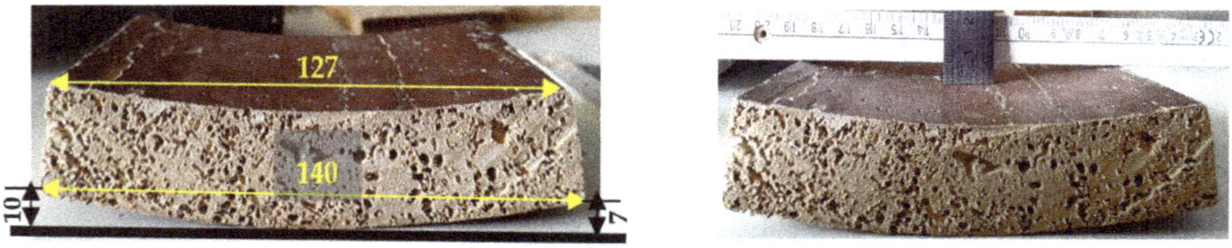

Figure 28. The test piece after 24 h of immersion in water.

Deformations following immersion in white spirit. Analyzing the results that were obtained after immersing the samples in white spirit, it can be seen that in the case of pieces with a thickness of 27 mm, the deformations are very reduced, almost insignificant. In the case of thin specimens, they show a tendency to deform in the direction opposite to the initial shape.

Deformations following immersion in acetone. Analyzing all the samples immersed in acetone, it is evident that the deformations produced by the solvent are non-unitary. It can be seen that in the case of thick specimens, more pronounced bends and fewer torsions occur, while, in the case of thin specimens, the bends are slightly smaller, but the torsions are more pronounced.

The curvatures are, on average, between 1.5 and 5 mm, while the twists have values of between 4 and 12 mm.

In sample 1 FP and sample 1P, a strong deformation appeared, which changed the sample from the initial state, with a convex curvature, to a reverse deformation, with a curvature in the concave plane. Point D was initially 6.75 mm; after immersion, it reached 9.23 mm. All this must also be correlated with the swelling ratio, as well as the effects on the polychromy, highlighted in Section 4.1, regarding absorption and swelling in acetone, as well as is shown in Figures 25 and 26, where the cracks produced in the polychromy layer are revealed.

5. Future Research Directions

The current research may extend to other solvents used in the restoration of wooden art objects. Dimensional changes and deformations produced during consolidation treatments with Paraloid B72 can also be studied comparatively.

6. Conclusions

This study focused on understanding the changes and deformations occurring in polychrome panels from heritage assets during restoration treatments.

The data obtained give us a better understanding of the mechanisms behind these changes. This information is essential for developing an intervention process that reduces or limits impairments without exacerbating them through the treatment response.

We wished to observe the changes produced by solvents on the wood, having a limited time and a reduced amount of solvent, respectively, at 10% of the volume of the tested pieces.

Wet treatments were applied to polychrome panels that show significant degradation caused by the attack of xylophagous insects; the treatments must be prepared and applied rigorously.

Liquid solutions used to perform such treatments can produce:

1. Dimensional changes—swelling during treatment and contraction after the evaporation and drying of the wood.
2. Deformations—concave/convex curvature and twists.

Changes and deformations occurred simultaneously throughout the composite of the work of art.

The elasticity of the components of the matrix of such cultural goods is not uniform and similar and, as such, behaves differently from the stresses and forces to which the goods are subjected during their treatments. Not knowing the limits and, especially, the impossibility of controlling the stresses, causes fissures, gaps, fractures, and damage to the art object.

Author Contributions: Conceptualization, A.M.-A. and C.Ș.I.; methodology, C.Ș.I.; software, A.M.-A. and C.Ș.I.; validation, A.L. and C.Ș.I.; formal analysis, C.Ș.I.; investigation, A.M.-A.; resources, A.M.-A.; data curation, A.L.; writing—original draft preparation, A.M.-A.; writing—review and editing, A.L. and C.Ș.I.; visualization, C.Ș.I.; supervision, C.Ș.I.; project administration, C.Ș.I.; funding acquisition, C.Ș.I. All authors have read and agreed to the published version of the manuscript.

Funding: This research received no external funding.

Institutional Review Board Statement: Not applicable.

Informed Consent Statement: Not applicable.

Data Availability Statement: Not applicable.

Acknowledgments: The authors thank the Transylvania University of Brasov for the technical and logistical support provided; the authors also thank the Research and Conservation Laboratory for Cultural Heritage RESTAURARE IONESCU CONSTANTIN, Sibiu, Romania.

Conflicts of Interest: The authors declare no conflict of interest.

References

1. Parliament of Romania—Law No. 182 of 25 October 2000 Regarding the Protection of the National Cultural Heritage Object. Published in the Official Gazette Number 10 of 8 January 2007. Available online: https://en.unesco.org/sites/default/files/rom_lege_182_romorof.pdf (accessed on 29 September 2023).
2. Code of Ethics Document Promoted by the European Confederation of Conservator-Restoration Organizations and Adopted by the General Assembly, Brussels, 03/01/2002 Professional Guide (II) Promoted by the European Confederation of Conservator-Restorer Organizations and Approved in Its General Assembly of 7 March 2003. Available online: https://www.academia.edu/2905696/Cod_deontologic_E_C_C_O_E_C_C_O_PROFESSIONAL_GUIDELINES (accessed on 20 April 2022).
3. Castelli, C. The Restoration of Panel Painting Supports Some Case Histories în Current Approaches to the Structural Conservation of Panel Paintings. In *The Structural Conservation of Panel Paintings: Proceedings of a Symposium at the J. Paul Getty Museum, 24–28 April 1995*; Getty Publications: Los Angeles, CA, USA, 1998; pp. 316–340.
4. Caggiani, M.C.; Cosentino, A.; Mangone, A. Pigments Checker version 3.0, a handy set for conservation scientists: A free online Raman spectra database. *Microchem. J.* **2016**, *129*, 123–132. [CrossRef]
5. Janas, A.; Fuster-López, L.; Andersen, C.K. Mechanical properties and moisture-related dimensional change of canvas paintings–canvas and glue sizing. *Herit. Sci.* **2022**, *10*, 160. [CrossRef]

6. Sargent, R. Evaluating dimensional stability in solid wood: A review of current practice. *J. Wood Sci.* **2019**, *65*, 36. [CrossRef]
7. Filipovici, J. *Studiul Lemnului*, 1st ed.; Didactică și Pedagogică: București, Romania, 1964; Volume 1, pp. 68–96.
8. Pescăruș, P. Proprietățile Fizice ale Lemnului. In *Tehnologia Prelucrării Lemnului*; Râmbu, I., Ed.; Tehnică: București, Romania, 1978; pp. 64–89.
9. Wadum, J. Microclimate Boxes for Panel Paintings in Approaches to the Structural Conservation of Panel Paintings. In *The Structural Conservation of Panel Paintings: Proceedings of a Symposium at the J. Paul Getty Museum, 24–28 April 1995*; Getty Publications: Los Angeles, CA, USA, 1998; pp. 497–524.
10. Ionescu, C.Ș. Stabilizarea și Consolidarea Suportului Lemnos Pentru Obiectele de Patrimoniu, Utilizând Materiale Clasice și Modern/Stabilization and Consolidation of Wooden Support for Heritage Objects, Using Classic and Modern Materials. Ph.D. Thesis, Transilvania University of Brașov, Brașov, România, 2020.
11. Avram, A.; Ionescu, C.S.; Lunguleasa, A. A consolidation of degraded lime wooden support from heritage objects using two types of consolidant. *BioResources* **2023**, *18*, 4580–4597. [CrossRef]
12. Arzola-Villegas, X.; Lakes, R.; Plaza, N.Z.; Jakes, J.E. Wood Moisture-Induced Swelling at the Cellular Scale—Ab Intra. *Forests* **2019**, *10*, 996. [CrossRef]
13. Zhan, T.; Lyu, J.; Eder, M. In situ observation of shrinking and swelling of normal and compression Chinese fir wood at the tissue, cell and cell wall level. *Wood Sci. Technol.* **2021**, *55*, 1359–1377. [CrossRef]
14. Callum, A.S. *Hill Wood Modification: Chemical, Thermal and Other Processes*; John Wiley & Sons Ltd.: Chichester, UK, 2006; pp. 19–44.
15. Rowell, R.M. *Handbook of Wood Chemistry and Wood Composites*; CRC Press: Boca Raton, FL, USA, 2005; Chapter 4; pp. 77–97.
16. Lunguleasa, A. *Wood Physics and Mechanics*; Editura Universiatii Transilvania: Brasov, Romania, 2007; pp. 18–43.
17. Tiryaki, S.; Bardak, S.; Aydin, A.; Nemli, G. Analysis of volumetric swelling and shrinkage of heattreated woods: Experimental and artificial neural network modeling approach. *Maderas Cienc. Technol.* **2016**, *18*, 477–492. [CrossRef]
18. Filipovici, J. *Studiul Lemnului*, 2nd ed.; Didactică și Pedagogică: București, Romania, 1965; Volume 2, p. 77.
19. Ispas, M. Experimental investigations on swelling pressure of natural and heat-treated ash wood. *Bull. Transilv. Univ. Brasov. Ser. II For. Wood Ind. Agric. Food Eng.* **2013**, *6*, 55.
20. Krauss, A. Swelling Pressure of Wood Determined from Hygro-Mechanical Creep Measurements. 2004. Available online: http://www.ejpau.media.pl/volume7/issue2/wood/art-01.html (accessed on 21 February 2023).
21. Zelinka, S.L.; Altgen, M.; Emmerich, L.; Guigo, N.; Keplinger, T.; Kymäläinen, M.; Thybring, E.E.; Thygesen, L.G. Review of Wood Modification and Wood Functionalization Technologies. *Forests* **2022**, *13*, 1004. [CrossRef]
22. Zelinka, S.L.; Glass, S.V.; Jakes, J.E. A solution thermodynamics definition of the fiber saturation point and the derivation of a wood–water phase (state) diagram. *Wood Sci. Technol.* **2016**, *50*, 443–462. [CrossRef]
23. Thybring, E.E.; Fredriksson, M. Wood Modification as a Tool to Understand Moisture in Wood. *Forests* **2021**, *12*, 372. [CrossRef]
24. Maher, Z.A. Evaluation of Moisture Content in Wood Fiber and Recommendation of the Best Method for Its Determination, Submitted in Partial Fulfillment of the Requirements for the Degree of Master of Science at Helwan University Ain Helwan, Cairo. 2006. Available online: https://maher2100.tripod.com/msc1.pdf (accessed on 21 February 2023).
25. Lahtela, V.; Kärki, T. Effects of impregnation and heat treatment on the physical and mechanical properties of Scots pine (*Pinus sylvestris*) wood. *Wood Mater. Sci. Eng.* **2016**, *11*, 217–227. [CrossRef]
26. Wu, G.; Shah, U.D.; Janeček, E.-R.; Burridge, H.C.; Reynolds, T.P.S.; Fleming, P.H.; Linde, P.F.; Ramage, M.H.; Scherman, O.A. Predicting the pore-filling ratio in lumen-impregnated wood. *Wood Sci. Technol.* **2017**, *51*, 1277–1290. [CrossRef]
27. Ionescu, C.Ș.; Lunguleasa, A.; Avram, A.; Spîrchez, C. Evaluation of the efficiency of the consolidation treatment with Paraloid B72, performed on artworks with degraded wood support. *MATEC Web Conf.* **2021**, *343*, 1–10. [CrossRef]
28. Avram, A.; Ionescu, C.Ș.; Lunguleasa, A. Some Methods for the Degradation-Fragility Degree Determination and for the Consolidation of Treatments with Paraloid B72 of Wood Panels from Icon-Type Heritage Objects. *Forests* **2022**, *13*, 801. [CrossRef]
29. Darwish, S.S.; El-Hadidi, N.M.N. The effect of solvents on the chemical composition of archaeological wood. In Proceedings of the International Conference on Giza through the Ages, Giza, Egypt, 4–6 March 2008.
30. Bossu, J.; Le Moigne, N.; Corn, S.; Trens, P.; Di Renzo, F. Sorption of water–ethanol mixtures by poplar wood: Swelling and viscoelastic behaviour. *Wood Sci. Technol.* **2018**, *52*, 987–1008. [CrossRef]
31. Chang, S.S.; Quignard, F.; Di Renzo, F.; Clair, B. Solvent polarity and internal stresses control the swelling behavior of green wood during dehydration in organic solution. *BioResources* **2012**, *7*, 2418–2430. [CrossRef]
32. Mantanis, G.I.; Young, R.A.; Rowell, R.M. Swelling of wood, Part 1. Swelling in water. *Wood Sci. Technol.* **1994**, *28*, 119–134.
33. Mantanis, G.I.; Raymond, A.Y.; Rowell, R.M. Swelling of compressed cellulose fiber webs in organic liquids. *Cellulose* **1995**, *2*, 1–22. [CrossRef]
34. Meier, P.; Kaps, T.; Kallavus, U. Swelling of pinewood (*Pinus sylvestris*) in binary aqueous solutions of organic substances. *Mater. Sci.-Medzg.* **2005**, *11*, 140–145.
35. Rowell, R.M.; Dale, E.W. Determination of dimensional stabilization of wood using the water soak method. *Wood Fiber Sci.* **1978**, *2*, 104–111. Available online: https://wfs.swst.org/index.php/wfs/article/view/1004/1004 (accessed on 24 September 2023).
36. Rafsanjani, A.; Stiefel, M.; Jefimovs, K.; Mokso, R.; Derome, D.; Carmeliet, J. Hygroscopic swelling and shrinkage of latewood cell wall micropillars reveal ultrastructural anisotropy. *J. R. Soc. Interface* **2014**, *11*. [CrossRef]

37. *International Standard ISO 3131-1975;* Wood Determination of Density for Physical and Mechanical Tests. International Organization for Standardization: Geneva, Switzerland, 1975.
38. *European Standard EN 13183-1/2002;* Moisture Content of a Piece of Sawn Timber. European Committee for Standardization: Brussels, Belgium, 2002.
39. *International Standard ISO 4859-1982;* Wood Determination of Radial and Tangential Swelling. International Organization for Standardization: Geneva, Switzerland, 1982.

applied sciences

MDPI

Article

Elemental Compositions of Wood Plastic Pellets Made from Sawdust and Refuse-Derived Fuel (RDF) Waste

Aujchariya Chotikhun [1,*], Rattikal Laosena [2], Jitralada Kittijaruwattana [1], Seng Hua Lee [3], Kanokorn Sae-Ueng [1], Charoen Nakason [1], Yutthapong Pianroj [1] and Emilia-Adela Salca [4]

[1] Faculty of Science and Industrial Technology, Prince of Songkla University, Surat Thani Campus, Mueang Surat Thani 84000, Surat Thani, Thailand; jitralada.k@psu.ac.th (J.K.); kanokorn.s@psu.ac.th (K.S.-U.); charoen.n@psu.ac.th (C.N.); yutthapong.p@psu.ac.th (Y.P.)

[2] Research and Development Center, Chalermkarnchana University, Mueang Sisaket 33000, Sisaket, Thailand; rattikalnearn@gmail.com

[3] Department of Wood Industry, Faculty of Applied Sciences, Universiti Teknologi MARA Pahang Branch Campus Jengka, Bandar Tun Razak 26400, Pahang, Malaysia; leesenghua@uitm.edu.my

[4] Faculty of Furniture Design and Wood Engineering, Transilvania University of Brasov, Universitatii 1, 500068 Brasov, Romania; emilia.salca@unitbv.ro

* Correspondence: aujchariya.c@psu.ac.th

Abstract: The purpose of this research was to investigate the production and properties of wood plastic pellets (WPP) made from rubberwood sawdust and refuse-derived fuel (RDF). WPP samples were tested for chemical and physical properties and compared to standard wood pellets. The results showed that when using RDF, the elemental compositions of WPP can affect the content of Zn, Cu, Pb, Cd, Cr, and As. In addition, RDF samples had a higher heating value of 21.19–22.09 MJ/kg. The physical properties of the samples revealed that they had a density of 1175–1286 kg/m^3, a mechanical durability of 98%, and a moisture content of 5.38–11.27%. According to the study's findings, these manufactured mixed pellets have the potential to be beneficial for alternative sustainable green energy as fuels. Moreover, using RDF, which comes from MSW, could help in global warming mitigation.

Keywords: wood plastic pellets; rubberwood sawdust; refuse-derived fuel; elemental compositions; chemical and physical properties

Citation: Chotikhun, A.; Laosena, R.; Kittijaruwattana, J.; Lee, S.H.; Sae-Ueng, K.; Nakason, C.; Pianroj, Y.; Salca, E.-A. Elemental Compositions of Wood Plastic Pellets Made from Sawdust and Refuse-Derived Fuel (RDF) Waste. *Appl. Sci.* **2023**, *13*, 11162. https://doi.org/10.3390/app132011162

Academic Editor: Giuseppe Lazzara

Received: 12 September 2023
Revised: 6 October 2023
Accepted: 9 October 2023
Published: 11 October 2023

1. Introduction

Due to fossil fuel reserves being depleted, the world is shifting toward the production of renewable energy. Biomass materials are one of the most intriguing alternatives due to their widespread availability. Because of their low bulk density, biomass materials are difficult to handle, store, and transport. Pelletization can densify the materials into pellets by using pressure and heat [1]. The use of raw materials, such as agricultural, forestry, and food waste, and surpluses from the agri-food industry has recently increased dramatically [2]. Wood pellets are a cost-effective, environmentally friendly alternative to traditional fossil fuels. In terms of calorific value, softwoods had a higher range between 19.66 and 20.36 MJ/kg, while hardwoods ranged between 17.63 and 20.81 MJ/kg [3]. Sawdust, planer shavings, and dry chips have been the primary source materials used in the production of wood pellets in Sweden since 2001.

Because of their widespread availability, alternative forms of biomass, such as bark and logging residues, have sparked interest as raw materials [4]. A previous study [5] investigated the physical, mechanical, and energy properties of wood pellets derived from three common tropical species: *Acacia wrightii*, *Ebenopsis ebano*, and *Havardia pallens*. According to the findings, the pellets obtained from the species have a high energy density, making them ideal for commercial and industrial heating applications. Aside from wood residues and uncommon species, a variety of biomasses, including almond shells, olive stones, coffee dregs, coffee husks, grape pomace, hazelnut shells, miscanthus, pine kernel shells, and switchgrass, were blended to produce high-quality pellets for industrial use [1]. Similarly, Carroll and Finnan [6] found that willow, miscanthus, wheat, barley, and rape straws were also used. Additionally, Limhengha et al. [7] showed the potential of empty fruit bunches as a biomass source for pellet production.

Thailand can grow plants all year, ensuring a steady supply of biomass feedstock. Thailand exported 172,441 tonnes of wood pellets to South Korea and Japan in 2019. Thai entrepreneurs want to increase wood pellet exports to Japan for power generation. According to the Kasikorn Research Center, Thailand will export 80,000–100,000 tonnes of wood pellets to Japan in 2020 [8]. The Thai government restricts the use of wood pellets to industrial heat production rather than electricity generation. In Thailand, biomass power plants use less expensive substitutes, such as palm shells, paddy, and bagasse, instead of wood pellets [9].

Rubberwood has the potential to produce a substantial amount of energy. It outperforms fruit fibers, rice husks, coconut husks, bagasse, and logging residues, and it even slightly outperforms wood residues. It does, however, have less potential than empty fruit bunches and palm shells. Rubberwood has lower ash and nitrogen content compared to oil palm biomass. There are numerous opportunities in Thailand for wood pellet production using para-rubber waste wood as a feedstock. Consumption of wood pellets can help reduce greenhouse gas emissions while also supporting the Thai government's efforts to use wood pellets for domestic purposes. Furthermore, there is a growing demand for wood pellets for export [10]. The global market for pellets is expanding. The output in 2008 was 9.8 T_g, with a global increase to 14.3 T_g in 2010 and an excess of 26 T_g in 2015 [11]. Because of the high demand for rubberwood in the wood product manufacturing industry, the quantity of rubberwood biomass available for energy production is relatively low compared to the abundance of oil palm biomass. Rubberwood biomass utilization as a fuel source is severely limited due to (1) systematic replanting, (2) extraction of all above-ground biomass up to 10 cm in diameter from the field, and (3) utilization of waste and residue from secondary milling activities in the panel products sector [12].

Pretreatment and blending diverse raw materials for co-pelletizing is a promising approach for improving biomass pellet quality, and co-pelletizing requires cost-effective, eco-friendly, and sustainable raw materials. Furthermore, in the coming years, co-pelletizing will be a viable option for increasing the competitiveness of large-scale biomass pellet fuel production [13]. Co-pelleting of woody and herbaceous crops; co-pelleting of crops containing high amounts of starch and oil, such as microalgae and peanut shells, with woody or herbaceous crops; mixed pelleting of different parts of the same crop, such as the bark and leaves of similar quality; and co-pelleting of biomass with solid waste, such as municipal sludge and paper waste, are examples of common co-pelleting combinations. The co-pelletization of non-biomass-based materials with biomass, such as household waste and inorganic additives, simplifies pelletizing and improves pellet fuel quality. The addition of 50% rubberwood sawdust and 50% lignite reduced the ash content by 50% [14].

Refuse-derived fuel (RDF) is a fuel substitute derived from waste management facilities. The use of RDF for heat and power production adheres to the European Union's waste hierarchy [15]. Materials such as refuse-derived fuel (RDF) fractions are used as fuel in cement or combined heat and power (CHP) plants. However, the low bulk density creates a number of transportation and storage issues [16]. In the case of biomass, these issues result in a decrease in pelletization. Because of its low bulk density, raw RDF has a high potential for densification operations. Furthermore, the negligible content of natural binders necessitates die redesign. Furthermore, studies have shown that temperature has a significant impact on pellet quality. For all variations of compaction pressure and channel diameter, the temperature condition of 120 °C produces the most durable pellets. This is due to plastics melting. Using appropriate operation combinations, RDF from municipal solid waste, packaging, wood, paper, and plastics could be more manageable and storable, with more predictable characteristics and specifications such as higher heating value and proximate and ultimate analyses of pellets [17]. RDF has gained value as an alternative fuel due to the vast quantity of non-recyclable combustible waste materials, which can replace traditional generation burning fuels [18,19].

The impact of fossil fuels (coal, oil, and gas) on global warming is highly unsustainable for developing economies worldwide, and they are the most significant contributor to global climate change [20]. The use of biomass as a biofuel can communicate the need for low-carbon energy while also reducing pollution, paving the way to carbon neutrality. As a result, the critical decision of using municipal solid waste and biomass as sustainable energy resources to manage energy supply and demand issues, as well as climate change issues, must be made [21,22].

Previous literary works have revealed a potential ability to co-pelletize. To the best of the authors' knowledge, there have been few studies on the co-pelletization of rubberwood and RDF. As a result, there is a limited understanding of the effects of co-pelletization on the performance of the produced pellets. Our previous study produced mixed pellets from rubberwood and RDF and tested their mechanical durability and calorific value (ultimate and proximate) [23]. Meanwhile, the goal of this research is to determine the physical properties, higher heating value (HHV), and elemental analysis of pellets made from a rubberwood and RDF mixture. The findings of this research could support the viability of mixed pellet fuel as a feedstock in thermal waste-to-energy technologies.

2. Materials and Methods

2.1. Sample Preparation and Pelletization

RDF was selected and separated from a landfill area in Pattalung Province, Thailand, where the municipal solid waste (MSW) used in this study had been landfilled for 2 years and managed by the Provincial Administrative Organization of Pattalung Province. Firstly, MSW was recovered and separated into combustible and incombustible materials, using a separating machine and manual selection by workers. Then, the combustible materials were converted by a machine into refuse-derived fuel (RDF) level 3, which only has fluff RDF, such as plastics, papers, and fibers. Eventually, the RDF wastes were shredded to a size suitable for pelletization.

Rubberwood sawdust, a waste product supplied by BNS Wood Industry Co., Ltd., a sawmill located at Mueang, Surat Thani, Thailand, was collected. The material was controlled through its moisture content reaching 10–15% in a laboratory oven. The oversized sawdust was screened using a sieve with 18 meshes before mixing with RDF.

The composition of sawdust/RDF materials with four different types of pellets was prepared in ratios of 100/0, 70/30, 60/40, and 50/50 by weight, respectively. A rotary drum mixer was used to completely mix the two materials for 5 min before pelletization. The mixed materials were pelleted in an electric flat die wood pellet mill, ZhengZhou Known Imp. & Exp. Co., Ltd, KN-D-200, Zhengzhou, Henan, China, with 7.5 hp (380 v), 50 Hz, with the extruded temperature of 90–110 °C and a pellet mill die of 6 mm to produce 10 kg of each condition. Five samplings of each condition were randomly collected for the testing (Figure 1), while the pellets test had five replications of each property.

2.2. Physical Properties of the Samples

Dimensions such as the diameter and length of the pellets were measured with precisions of 0.01 mm. The density of samples was calculated by mass/volume, where pellets were weighed with precisions of 0.001 g. Mechanical durability testing of the pellets was carried out in accordance with the procedure outlined in EN 15210-1 [24]. The water absorption (WA) of the samples was evaluated with the water soaking method. The samples were soaked in distilled water in a 250 mL beaker with 100 mL of water for 5 min. After 24 h, the samples were taken out and drenched through filter paper—which had a pore size of 11 μm and was 125 mm in diameter—for 10 min [24]. The WA samples were measured by weight with precision of 0.01 g using the following equation:

$$WA = (M_{after} - M_{before})/M_{before} \times 100 \quad (1)$$

where WA is the water absorption of the pellet (%); M_{after} is the mass of the pellet after soaking (g); M_{before} is the mass of the pellet before soaking (g).

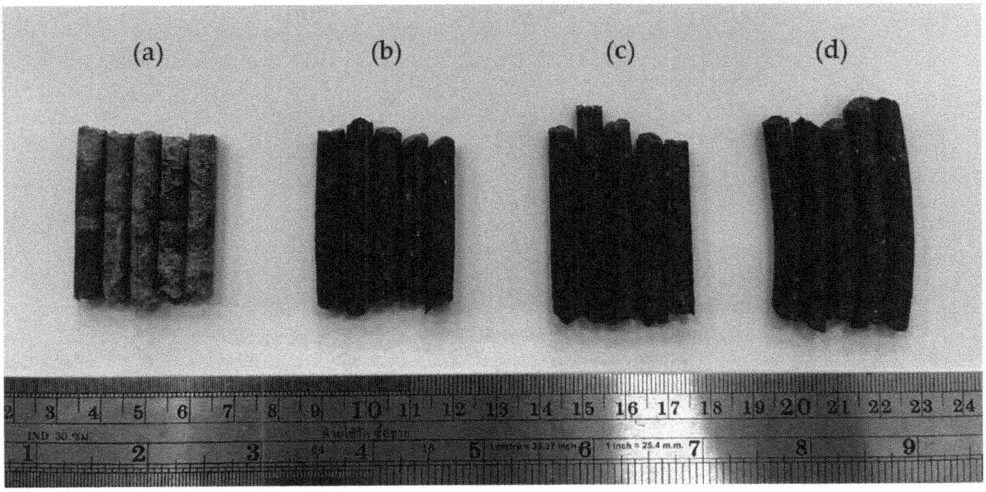

Figure 1. The completely randomized samples of each condition manufactured with sawdust/RDF ratios, (**a**) 100/0, (**b**) 50/50, (**c**) 60/40, and (**d**) 70/30.

2.3. Microstructure Evaluation of the Samples

A scanning electron microscope (SEM), FEI Quanta 250, Waltham, MA, USA, was used to image the samples. Before testing, the samples were coated with a thin layer of gold. The images were taken with a SEM set to high vacuum (HV) mode at 15 kV. Pellet cross-sections and longitudinal sections were photographed.

2.4. Proximate Analysis of the Samples

The proximate analysis of samples typically determined the moisture content (MC), volatile matter (VM), ash content (AC), and fixed carbon (FC) as a percentage. The percentage of fixed carbon was determined by using the following equation:

$$FC = 100 - (MC + VM + AC)\ \% \tag{2}$$

where FC is the fixed carbon (%); MC is the moisture content (%); VM is the volatile matter (%); and AC is the ash content (%).

2.5. Higher Heating Value of the Samples

The higher heating value (HHV) or gross calorific value (GCV) of pellet samples was determined by using an isoperibol bomb calorimeter, Leco AC500, St. Joseph, MI, USA. The measurements were taken in five replicates, and the results are given as means with standard deviations in MJ/kg following ASTM D 5865-13 [25]. The HHV of the samples as a function of fixed carbon (FC, wt%) was calculated using the following equation:

$$HHV = 0.196(FC) + 14.119 \tag{3}$$

for which the correlation coefficient was 0.9997.

2.6. Ultimate Analysis of the Samples

The samples were analyzed using the ASTM D5373-93 (1997) [26] procedure to determine the contents of elemental carbon (C), hydrogen (H), nitrogen (N), and sulfur (S) using a Perkin Elmer, 2400 Series II CHNS/O analyzer, Waltham, MA, USA.

2.7. Elemental Analysis of Samples

The elemental analysis of samples finally analyzed their compositions. The potassium (K), sodium (Na), cadmium (Cd), zinc (Zn), copper (Cu), lead (Pb), and chromium (Cr) content was determined by using a high-performance flame AA spectrometer, PinAAcle 900F, Perkin Elmer, USA. The chlorine (Cl) content was determined by using a 785 DMP Titrino, Metrohm, Herisau, Switzerland. The arsenic (As) content was determined by using a FIAS 100, Perkin Elmer, USA. Ultimately, mercury (Hg) was determined using a FIMS 100, Perkin Elmer, USA.

2.8. FTIR-ATR Spectral Analysis of Samples

The FTIR spectroscopy of samples was performed using a Spectrum Two FT-IR Spectrometer (DTGS Detector), PerkinElmer, Llantrisant, UK, in order to compare the spectra of each pellet condition at 400–4000 cm^{-1} resolution. The FTIR analysis was conducted quantitatively, with consideration given to the shape of the spectra at specific peaks and the intensity of the FTIR graph.

2.9. Data Analysis

A completely randomized design of sample types was used for this experiment. The analysis of variance (ANOVA) was used to determine the significant differences between the four types of wood pellet specimens, and Duncan's multiple range tests were used for additional analysis using SPSS Statistics version 22, IBM SPSS Statistics for Windows, Armonk, NY, USA. A p-value of 0.05 was used as the level of confidence.

3. Results and Discussion

3.1. Physical Characteristics of Pellets

The mean values of diameter, length, density, and mechanical durability of the samples are shown in Table 1. The diameter of the pellets did not differ much, ranging from 6.12 to 6.19 mm. The pellet with the addition of RDF was slightly bigger than the control sample (WPP). It seems that the size of the pellets increased with the increasing ratio of RDF. Additionally, the pellets with the addition of RDF were longer than those with WPP. However, WPP had a higher density than the RDF samples. The density of WPP was 1.29 g/cm^3, while the RDF samples had a density ranging from 1.12 to 1.24 g/cm^3. This experiment used mixing ratios by weight, and by adding RDF, the density decreased compared to the wood pellet; however, having a higher RDF ratio could increase the density of WPP. The highest density was observed in the sample with 50% RDF addition. The findings were in line with the study by Laosena et al. [23]. When the pellets had a higher RDF content, the water absorption (WA) of the samples was lower. Because plastic is a hydrophobic material, mixing RDF with the pellet can reduce WPP's moisture uptake. According to this study, having a 50/50 (Wood/RDF) ratio reduced the WA of the sample by 57%. The results showed that the WA values of WPP soaked in water for 5 min ranged from 4.17 to 6.83%, while wood pellets had a value of 9.68% [27].

Table 1. Physical characteristics of pellets.

Sample Type (Sawdust/RDF)	Sample Type	Diameter (mm) Mean		Length (mm) Mean		Density (g/cm^3) Mean		Water Absorption (%) Mean		Mechanical Durability (%) Mean	
100/0	Control	6.12	±(0.09)	36.48	±(0.44)	1.29 ^{d}	±(0.08)	9.68 ^{d}	±(0.42)	98.48 ^{a}	±(0.30)
70/30	70/30	6.18	±(0.01)	42.81	±(0.77)	1.12 ^{a}	±(0.04)	6.83 ^{c}	±(0.59)	98.26 ^{a}	±(0.06)
60/40	60/40	6.15	±(0.04)	40.81	±(1.12)	1.18 ^{b}	±(0.15)	5.56 ^{b}	±(0.28)	98.87 ^{b}	±(0.04)
50/50	50/50	6.19	±(0.03)	41.29	±(1.35)	1.24 ^{c}	±(0.12)	4.17 ^{a}	±(0.50)	99.06 ^{b}	±(0.10)

Numbers in parentheses are standard deviation values. Mean values with the different letters are significantly different at $p < 0.05$.

Meanwhile, the WPP had a mechanical durability of 98.48%. The results showed that the samples produced in this study are of high quality, with a mechanical durability value greater than 96%. However, the mechanical durability of the pellets decreased initially when 30% RDF was added but increased as the RDF ratio increased. The highest mechanical durability of 99.06% was recorded in the samples with 50% RDF addition. This phenomenon may be due to the strong association with the density of the pellets, as reported by Jewiarz et al. [16]. Pellets with 50% RDF addition had the highest density, and therefore, this may be the reason they showed the greatest durability among the tested samples.

3.2. SEM Images of Pellets

Figure 2 depicts the SEM images of rubberwood samples (Figure 2a,b) and the pellet made of wood and RDF at a 60/40 ratio (Figure 2c,d). The textures of the pellets differed from one another. The cross-section of the rubberwood pellet (Figure 2a) was coarse, revealing the wood fibers, whereas the longitudinal section (Figure 2b) was smooth. The pellet with RDF addition, on the other hand, showed the finished blending with smooth areas on the pellet's cross-section (Figure 2c) and longitudinal section (Figure 2d). The smooth surface indicates lower porosity, and thus, the water absorption of RDF pellets was lower than that of pellets made entirely of wood.

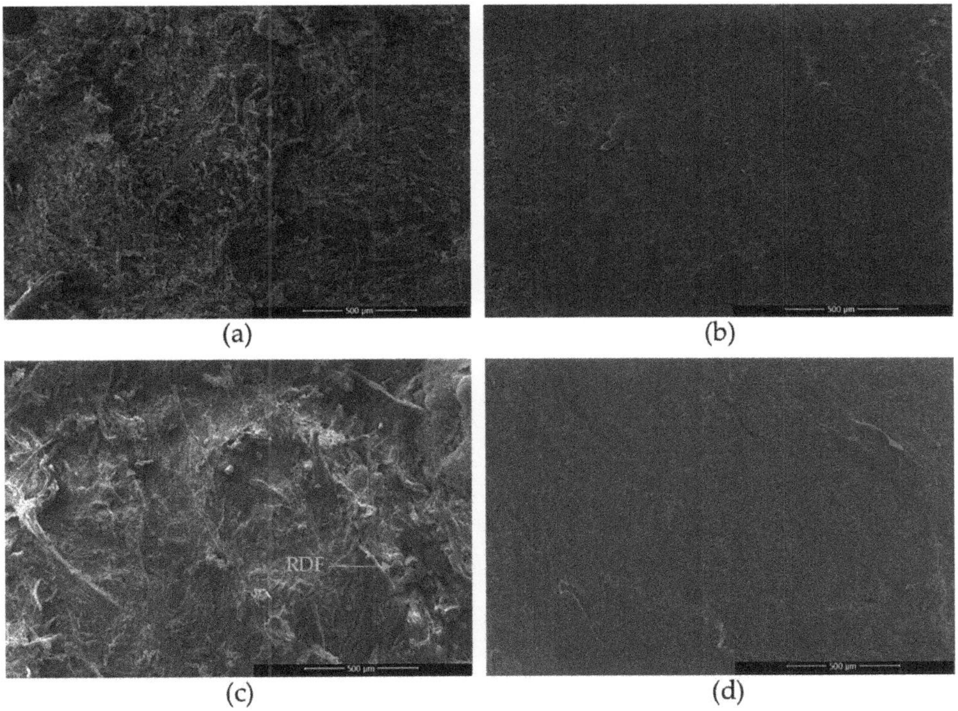

Figure 2. The SEM images of (**a**) cross-section of wood pellet, (**b**) longitudinal section of wood pellet, (**c**) cross-section of wood/RDF pellet, and (**d**) longitudinal section of wood/RDF pellet.

3.3. Proximate Analysis and Higher Heating Values (HHV) of Pellets

Table 2 shows the results of the experiment's proximity analysis and higher heating values (HHV). The moisture content of the samples ranged from 5.38 to 11.27%. Compared to the RDF samples (65.17 to 68.17%), WPP had the highest volatile matter of 72.11%. Meanwhile, the fixed carbon of pellets decreased when RDF was added. Furthermore, the ash content of the pellets was unsatisfactory in this experiment, with the control pellet

(2.54%) having less ash content (9.85–13.86%), which was 3.9–5.4 times lower than that of RDF samples. However, RDF reported an ash content of 11–13.4% [28]. The increase in the RDF ratio increased the volatile matter and ash content of the pellets while decreasing the fixed carbon content. Kramens et al. [28] conducted a similar study using pinewood and non-recyclable textile materials. On the other hand, the HHV of the pellets produced in this study ranged between 19.40 and 22.09 MJ/kg, whereas the HHV of typical RDF values were reported at 13–20 MJ/kg [29]. The results showed that adding RDF to the pellet significantly increased its HHV. The highest higher heating values of 22.09 MJ/kg were observed in the pellet with 50% RDF, while values of 19.40, 21.19, and 21.83 MJ/kg resulted in 100/0, 70/30, and 60/40 ratios, respectively. A previous investigation established that wood pellets made from mixed wood species had HVV of 19.85–20.45 MJ/kg [30], whereas an ultimate analysis of RDF in Europe reported an average HVV of 20.79 MJ/kg [31]. The highest HHV found in 50% RDF samples might be due to their lower MC and higher volatile matter and ash content. Ozkan et al. [32] demonstrated that the HHV of RDF pellets were inversely proportional to the MC but positively proportional to the volatile matter, ash content, and fixed carbon content.

Table 2. Proximate analysis and higher heating values of samples.

Sample	Proximate Analysis								HHV	
Type (Sawdust/RDF)	MC Mean (%)		VM Mean (%)		AC Mean (%)		FC Mean (%)		(MJ/kg) Mean	
100/0	9.57 [b]	±(0.26)	72.11 [d]	±(0.31)	2.54 [a]	±(0.09)	15.78 [b]	±(0.56)	19.40 [a]	±(0.07)
70/30	11.27 [d]	±(0.31)	65.87 [b]	±(0.36)	9.85 [b]	±(0.11)	13.01 [a]	±(0.45)	21.19 [b]	±(0.09)
60/40	10.48 [c]	±(0.06)	65.17 [a]	±(0.36)	11.86 [c]	±(0.15)	12.48 [a]	±(0.40)	21.83 [c]	±(0.14)
50/50	5.38 [a]	±(0.35)	68.17 [c]	±(0.49)	13.86 [d]	±(0.16)	12.59 [a]	±(0.55)	22.09 [d]	±(0.08)

Numbers in parentheses are standard deviation values. Mean values with the different letters are significantly different at $p < 0.05$.

3.4. Ultimate Analysis Values of Pellets

Table 3 shows the percentages of carbon (C), hydrogen (H), nitrogen (N), and sulphur (S) in the control and RDF samples tested. The control samples had C, H, N, and S content of 51.20%, 8.01%, 0.27%, and 0.60%, respectively. In the case of RDF samples, the C content ranged from 47.46 to 49.87%. When RDF was added, the C content decreased, but it increased as the RDF ratio increased. On the contrary, the H, N, and S contents of the RDF samples were greater than those of control samples. Generally, the C content of the RDF samples produced in this study was higher than the RDF pellets produced by Kobyashi et al. (42–44%) [33] but comparable to those produced in Italy (49%) [34].

Table 3. Ultimate analysis values of samples.

Sample	Sample	Ultimate Analysis (%)							
Type (Sawdust/RDF)	Type	C Mean		H Mean		N Mean		S Mean	
100/0	Control	51.20 [a]	±(7.24)	8.01 [a]	±(1.13)	0.27 [a]	±(0.03)	0.60 [a]	±(0.03)
70/30	70/30	47.46 [a]	±(0.27)	10.21 [b]	±(1.10)	0.47 [b]	±(0.03)	0.91 [b]	±(0.02)
60/40	60/40	47.67 [a]	±(0.34)	11.45 [b]	±(0.69)	0.50 [b]	±(0.13)	1.38 [c]	±(0.15)
50/50	50/50	49.87 [a]	±(1.58)	10.38 [b]	±(1.00)	0.69 [c]	±(0.04)	1.04 [b]	±(0.08)

Numbers in parentheses are standard deviation values. Mean values with the different letters are significantly different at $p < 0.05$.

3.5. Elemental Analysis Values of Pellets

Table 4 reports the ten elemental compositions of the pellets. According to Table 4, there was no significant difference in the K and Hg content between WPP samples and RDF samples. Meanwhile, the remaining elemental compositions varied greatly. For instance,

RDF samples had significantly higher Na, Cl, Zn, Cu, Pb, Cd, Cr, and As content compared to that of WPP samples. Additionally, ANOVA testing showed a significant difference between groups and within groups of the sample type, as displayed in Table 5. Generally, pellets with the highest RDF ratio (50%) had the highest value of all elemental compositions. RDF is a fuel made from various wastes, including municipal solid waste (MSW) and industrial waste. MSW is primarily composed of various types of waste, such as plastics, textiles, rubber, and foam, which contain varying amounts of elemental components [35]. As a result, it is understandable that the RDF samples had more elemental components than the WPP samples.

Table 4. Elemental analysis values of samples.

Sample Type (Sawdust/RDF)	Elemental Analysis									
	K Mean (%)	Na Mean (%)	Cl Mean (%)	Zn Mean (mg/kg)	Cu Mean (mg/kg)	Pb Mean (mg/kg)	Cd Mean (mg/kg)	Cr Mean (mg/kg)	As Mean (mg/kg)	Hg Mean (mg/kg)
100/0	0.228 [a] ±(0.01)	0.016 [a] ±(0.001)	0.017 [a] ±(0.001)	10.190 [a] ±(1.78)	5.475 [a] ±(0.94)	27.163 [a] ±(3.05)	0.001 [a] ±(0.00)	19.014 [a] ±(1.79)	0.180 [a] ±(0.01)	0.0001 [a] ±(0.00)
70/30	0.237 [a] ±(0.09)	0.101 [b] ±(0.003)	0.092 [b] ±(0.004)	328.492 [b] ±(16.90)	61.763 [b] ±(5.67)	57.449 [bc] ±(7.82)	1.517 [b] ±(0.20)	44.014 [b] ±(2.22)	2.306 [b] ±(0.16)	0.0001 [a] ±(0.00)
60/40	0.234 [a] ±(0.01)	0.122 [c] ±(0.002)	0.116 [c] ±(0.003)	380.599 [c] ±(7.74)	70.076 [b] ±(11.09)	48.275 [b] ±(7.51)	2.287 [c] ±(0.14)	64.274 [c] ±(3.45)	2.927 [c] ±(0.19)	0.0001 [a] ±(0.00)
50/50	0.237 [a] ±(0.01)	0.144 [d] ±(0.004)	0.125 [d] ±(0.002)	400.015 [d] ±(12.27)	70.496 [b] ±(3.43)	61.996 [c] ±(5.34)	3.360 [d] ±(0.34)	64.735 [c] ±(5.34)	2.725 [c] ±(0.34)	0.0001 [a] ±(0.00)

Numbers in parentheses are standard deviation values. Mean values with the different letters are significantly different at $p < 0.05$.

Table 5. The ANOVA testing of some elemental compositions of samples.

		Sum of Squares	df	Mean Square	F	Sig.
Zn	Between Groups	498,363.018	3	166,121.006	1063.927	0.000
	Within Groups	2498.231	16	156.139		
	Total	500,861.250	19			
Cu	Between Groups	14,643.652	3	4881.217	92.952	0.000
	Within Groups	840.214	16	52.513		
	Total	15,483.865	19			
Pb	Between Groups	3586.758	3	1195.586	24.576	0.000
	Within Groups	778.374	16	48.648		
	Total	4365.132	19			
Cd	Between Groups	29.941	3	9.980	177.785	0.000
	Within Groups	0.898	16	0.056		
	Total	30.839	19			
Cr	Between Groups	7004.878	3	2334.959	253.536	0.000
	Within Groups	147.353	16	9.210		
	Total	7152.231	19			
As	Between Groups	23.944	3	7.981	112.429	0.000
	Within Groups	1.136	16	0.071		
	Total	25.080	19			

3.6. FTIR Spectra of Pellets

Figure 3 shows the changes in function groups of different wood/RDF pellet samples. In 100/0 pellet samples with 100% wood composition, the broad absorption near the peaks around 3300–3400 cm^{-1} and 2800–3000 cm^{-1} corresponds to the hydrogen-bonded O-H stretch and C-H stretch, respectively. These structures are commonly found in lignocellulosic material [36], which is rubberwood in this case. After mixing with RDF, two new sharp peaks at around 2800–3000 cm^{-1} can be seen. These peaks could correspond to the $C\text{-}CH_3$ groups [37], which may belong to the plastic waste in the RDF. The peaks, including those around 3000–3400 cm^{-1}, become less intense as the proportion of RDF increases, most likely due to the successful cross-linking during rubberwood and RDF co-pelletization, especially when RDF is added at a higher proportion.

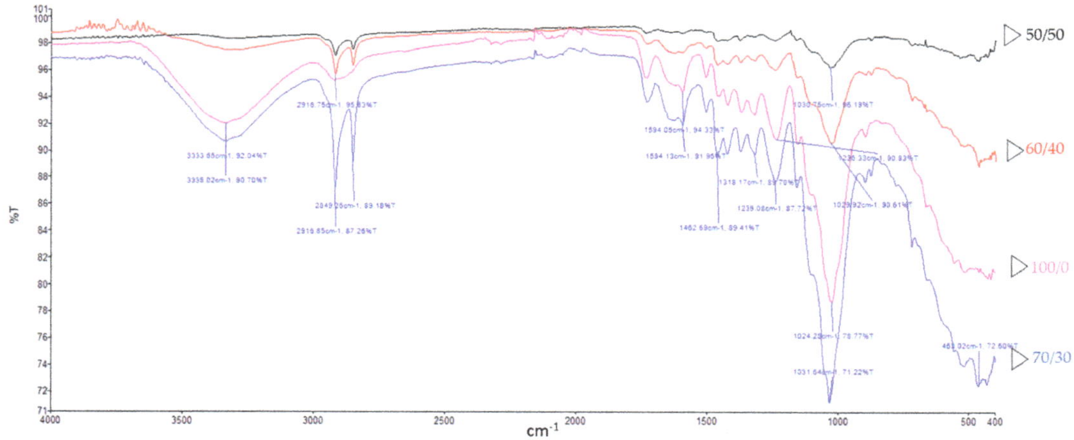

Figure 3. The FTIR spectra of samples with the 100/0, 70/30, 60/40, and 50/50 ratios of wood/RDF.

4. Conclusions

This study provided a comprehensive foundation for wood plastic pellets made from RDF and rubberwood sawdust. Specimen elemental compositions were compared with the standard. Furthermore, the information on their properties revealed a significant difference between each component ratio at a 95% confidence level. The HHV of RDF samples increased significantly, while the ash content needs to be reduced. However, using RDF derived from MSW as an alternative sustainable green energy fuel could aid in global warming mitigation. Therefore, it can be concluded that wood plastic pellets have the potential to be used as a biofuel in a clean industry. Moreover, this product may also be useful for composite materials, such as building materials, which previous studies had examined in order to promote plastic waste use as a substitute for concrete components, which improve the mechanical properties of the materials [38,39]. In a further study, the properties of light weight concrete made from WPP and Portland cement will be investigated in order to attain green materials.

Author Contributions: Conceptualization, A.C. and R.L.; Data curation, A.C.; Formal analysis, R.L., K.S.-U. and J.K.; Investigation, R.L., J.K. and K.S.-U.; Methodology, A.C.; Project administration, A.C.; Resources, J.K. and K.S.-U.; Supervision, E.-A.S., S.H.L., Y.P. and C.N.; Writing—Original draft, K.S.-U., J.K. and A.C.; Writing—Review and editing, C.N., S.H.L., E.-A.S. and A.C. All authors have read and agreed to the published version of the manuscript.

Funding: This project was funded by a National Research Council of Thailand (NRCT) grant allocated to Prince of Songkla University, Grant Number N71A660396, and the APC was also funded. This research was also funded by a grant from the Interdisciplinary Graduate School of Energy Systems (IGS-Energy), Prince of Songkla University, allocated to Rattikal Laosena.

Institutional Review Board Statement: Not applicable.

Informed Consent Statement: Not applicable.

Conflicts of Interest: The authors declare that there are no conflict of interest regarding the publication of this paper.

References

1. García, R.; Gil, M.V.; Rubiera, F.; Pevida, C. Pelletization of Wood and Alternative Residual Biomass Blends for Producing Industrial Quality Pellets. *Fuel* **2019**, *251*, 739–753. [CrossRef]
2. Zawiślak, K.; Sobczak, P.; Kraszkiewicz, A.; Niedziółka, I.; Parafiniuk, S.; Kuna-Broniowska, I.; Tanaś, W.; Żukiewicz-Sobczak, W.; Obidziński, S. The Use of Lignocellulosic Waste in the Production of Pellets for Energy Purposes. *Renew. Energy* **2020**, *145*, 997–1003. [CrossRef]
3. Telmo, C.; Lousada, J. Heating Values of Wood Pellets from Different Species. *Biomass Bioenergy* **2011**, *35*, 2634–2639. [CrossRef]
4. Lehtikangas, P. Quality properties of pelletised sawdust, logging residues and bark. *Biomass Bioenergy* **2001**, *20*, 351–360. [CrossRef]
5. Artemio, C.P.; Maginot, N.H.; Serafín, C.U.; Rahim, F.P.; Guadalupe, R.Q.J.; Fermín, C.M. Physical, mechanical and energy characterization of wood pellets obtained from three common tropical species. *PeerJ* **2018**, *6*, e5504. [CrossRef]
6. Carroll, J.P.; Finnan, J. Physical and chemical properties of pellets from energy crops and cereal straws. *Biosyst. Eng.* **2012**, *112*, 151–159. [CrossRef]
7. Limhengha, S.; Mahathaninwong, N.; Chucheep, T.; Karrila, S.; Tipayanon, T. Making Blends of Agarwood Waste with Empty Palm Bunches or Rubber Wood Sawdust for Pelletized Biofuels. *BioResources* **2021**, *16*, 2971–2986. [CrossRef]
8. Kasikorn Research Center Market. Opportunity of Wood Pellet in Thailand. (n.d.). Available online: https://kasikornresearch.com/th (accessed on 6 June 2023).
9. Asian Wood Pellet Producer & Market Snapshot | Biomassmagazine.com. (n.d.). Available online: https://biomassmagazine.com/articles/13880/asian-wood-pellet-producer-market-snapshot (accessed on 7 June 2023).
10. Saosee, P.; Sajjakulnukit, B.; Gheewala, S.H. Life cycle assessment of wood pellet production in Thailand. *Sustainability* **2020**, *12*, 6996. [CrossRef]
11. Thrän, D.; Schaubach, K.; Peetz, D.; Junginger, M.; Mai-Moulin, T.; Schipfer, F.; Olsson, O.; Lamers, P. The Dynamics of the Global Wood Pellet Markets and Trade—Key Regions, Developments and Impact Factors. *Biofuels Bioprod. Biorefin.* **2018**, *13*, 267–280. [CrossRef]
12. Ratnasingam, J.; Ramasamy, G.; Wai, L.T.; Senin, A.L.; Muttiah, N. The prospects of rubberwood biomass energy production in Malaysia. *BioResources* **2015**, *10*, 2526–2548. [CrossRef]
13. Li, W.; Wang, M.; Meng, F.; Zhang, Y.; Zhang, B. A Review on the Effects of Pretreatment and Process Parameters on Properties of Pellets. *Energies* **2022**, *15*, 7303. [CrossRef]
14. Kan, R.; Kaosol, T.; Tekasakul, P. Characterization and elemental composition of lignite and rubber wood sawdust pellets. *Eng. Appl. Sci. Res.* **2016**, *43*, 259–262.
15. Vounatsos, P.; Agraniotis, M.; Grammelis, P.; Kakaras, E.; Skiadi, O.; Zarmpoutis, T. Refuse-derived fuel classification in a mechanical–biological treatment plant and its valorization with techno-economic criteria. *Int. J. Environ. Sci. Technol.* **2015**, *12*, 1137–1146. [CrossRef]
16. Jewiarz, M.; Mudryk, K.; Wróbel, M.; Frączek, J.; Dziedzic, K. Parameters affecting RDF-based pellet quality. *Energies* **2020**, *13*, 910. [CrossRef]
17. Buekens, A. Refuse-Derived Fuel. In *Incineration Technologies*; SpringerBriefs in Applied Sciences and Technology; Springer: New York, NY, USA, 2013. [CrossRef]
18. Akdağ, A.S.; Atimtay, A.; Sanin, F.D. Comparison of Fuel Value and Combustion Characteristics of Two Different RDF Samples. *Waste Manag.* **2016**, *47*, 217–224. [CrossRef]
19. Januszewski, J.; Brzezińska, D. RDF Fire and Explosion Hazards at Power Plants. *Sustainability* **2021**, *13*, 12718. [CrossRef]
20. Khawaja, A.S.; Zaheer, M.A.; Ahmad, A.H.; Mirani, A.A.; Ali, Z. Advances in Limitations and Opportunities of Clean Biofuel Production to Promote Decarbonization. *Fuel* **2023**, *342*, 127662. [CrossRef]
21. Hansen, T. Stranded Assets and Reduced Profits: Analyzing the Economic Underpinnings of the Fossil Fuel Industry's Resistance to Climate Stabilization. *Renew. Sustain. Energy Rev.* **2022**, *158*, 112144. [CrossRef]
22. Porichha, G.K.; Hu, Y.; Rao, K.T.V.; Xu, C.C. Crop Residue Management in India: Stubble Burning vs. Other Utilizations including Bioenergy. *Energies* **2021**, *14*, 4281. [CrossRef]
23. Laosena, R.; Palamanit, A.; Luengchavanon, M.; Kittijaruwattana, J.; Nakason, C.; Lee, S.H.; Chotikhun, A. Characterization of Mixed Pellets Made from Rubberwood (*Hevea brasiliensis*) and Refuse-Derived Fuel (RDF) Waste as Pellet Fuel. *Materials* **2022**, *15*, 3093. [CrossRef]
24. *EN 15210-1*; Solid Biofuels—Determination of Mechanical Durability of Pellets and Briquettes Part 1: Pellets. British Standards Institution: London, UK, 2010.
25. *ASTM D 5865-13*; Standard Test Method for Gross Calorific Value of Coal and Coke. ASTM International: West Conshohocken, PA, USA, 2013.
26. *ASTM D5373-93*; Standard Test Methods for Instrumental Determination of Carbon, Hydrogen, and Nitrogen in Laboratory Samples of Coal and Coke. ASTM International: West Conshohocken, PA, USA, 1997.
27. Chotikhun, A.; Kittijaruwattana, J.; Pianroj, Y.; Tor, O.; Birinci, E.; Hengniran, P.; Lee, S.H. Some properties of white and torrefied pellets obtained from oil palm trunk as raw material. *BioResources* **2022**, *17*, 6818–6831. [CrossRef]
28. Paszkowski, J.; Domański, M.; Caban, J.; Zarajczyk, J.; Pristavka, M.; Findura, P. The Use of Refuse Derived Fuel (RDF) in the Power Industry. *Agric. Eng.* **2020**, *24*, 83–90. [CrossRef]

29. Kramens, J.; Vīgants, E.; Kanukuntla, S.P.; Goljandin, D.; Glušņova, J. Study of the application of pellets from textile material waste and biomass mixture in industrial and residental heating systems. In *Environment. Technologies. Resources, Proceedings of the International Scientific and Practical Conference, Rezekne, Latvia, 15–16 June 2023*; Rezekne Academy of Technologies: Rezekne, Latvia, 2023; Volume 1, pp. 86–92.
30. Ilari, A.; Pedretti, E.F.; De Francesco, C.; Duca, D. Pellet Production from Residual Biomass of Greenery Maintenance in a Small-Scale Company to Improve Sustainability. *Resources* **2021**, *10*, 122. [CrossRef]
31. Vounatsos, P.; Atsonios, K.; Itskos, G.; Agraniotis, M.; Grammelis, P.; Kakaras, E. Classification of Refuse Derived Fuel (RDF) and Model Development of a Novel Thermal Utilization Concept Through Air-Gasification. *Waste Biomass Valor.* **2016**, *7*, 1297–1308. [CrossRef]
32. Ozkan, K.; Işık, Ş.; Günkaya, Z.; Özkan, A.; Banar, M. A heating value estimation of refuse derived fuel using the genetic programming model. *Waste Manag.* **2019**, *100*, 327–335. [CrossRef]
33. Kobyashi, N.; Itaya, Y.; Piao, G.; Mori, S.; Kondo, M.; Hamai, M.; Yamaguchi, M. The behavior of flue gas from RDF combustion in a fluidized bed. *Powder Technol.* **2005**, *151*, 87–95. [CrossRef]
34. Galvagno, S.; Casciaro, G.; Casu, S.; Martino, M.; Mingazzini, C.; Russo, A.; Portofino, S. Steam gasification of tyre waste, poplar, and refuse-derived fuel: A comparative analysis. *Waste Manag.* **2009**, *29*, 678–689. [CrossRef]
35. Nasrullah, M.; Vainikka, P.; Hannula, J.; Hurme, M.; Oinas, P. Elemental balance of SRF production process: Solid recovered fuel produced from municipal solid waste. *Waste Manag. Res.* **2016**, *34*, 38–46. [CrossRef]
36. Miklecic, J.; Jirous-Rajkovic, V.; Antonovic, A.; Spanic, N. Discolouration of thermally modified wood during simulated indoor sunlight exposure. *BioResources* **2011**, *6*, 434–446. [CrossRef]
37. Sarker, M.; Rashid, M.M.; Molla, M. Hydrocarbon fuels produced from municipal solid waste plastics. *J. Environ. Sci. Eng.* **2011**, *5*, 446–452.
38. Nafees, A.; Althoey, F.; Khan, S.; Sikandar, M.A.; Alyami, S.H.; Rehman, M.F.; Javed, M.F.; Eldin, S.M. Plastic Concrete Mechanical Properties Prediction Based on Experimental Data. *Case Stud. Constr. Mater.* **2023**, *18*, e01831. [CrossRef]
39. Nibudey, R.; Nagarnaik, P.; Parbat, D.; Pande, A. A Model for Compressive Strength of PET Fiber Reinforced Concrete. *Am. J. Eng. Res. (AJER)* **2013**, *2*, 367–372.

applied sciences

MDPI

Article

Lightweight Solid Wood Panels Made of Paulownia Plantation Wood

Marius Cătălin Barbu [1,2], Helmut Radauer [1], Alexander Petutschnigg [1,3], Eugenia Mariana Tudor [1,2,*] and Markus Kathriner [4]

1 Green Engineering and Circular Design Department, Salzburg University of Applied Sciences, Markt 136a, 5431 Kuchl, Austria; cmbarbu@unitbv.ro (M.C.B.); helmut.radauer@fh-salzburg.ac.at (H.R.); alexander.petutschnigg@fh-salzburg.ac.at (A.P.)
2 Faculty of Furniture Design and Wood Engineering, Transilvania University of Brașov, B-dul. Eroilor nr. 29, 500036 Brasov, Romania
3 Institute of Wood Technology and Renewable Materials, University of Natural Resources and Life Sciences (BOKU), Konrad Lorenz-Straße 24, 3340 Tulln an der Donau, Austria
4 Controlling Department, Paris Lodron Universität Salzburg, Residenzplatz 9, 5020 Salzburg, Austria; markus.kathriner@plus.ac.at
* Correspondence: eugenia.tudor@fh-salzburg.ac.at

Abstract: Light Paulownia seamless-edged glued solid wood panels (SWPs), single-layered and three-layered, were analyzed in this study. Both panel types were calibrated at a thickness of 19 mm, a dimension very often in demand on the SWP market, but produced with other wood species (for example, spruce, pine, larch and fir). The panels were bonded with melamine-urea formaldehyde, polyurethane and polyvinyl acetate resins. The panels were tested for their physical (density) and mechanical (modulus of rupture, modulus of elasticity, compressive shear strength and wood breakage rate) properties. For the single-layered panels, the mechanical and physical properties did not differ significantly and were similar to massive Paulownia wood. For the three-layered panels, the adhesive application of polyurethane influenced positively all SWP properties. Considering the differences in density, these composites failed to achieve the performance of one- and single-layered panels made of spruce. The results of these findings recommend Paulownia SWPs to be used as lightweight and sustainable core materials in sandwich structures for the furniture and packaging industry, sport articles or non-load-bearing constructions.

Keywords: plantation wood lumber; Paulownia; lightweight solid wood panels

Citation: Barbu, M.C.; Radauer, H.; Petutschnigg, A.; Tudor, E.M.; Kathriner, M. Lightweight Solid Wood Panels Made of Paulownia Plantation Wood. *Appl. Sci.* **2023**, *13*, 11234. https://doi.org/10.3390/app132011234

Academic Editor: Manoj Gupta

Received: 12 September 2023
Revised: 9 October 2023
Accepted: 11 October 2023
Published: 12 October 2023

1. Introduction

Lightweight wood materials offer enormous potential for resource-efficient construction systems considering functional, environmental and economic factors [1]. The structure of these products can consist of a lighter material and/or use less material (e.g., hollow construction) [2]. These do not only imply the lower weight of structures, but ensure the load-oriented use of materials, aiming towards more energy efficiency due to decreased thermal conductivity [3]. Lightweight materials are a sine qua non of aircraft performance to allow greater range and speed, where low weight in relation to strength is needed. The emergence of still-lighter materials in the building and furniture sectors plays a significant role in the development of next-generation products [2]. Considering material savings and reduced weight, furniture is expected to use lightweight panels following designers' recommendations or for portable furniture (for do-it-yourself purposes). Experts expect that hutch furniture will consist no longer 100% of chipboard, but of "hybrid construction", where the optimal material is used depending on the functional requirements. The fact that light furniture is more likely to be associated with limited durability than heavier furniture speaks to a certain reluctance on the part of end consumers [4]. There are some ways to reduce the weight of composites, e.g., the use of low-density wood species, lower compaction of the wooden mat and filling the core layers of panels with low-density materials (foam or honeycomb structures) for sandwich structures. It is important to note that lightweight wood products are primarily used for applications where structural strength is not a primary concern. These lightweight solutions should consider tailored manufacturing, machinability (connections and the lamination of the edges) and performance (the physicomechanical properties of the composites) [5]. As a consequence of weight reduction, the transportation and material costs are diminished and ecological aspects as efficiency, consistency and sufficiency are considered [4].

Sustainable raw materials for the manufacture of solid wood panels are of great interest and highly demanded nowadays [6]. Solid wood panels (SWPs), single- and multi-layer, are used for both indoor and outdoor applications [4]. Either as an outer covering, with aesthetic quality and original particularities, or for use indoors in furniture construction [7], the request for SWPs develops constantly. The increased interest in prefabricated building requires the use of high-quality SWPs [8]. In the context of significant pressure for carbon emission reduction, prefabricated wood modules exhibit lower footprints and garner widespread interest [9]. For exterior applications, e.g., timber cladding, the wood species selected include Douglas fir (*Pseudotsuga menziesii* (Mirb.) Franco), European larch (*Larix decidua* Mill.), Siberian larch (*Larix sibirica* Ledeb.) and European oak (*Quercus robur* L.). Coniferous species such as fir (*Abies alba* Mill.), spruce (*Picea abies* L. (Karst)) or pine (*Pinus sylvestris* L.) are considered only partly durable for use unless they are treated for enhanced protection or preservation [10]. Durability is very important also for structural timber. Considering the shortage of this lignocellulosic resource, alternative raw materials from short-rotation plantations can be considered for the manufacture of SWPs.

Lighter wood species such as balsa (*Ochroma lagopus* Sw.) have been used since the logs in the FSC-certified plantations in Central America have reached harvestable diameters. Special lightweight fittings are usually not necessary in weight-sensitive furniture and vehicle interior fittings. If higher shear strengths are to be achieved, balsa brainwood panels can be used (e.g., for the rotor blades of wind turbines). For many applications, balsa wood products are too expensive, so lightweight wood species such as Paulownia could represent an alternative to balsa [4].

Paulownia, also known as the princess tree, emperor tree, miracle tree or the European balsa [11], has been cultivated on the old continent more extensively in recent decades [12]. Koman and Feher (2020) [13] reported about at least nine sub-species of Paulownia, of which *Paulownia tomentosa*, *Paulownia elongata* and *Paulownia fortunei* (and their hybrids) are most prevalent [14]. It is worth mentioning that short-rotation plantations are not considered forests when the rotation time is less than 20 years [15]. One of the main properties of Paulownia is its low density, which can range from 250 to 350 kg/m^3 [11], and other mechanical properties such as modulus of rupture (on average 40 MPa), modulus of elasticity (on average 4400 MPa), and compressive strength (on average 25 MPa) or tensile strength (on average 38 MPa) for Paulownia Clone In Vitro 112 [13], Paulownia COTE-2 [16] and *Paulownia tomentosa* × *elongata* from European plantations [11]. But similar to other fast-growing wood species, Paulownia has low performance in its mechanical properties. One of the limitations of the industrial usage of Paulownia timber is due to the influence of its low density on its mechanical properties [14]. This lightweight wood species can be used for the manufacture of particleboard using raw material from Turkey [17], a combination of bio-waste cotton stalks (*Gossypium hirsutum* L.) and Paulownia wood particles [18], using raw material from Portugal [19] or in non-load-bearing constructions [20]. Moreover, Paulownia represents a solution to replace expensive and rare tropical low-weight/density wood species (i.e., balsa (*Ochroma pyramidale* Cav. ex Lam) [11].

To the best of our knowledge, solid wood Paulownia panels were studied hitherto in [21] regarding their microscopic changes under thermal compression. The results of this study showed that Paulownia wood resembles balsa with respect to its reduced density and the occurrence of a hexagonal cell shape in its microstructure. The authors emphasise the idea that Paulownia can replace balsa wood as a core material in sandwich structures. It should be mentioned that SWPs with this cellular structure (e.g., honeycomb, Dendrolight®, Bionic Delta, Steinakirchen am Forst, Austria) as industrial established products are not the object of this study. The aim of this study is to analyse the properties of single- and three-layered solid wood panels made of European Paulownia plantations' wood and to compare them with similar lightweight commercially available SWPs.

2. Materials and Methods

The Paulownia European plantation timber (*Paulownia tomentosa* × *elongata*) was provided by Glendor Holding GmbH (Kilb, Austria) and sourced from 5- to 7-year-old trees. The plantation wood from Petrinja, Croatia, was delivered by Moserholz GmbH (Pettenbach, Austria) dried (m.c. = 12%) as rough-sawn lumber of approximately 100 cm length and 20–30 cm width and with a thickness of 20–25 mm. Prior to testing, the raw material was conditioned to constant weight at 20 °C and 65% relative air humidity for at least 14 days, until constant weight was achieved.

For the manufacture of single- and three-layered SWPs with a thickness of 19 mm, two types of lamellae with the following dimensions were cut with a circular saw and planed: 112 × 3.2 × 2.2 cm for the single-layer SWPs and 98 × 6.6 × 1 cm for the 3-layer SWPs. A lamella width of 32 mm (single-layered board) was opted for according to the dimension of the core layer of a standard industrial blockboard from Moralt AG (Hausham, Germany). A lamella width of 66 mm (3-layered board) was chosen after optimising the available raw material in order to obtain the maximum yield.

Both the one- and three-layered SWPs were glued using polyvinylacetate (PVAc) type Pattex® PV/H, category D3 (Pattex, Düsseldorf, Germany); melamine urea formaldehyde (MUF), type Dynea Prefere 4564 with hardener Prefere 5013, at a ratio of 100:8 (Dynea Austria GmbH, Krems, Austria) and polyurethane Kleiberit® PUR 501.0 (Kleiberit SE & Co. KG, Weingarten, Germany). The adhesive amount was as follows: for the PVAc, 330 g/m^2, for the MUF, 340 g/m^2 and 210 g/m^2 for the PUR.

The single-layered SWPs were manufactured with Paulownia lamellae with 12% moisture content [22]. The lamellae were previous glued with PVAc, MUF and PUR and consequently cold-pressed at 0.68 MPa with screw clamps for two hours (Figure 1).

Figure 1. Pressing a single-layered Paulownia solid wood panel with clamps, where the lamellae were longitudinally bonded (95 × 44.6 × 2.2 cm).

The single-layer panels were calibrated using a wide belt sander, grit 60, to a thickness of 19 mm, a width of 446 mm and a length of 950 mm.

The individual layers of the 3-layer SWP were formatted to 920 × 920 mm and calibrated. The top layers (faces) had a thickness of 5.75 mm and the middle layer (core) 7.5 mm, considering the model of a standard spruce 3-layer panel from Binderholz GmbH (St. Georgen, Austria) of 19 mm thickness (2 × 5.75 + 7.5 mm).

The adhesive was applied to the individual layers of the 3-layer SWP with a notched trowel of tooth size B1 for all the adhesive types (Table 1). The 3-layer SWPs were pressed at 60 °C for 15 min in a veneer press OTT (Lambach, Austria) with a pressure of 0.6 N/m^2 for all panel types (bonded with MUF, PUR and PVAc) (Figure 2).

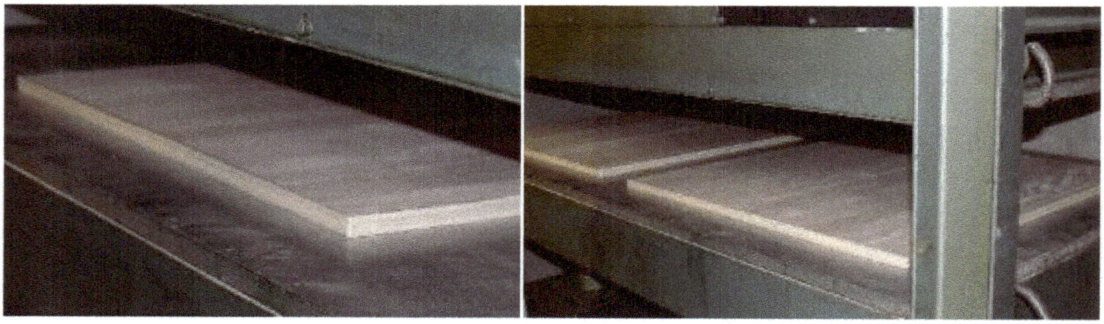

Figure 2. Pressing of a 3-layer SWP (92 × 92 × 1.9 cm) bonded with PVAc (**left**); pressing of 3-layer SWP bonded with MUF and PUR (**right**).

For the manufacture of SWPs, lamellae with different thicknesses (32 and 66 mm) were used, as well as distinct amounts of adhesive (Table 1):

Table 1. Design of experiments for the single- and three-layered Paulownia SWPs.

Board Type	Adhesive Type	Lamellae Width (mm)	Adhesive Application (g/m^2)
Single-layered	MUF	32	160
Single-layered	PUR	32	180
Single-layered	PVAc	32	150
Three-layered	MUF	66	340
Three-layered	PUR	66	210
Three-layered	PVAc	66	330

The testing specimens were prepared from the 1-layer and 3-layer SWPs according to EN 326-1:2005 [23] and then stored in a standard climate (20 °C and 65% relative air humidity).

For the density determination of the 1- and 3-layer SWPs (Figure 3), 6 test specimens were prepared for each type of board according to EN 323:2005 [24]. The mechanical properties of 3-point flexural strength (MOR) and modulus of elasticity (MOE) in bending were determined from 6 test specimens per board according to EN 310:2005 [25]. In the case of the 3-layer SWPs, 6 test specimens for each board were tested parallel and perpendicular to the grain. To test the quality of the bond for SWP 1 (24 h water storage), 10 test specimens per board were used to determine the compressive shear strength according to EN 13354:2009 [26], as depicted in Figure 4.

Figure 3. PVAc adhesive application on a single-layered SWP before layering (**left**) and three-layered SWP (92 × 92 × 1.9 cm) bonded with MUF, PUR and PVAc (**right**).

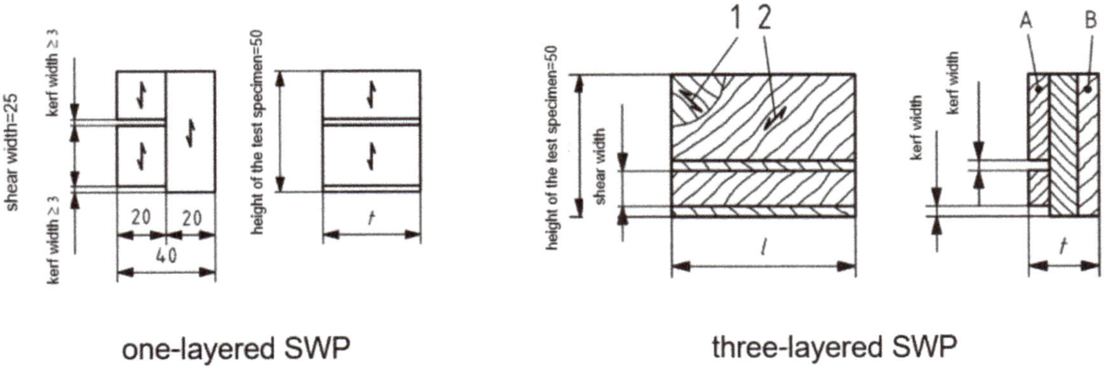

Figure 4. Cutting plan for the compressive shear strength test of single- and three-layered Paulownia SWPs (the values are in millimetres).

3. Results and Discussion

In this section are presented the density, bending strength, modulus of elasticity, compressive shear strength and wood breakage of the samples from single- and three-layered SWPs tested parallel and perpendicular to the grain.

3.1. Density

Table 2 and Figure 5 show the values of the raw density of the Paulownia boards, single- and three-layered. Similar average values of 258 kg/m^3 were measured for the single-layered boards bonded with MUF and PVAc, but with different coefficients of variation (CVs) of 0.055 for the panel glued with MUF and 0.019 for the panel with PVAc adhesive application. The single-layered panel bonded with PUR had a density of 249 kg/m^3, and a CV of 0.059.

In the case of the three-layered boards, the density [24] exhibited values of about 280 kg/m^3 with a minimum of 264 kg/m^3 and a maximum of 300 kg/m^3, with the smallest CV for the panel with MUF application (0.023), followed by the panel with PVAc application (0.038) and the highest CV of 0.042 for the panel bonded with PUR. It can be observed that the density values for the single-layered panels are consistent with the raw density of Paulownia wood in results from studies concerning plantation wood from Spain, Serbia and Bulgaria [11,14] and Hungary [13,27]. The three-layered SWPs had increased density due to the adhesive application and there was no difference regarding the extraction of samples parallel or perpendicular to the direction of production/or the grain.

Table 2. Density of single- and three-layered SWPs made of Paulownia lamellae.

Board Type	Adhesive Type	Lamellae Width (mm)	Density Mean Value kg/m^3	Density Min/Max Value, kg/m^3
1-layer SWP	MUF	32	258 ^{a} (14.04)	234/276
1-layer SWP	PUR	32	249 ^{a} (14.86)	236/280
1-layer SWP	PVAc	32	258 ^{a} (5.13)	251/264
3-layer SWP ‖	MUF	66	277 ^{b} (6.45)	265/284
3-layer SWP ⊦	MUF	66	277 ^{b} (6.45)	265/284
3-layer SWP ‖	PUR	66	283 ^{b} (12.00)	270/300
3-layer SWP ⊦	PUR	66	283 ^{b} (12.00)	270/300
3-layer SWP ‖	PVAc	66	280 ^{b} (10.73)	264/298
3-layer SWP ⊦	PVAc	66	280 ^{b} (10.73)	264/298

(^{a,b} values with the same letter are not significantly different: ANOVA, Post Hoc Tukey HSD, α = 0.05) (standard deviation for MOR and MOE for average/mean value in parentheses).

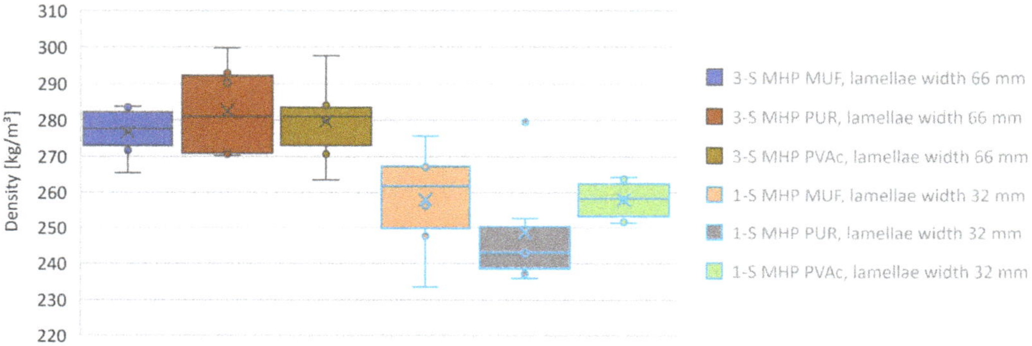

Figure 5. Density of single- and three-layered Paulownia SWPs (n = 6).

3.2. Three-Point Flexural Modulus of Rupture and Modulus of Elasticity

The elastic properties of the Paulownia SWPs are summarised in Table 3. For the single-layered SWPs (Figure 6, left), there is no significant difference between the values of modulus of rupture (MOR) for panels with three types of adhesive application. A clear difference is seen in the CV of the panel with PUR adhesive application, 0.117, compared with 0.082 (for MUF adhesive application) and 0.078 (for PVAc adhesive application).

Table 3. Modulus of rupture and modulus of elasticity of the single- and three-layered Paulownia SWPs.

Board Type	Adhesive Type	Lamellae Width mm	MOR Average Value, N/mm^2	MOR Min/Max Values N/mm^2	MOE Average Value, N/mm^2	MOE Min/Max Values, N/mm^2
1-layer SWP	MUF	32	33 ^{a} (2.72)	29/38	3733 ^{a} (334)	3283/4271
1-layer SWP	PUR	32	32 ^{a} (3.75)	29/40	3661 ^{a} (443)	3179/4528
1-layer SWP	PVAc	32	32 ^{a} (2.15)	29/36	3539 ^{a} (223)	3134/3819
3-layer SWP ‖	MUF	66	28 ^{b} (4.17)	23/34	3998 ^{b} (333)	3736/4507
3-layer SWP ⊦	MUF	66	8 ^{c} (1.42)	6/10	541 ^{c} (43)	463/595
3-layer SWP ‖	PUR	66	36 ^{a} (3.25)	31/41	4604 ^{d} (326)	4256/5212
3-layer SWP ⊦	PUR	66	10 ^{c} (0.93)	9/11	515 ^{c} (48)	454/587
3-layer SWP ‖	PVAc	66	29 ^{b} (0.66)	29/31	4071 ^{d} (322)	3485/4374
3-layer SWP ⊦	PVAc	66	9 ^{c} (1.34)	7/11	461 ^{c} (48)	422/462

(^{a,b,c,d} values with the same letter are not significantly different: ANOVA, Post Hoc Tukey HSD, α = 0.05) (standard deviation for MOR and MOE for average/mean value in parentheses).

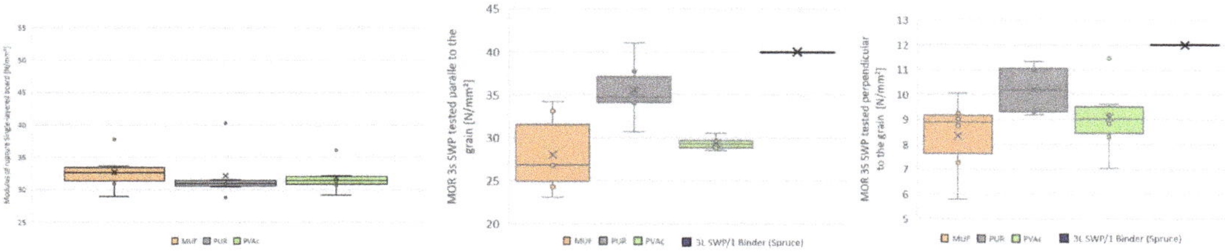

Figure 6. Modulus of rupture of the single-layered panels (**left**) and three-layered boards tested parallel to the grain (**middle**) and perpendicular to the grain (**right**).

The MOR [25] was tested for the three-layered SWPs parallel and perpendicular to the grain (Figure 6, middle and right). The MOR values for the samples tested perpendicular to the grain were at least three times lower compared to the other testing direction, parallel to the grain. The highest MOR (Figure 7) was achieved by the boards with PUR application (36 N/mm^2). These boards had similar CVs (about 0.09) when tested in both directions. Significant differences in the CV were determined for the panels bonded with PVAc, 0.022 (parallel to the grain) and 0.147 (perpendicular to the grain).

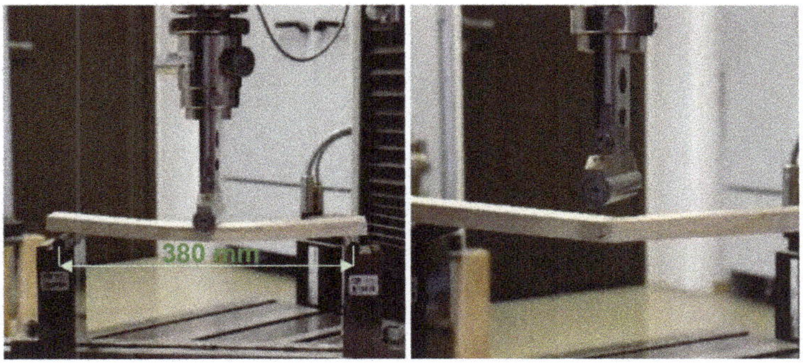

Figure 7. Testing of the 3-point bending strength of a Paulownia single-layered SWP.

The values for modulus of elasticity [25] are similar for the single-layer SWPs. Again, MUF, PUR and PVAc (Figure 8) have no significant influence on the edge-gluing of a single-layered SWP. The knot-free structure of the Paulownia lamellae [28], straight-grained [21], determines the relative related values for MOE (from 3500 to 3700 N/mm^2) for the single-layered SWPs. The CVs are different when the lamellae are bonded with MUF (0.089), PUR (0.121) and PVAc (0.063).

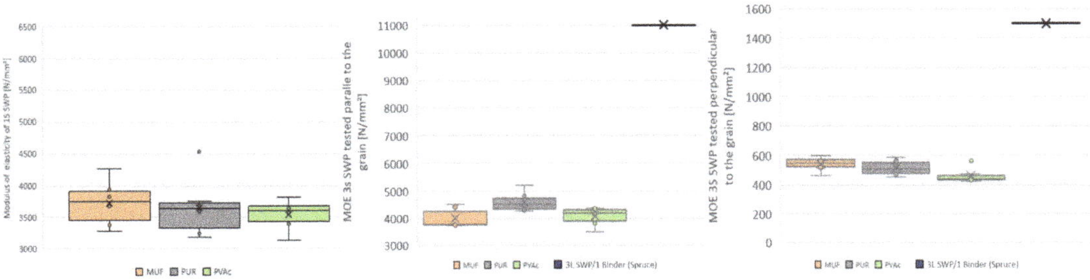

Figure 8. Modulus of elasticity of the single-layered panels (**left**) and three-layered boards tested parallel to the grain (**middle**) and perpendicular to the grain (**right**).

For the three-layered SWPs, the MOE was higher when the samples were tested parallel to the grain and about eight times lower when the testing was performed perpendicular to the grain. For the MUF bonding, the CVs were almost equal in both testing directions (0.09). For PUR bonding, the CVs were 0.070 (parallel) and 0.093 (perpendicular). Higher CV differences were determined for the bonding with PVAc, namely 0.079 (parallel) and 0.104 (perpendicular).

A comparison of the MOR and MOE of Paulownia SWPs with other lightweight wood products (made of balsa, for example) is relatively difficult, as balsa is used in sandwich structures with glass-fibre reinforcements [29] and is mostly composed as end-grain in a setup that consists of two face sheets (skins) and a core structure (e.g., balsa) [30].

3.3. Compressive Shear Strength and Wood Breakage

The highest compressive shear strength [26] after 24 h of water storage (Figure 9) was measured for the single-layered SWPs (about 3 N/mm^2), compared to 2 N/mm^2 for the three-layered SWPs bonded with PUR and PVAc (Table 4, Figure 10).

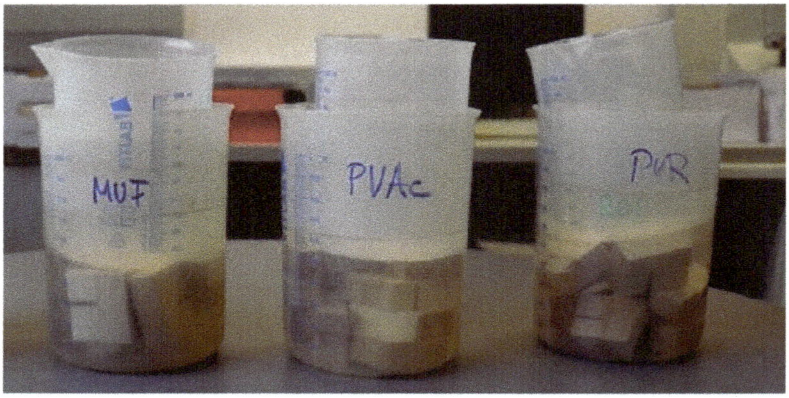

Figure 9. Water storage (for 24 h) for the SWP samples before a compressive shear test.

Table 4. Compressive shear strength and wood breakage rate of the single- and three-layered Paulownia SWPs.

Panel Type	Adhesive Type	Lamellae Width (mm)	Compressive Shear Strength Average Value, N/mm^2	Compressive Shear Strength Min/Max Value, N/mm^2	Wood Breakage Rate Average Value %	Wood Breakage Rate Min/Max Value %
1-layer SWP	MUF	32	3.16 ^{a} (0.58)	1.81/3.84	95 ^{a} (5)	85/100
1-layer SWP	PUR	32	2.92 ^{a} (0.32)	2.27/3.3	97.5 ^{a} (5.12)	85/100
1-layer SWP	PVAc	32	2.71 ^{b} (0.34)	2.21/3.26	79 ^{b} (15.3)	50/100
3-layer SWP ‖	MUF	66	1.62 ^{c} (0.33)	1.00/2.06	99 ^{a} (2)	95/100
3-layer SWP ⊦	MUF	66	1.62 ^{c} (0.33)	1.00/2.06	99 ^{a} (2)	95/100
3-layer SWP ‖	PUR	66	2.03 ^{b} (0.17)	1.59/2.23	98.5 ^{a} (2.29)	95/100
3-layer SWP ⊦	PUR	66	2.03 ^{b} (0.17)	1.59/2.23	98.5 ^{a} (2.29)	95/100
3-layer SWP ‖	PVAc	66	1.9 ^{b} (0.25)	1.5/2.62	96 ^{a} (4.36)	85/100
3-layer SWP ⊦	PVAc	66	1.9 ^{b} (0.25)	1.5/2.62	96 ^{a} (4.36)	85/100

(^{a,b,c} values with the same letter are not significantly different: ANOVA, Post Hoc Tukey HSD, α = 0.05) (standard deviation for MOR and MOE for average/mean value in parentheses).

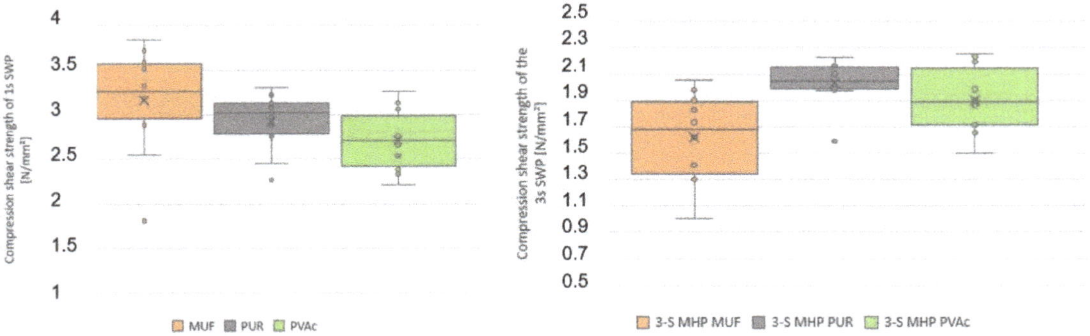

Figure 10. Compressive shear strength of the single-layered panels (**left**) and three-layered boards (**right**).

The compressive shear strength (Figures 10 and 11) of the multi-layered panels glued with MUF was 20% lower. The coefficients of variance in this case were the lowest when the lamellae were glued with MUF (0.023 for single-layered boards and 0.082 for three-layered boards) and PVAc (0.038 for single-layered boards and 0.078 for three-layered boards). For the MUF gluing, the CVs were 0.042 (single-layered boards) and 0.117 (three-layered boards). These results are consistent with the findings of [31], which studied the shear strength of balsa end-grain solid wood panels for densities ranging from 200 to 300 kg/m^3. Although the orientation of the balsa blocks was different compared to the direction of the production of the Paulownia SWPs, the values of the shear strength for balsa panels, including an interval of 0.5 to 2 N/mm^2 for both the parallel and perpendicular grain orientations of the testing specimens, are in line with the results obtained for the single- and three-layered Paulownia SWPs. A major difference, however, is in the panel thickness, which was 5-fold higher for the balsa SWPs. Other solid wood composite panels with a thickness of 20 mm, with a mixture of balsa blocks (flat-grain and end-grain) with a side reinforcement of Brazilian fern wood (*Schizolobium parahyba* (Vell.) Blake), are an industrial product with a shear strength of 4.1 N/mm^2 [32].

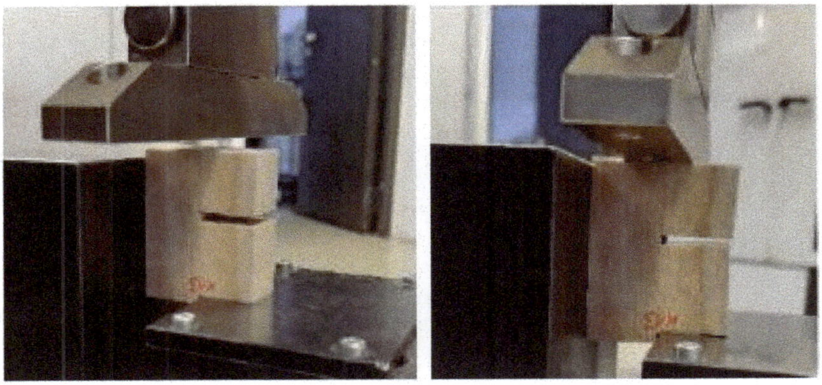

Figure 11. Testing of the compressive shear strength of a single-layered Paulownia SWP.

The wood breakage rate (Figure 12) after testing the compressive shear strength was relatively high for all panel types (more than 95%), with a smaller rate of 80% for the panels bonded with PVAc (Table 4). This is the case when the stress is perpendicular to the longitudinal axis and wood breakage may develop as intercellular failure due to cells being split up and peeling. Cracks are multiplied via the compound middle lamella (which is also called intercellular fracturing) [33].

Figure 12. Wood breakage (98%) of a sample bonded with PUR after testing of compressive shear strength.

4. Conclusions

Paulownia SWPs can be processed well with conventional woodworking machines. The basic prerequisites are well-sharpened tools for the production of the lamellae and SWPs, as well as the appropriate settings of the processing machines. Due to the wide annual rings in Paulownia wood, the variance of testing results for mechanical properties such as modulus of elasticity, bending strength and compression shear test, both positive and negative, is increased. Single-layered SWPs have similar properties even if they are bonded with MUF, PUR or PVAc. In the case of three-layered panels, the use of PUR adhesive influences positively the mechanical properties of SWPs. The presence of PUR-based adhesive joints may influence the elasticity and the shear response of the panels.

In order to meet the demand for ecologically compatible binders and the reduction of emissions from adhesives, it is recommended to carry out further series of tests with protein-based binders, such as casein-based adhesives. As further research, it is recommended to carry out further tests on Paulownia solid wood panels with regard to surface treatment, in terms of surface treatment, moisture behaviour and creep behaviour too. Another aspect to consider for the bonding of Paulownia wood can be the examination of the wood surfaces to be bonded. Different production methods are possible for industrially produced lamellae. These can be cut or split and then further processed in this form, or still be planed or sanded. These surface properties influence the quality of the bonding and offer further possibilities for various test arrangements.

The use of Paulownia wood to produce one-layer and three-layer MHPs is certainly possible. The results obtained suggest, among other things, their use in lightweight construction, and they could find a variety of applications in the furniture and door industry, in interior construction or in means of transport. In Europe, however, the availability of Paulownia sawn timber or lamellae for the production of MHPs is still an open question. Furthermore, the issue of the market preparation and market introduction of MHPs made of Paulownia wood still needs to be considered.

Paulownia, the so-called wood aluminium, can serve in sandwich structures as a lightweight core material for three-layered laminates with thin faces. Paulownia wood from European plantations resembles balsa wood and can successfully replace this lightweight wood species in sandwich structures as the core layer or in engineered products such as three-layered solid wood panels. In this way, the footprint of the end product is considerably reduced and the production costs are diminished by using Paulownia instead of balsa wood.

Author Contributions: Conceptualization, E.M.T., H.R. and M.K.; methodology, H.R. and M.K.; validation, M.C.B., E.M.T. and A.P., formal analysis, A.P.; investigation, H.R. and M.K.; resources, H.R. and M.K.; data curation, E.M.T.; writing—original draft preparation, E.M.T.; writing—review and editing, E.M.T.; visualization, M.C.B.; supervision, A.P.; project administration, M.C.B. All authors have read and agreed to the published version of the manuscript.

Funding: This research received no external funding.

Institutional Review Board Statement: Not applicable.

Informed Consent Statement: Not applicable.

Data Availability Statement: Not applicable.

Acknowledgments: The authors would like to thank to Arien Crul, Glendor Holding GmbH (Kilb, Austria), for providing all the Paulownia lumber from Petrinja, Croatia; DI (FH) Thomas Harreither, Moserholz GmbH (Pettenbach, Austria), for kiln-drying the Paulownia lumber and Ing. T. Wimmer, for his support with the testing of the Paulownia samples.

Conflicts of Interest: The authors declare no conflict of interest.

References

1. Nilsson, J.; Ormarsson, S.; Johansson, J. Moisture-related distortion and damage of lightweight wood panels—Experimental and numerical study. *J. Indian Acad. Wood Sci.* **2017**, *14*, 99–109. [CrossRef]
2. Barbu, M.C. Evolution of lightweight wood composites. *Pro Ligno* **2015**, *11*, 21–26.
3. Švajlenka, J.; Kozlovská, M. Perception of the Efficiency and Sustainability of Wooden Buildings in the Use Phase of the Building. In *Efficient and Sustainable Wood-Based Constructions*; Švajlenka, I.J., Kozlovská, I.M., Eds.; Springer International Publishing: Cham, Switzerland, 2022; pp. 35–57. ISBN 978-3-030-87574-9.
4. Paulitsch, M.; Barbu, M.C. *Holzwerkstoffe der Moderne*; DRW-Verlag: Leinfelden-Echterdingen, Germany, 2015; ISBN 3871818917.
5. Monteiro, S.; Martins, J.; Magalhães, F.D.; Carvalho, L. Lightweight Wood Composites: Challenges, Production and Performance. In *Lignocellulosic Composite Materials*; Kalia, S., Ed.; Springer International Publishing: Cham, Switzerland, 2018; pp. 293–322. ISBN 978-3-319-68695-0.
6. May, N.; Günther, E.; Haller, P. The sustainable use of wood as a regional resource—An ecological assessment of common and new processing technologies for wood poles. *Sustain. Manag. Forum Nachhalt.* **2019**, *27*, 177–201. [CrossRef]
7. Ab-Latib, H.; Choon Liat, L.; Ratnasingam, J.; Law, E.L.; Abdul Azim, A.A.; Mariapan, M.; Natkuncaran, J. Suitability of paulownia wood from Malaysia for furniture application. *BioResources* **2020**, *15*, 4727–4737. [CrossRef]
8. Wilk, K. The Use of a Prefabricated Mass Timber Structure in the Design of Single-Family Houses in Terms of Sustainable Development. In *Environmental Challenges in Civil Engineering II*; Zembaty, Z., Perkowski, Z., Beben, D., Massimino, M.R., Lavan, O., Eds.; Springer International Publishing: Cham, Switzerland, 2023; pp. 226–233. ISBN 978-3-031-26878-6.
9. Tenório, M.; Branco, J.M.; Silva, S.M. Application of a Prefabricated Wooden-Based System for Collective Buildings in a Four-Storey Portuguese Building. In *Sustainable and Digital Building*; Gaspar, F., Mateus, A., Eds.; Springer International Publishing: Cham, Switzerland, 2023; pp. 183–193. ISBN 978-3-031-25794-0.
10. Hill, C.; Kymäläinen, M.; Rautkari, L. Review of the use of solid wood as an external cladding material in the built environment. *J. Mater. Sci.* **2022**, *57*, 9031–9076. [CrossRef]
11. Barbu, M.C.; Buresova, K.; Tudor, E.M.; Petutschnigg, A. Physical and Mechanical Properties of *Paulownia tomentosa* × *elongata* Sawn Wood from Spanish, Bulgarian and Serbian Plantations. *Forests* **2022**, *13*, 1543. [CrossRef]
12. Jakubowski, M. Cultivation Potential and Uses of Paulownia Wood: A Review. *Forests* **2022**, *13*, 668. [CrossRef]
13. Koman, S.; Feher, S. Physical and mechanical properties of Paulownia clone in vitro 112. *Eur. J. Wood Prod.* **2020**, *78*, 421–423. [CrossRef]
14. Barbu, M.C.; Tudor, E.M.; Buresova, K.; Petutschnigg, A. Assessment of Physical and Mechanical Properties Considering the Stem Height and Cross-Section of *Paulownia tomentosa* (Thunb.) Steud. × *elongata* (S.Y. Hu) Wood. *Forests* **2023**, *14*, 589. [CrossRef]
15. Nelis, P.A.; Michaelis, F.; Krause, K.C.; Mai, C. Kiri wood (*Paulownia tomentosa*): Can it improve the performance of particleboards? *Eur. J. Wood Prod.* **2018**, *76*, 445–453. [CrossRef]
16. Lachowicz, H.; Giedrowicz, A. Charakterystyka jakości technicznej drewna paulowni COTE−2. *Sylwan* **2020**, *164*, 414–423. [CrossRef]
17. Kalaycioglu, H.; Deniz, I.; Hiziroglu, S. Some of the properties of particleboard made from paulownia. *J. Wood Sci.* **2005**, *51*, 410–414. [CrossRef]
18. Khanjanzadeh, H. Utilization of bio-waste cotton (*Gossypium hirsutum* L.) stalks and underutilized paulownia (*Paulownia fortunie*) in wood-based composite particleboard. *Afr. J. Biotechnol.* **2012**, *11*, 8045–8050. [CrossRef]
19. Esteves, B.; Aires, P.; Sen, U.; Da Gomes, M.G.; Guiné, R.P.F.; Domingos, I.; Ferreira, J.; Viana, H.; Cruz-Lopes, L.P. Particleboard Production from *Paulownia tomentosa* (Thunb.) Steud. Grown in Portugal. *Polymers* **2023**, *15*, 1158. [CrossRef] [PubMed]

20. Huber, C.; Moog, D.; Stingl, R.; Pramreiter, M.; Stadlmann, A.; Baumann, G.; Praxmarer, G.; Gutmann, R.; Eisler, H.; Müller, U. Paulownia (*Paulownia elongata* S.Y. Hu)—Importance for forestry and a general screening of technological and material properties. *Wood Mater. Sci. Eng.* **2023**, *18*, 1663–1675. [CrossRef]
21. Dogu, D.; Tuncer, F.D.; Bakir, D.; Candan, Z. Characterizing Microscopic Changes of Paulownia Wood under Thermal Compression. *BioResources* **2017**, *12*, 5279–5295. [CrossRef]
22. *EN 322:1993*; Wood-Based Panels—Determination of Moisture Content. European Committee for Standardization: Brussels, Belgium, 1993.
23. *EN 326-1:2005*; Wood-Based Panels—Sampling, Cutting and Inspection—Part 1: Sampling and Cutting of Test Pieces and Expression of Test Results. European Committee for Standardization: Brussels, Belgium, 2005.
24. *EN 323:2005*; Wood-Based Panels—Determination of Density. European Committee for Standardization: Brussels, Belgium, 2005.
25. *EN 310:2005*; Wood-Based Panels; Determination of Modulus of Elasticity in Bending and of Bending Strength. European Committee for Standardization: Brussels, Belgium, 2005.
26. *EN 13354:2009*; Solid Wood Panels (SWP)—Bonding Quality—Test Method. European Committee for Standardization: Brussels, Belgium, 2009.
27. Koman, S.; Feher, S.; Vityi, A. Physical and mechanical properties of *Paulownia tomentosa* wood planted in Hungaria. *Wood Res.* **2017**, *62*, 335–340.
28. Fos, M.; Oliver-Villanueva, J.-V.; Vazquez, M. Radial variation in anatomical wood characteristics and physical properties of *Paulownia elongata* × *Paulownia fortunei* hybrid Cotevisa 2 from fast-growing plantations. *Eur. J. Wood Prod.* **2023**, *81*, 819–831. [CrossRef]
29. Goodrich, T.; Nawaz, N.; Feih, S.; Lattimer, B.Y.; Mouritz, A.P. High-temperature mechanical properties and thermal recovery of balsa wood. *J. Wood Sci.* **2010**, *56*, 437–443. [CrossRef]
30. Galos, J.; Das, R.; Sutcliffe, M.P.; Mouritz, A.P. Review of balsa core sandwich composite structures. *Mater. Des.* **2022**, *221*, 111013. [CrossRef]
31. Osei-Antwi, M.; de Castro, J.; Vassilopoulos, A.P.; Keller, T. Shear mechanical characterization of balsa wood as core material of composite sandwich panels. *Constr. Build. Mater.* **2013**, *41*, 231–238. [CrossRef]
32. Bcomp. *bCores Qx + Sb*; Bcomp: Fribourg, Switzerland, 2023. Available online: https://pdf.nauticexpo.com/pdf/bcomp-ltd/bcores-qx-sb/51016-104454.html#open546181 (accessed on 11 September 2023).
33. Karinkanta, P.; Ämmälä, A.; Illikainen, M.; Niinimäki, J. Fine grinding of wood—Overview from wood breakage to applications. *Biomass Bioenergy* **2018**, *113*, 31–44. [CrossRef]

applied sciences

MDPI

Article

Properties of Un-Torrefied and Torrefied Poplar Plywood (PW) and Medium-Density Fiberboard (MDF)

Cosmin Spîrchez [1,*], Aurel Lunguleasa [1], Carmen-Mihaela Popescu [2], Anamaria Avram [1] and Constantin Stefan Ionescu [1]

[1] Faculty of Furniture Design and Wood Engineering, Transilvania University of Brasov, 500036 Brasov, Romania; lunga@unitbv.ro (A.L.); anamaria.avram@unitbv.ro (A.A.); ionescu.constantin.stefan@unitbv.ro (C.S.I.)

[2] Petru Poni Institute of Macromolecular Chemistry, 700487 Iasi, Romania; mihapop@icmpp.ro

* Correspondence: cosmin.spirchez@unitbv.ro

Abstract: In a context where there is an increasing need for thermal treatments of wooden products, the current research contributes a description of the torrefaction treatment of two of the composite wood materials available on the international market. The present paper presents the importance of the torrefaction process for poplar plywood and medium-density fiberboard. In this paper, the positive aspects of the torrefaction process (decrease in water absorption, thickness swelling and shrinkage, and color) but also the negative aspects of mechanical resistance to static bending are presented. Poplar plywood (PW) and medium-density fiberboard (MDF) panels, with the initial dimensions of 2000 × 1250 mm, were used. From these, 300 × 300 mm samples were cut and torrefied using two different temperatures (170 and 190 °C) and two different periods (for 1 and 2 h). After the treatment, the samples were cut in different sizes (as necessary for each type of evaluation method) from different zones of the panels and used to evaluate the water absorption and thickness swelling, to determine their modulus of rupture, roughness, and color changes. The obtained results emphasize that the mass loses increase at high temperature as the main disadvantageous characteristics of torrefaction. Also, while the calorific power increases with the increase in the parameters of the torrefaction regime, the hygroscopicity and some mechanical properties of the material simultaneously decrease.

Keywords: water absorption; modulus of rupture; roughness; medium-density fiberboard

Citation: Spîrchez, C.; Lunguleasa, A.; Popescu, C.-M.; Avram, A.; Ionescu, C.S. Properties of Un-Torrefied and Torrefied Poplar Plywood (PW) and Medium-Density Fiberboard (MDF). *Appl. Sci.* **2023**, *13*, 11950. https://doi.org/10.3390/app132111950

Academic Editor: Giuseppe Lazzara

Received: 23 September 2023
Revised: 28 October 2023
Accepted: 29 October 2023
Published: 1 November 2023

1. Introduction

The main properties of wood vary between forest species, but also between the same wood species; thus, the properties of wood cannot be controlled during different processing stages. Moreover, there are differences between wood-based composite materials, which are made of technical veneer (plywood), fibers (wood fiber boards), and chips (wood chip boards—PB—and long oriented chip boards—OSB), from fine sawdust (WPC boards) and timber (GLULAM) to wood chips (boards with mineral binders).

In Europe, the consumption of wood and wood-based composite materials has increased year by year. The demand for these materials is differentiated from country to country. Even if the increase in people's needs was not the same in different areas of the globe, there was a total increase in the level of the consumption of wood and wood-based composite materials. So, the consumption of composite materials recorded an increase up to the level of 2010, followed by a stagnation period for wood chipboards and plywood and an increase in oriented strand boards (OSB) and medium-density fiberboards (MDF) [1–5].

Wood-based composite materials are used for applications in the furniture industry, construction, aviation, and musical instruments, showing many advantages over other materials, such as high temperature resistance, stiffness, very good durability, strengths, and higher rigidity and adhesion, in relation to metals or plastics [2,6]. Considering the quality coefficients (the ratio between strength and density), sometimes, wood-based composite materials come close to the strongest materials, such as steel, cast iron, and Kevlar [7,8]. The main advantage of these composite materials exceeds the area of their superior properties, through the use of residues obtained from the exploitation of forests, vines, and orchards; from the processing of round wood in the form of logs and other wood materials; but also from the use of wood products taken out of use (cabins, wooden buildings, temporary civil constructions, furniture, crates, etc.). From this point of view, wood-based boards are renewable, by the fact that, at certain time intervals, a base of raw material is created, through the collected remains and wood waste. Aside from all these advantages, composite materials also have the disadvantage of the use of adhesives for gluing, which are non-ecological materials with serious impact on human health [9]. And yet, through using natural adhesives, the attractiveness of these composite products increases significantly, becoming essential to current human activities.

The quality of composite materials can be evaluated through observing their physico-mechanical properties through following specific tests, as well as through micro- and macroscopic evaluations of fibers, chops, veneers, sawdust, chips, or timber [1].

Two of the most important wood-based composite materials are plywood and medium-density fiberboard. Plywood is a very durable composite material; it is waterproof and is heavier than MDF [4]. Plywood has very good strength in relation to its weight and presents superior rigidity on both of it sides [10–19]. It can be used as a structural material, in which case its dimensions and properties must be precisely calculated [20]. The mechanical resistance of plywood decreases with increasing humidity and water [21–24]. Dimensional changes in longitudinal and transverse direction are on average 0.015% for every 1% change in its humidity [13,17].

Medium-density fiberboards are produced from a series of lignocellulosic fibers, following a thermal and mechanical process of defibration that uses the combined action of heat with mechanical energy to break the bonds between cells [6,23]. Depending on the manufacturing process of MDF, fibers with a high degree of fibrilization are used for the wet process, and long and thin fibers are used for the dry process [9,25–31]. To obtain MDF, small-diameter fibers are used, so that a degree of compressibility is achieved without additional energy consumption [29]. Plywood and medium-density fiberboard are dimensionally stable, as they have a density of about 750–800 kg/m^3, compared to wood chipboards with a density of 650–750 kg/m^3 [11,20]. Medium-density fiberboard is used as a raw material for high-quality furniture, having a much denser composition than wood chipboards [31–34].

MDF and plywood do not have moisture resistance, but painted or laminated, they behave well in wet or humid environments [8,34–36]. In order to increase the moisture resistance of the two composite materials and to extend their field of use in humid environments, heat treatment at high temperatures between 170–220 °C in an oxygen-poor environment, or even at higher temperatures under a nitrogen environment, can be used [37–40]. This treatment is called torrefaction [26]. Through the torrefaction process, the molecular structure of the wood is changed, its dimensional stability is improved by 30–50%, the hygroscopicity is reduced, and, implicitly, the calorific value is increased [41–43]. In addition to the advantages highlighted above, torrefied wood-based composites (MDF and PW) have also some disadvantages, like a slight decrease in density and mechanical resistance [26,30].

The aim of this work was to study two of the properties of the most used wood-based composites, namely plywood (PW) and medium-density fiberboard (MDF), to reduce water absorption, in order to use them in wet environments such as bathrooms and kitchens. They were thermally treated at high temperatures, and for comparison, unmodified PW and MDF were used.

2. Materials and Methods

Poplar plywood (PW) and medium-density fiberboard (MDF) panels, with the initial dimensions of 2000 × 1250 mm, were used. These are commercially available panels and were purchased from Compozite S.R.L. Brasov, Romania. The poplar PW consists of seven veneer sheets, glued using ureo-formaldehyde resin, with a moisture content of 8%. The MDF had a moisture content 9%. The calculated density for the poplar PW was 671.98 kg/m^3, and for MDF, it was 752.48 kg/m^3. From these, 30 pieces of 300 × 300 mm were cut with a circular saw in order to be used for the torrefaction process. Before and after the modification, the samples were weighed with a Kern 250 scale (Düsseldorf, Germany), and their thickness was measured with an electronic caliper.

Torrefaction process. After conditioning, the samples were placed in a laboratory oven type Memmert (Berlin, Germany) for drying at 105 ± 3 °C for 10 h. Then, they were weighed, and after another hour of drying, the thickness of the samples was determined. The torrefaction process was performed, at temperatures of 170 °C (T1) and 190 °C (T2), for 1 h (t1) and 2 h (t2), respectively. Through combining times with temperatures, there were 4 types of thermal hardening regimes, namely T1t1, T1t2, T2t1, and T2t2.

Firstly, in order to determine the main parameter of torrefaction process, the mass loss was calculated, using the following equation (Equation (1)):

$$M_l = \frac{m_i - m_f}{m_i} \cdot 100\ [\%] \tag{1}$$

where m_i is the initial mass of the sample, before torrefaction; m_f is the final mass of the sample.

Water absorption. Samples measuring 50 × 50 mm were cut from different zones of the panel and used to determine the water absorption and thickness swelling. For each sample, 5 replicates were used.

The untreated and torrefied samples were dried in an oven at 105 °C until constant weight was observed. After that, the samples were conditioned at 65% RH and 20 °C and then placed in an immersion tank with distilled water at room temperature (of 20 °C), 20 mm below the water level. The samples were placed horizontally on special grids. After 2 and 24 h, the samples were extracted, the excess of water was removed with filter paper, and the mass and their dimensions were measured again. The experiment was realized according to the EN 317:1993 standard [19].

Water absorption was determined using the following equation (Equation (2)):

$$A_i = \frac{m_i - m_{0i}}{m_{0i}} \cdot 100\ [\%] \tag{2}$$

where m_i is the mass of the sample for each period of time (2 or 24 h), m_{0i} is the initial dry mass of the sample, and i = 2 or 24 h. Customizing equation (2) for immersion durations of 2 and 24 h, the following two equations are obtained (Equation (3)):

$$A_{2h} = \frac{m_{2h} - m_{02h}}{m_{02h}} \cdot 100;\ A_{24h} = \frac{m_{24h} - m_{024h}}{m_{024h}} \cdot 100\ [\%] \tag{3}$$

where A_{2h} is the water absorption percentage after 2 h of immersion in water; m_{2h} is the mass of the samples after 2 h of immersion in water; m_{02h} is the mass of absolutely dry samples for the case of samples immersed for 2 h in water; A_{24h} is the water absorption percentage after 24 h immersion in water; m_{24h} is the mass of the samples immersed for 24 h in water; m_{024h} is the mass of dry samples for the case of immersion in water for a period of 24 h.

The thickness swelling of the samples was determined according to the variation in the thickness of the samples before and after immersion, using the data obtained during the measurements mentioned above and the following equation (Equation (4)):

$$S_i = \frac{g_i - g_{0i}}{g_{0i}} \cdot 100\ [\%] \tag{4}$$

where S_i represents the swelling percentage in the thickness direction; g_i is the thickness of the samples at time i (2 or 24 h); g_{0i} is the initial thickness after drying, at time i (2 or 24 h).

Customizing the previous relationship (4) for immersion periods of 2 h and 24 h, the following two calculation formulas were obtained (Equation (5)):

$$S_{2h} = \frac{g_{2h} - g_{02h}}{g_{02\ h}} \cdot 100\ [\%]\ S_{24h} = \frac{g_{24h} - g_{024h}}{g_{024h}} \cdot 100\ [\%] \tag{5}$$

where S_{2h} is the percentage of swelling of the thickness of the samples after 2 h of immersion in water; g_{2h} is the thickness recorded after 2 h immersion; g_{02h} is the thickness of the absolutely dry samples before immersion in water for a period of 2 h; S_{24h} is the swelling percentage of the thickness of the samples after 24 h of immersion in water; g_{24h} is the thickness of wet samples, after immersion for a period of 24 h; g_{024h} is the thickness of the absolutely dry samples before immersion in water for a period of 24 h.

Modulus of rupture (MOR). These tests were performed according to EU standard EN 310: 1993 [44]. To determine the modulus of rupture, samples measuring 50 × 360 mm (width × length) were processed from poplar PW, and measuring 50 × 400 mm (width × length) from MDF. From each type of wooden composites (PW and MDF), 10 replicates were used, including the addition of 10 new control replicates. Before the measurements, the samples were conditioned at 20 °C and 65% relative humidity. To measure the deformation, a minimum force of 200 N and a maximum one of 600 N were applied. The MOR was calculated using the following equation (Equation (6)):

$$MOR = \frac{3}{2}\frac{P_{max} \cdot l}{b \cdot g^2}\ \left[N/mm^2\right] \tag{6}$$

where *MOR* (N/mm^2) is the modulus of rupture; P_{max} (*N*) is the maximum breaking force of the sample; l (*mm*) is the distance between supports; b (*mm*) is the width of the samples of each type of wood-based composite; g (*mm*) is the thickness of the samples. A schematic representation of the determination of modulus of rupture is presented in Figure 1.

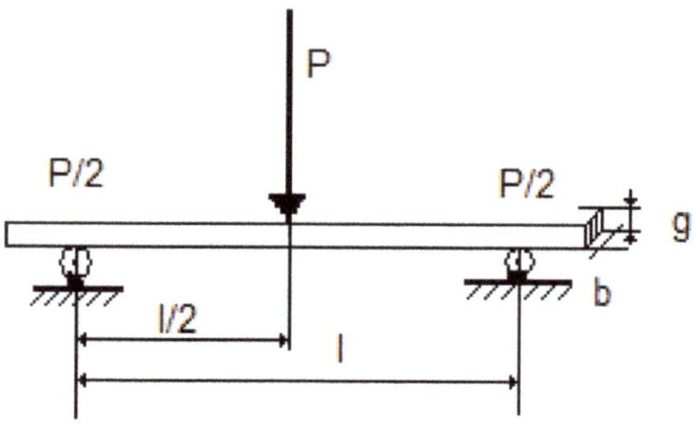

Figure 1. Determination of modulus of rupture.

Roughness of PW and MDF. To determine the roughness, 30 samples with dimensions of 50 × 50 mm (width × length) from poplar PW and MDF were used. The roughness was expressed through three parameters: the arithmetic mean of irregularities (R_a), the mean height of irregularities (R_z), and the mean square root (R_q).

For the measurements, the Mitutoyo Surftest SJ-301 instrument (Mitutoyo company, Kawasaki, Japan) was used. The cut-off length was 2.5 mm, the sampling length was 12.5 mm, and the detection tip radius was 5 µm in the surface roughness measurements.

Color measurements of PW and MDF. A portable colorimeter type Tes-135 A, manufactured by Tes Electrical Electronic Corp (Taipei, Taiwan), was used to determine the color, using the CIELab system. The L* axis represents lightness and has the value 0 for black and 100 for white, with values in between being a series of shades of grey. The a* axis refers to the green–red opposition, with negative values towards green and positive values towards red. The b* axis quantifies the blue–yellow opposition, with negative values towards the blue zone and positive values towards the yellow zone [45]. Figure 2 defines the entire CIELab colorimetric space with the 3 components L, a, and b. From the L*, a*, and b* values, the color uniformity was calculated as the difference in the lightness and chromaticity parameters between heat-treated and control (untreated) poplar PW and MDF samples.

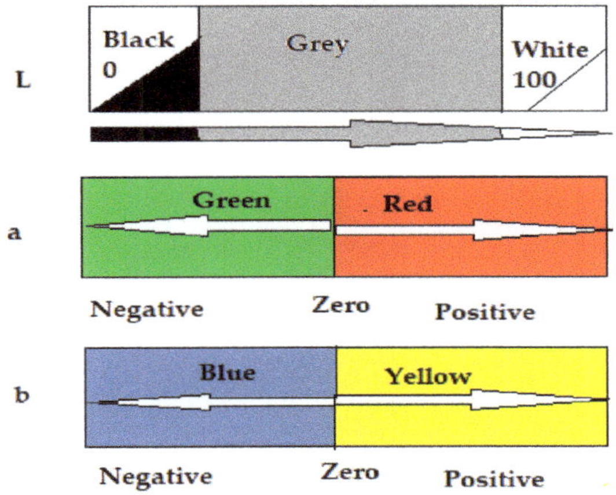

Figure 2. The CIELab color space.

Based on the direct observations on the analyzed parts, it is predicted that after the heat treatment of torrefaction, the luminance will migrate towards black (i.e., it will decrease), the coefficient a* will migrate towards green (i.e., it will decrease slightly), while the coefficient b* will migrate towards yellow (i.e., it will increase slightly).

3. Results

3.1. Mass Losses of Torrefaction Process

As a first evaluation for the torrefaction process, the mass loss of all samples was calculated. The variation in the mass losses for the treated samples for different periods and temperatures can be observed in Figure 3. The regression curves that best highlight this are the polynomial ones. As can be seen, the mass loses present higher values for poplar PW compared to MDF, and this is due to its layered structure. An exception is observed the T_2t_2 treatment regime, where the mass loss for MDF is almost similar, with a slight increase, compared to poplar PW. The maximum mass loss for poplar PW was 7.65% for a torrefaction treatment at 170 °C and 2 h, and 11.50% for a torrefaction treatment at 190 °C and 2 h. The maximum mass loss for MDF was 6.86% for the treatment at 170 °C and 2 h, and 13.85% for the treatment at 190 °C and 2 h.

3.2. Water Absorption after 2 and 24 h of Immersion

The water absorption percentage depends on the structure of the two materials (PW and MDF) [12]. Thus, for poplar PW, it is observed that the water absorption for 2 h was higher than the control sample in the case of the T_1t_1 regime (with 72.73%), while in the case of the T_2t_1 regime, the water absorption for 2 h was lower than the control sample (with 14.62%).

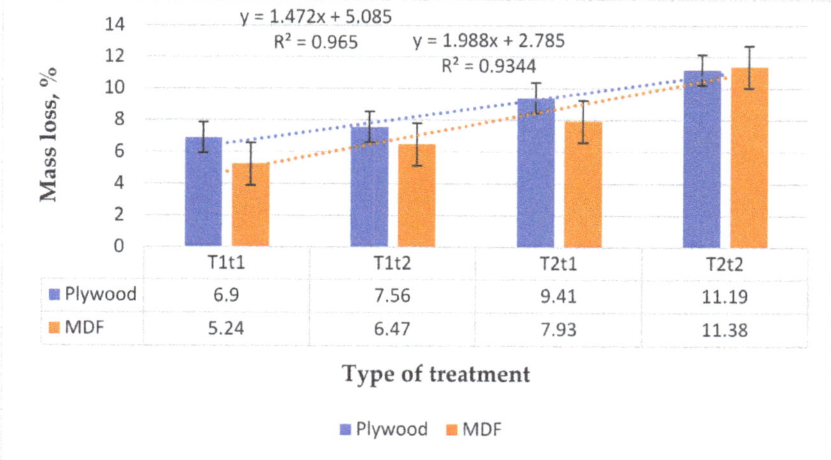

	T1t1	T1t2	T2t1	T2t2
Plywood	6.9	7.56	9.41	11.19
MDF	5.24	6.47	7.93	11.38

Figure 3. Mass loss in poplar PW and MDF: T_1t_1—torrefaction treatment at 170 °C for 1 h; T_1t_2—torrefaction treatment at 170 °C for 2 h; T_2t_1—torrefaction treatment of at 190 °C for 1 h; T_2t_2—torrefaction treatment at 190 °C for 2 h.

The water absorption percentage after 24 h was higher than the control sample for both regimes (T_1t_1 and T_2t_1) (with 76.79% and 9.07%, respectively). This evaluation was validated by the two polynomial curves, with the Pearson coefficient R^2 =1 (see Figure 4a).

The maximum water absorption for poplar PW control sample was 31.8% after 2 h and 54.76% after 24 h. The maximum water absorption for torrefied poplar (PW) in the T_1t_1 regime after 2 h was 48.53%, and after 24 h, it was 99.65%, respectively. The maximum water absorption for torrefied poplar PW in the T_2t_1 regime after 2 h was 23.19%, while after 24 h, it was 56.8%.

The maximum water absorption percentage for the control MDF was 8.66% after 2 h and 30.89% after 24 h. The maximum water absorption for torrefied MDF in the T_1t_1 regime after 2 h was 9.07%, and after 24 h, 34.92%. The maximum water absorption for torrefied MDF in the T_2t_1 regime after 2 h was 5.55%, while after 24 h, it was 26.71%.

For MDF, it is observed that the water absorption after 2 h was higher than for the control sample in the case of the T_1t_1 regime (with 3.61%), while in the case of the T_2t_1 regime, the water absorption was lower than the control sample (with 31.53%). In the case of water absorption for 24 h, the same trend was maintained as in the case of water absorption for 2 h (with 9.17% for the T_1t_1 regime and 16.81% for the T_2t_1 regime). Polynomial curves, with Pearson coefficient R^2 = 1, highlighted this trend of water absorption (Figure 4b). Through comparing the used regimes, it was noticed that in all working regimes, poplar PW absorbs more water compared to MDF.

For the control poplar PW, the standard deviation for water absorption for 2 h was 3.97, while for water absorption for 24 h, it was 4.03, respectively. For poplar PW treated at 170 °C, the standard deviation for water absorption for 2 h was 1.17, while for water absorption for 24 h, it was 1.31, respectively. For poplar PW treated at 190 °C, the standard deviation for water absorption for 2 h was 2.09, while for water absorption for 24 h, it was 2.76, respectively.

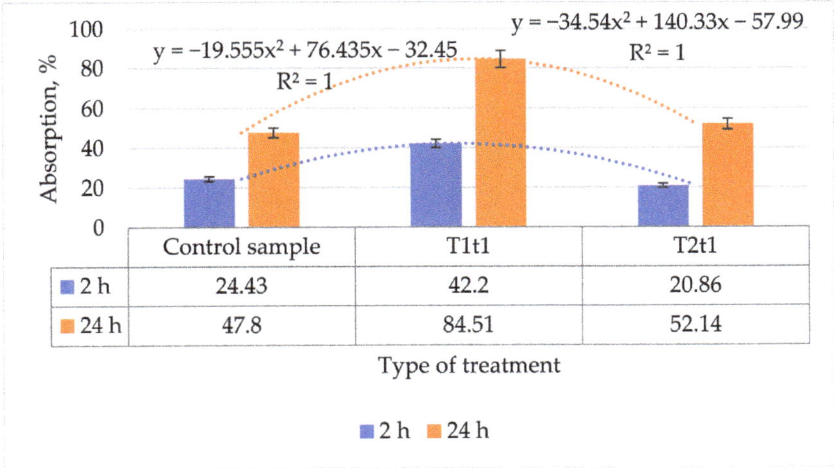

(**a**)

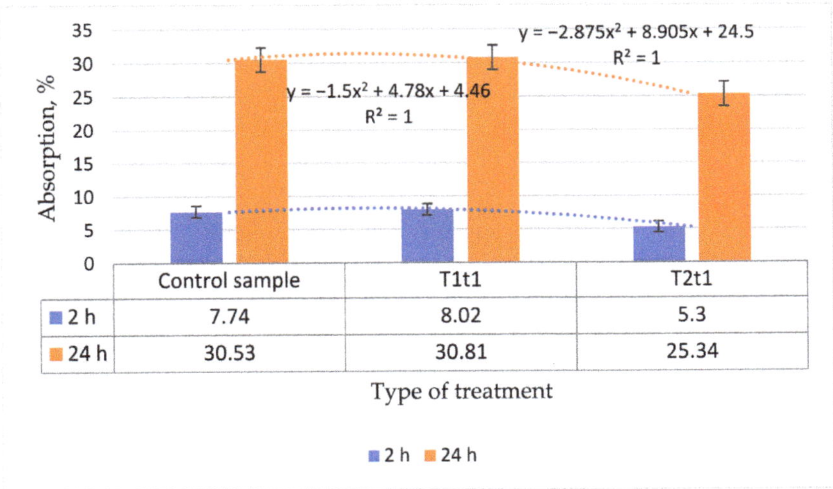

(**b**)

Figure 4. Water absorption percentage for (**a**) poplar PW and (**b**) MDF.

In the case of the control MDF, the standard deviation for water absorption for 2 h was 1.71, while for water absorption for 24 h, it was 2.92, respectively. For MDF treated at 170 °C, the standard deviation for water absorption for 2 h was 1.82, and for water absorption for 24 h, it was 2.94, respectively. For MDF treated at 190 °C, the standard deviation for water absorption for 2 h was 1.18, while for water absorption for 24 h, it was 2.69.

3.3. Thickness Swelling after Immersion in Water

The dimensional stability of composite materials (PW and MDF) was exemplified by the parameter of thickness swelling after total immersion in distilled water. Only the swelling in thickness was used, because the swelling in the longitudinal direction was very small, almost imperceptible. This behavior is very well known from the properties of solid wood, where the ratio between the two directions is 1: (10–20).

For poplar PW, the thickness swelling after 2 h was lower than the control sample, for the two work regimes (17.48% for T_1t_1 and 12.02% for T_2t_1). The same trend was preserved for thickness swelling when the samples were immersed in water for 24 h (6.77% for T_1t_1

and 5.53% for T_2t_1, respectively) (see Figure 5a). This means that the swelling in thickness is different for the two composite materials studied and is determined by the nature of each composite material, respectively, by the layered nature of the poplar PW and the agglomerated–fibrous nature of the MDF.

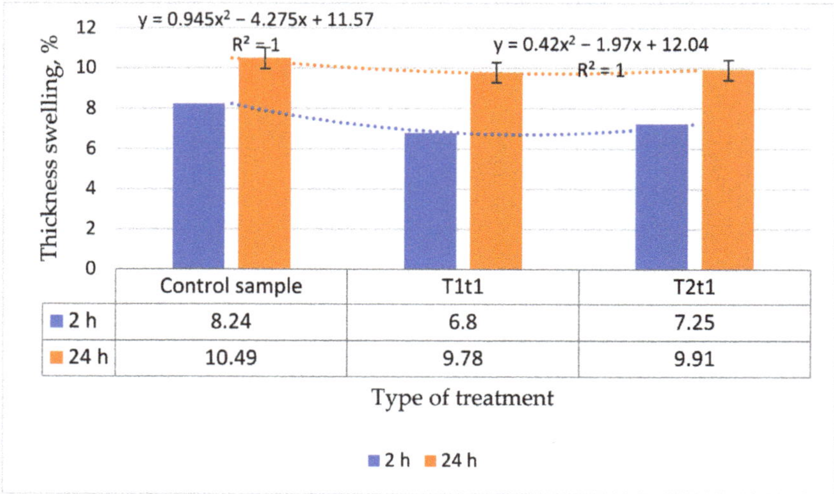

(**a**)

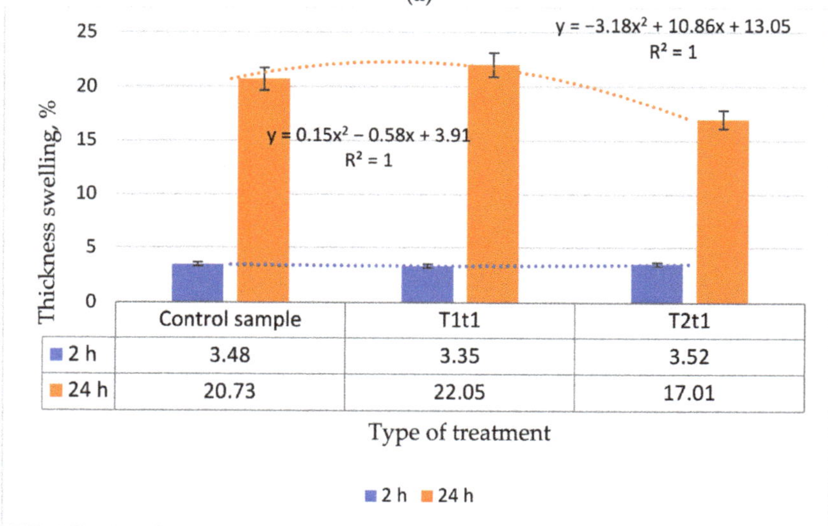

(**b**)

Figure 5. Thickness swelling of (**a**) poplar PW and (**b**) MDF; 2 h—2 h immersions in water; 24 h—24 h immersions in water; T_1t_1—treatment regime at 170 °C for 1 h; T_2t_1—treatment regime at 190 °C for 1 h.

The thickness swelling of the treated MDF in the T_1t_1 regime after 2 h immersion in water was lower than the values recorded for the control sample (3.74%) and higher than the control sample for the treated MDF in the T_2t_1 regime (with 1.14%) (Figure 5b). For treated MDF in the T_1t_1 regime, the thickness swelling after 24 h immersion in water was higher than the recorded values for the control sample (with 6.36%) and lower for the treated MDF in the T_2t_1 regime (with 17.95%), as Figure 5b shows us.

The variation in thickness swelling in poplar PW and MDF was evidenced by polynomial curves with Pearson coefficient (R) = 1. The maximum thickness swelling for poplar PW related to the control sample was 9.46% after 2 h and 11.77% after 24 h immersion in water. The maximum thickness swelling for treated poplar PW in the T_1t_1 regime was 10.71% after 2 h and 13.32% after 24 h. The maximum thickness swelling for treated poplar PW in the T_2t_1 regime was 8.06% after 2 h and 10.34% after 24 h.

The maximum thickness for MDF related to the control sample was 4.65% after 2 h and 23.14% after 24 h. The maximum thickness swelling for treated MDF in the T_1t_1 regime was 3.93% after 2 h, while after 24 h, it was 26.95%. The maximum thickness swelling for treated MDF in the T_2t_1 regime was 3.49% after 2 h, while after 24 h, it was 18.81%.

For the control poplar PW, the standard deviation for thickness swelling after immersion in water for 2 h was 1.68, and 2.04 after immersion for 24 h. For poplar PW treated at 170 °C, the standard deviation for the thickness swelling was 1.82 after immersion in water for 2 h, and 1.90 after immersion in water for 24 h, respectively. For poplar PW treated at 190 °C, the standard deviation for the thickness swelling was 1.73 after immersion in water for 2 h, and 1.98 after immersion in water for 24 h.

For the control MDF, the standard deviation for the thickness swelling was 1.79 after immersion in water for 2 h and 2.09 after immersion in water for 24 h. For MDF treated at 170 °C, the standard deviation for the thickness swelling was 1.82 after immersion in water for 2 h and 2.15 after immersion in water for 24 h, respectively. For MDF treated at 190 °C, the standard deviation for the thickness swelling was 1.36 after immersion in water for 2 h, and 2.75 after immersion in water for 24 h respectively.

3.4. Modulus of Rupture (MOR)

Another important property of composite materials is the resistance to bending strength or modulus of rupture (MOR). For poplar PW, the bending resistance decreases by 45.27% in the case of torrefied samples in the T_1t_2 regime and by 24.52% in the case of the T_2t_2 regime, compared to the control sample, as can be observed from Figure 6. This means that the increase in temperature negatively influences the resistance of this composite material, all due to the irreversible changes occurring in the structure of the wood during treatment at high temperatures. For MDF, the bending resistance decreases by 13.06% for the torrefied samples with the T_1t_2 regime and by 32.51% for the torrefied samples with the T_2t_2 regime, compared to control sample. The regression curves that highlighted the decrease in bending resistance for poplar PW and MDF were the polynomial ones of the second degree, with Pearson coefficient $R^2 = 1$.

The gluing quality is one of the most effective factors related to modulus of rupture of poplar PW because the improper gluing of veneer sheets in plywood manufacturing can cause the delamination of layers, and such samples exhibit a low modulus of rupture value. The maximum modulus of rupture for the control sample of poplar PW was 70.06 N/mm^2. This value falls within the limits obtained by other researchers (95.65 N/mm^2 for poplar plywood and 117.37 N/mm^2 in the case of MDF) [39,46]. The maximum modulus of rupture for torrefied poplar PW at 170 °C was 64.39 N/mm^2, and 40.47 N/mm^2 for torrefied poplar PW at 190 °C. The maximum modulus of rupture for the control MDF sample was 46.46 N/mm^2, while the maximum modulus of rupture for torrefied MDF at 170 °C was 45.86 N/mm^2, and for torrefied MDF at 190 °C, it was 31.27 N/mm^2.

3.5. Roughness of PW and MDF

Roughness is an important parameter because it characterizes the quality of the processed surfaces. The main parameters of roughness characterization are R_a (arithmetic mean of irregularities), R_z (average height of irregularities), and R_q (mean square root). It is noted that regarding poplar PW parallel to fibers and MDF, the optimal regime of roughness is for the torrefaction of the samples at 170 °C (see Figures 7 and 8). The surface quality of solid wood products is one of the most important properties influencing further manufacturing processes such as finishing or the strength of adhesive joint.

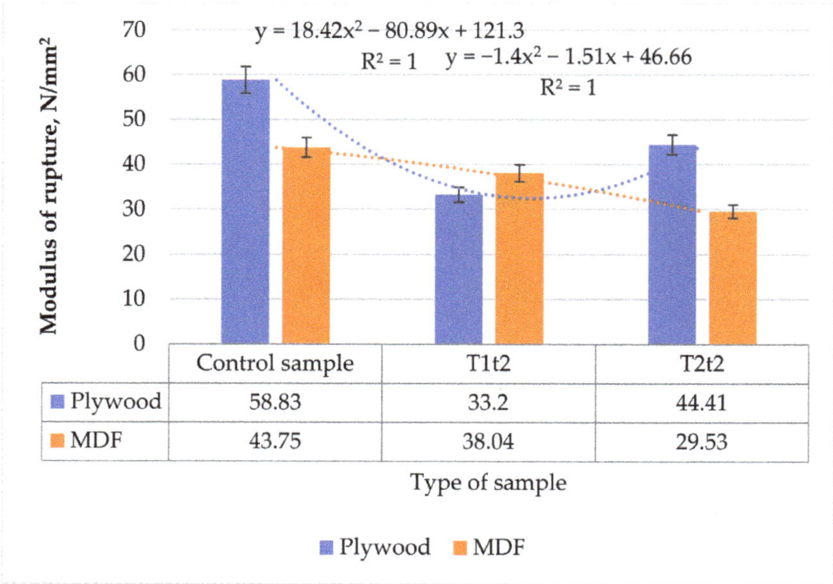

Figure 6. Modulus of rupture for poplar PW and MDF.

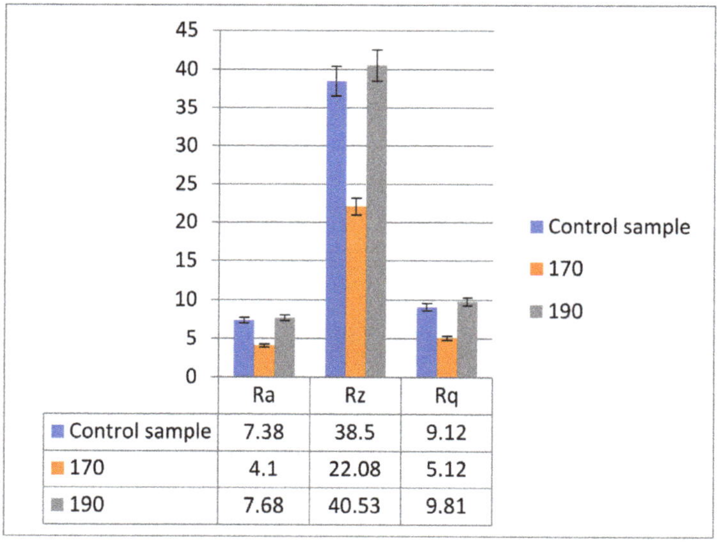

Figure 7. Roughness of poplar PW parallel to the surface of the fiber.

Regarding the measurements for poplar PW perpendicular to the fibers, the torrefaction treatment does not influence the roughness for the samples treated at 170 °C (see Figure 9).

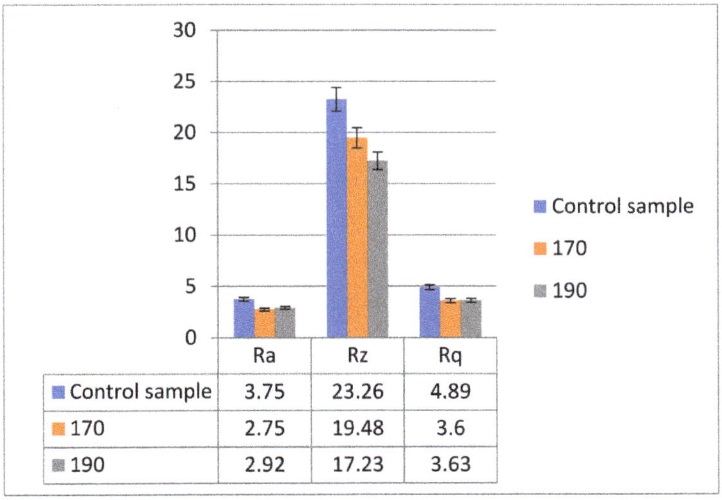

Figure 8. Roughness of MDF.

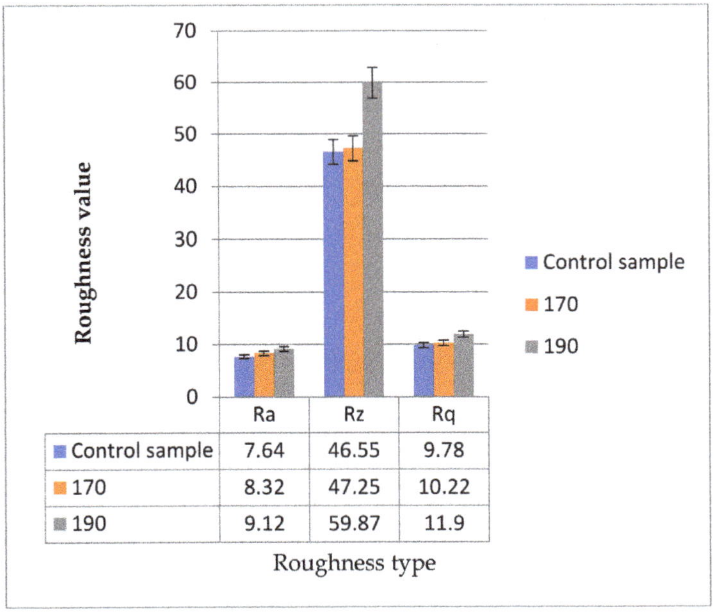

Figure 9. Roughness of poplar PW perpendicular to the fibers.

Also, we observed that the roughness values obtained for poplar PW perpendicular to the fiber direction are much higher than those measured in the parallel direction.

3.6. CIELab Colour Results

The results regarding the color values in the CIELab space are presented in Figure 10. Firstly, regarding the difference in the brightness of the two analyzed composite materials, poplar PW is whiter than MDF (values decrease from 25.07 to 22.4).

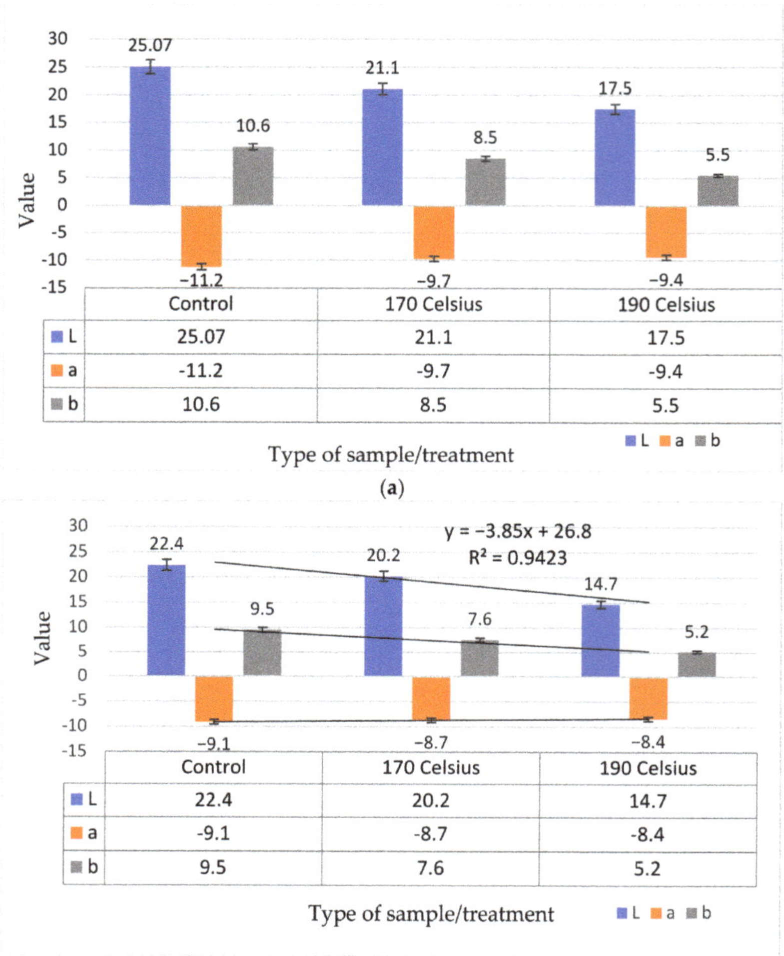

Figure 10. Color measurements for (**a**) poplar PW and (**b**) MDF.

Further, all values decrease after heat treatment regardless of the type of composite material analyzed. This means that the two composites approach black (luminance L is lower), the green shade decreases and approaches the shade of red (coefficient a is lower), without entering this field, and the yellow shade (positive component b) approaches blue shade without entering this field. The temperature had a much more remarkable impact on color changes than the duration of the torrefaction process. These results are in agreement with the results obtained by other researchers [33], where a more remarkable effect of temperature than time of torrefied poplar PW on color change was confirmed.

For poplar PW, L* values decreased from 21.1 (for 170 °C) to 17.5 (for 190 ° C), and a* values decreased from −9.7 (for 170 °C) to −9.4 (for 190 °C), while b* values decreased from 8.5 (for 170 °C) to 5.5 (for 190 °C). At the same time, for MDF, L* values decreased from 20.2 (for 170 °C) to 14.7 (for 190 °C), and a* values decreased from −8.7 (for 170 °C) to −8.4 (for 190 °C), while b* values decreased from 7.6 (for 170 °C) to 5.2 (for 190 °C).

The control poplar PW and torrefied samples at 170 °C and 190 °C are presented in Figure 11.

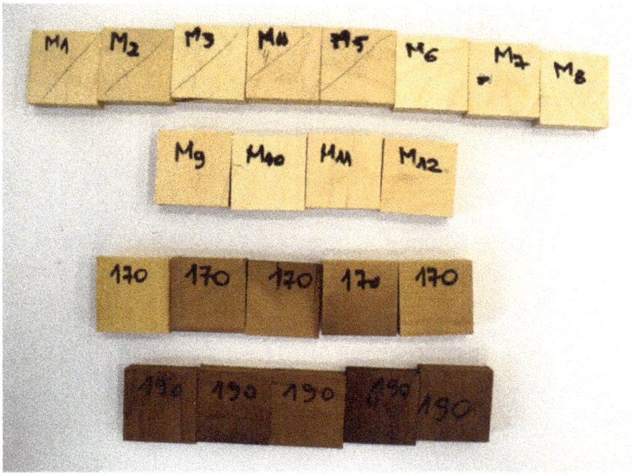

Figure 11. Images of control and torrefied poplar PW samples at 170 °C and 190 °C.

In Figure 12, the control and torrefied MDF sample at 170 °C and 190 °C are presented.

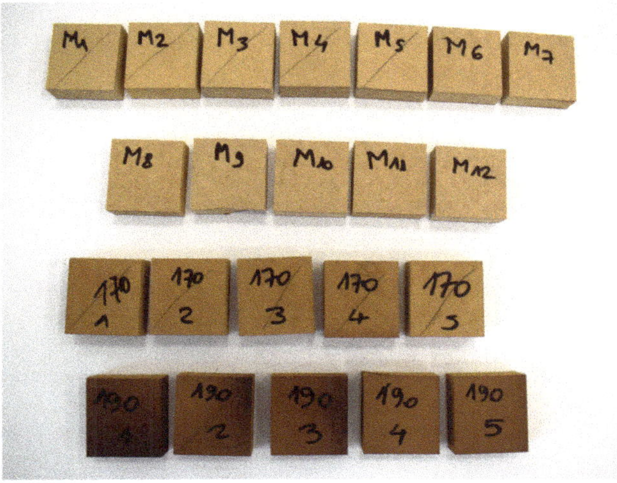

Figure 12. Images of control and torrefied MDF samples at 170 °C and 190 °C.

4. Discussion

The most commonly used furniture materials are plywood and particleboard; the latter one includes oriented strand board (OSB), as well as a low-density fiberboard (LDF), medium-density fiberboard (MDF), or high-density fiberboard (HDF) [5,18]. Also, thermally modified wood has numerous applications because of the enhancement in its properties (biotic resistance and color) [32].

Aydin and coworkers [33] tested the roughness of alder and beech veneers that were thermally modified at 110 °C and 180 °C. The obtained results showed a slight increase in surface roughness. The surface roughness of thermally modified wood is dependent on many factors, such the properties of the wood or the parameters used for the thermal modification process.

Thermally modified plywood decreased in surface roughness as the modification temperature was increased to 170 °C [34], while another study [35] outlined that thermal modification at 120 °C does not influence the surface roughness of pine wood. It is noted that increasing the temperature to 200 °C for 2h reduced R_a values by 6% and increased the R_z value by more than 7%.

In many studies on the influence of thermal modification on wood properties, it is recorded that thermal modifications lead to a decrease in the equilibrium moisture content [5], a decrease in the surface hydrophilicity of the treated material [16], an increase in dimensional stability [17], and a change of color [27,47].

The advantages of torrefying wood composite are the following: it improves its dimensional stability by 30–50%, it produces a pretty color, it decreases the hygroscopicity, and it increases the calorific value [7,33].

Another study [36] mentioned that there is good-quality adhesion between pieces of thermally modified wood when cold-setting melamine-urea-formaldehyde adhesive is used [36]. Further, Wang and its collaborators [41] shows the fact that the use of hot-setting melamine-urea-formaldehyde adhesive in the production of boards is important for increased air humidity conditions or outside applications. The variation in the torrefaction's effects on depth is reduced, especially for samples with small thickness up to 20 mm [12,15]. Medium-density fiberboard has an advantage compared to timber, chipboard, and high-density fiberboard [14,22], because of its large and flat surface due to fine and uniform fibers.

The heat treatment provided a darkening of the surface of the poplar PW. At 170 °C, the changes in color are very low. This is because the important change in the samples was caused by the loss of water. The heat treatment of poplar PW and MDF samples reduces the lightness of color L* and the chromatic coordinates a* and b*.

During the high-temperature torrefaction process (190 °C), the color changes of poplar PW and MDF were more pronounced. L* decreased at all treatment temperatures, indicating that the color became darker.

5. Conclusions

Through the torrefaction process of composite materials (poplar plywood and medium-density fiberboard), it was observed that their density decreases, and water absorption and thickness swelling show good characteristics.

The mass loss increased with the increase in heat treatment duration and temperature.

In terms of modulus of rupture, its values decreased due to high-temperature torrefying processes. Thus, for both studied series of samples, poplar PW and MDF, the modulus of rupture decreased for all analyzed torrefaction regimes compared to the control sample.

Due to the fact that torrefied composite materials are less hydrophilic and more dimensionally stable, they can be used in the manufacture of furniture in humid conditions (kitchen, bathroom, etc.).

For plywood, it was observed that the water absorption after 2 h was higher than the control sample in the case of the T_1t_1 regime, while in the case of the T_2t_1 regime, the water absorption after 2 h was lower than for the control sample.

For MDF, it was observed that the water absorption after 2 h was higher than that for the control sample in the case of the T_1t_1 and T_2t_1 regimes, while in the case of the T_2t_1 regime, the water absorption was lower than the control sample.

The novelty of this study resides in the torrefaction process of composite panels. Companies should have additional torrefaction systems. The later ones are useful when a company has special requests from customers. Moreover, quality control institutions should include in their evaluation systems the quality assessment of these products, as well as their applicability domain.

Author Contributions: Conceptualization, C.S. and A.L.; methodology, C.S.; software, A.A.; validation, C.S., C.-M.P. and C.S.I.; formal analysis, C.S.I.; investigation, A.L.; resources, C.S.; data curation, C.-M.P.; writing—original draft preparation, A.A.; writing—review and editing, C.S. and C.-M.P.; visualization, A.L.; supervision, C.S.; project administration, C.S.; funding acquisition, C.S. All authors have read and agreed to the published version of the manuscript.

Funding: The APC was funded by Transilvania University of Brasov.

Institutional Review Board Statement: Not applicable.

Informed Consent Statement: Not applicable.

Data Availability Statement: Not applicable.

Conflicts of Interest: The authors declare no conflict of interest.

References

1. Thybring, E.E.; Fredriksson, M. Wood and Moisture. In *Springer Handbook of Wood Science and Technology*; Springer Handbooks; Niemz, P., Teischinger, A., Sandberg, D., Eds.; Springer: Cham, Switzerland, 2023. [CrossRef]
2. Windeisen, E.; Bächle, H.; Zimmer, B.; Wegener, G. Relations between chemical changes and mechanical properties of thermally treated wood. *Holzforschung* **2009**, *63*, 773–778. [CrossRef]
3. Aydin, I.; Demirkir, C.; Colak, S.; Salcă, E.A. The Effect of Veneers Roughness on Bonding and Some Mechanical Properties of Plywood. *Pro Ligno* **2013**, *9*, 41–49.
4. Arruda, L.; Del Menezzi, C. Properties of a laminated wood composite produced with thermomechanically treated veneers. *Adv. Mater. Sci. Eng.* **2016**, *2016*, 8458065. [CrossRef]
5. Bakar, B.F.A.; Hiziroglu, S.; Tahir, P.M. Properties of some thermally modified wood species. *Mater. Des.* **2013**, *43*, 348–355. [CrossRef]
6. Bekhta, P.; Prozsyk, S.; Lis, B.; Krystofiak, T. Gloss of thermally densified alder (*Alnus glutionosa Goertn*), beech (*Fagus sylvatica* L.), birch (*Betula verrucosa Ehrh*), and pine (*Pinus sylvestris* L.) wood veneers. *Eur. J. Wood Wood Prod.* **2014**, *72*, 799–808. [CrossRef]
7. Garcia Perez, A.; Salcă, E.A.; Maldonado, I.B.; Hiziroglu, S. Evaluation of surface quality of medium density fiberboards (MDF) and particleboards as function of weathering. *Pro Ligno* **2012**, *8*, 10–17.
8. Işleyen, U.; Karamanoğlu, M. The influence of machining parameters on surface roughness of MDF in milling operation. *BioResources* **2019**, *14*, 3266–3277. [CrossRef]
9. Jehlickova, B.; Morris, R. Effectiveness of policy instruments for supporting the use of waste wood as a renewable energy resource in the Czech Republic. *Energy Policy* **2007**, *35*, 577–585. [CrossRef]
10. Murata, K.; Watanabe, Y.; Nakano, T. Effect of thermal treatment of veneer on formaldehyde emissions of poplar plywood. *Materials* **2013**, *6*, 410–420. [CrossRef] [PubMed]
11. Borrega, M.; Karenlampi, P.P. Higroscopicity of heat-treated Norway spruce (*Picea abies*) wood. *Eur. J. Wood Wood Prod.* **2010**, *68*, 233–235. [CrossRef]
12. Bridgwater, A.V. Review of fast pyrolysis of biomass and product upgrading. *Biomass Bioenergy* **2012**, *38*, 68–94. [CrossRef]
13. Candan, Z.; Hiziroglu, S.; McDonald, A.G. Surface quality of thermally compressed douglas fir veneer. *Mater. Des.* **2010**, *31*, 3574–3577. [CrossRef]
14. *CEN/TS 15679:2011*; Thermal Modified Timber-DEFINITIONS and characteristics. European Committee for Standardization (CEN): Brussels, Belgium, 2011.
15. Chen, W.H.; Cheng, W.Y.; Lu, K.M.; Wuang, Y.P. An evaluation on improvement of pulverized biomass property for solid through torrefaction. *Appl. Energy* **2011**, *11*, 3636–3644. [CrossRef]
16. Chen, W.H.; Ye, S.C.; Sheen, H.K. Hydrothermal carbonization of sugarcane bagasse via wet torrefacton in Association with Microwave Heating. *Bioresources Tecnol.* **2012**, *118*, 195–203. [CrossRef] [PubMed]
17. Demirbas, A. Biomass resource facilities and biomass conversion processing for fuels and chemicals. *Energy Convers. Manag.* **2001**, *42*, 1357–1378. [CrossRef]
18. Derek, W.T.; Robert, A.K.; Philip, D.E. Thermal modification of color in red alder veneer. Effects of temperature, heating time and wood type. *Wood Fiber Sci.* **2005**, *37*, 653–661.
19. *EN 317:1993*; Particleboards and Fibreboards-Determination of Swelling in Thickness after Immersion in Water. European Committee for Standardization: Brussels, Belgium, 1993.
20. *EN 636:2003*; Plywood-Specifications. European Committee for Standardization: Brussels, Belgium, 2003.
21. Feher, S.; Kolman, S.; Borcsok, Z.; Taschner, R. Modification of hardwood veneers by heat treatment for enhanced colors. *Bioresources* **2014**, *9*, 3456–3465. [CrossRef]
22. Grîu (Dobrev), T. The Appreciation and Increasing of Calorific Value for Wooden Biomass. Ph.D. Thesis, Transilvania University of Brasov, Brasov, Romania, 2014.
23. Kocaefe, D.; Poncsak, S.; Dor'e, G.; Younsi, R. Effect of the heat treatment on the wettability of white ash and soft maple by water. *Holz Als Roh-Werkst.* **2008**, *66*, 355–361. [CrossRef]
24. Korkut, S.; Bektas, I. The effects of heat treatment on physical properties of fir (*Abies bornuellriana Mattf*) and scots pine (*Pinus sylvestris* L.) wood. *For. Prod. J.* **2008**, *58*, 95–99.
25. Kukla, M.; Wargula, L. Wood-based boards mechanical properties and their effects on the cutting process during shredding. *BioResources* **2021**, *16*, 8006–8031. [CrossRef]
26. Laskowska, A.; Kozakiewicz, P.; Zbiec, M.; Zaton, P.; Olenska, S.; Beer, P. Surface characteristics of scots pine veneers produced with a peeling process in industrial conditions. *BioResources* **2018**, *13*, 8342–8357. [CrossRef]
27. Lovrić, A.; Zdravković, V.; Furtula, M. Influence of thermal modification on colour of poplar (Populus X Euramericana) rotary cut veneer. *Wood Res.* **2014**, *59*, 661–670.

28. Lovrić, A.; Zdravković, V.; Popadic, R.; Milic, G. Properties of plywood boards composed of thermally modified and non-modified poplar veneer. *Bioresources* **2017**, *12*, 8581–8594. [CrossRef]
29. Lunguleasa, A.; Spîrchez, C. The characteristics of thermally treated MDF Panels. *Bull. Transilv. Univ. Bras.* **2015**, *8*, 55–60.
30. Moya, R.; Tenorio, C. Fuel wood characteristics and its relation with extractives and chemical properties of ten fast-growth species in Costa Rica. *Biomass Bioenergy* **2013**, *56*, 14–21. [CrossRef]
31. Nazerian, M.; Dahmardeh, M. Physical and mechanical properties of laminated veneer lumber manufactured by poplar veneer. *J. Agric. Sci. Technol.* **2011**, *1*, 1040–1045. [CrossRef]
32. Nazerian, M.; Ghalchno, M.D.; Kashkooli, C. Effect of wood species, amount of juvenile wood and heat treatment on mechanical and physical properties of laminated veneer lumber. *J. Appl. Sci.* **2011**, *11*, 980–987. [CrossRef]
33. Aydin, G.; Colakoglu, S.; Colak, S.; Demirkir, C. Effects of moisture content on formaldehyde emissions and mechanical properties of plywood. *Build Environ.* **2006**, *10*, 1311–1316. [CrossRef]
34. Ozlusoylu, I.; Istek, A. Effects of surface lamination process parameters on medium density fiberboard (MDF) properties. *Bioresources* **2023**, *18*, 767–777. [CrossRef]
35. Tasdemir, C.; Hiziroglu, S. Achieving cost efficiency through increased inventory leanness: Evidences from oriented strand board (OSB) industry. *Int. J. Product. Econ.* **2019**, *208*, 412–437. [CrossRef]
36. Percin, O.; Peker, H.; Atilgan, A. The effect of heat on the some physical and mechanical properties of beech (Fagus Orientalis Lipsky) wood. *Wood Res.* **2016**, *61*, 443–456.
37. Salcă, E.A.; Kobori, H.; Inagaki, T.; Kojima, Y.; Suzuki, S. Effect of heat treatment on colour changes of black alder and beech veneers. *J. Wood Sci.* **2016**, *62*, 297–304. [CrossRef]
38. Schnabel, T.; Zimmer, B.; Petutschings, A.J.; Schonberger, S. An approach to classify thermally modified hardwoods by color. *For. Prod. J.* **2007**, *57*, 105–110.
39. Sedlecky, M. Surface roughness of medium-density fiberboard (MDF) and edge-glued panel (EGP) after edge milling. *BioResources* **2017**, *12*, 8119–8133. [CrossRef]
40. Shulga, G.; Betkers, T.; Brovkina, J.; Aniskevicha, O.; Ozolins, J. Relationship between composition of the lignin-based interpolymer complex and its structuring ablity. *Envir. Eng. Manag. J.* **2008**, *7*, 397–400. [CrossRef]
41. Wang, M.; Wang, X.; Li, L.; Ji, H. Fire performance of plywood treated with ammonium polyphosphate and 4A Zeolite. *BioResources* **2022**, *9*, 4934–4945. [CrossRef]
42. Zdravković, V.; Lovrić, A.; Stanković, B. Dimensional stability of plywood panels made from thermally modified poplar veneers in the conditions of variable air humidity. *Drv. Ind.* **2013**, *64*, 175–181. [CrossRef]
43. Wikipedia—The Free Encyclopedia, CIELAB Color Space. Available online: https://en.wikipedia.org/wiki/CIELAB_color_space (accessed on 12 August 2022).
44. *EN 310:1993*; Determination of Modulus of Elasticity in Bending and of Bending Strength. European Committee for Standardization: Brussels, Belgium, 1993.
45. Teuch, O.; Hofeanuer, A.; Troger, F.; From, J. Basic properties of specific wood based materials carbonised in a nitrogen atmosphere. *Wood Sci. Technol.* **2004**, *38*, 323–333. [CrossRef]
46. Percin, O.; Uzun, O. Physical and mechanical properties of laminated wood made from heat-treated scotch pine reinforced with carbon fiber. *BioResources* **2023**, *18*, 5146–5164. [CrossRef]
47. Bekhta, P.; Niemz, P. Effect of high temperature on the Change in Color, Dimensional stability and mechanical properties of spruce wood. *Holzforschung* **2003**, *57*, 539–546. [CrossRef]

applied sciences

Article

Date-Palm-Based Sustainable Hybrid Composite with Cotton and Kevlar Fibre Participation

Octavia Zeleniuc *, Mohammad Hassan Mazaherifar, Camelia Coşereanu and Andrei Suciu

Faculty of Furniture Design and Wood Engineering, Transilvania University of Brasov, 29 B-dul Eroilor, 500036 Brasov, Romania; mohammad.mazaherifar@unitbv.ro (M.H.M.); cboieriu@unitbv.ro (C.C.); andrei-a.suciu@student.unitbv.ro (A.S.)

* Correspondence: zoctavia@unitbv.ro

Abstract: This research aims to evaluate the physical and mechanical performance of three types of hybrid composites made of date palm (*Phoenix dactylifera* L.) (DP), additional layers of cotton (DP/C) and Kevlar fibres (DP/K). The fibres were formed into flat sheets and employed as reinforcement layers embedded in a polyester matrix. Three-layer and five-layer hybrid composites were created using the hand layup method. The layers have alternative longitudinal–transversal orientation. The composites were investigated for density, thickness swelling (TS), water absorption (WA), flexural strength and modulus of elasticity (MOE) properties. Moreover, the composites were subjected to cycles of water immersion, freezing and drying, and the changes in mass and mechanical performance were analysed before and after the cyclic testing. The hybrid composite with Kevlar as the inner layers displayed better physical and mechanical properties when compared to the other two hybrid composites. A stereo-microscopic investigation revealed that poor adhesion between the layers of composites contributed to a reduction in the mechanical properties of DP/C and DP hybrid composites. The DP/C composite had the highest thickness swelling and water absorption, with the water uptake more pronounced than in the cases of the other composites. The hybridisation of date palms with Kevlar fibres improved the properties of the hybrid composites.

Keywords: hybrid composite; date palm; cotton; Kevlar; polyester resin; cyclic testing; mechanical testing

Citation: Zeleniuc, O.; Mazaherifar, M.H.; Coşereanu, C.; Suciu, A. Date-Palm-Based Sustainable Hybrid Composite with Cotton and Kevlar Fibre Participation. *Appl. Sci.* **2024**, *14*, 1008. https://doi.org/10.3390/app14031008

Academic Editor: Victor Franco Correia

Received: 4 December 2023
Revised: 15 January 2024
Accepted: 19 January 2024
Published: 24 January 2024

1. Introduction

The development of wood composites was driven by the increase in the cost of logs, the promotion of sustainable management of wood raw materials, and market demands [1,2]. There is an increasingly convincing argument for sustainability in the selection process of materials related to the fields of use [2]. Therefore, the increase in demand for sustainable and renewable materials has brought the possibility of using natural fibres in the structure of the composites to the attention of specialists, due to their low cost, light weight, biodegradability and non-hazardous characteristics [3,4]. Natural fibres such as sisal, flax, jute, kenaf, hemp and ramie were employed as reinforcing materials in polymer-based matrices [5–9] to surpass negative impacts on the environment and resource availability issues [10]. Among them, date palm (*Phoenix dactylifera*) (DP) midrib long fibres are possible raw materials for the design and production of lignocellulosic composites.

The date palm is known for its fruits but, from an economic point of view, represents an important source of income for local farmers and the communities where it grows; it also provides the raw material for furnishing, housing and lots of handcrafts [11]. Date palm trees are found in countries like Egypt (1,352,950 metric tons), Saudi Arabia (1,078,300 metric tons), Iran (1,023,130 metric tons), the UAE (775,000 metric tons) and Algeria (710,000 metric tons) [12]. There are more than 120 million date palm trees in the world, and they are distributed in 94 countries [13]. Date palm leaves are considered waste materials, and they are extensively used for the production of natural fibres [14]. Each date palm is pruned, generating up to 35 kg of biomass per tree, of which the leaves weigh 20 kg [15,16], representing about

4,200,000 tons of natural fibres and an important amount of natural and renewable resources, 20 and 10 times more than hemp and sisal production, respectively, which are commonly used as natural fibres in the structure of composites [17]. With lower cellulose content and lower density than sisal and hemp, date palm fibres provide better wettability resistance compared to the other two natural fibres and contribute to lower-weight composites suitable for automotive and space applications [17]. Date palm fibres can be extracted from the leaves and trunk, and each part's fibres have different properties, which can be modified by applying chemical and physical treatments to control ash, extractive content and surface bonding [18]. The fibre extraction technique is an important factor in obtaining fibres with high-performance properties, and it can be achieved biologically (by retting and enzymatic treatment), chemically (by alkaline treatment, specifically with NaOH) or mechanically (through the decorticator mechanism) [19].

Studies have investigated the possibility of using date palms in the structure of composites. In the literature [11], the improvement of flexural strength properties and bending resistance in concrete and cement with the addition of date palm fibres was reported. Fibres from different parts of the date palm tree with different shapes and sizes were used by researchers to manufacture wood-based composites: trunk and rachis particles were used for the production of particleboard [20]; date palm waste trunk was used for the production of blockboard and battenboard [21]; and date palm pruning residues were used to manufacture MDF [22]. Several research works focused on using this resource in the structure of polymer-reinforced composites. Thus, small fibres with lengths between 1 mm and 1.3 mm and diameters of a maximum of 40 μm were extracted from date palm rachis, trunk and petioles and used to reinforce an HDPE matrix [14]. Short fibres extracted from leaves [23] were used for polyester-reinforced composites, with a bending strength between 28.8 N/mm^2 and 58.1 N/mm^2. Leaf fibres ground into small particles and mixed with wood flour were used to reinforce polyester, creating a polyester/woven flax fibre sandwich biocomposite matrix [24]. Fibres extracted from date palm stems and polyester resin were used to form treated date palm fibre polyester-reinforced composites in a die with dimensions of 125 mm × 90 mm × 20 mm, applying the hand layup method; they were tested for erosion and water absorption and showed severe deterioration of both the matrix and date palm fibres, involving matrix micro-cracking, fibre pulling and fracture [25].

Date palm midribs have also had the attention of researchers. With bending strengths in the range of 80–90 N/mm^2 and compression strengths between 35 N/mm^2 and 42 N/mm^2, they are close to the mechanical performance of the red European pine [26]. Long textile fibres that could be converted into woven fabrics or nonwoven mats using a combined alkaline/-mechanical treatment were, for the first time, extracted from date palm midribs without testing them as reinforcements for polymer composites [27]. Using date palm midribs and the extraction method mentioned in [27], an oriented, self-bonded mat without chemical or synthetic materials was obtained in the previous research and used to reinforce a hybrid epoxy composite [28]. The present research is a continuation of the investigation on the potential of using these long textile fibres in self-bonded mats as reinforcement in laminate hybrid polyester composites.

The contribution of other natural fibres was also investigated in the laminate structures of date-palm-midrib-reinforced composites, showing that the flexural strength and flexural modulus of elasticity were improved with the contribution of *Calotropis Procera* fibres with low weight and diameters between 0.015 mm and 0.022 mm [28]. A similar natural fibre to be tested in such structures could be cotton. Cotton is a textile fibre obtained from seeds through ginning. Cotton fibres have a matte sheen, and their yellowish-white colour is due to the natural pigments they contain. As a natural fibre, cotton is known for its properties such as low density, good heat conductivity, high flexibility, resilience, elasticity and renewability [29]. Despite the fact that natural fibres are biodegradable, easy to recycle, readily available, and have low density and low cost, their use is limited due to their tendency to absorb moisture, have low adhesion with synthetic components, poor wettability [30] and lack of compatibility with the matrix [31]. The hybridization of natural fibres with synthetic fibres is necessary to increase their performance, which is difficult to obtain from individual natural fibres [32]. As long as date palm midrib fibres show comparable properties to those of other natural fibres such as sisal, hemp, and flax [27], possible hybrid Kevlar/natural-fibre-reinforced polymer composites can be designed considering the results obtained by other researchers [31]. Kevlar is a strong material, with high rigidity modulus, toughness and thermal stability, belonging to the aramid fibres. It has a wide range of applications, including in aerospace engineering (such as aircraft wing and fuselage structure), bulletproof vests, face shields, ships and body armour [31,33].

The chemical bonding between the matrix and the natural fibres plays an important role in the performance of the composite, with a notable effect on its mechanical properties [34]. Except epoxy resin, which exhibits excellent properties due to its high crosslinking characteristics, adequate strength and dimensional stability [35] and has been often used in the structure of date-palm-reinforced composites [28,36–38], other polymers like polyamide, polyurethane or polyester could be used as matrices in composites with natural fibre reinforcement.

Date palm long fibrillated fibres obtained recently using a combined alkaline–mechanical extraction process of delignification and fibrillation from midribs [27,28] are not sufficiently studied in textile applications, nor as reinforcement for polymer composites. The present study deals with the use of these fibres as reinforcement in polyester composite. The influences of cotton, as a natural and soft fibre, and Kevlar, as a synthetic and strong fibre, were also investigated regarding the bending strength properties, resistance to water and temperature changes in terms of freezing and drying. The results of the present study are oriented towards the finding of an appropriate application of the proposed composites, intended to be used in a humid environment.

2. Materials and Methods

2.1. Materials

2.1.1. Wood Fibres

Date Palm (*Phoenix dactylifera* L.) Fibres (DP)

For the present research, the raw material needed for the extraction of the date palm fibres, was obtained from the date palm groves of Hormozgan province (Rudan city, Iran). The fibres were extracted from the midribs. The midrib is the only part from where the longest fibres can be extracted using the chemical method. The stages of the process for fibre extraction follow the procedure presented by [19,27,28], and they are described in Figure 1.

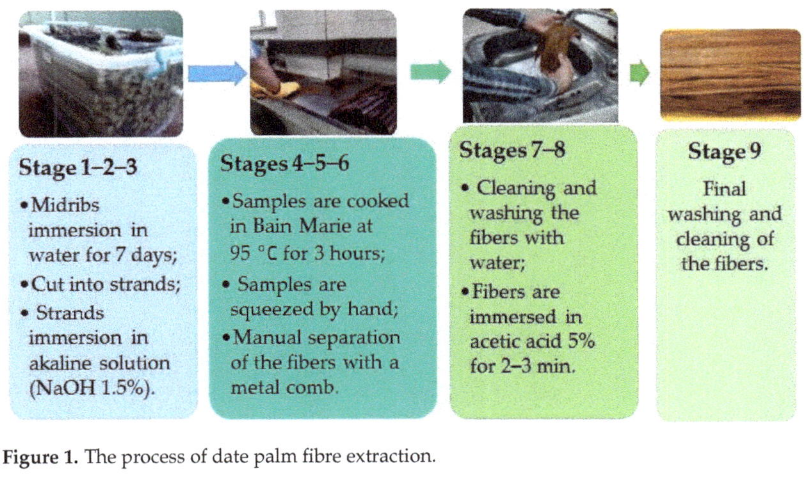

Figure 1. The process of date palm fibre extraction.

The extracted date palm fibres were longitudinally oriented and arranged manually into flat sheets with dimensions of 350 mm × 350 mm (Figure 2a). The measurement made under the stereo-microscope NIKON SMZ 18-LOT2 (Nikon Corporation, Tokyo, Japan) with 240× magnification on a bundle of fibres indicated diameter values between 0.22 mm and 0.37 mm [28].

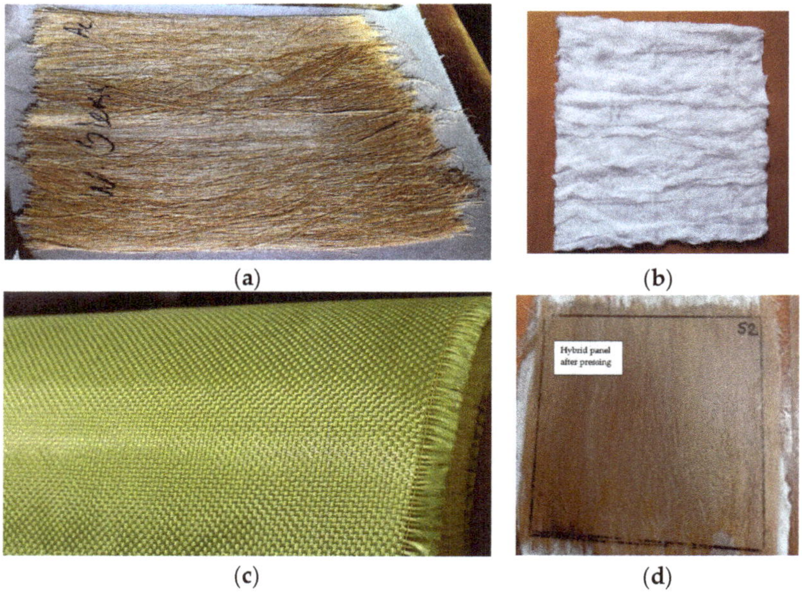

Figure 2. Sheets of reinforcement fibres: (**a**) date palm; (**b**) cotton; (**c**) Kevlar; (**d**) experimental hybrid composite.

Cotton Fibres

Cotton has a crystalline and fibrillary structure and has the potential to be used in reinforced composites due to its exceptional strength [29]. Cotton wool for medical applications was used in this experiment. The cotton wool was purchased from the Romanian market (Optimum medical, Bucharest, Romania). The cotton fibres were extracted by shredding the wool into a thin mat with a size of 350 mm × 350 mm (Figure 2b). The cotton fibre diameters were measured using the stereo-microscope NIKON SMZ 18-LOT2 (Nikon Corporation, Tokyo, Japan) with 240× magnification and varied between 0.008 mm and 0.02 mm.

Kevlar Fibres

Kevlar is a heat-resistant para-aramid synthetic fibre with a molecular structure of many inter-chain bonds that make it incredibly strong. Kevlar ARAMID fabric VC1 AG170, PLAIN, 170 g/m^2, width 120 cm, filament around 0.012 mm, and density of 1.44 g/cm^3 purchased from the Romanian market (Best Tools company Brasov, Brasov, Romania) and produced by the Vosschemie company (Uetersen, Germany) was used in this research as sheets of 350 mm × 350 mm (Figure 2c). This fabric is bidirectional with 2/2 twill weave. It is suitable for use in polyester, vinyl ester and composite materials: it has low weight, high elasticity and good tensile strength fibre characteristics, making it suitable to be exposed to weather and water. It has the following characteristics: number of threads 6.8 ± 2% threads/cm; tensile strength 3154 N/mm^2, modulus of elasticity 108,000 N/mm^2, fabric thickness approx. 0.3 mm; can be processed using the hand lay-up method; recommended to be used for reinforcement polyester, vinyl ester and epoxy resins.

2.2. Composite Design

The date palm/cotton/Kevlar fibre-reinforced polyester hybrid composites were manufactured in laboratory conditions. The two-component polyester resin VISCOVOSS AZUR SUPER+ type was used as the matrix and was purchased from the Romanian market (Best Tools company Brasov, Brasov, Romania). This resin has a density of 1.1 g/cm^3 and a viscosity of 900–1100 mPa·s at 20 °C (Brookfield, Toronto, ON, Canada) as specified in the technical sheet. It is an unsaturated, pre-accelerated, ambient-temperature-curing polyester resin with low styrene emissions. Methyl ethyl ketone peroxide (MEKP) was used as the hardener and was added to the base component resin with a participation rate of 3% of the resin weight.

Natural fibres (DP and C) and synthetic fibres (K) were used to reinforce the polyester matrix and to manufacture the hybrid composites. The design of the three types of composites is presented in Table 1.

Table 1. The design of experimental hybrid composites.

Code	Structure	Design [1]	Date Palm Fibres Weight Fraction (%)	Cotton/Kevlar Fibres Weight Fraction (%)
Date palm (DP)	3 layers (DP-DP-DP)	Alternating parallel–perpendicular orientation of layers	20	-
Date palm/cotton (DP/C)	5 layers (DP-C-DP-C-DP)	Alternating parallel–perpendicular orientation of layers	17	2.17
Date palm/Kevlar (DP/K)	5 layers (DP-K-DP-K-DP)	Alternating parallel–perpendicular orientation of layers	15.6	7.35

[1] The layers for the faces were oriented longitudinally in each structure.

Each layer of the composites was firstly weighed with an accuracy of 0.01 g, and the sheets with similar weight were used for the faces in the composite structures. The average weight of the layers was as follows: the DP layers had 33.26 g for the faces and 36.81 g for the core; the cotton layers had 6.6 g; and the Kevlar layers had 24.35 g. The proportion of the fibres in the polyester resin matrix was 20 wt% for DP, 19.17 wt% for DP/C and 22.95 wt% for DP/K (Table 1). The proportion of the cotton and Kevlar fibres from the total volume of the panels is cotton fibres 9.76 vol% for DP/C and Kevlar fibres 8.18 vol% for DP/K.

The polyester resin amount was calculated by multiplying the weight of each layer as follows: five times for the date palm fibres, seven times for the cotton fibres and three times for the Kevlar fibres. These rates were established according to the recommendations of the manufacturer for the synthetic fibres and after preliminary determinations for the date palm fibres and natural fibres similar to cotton [28]. The resin was applied via brush on each individual sheet of fibres. The working temperature was about 24 °C, and the working time for applying the resin was up to 30 min to avoid the stickiness effect as recommended in the technical sheet. On the top and bottom of the sandwich hybrid composite, white silicon baking paper was used to avoid sticking to the support where the sandwich was formed (Figure 2d). Each of the laminate composites was placed between two blockboard panels, and a weight of 20 kg was used for cold pressing the composites at a pressure of 0.019 bar for three days. Fully curing this resin needs at least 48 h, according to its technical sheet. The amounts of resin were, as follows: 490 g (DP structure), 595 g (DP/C structure), 660 g (DP/K structure). The method applied to manufacture the experimental hybrid composites is presented in Figure 3.

1. Selecting the sheets for the faces and core and weighing them.
2. Applying the resin using the hand layup method.
3. Positioning the core layer perpendicular to the faces.

4. The sandwich composite was placed between two sheets of silicone paper.
5. Cold pressing under a weight of 20 kg.
6. Final hybrid composite

Figure 3. The manufacturing process of the hybrid composites.

Three replicates of each hybrid composite type (DP, DP/C, DP/K) were manufactured. After six days of conditioning at 23 °C and 65% relative humidity of the air, the experimental panels were sized at 300 mm × 300 mm. The test specimens were cut according to the specific standard requirements for the evaluation of the physical properties, such as density, moisture content, thickness swelling (TS), water absorption (WA) and mechanical properties related to flexural strength (MOR) and modulus of elasticity (MOE).

2.3. Testing

The density was determined according to EN 323:1996 [39], the moisture content according to EN 322:1996 [40], and the thickness swelling (TS) and water absorption (WA) according to EN 317:1996 [41], with specimens with dimensions of 50 mm × 50 mm (length × width). The samples were immersed in distilled water at a temperature of 20 °C for 24 h. The thickness and the mass of the samples were recorded before and after immersion in water at 2, 24, 48, 72, 96, 120, 144, 168 192 and 216 h. An electronic calliper with an accuracy of 0.01 mm and the electronic scale with the accuracy of 0.01 g were used to measure the thickness and to weigh the samples. The purpose of this test was to assess the water resistance and the dimensional stability of the hybrid composites. The flexural strength (MOR) and modulus of elasticity (MOE) were evaluated according to EN 310:1996 [42], the tests being carried out with the help of a Zwick/Roell Z010 universal testing machine (Ulm, Germany). Six samples were used for each type of hybrid composites and each test.

The hybrid structures were designed to resist the humid environment. Thus, in addition to the water immersion test, they were also subjected to cyclic moisture resistance tests according to EN 321:2003 [43]. The experimental panels of 300 mm × 300 mm were subjected to three exposure cycles consisting of immersion in water followed by freezing and then drying at a high temperature. After cyclic testing, their thickness, mass, MOR and MOE were measured.

A water bath at a temperature of 20 ± 1 °C was used for the water immersion of panels in the first cycle for 24 h. The other two cycles were carried out in a test chamber of HL type [DongGuan City, China] set to the 24 h freezing cycle at temperatures between −12 °C and −25 °C and drying for 72 h at 70 ± 2 °C. Three cycles were performed per edge of each composite. After completing these cycles, the panels were conditioned at 23 °C and 65% relative humidity of air until constant mass. The changes in thickness, mass and flexural properties were tested and compared to those obtained before cyclic testing. The increase in thickness (dT) and the mass decrease (dM) of panels after cyclic testing were calculated with the following equations:

$$\mathrm{dT} = \frac{tf - ti}{ti} \times 100 \quad (1)$$

where

tf—the final thickness after cyclic testing, in mm;

ti—the initial thickness before cyclic testing, in mm.

$$\mathrm{dM} = \frac{Mf - Mi}{Mi} \times 100 \quad (2)$$

where

Mf—the final mass after cyclic testing, in g;

Mi—the initial mass before cyclic testing, in g.

A microscope investigation of the samples before and after cyclic testing was carried out using a NIKON SMZ 18-LOT2 (Nikon Corporation, Tokyo, Japan) stereo-microscope. This allowed for a better visualization of the structures and of the adhesion between the component layers. The microscopic analysis was made on the cross sections of the experimental hybrid composites with 30× magnification. Six samples from each experimental composite, were prepared to perform the test.

2.4. Statistical Analysis

The values of the physical properties (density, TS, WA) and mechanical properties (MOR and MOE, before and after cyclic testing) were compared individually. A single-factor analysis of variance (ANOVA) was performed with Minitab Statistical Software Version 21.1.0 at a 95% confidence level and a significance level of 0.05 ($p < 0.05$). The analysis shows how the properties of the hybrid composites were affected by the presence of cotton and Kevlar in the structure. Furthermore, Tukey's test was utilized to specify the significant differences created by the fibres.

3. Results and Discussion

3.1. Density and Moisture Content

The density and the moisture content of the experimental hybrid composites are presented in Figure 4.

The highest density value was obtained for the panels with date palm fibres (1012 kg/m^3) due to the fact that the quantity of resin applied was about 30% greater than the resin amount applied on the layers of the other two composites. The density of composites was not influenced by the densities of the fibres, which were almost in the same range of values: 1.44 g/cm^3 for Kevlar fibres, between 0.9 g/cm^3 and 1.2 g/cm^3 for date palm fibres and 1.5 g/cm^3 for cotton fibres [17]. The density of DP with polyester was about 44% lower than similar composites with epoxy resin as the matrix [44]. Both resins are generally used as matrices for natural fibre composites that cure at environmental temperatures. The lower density of the hybrid composites with polyester could be an advantage in applications where lightness is essential, such as in aerospace and automotive applications.

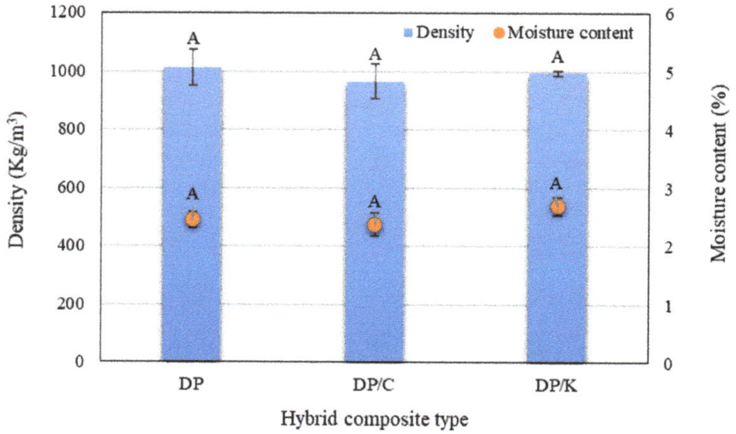

Figure 4. Density and moisture content (average values) (A—variables with the same letter mean that the difference is not statistically significant).

Furthermore, there were no big differences between the moisture content values, these being similar to those found in the literature [45].

According to the statistical examination, the investigation revealed the absence of statistically significant distinctions in both density and moisture content values across all three types of hybrid composites, with findings established at a 95% confidence level. The computed p-values and F-values for density and moisture content were 0.354, 0.182, 1.17 and 1.97, respectively. Additionally, the outcomes derived from the Tukey test corroborated with the aforementioned findings, categorizing all three variants of hybrid composites into identical groups concerning both density and moisture content.

3.2. Thickness Swelling (TS) and Water Absorption (WA)

The values obtained for TS and WA during the period of 216 h (9 days) of immersion in water are presented in Figures 5 and 6. By analysing the thickness swelling values, it can be seen that after 24 h and 216 h, the lowest values were obtained for the composites with Kevlar (1.22% and 2.3%, respectively) followed by those with cotton (1.01% and 3.0%, respectively) and date palm fibres (2.74% 5.42%, respectively) (Figure 5).

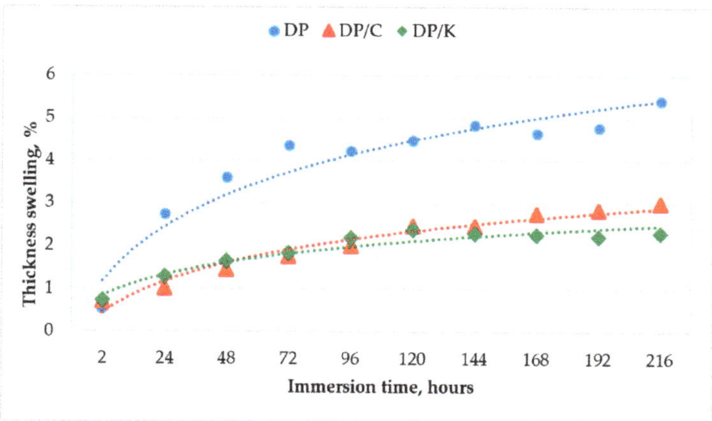

Figure 5. The dynamics of swelling in thickness for the 3 structures (DP, DP/C and DP/K).

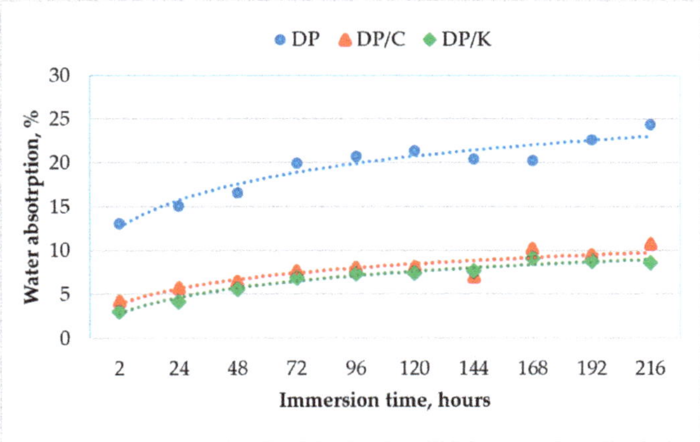

Figure 6. The dynamics of water absorption for the 3 structures (DP, DP/C and DP/K).

In the first 24 h, the thickness increased at about 0.02%/hour for the composites DP/C and DP/K and at about 0.09%/hour for DP. These results can be explained by the sensitivity of date palm fibres to water absorption due to their hydrophilic character, as was also observed by [46]. Another cause may be related to the presence of voids and pores in the structure, due to uneven application of the adhesive on the surfaces of the layers.

For all composites, the water uptake increased more rapidly in the first 72 h; then, until the end of the period, the dynamic of the water uptake decreased. This is in agreement with [47,48] (Figure 6).

The rate of swelling and absorption was more pronounced for the DP composite compared to the other two. The presence of high voids in DP specimens increased the water absorption, as water was trapped in the voids. After 24 h and 216 h, the water absorptions were 15% and 24.3%, respectively, for DP; 5.6% and 10.8%, respectively, for DP/C; and only 4.2% and 8.6%, respectively, for DP/K. Reinforcement with Kevlar fibres improved the performance in terms of water uptake by 20% compared to DP/C and by about 60% compared to DP.

However, the addition of the cotton and Kevlar fibres in the structure of the hybrid composite significantly ($p < 0.05$) affected both TA and WA after 24 h ($p < 0.01$) and 216 h, for which this influence was more evident, since $p < 0.009$. This explains the better interface between polyester resin with cotton and Kevlar fibres.

In Table 2, the equations that describe the dynamic of thickness swelling and water absorption during the period of testing are presented. The variation in TS and WA over time is expressed by a logarithmic curve with a high correlation factor ($R^2 > 0.9$).

Table 2. Equations describing the dynamics of TS and WA.

Code	TS	WA
DP	$y = 1.8459\ln(x) + 1.1636$ $R^2 = 0.9226$	$y = 4.5477\ln(x) + 12.594$ $R^2 = 0.911$
DP/C	$y = 1.067\ln(x) + 0.4358$ $R^2 = 0.9608$	$y = 2.5452\ln(x) + 3.8948$ $R^2 = 0.8393$
DP/K	$y = 0.751\ln(x) + 0.7491$ $R^2 = 0.9186$	$y = 2.6858\ln(x) + 2.7968$ $R^2 = 0.9659$

The natural fibres' cell walls contain free hydroxyl (OH) groups, which are a favourable medium for water absorption, leading to the swelling of fibres within the matrix. As a consequence, the interfacial adhesion between the matrix and fibres can be destroyed and voids or cracks can occur, negatively influencing the mechanical properties and performance of the composites [49,50].

These disadvantages of natural fibres can be reduced or removed by hybridization with synthetic fibres, as in this case with Kevlar, which led to higher resistance to moisture and better aging properties.

3.3. Resistance to Moisture

The test of resistance to moisture revealed the performance of the hybrid composites when exposed to different climatic conditions. The changes in thickness, mass and mechanical properties were analysed. At the end of the cycles (water immersion, freezing and drying), no visible changes were observed on the surfaces of the specimens; however, a slight increase in thickness (dT) of all composites was found, more obviously so for DP/C, at about 6%, compared to 4.8% for DP and 4% for DP/K (Figure 7).

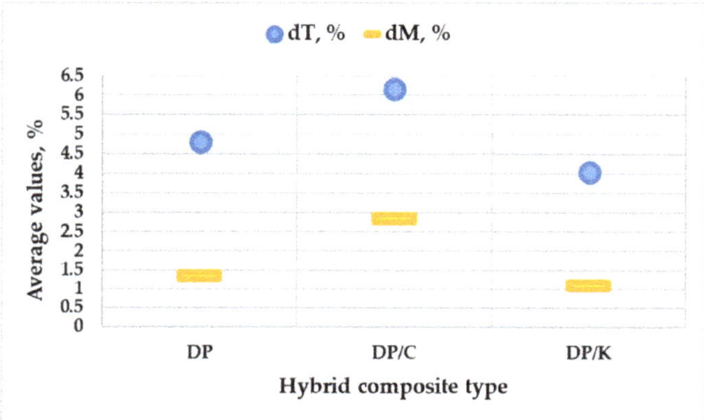

Figure 7. The differences in thickness (dT) and mass (dM) after cyclic testing.

A slight decrease in mass of the composites (dM) was noticed, in the range of 1–2.8% (Figure 7). Similar findings were reported by [51] for hybrid composites made with kenaf and glass fibres after freeze-drying exposure.

The increase in thickness was caused mainly by the hydrophilic character of natural fibres due to the presence of hemicellulose [52] and, in terms of other factors, by the small voids produced inside of structures because of the poor impregnation of date palm fibres into the polyester resin compared to the Kevlar composite (DP/K). Tensile forces can occur during freezing-drying in the polymer matrix, due to the expanding fibre volume in the saturated state after water immersion. These lead to some micro-cracks or voids in the structure (Figure 8d). These voids are more evident in the composites DP/C and DP which had greater decreases in mass compared to the DP/K structure.

DP/C and DP registered decreases in mechanical properties after cyclic testing of about 25% and 6.5%, respectively, for MOE and of about 13% and 5.8%, respectively, for MOR compared to the values obtained before testing. The substantial strength losses occurred in the composite structures, induced by degradation through freeze-drying mechanisms, which cause losses in bonding at the fibre–matrix interface and via the micro-cracks produced by internal stresses [46,51,53]. DP/K performed better, Kevlar fibre being slightly stronger at low temperatures.

After cyclic testing, the flexural strength and modulus of elasticity of DP/K registered slight decreases of about 1.4% and 2%, respectively (Figures 7 and 8).

In the literature [54,55], the effect of a freeze–thaw environment has been studied on carbon-fibre- or glass-fibre-reinforced polymer composites, reporting insignificant effects on shear, axial and flexural properties.

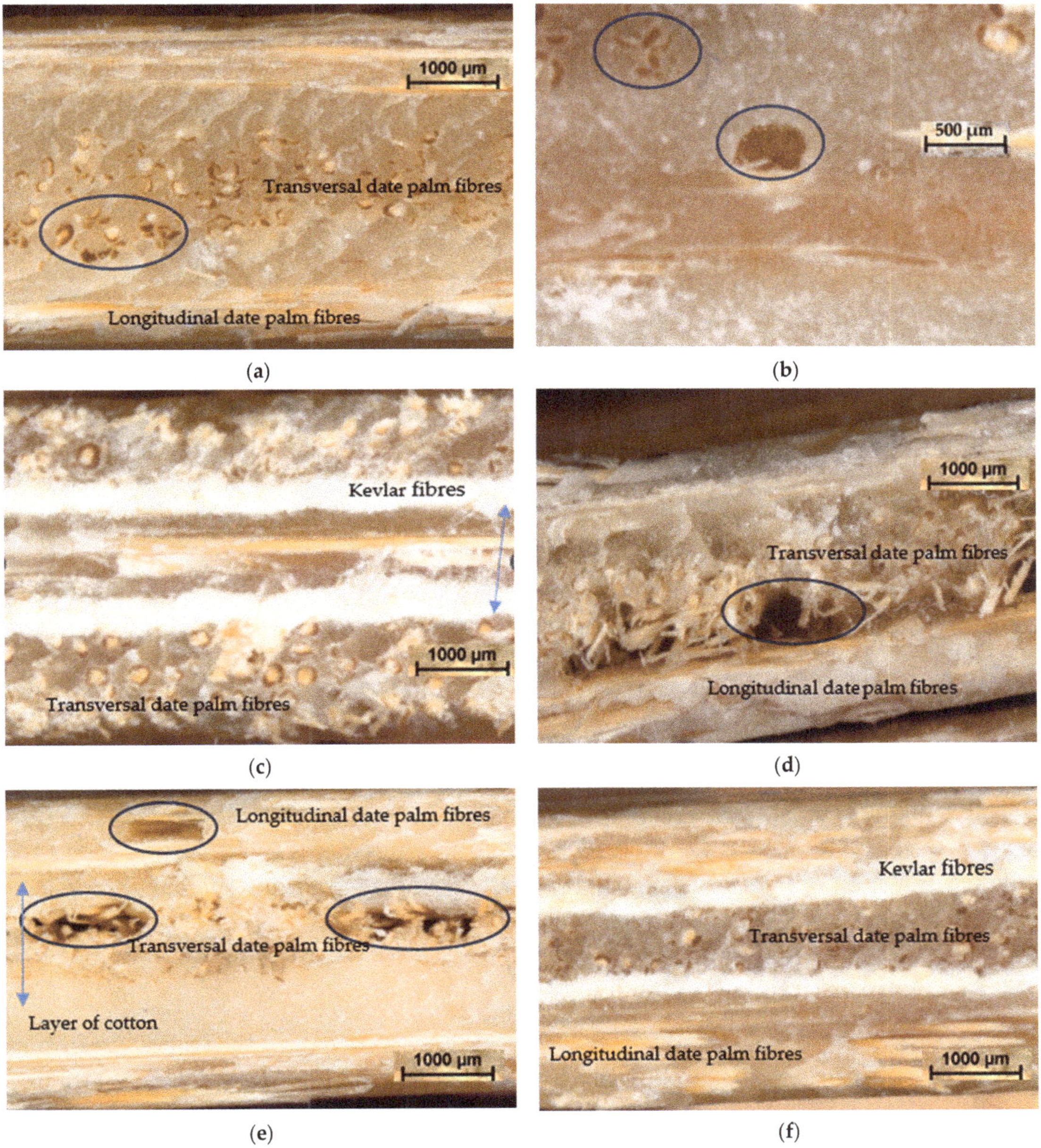

Figure 8. The microscopic investigation of the hybrid composites' structures: (**a**) DP before cyclic testing; (**b**) DP/C: detail in the cotton layer before cyclic testing (60× magnification); (**c**) DP/K: compact structure before cyclic testing; (**d**) DP after cyclic testing; (**e**) DP/C after cyclic testing; (**f**) DP/K after cyclic testing. The circled areas represent the voids inside structures.

3.4. Microscopic Investigation

A microscopic analysis was made on cross-sections of the hybrid composites before and after cyclic testing (CT). Some small voids in the date palm middle layer (cross-section) (Figure 8a) and in the cotton layer (Figure 8b) in the structure DP/C can be seen before the cyclic testing. These influenced negatively the behaviour of the structures, decreasing their resistance to water. The impregnation with resin of the cotton layers was difficult and as a consequence, after cyclic testing, voids occurred in some areas in the vicinity of the date palm fibres, as can be seen in Figure 8e.

Larger voids were observed also in a few DP specimens at the interface area located between date palm fibres and the polyester resin matrix after CT. This is due to the poor wettability between the fibres and the matrix, leading to insufficient impregnation. This non-continuity in the structure caused a decrease in strength after CT, as can be observed in Figures 9 and 10. This phenomenon was observed also by [56] after freeze-thaw exposure, which induced deterioration of natural fibre composites.

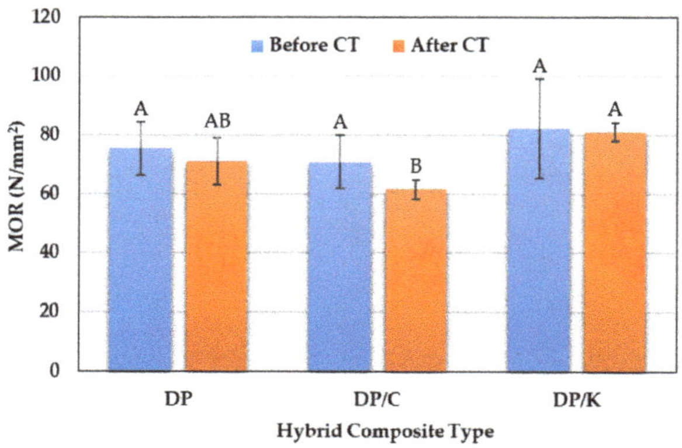

Figure 9. Comparative values of MOR for the 3 structures before and after exposure to cyclic testing (CT) (variables with the same letter mean that the difference is not statistically significant and those with different letters are statistically significant).

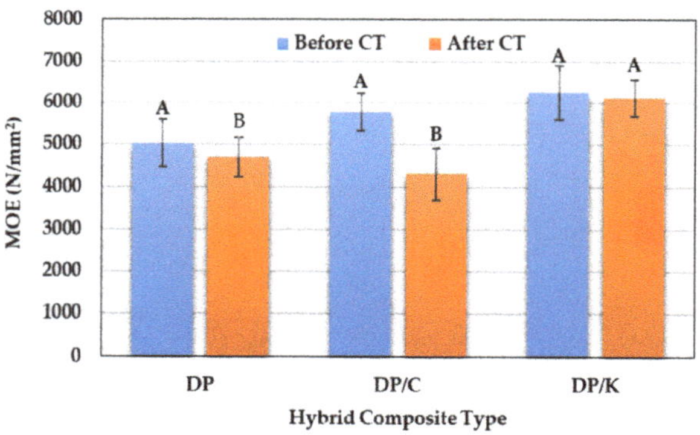

Figure 10. Comparative values of MOE for the 3 structures before and after exposure to cyclic testing (CT) (variables with the same letter mean that the difference is not statistically significant and those with different letters are statistically significant).

3.5. Mechanical Properties

In Figures 9 and 10 are presented the flexural strength (MOR) and modulus of elasticity (MOE) values before and after cyclic testing.

The DP/K hybrid composite exhibited the highest values of flexural strength and modulus of elasticity: 82.24 N/mm^2 and 6248 N/mm^2, respectively. Similar results were found by [30], who also confirmed that hybrid composites with Kevlar show higher mechanical properties.

Lower values of MOR and MOE were achieved for DP/C and DP composites, the values ranging between 70.78 N/mm^2 and 75.4 N/mm^2 and between 5774 N/mm^2 and 5028 N/mm^2, respectively. The natural fibre composites exhibit inferior mechanical properties compared to synthetic fibres, and in the hybrid composites, these properties mainly depend on the binding ability between the matrix and the fibres [57]. DP/C had the lowest density and the adhesive amount applied to each layer was about 40% lower than that applied to DP; consequently, the flexural strength decreased by 14% compared to DP. The cotton fibres did not improve the strength performance of the hybrid composite, due to weaker fibre–matrix interfacial bonding. After cyclic testing, the mechanical properties were analysed and no substantial changing was observed in the case of DP/K. The resin was much easier to apply to the Kevlar sheets and the resin amount was greater than in the case of DP/C. The resin penetrated better in all layers and no voids were observed in the structure (Figure 8f). In the case of the other two composites (DP and DP/C), when subjected to water immersion (first cycle), the softening of fibres occurred, as hydroxyl groups contained in date palm and cotton absorbed water. Then, the expansion of water molecules and contraction of the polymer matrix happened in the freeze cycle, leading to the degradation of the bonding between the fibres and the matrix, further influencing the mechanical properties. DP/C registered a decrease in MOR and MOE by 24% and 30%, respectively, and DP registered decreases by 12% and 23%, respectively, compared to the DP/K composite.

The flexural strength (MOR) and modulus of elasticity (MOE) before cyclic testing were not influenced by cotton and Kevlar fibres in the structures, as a statistical analysis displays no significant effect ($p < (0.2–0.6)$). However, after the cyclic test, the thickness swelling, MOR and MOE were statistically significantly affected by Kevlar layers ($p < (0.01–0.004)$), whereas cotton layers had no significant effect on MOR and MOE after cyclic testing ($p < (0.1–0.4)$). This demonstrates a better interface between polyester and Kevlar fibres compared to cotton fibre, the behaviour of which is influenced by its hydrophilic properties.

Due to their low density and good performance in the freeze-dry test, the experimental hybrid composites with Kevlar (DP/K) could be alternative materials for metal, which in the marine condition is susceptible to corrosion and cracks in time. On the other hand, due to their high flexural strength (75–82 N/mm^2) and minimal changes in thickness after water immersion for 216 h and cyclic testing (between 2.3% and 4%), structures DP/K is more reliable for application in construction for floors or roof tiles where moisture resistance is required. However, related to future requirements for sustainable resources and low-impact products, the combination of natural fibres in a hybrid product like DP/C could be an effective solution for indoor partition panels or floor tiles where high flexural strength of date palm fibres with the good impact properties of cotton fibres could be combined.

Durability is an important issue for hybrid composites; thus, further research needs to be carried out to improve the interfacial bonding between fibres and the matrix to increase the mechanical properties and the performance with regard to moisture, considering that natural fibres can replace synthetic fibres in a wide range of outdoor applications.

4. Conclusions

Oriented self-bonded mats of date palm long fibrillated fibres extracted from midribs are very suitable as reinforcement for polyester composites in the form of laminates, proving to have a good adhesion with polyester resin.

The positive influence of the two additional layers of cotton and Kevlar mats in the structure was noticed in a decrease in water absorption and thickness swelling.

The cyclic test consisting of immersion into water, freezing and drying negatively affected the initial mechanical properties of the composites, especially for the structures with two additional layers of cotton, and less affected the structures with Kevlar mats.

The cracks and voids revealed by the microscopic investigation on the DP and DP/C structures after the cyclic test do not recommend these structures to be used in very humid climates with alternating positive and negative temperatures. Instead, the participation of Kevlar mats in the laminate structure of the date-palm-reinforced polyester composite brings advantages of higher resistance to water and improved mechanical performance, even after immersion in water and being subjected to abrupt differences in temperature. For this reason, the hybrid composite with Kevlar fibres (DP/K) exhibits promising characteristics to be recommended for outdoor applications, in the marine environment, and also in the construction of floors or roof tiles, where moisture resistance is required.

Further studies are envisaged to evaluate other properties such as impact resistance, behaviour in salt water, and the addition of other natural fibres in order to improve the performance of date-palm-based sustainable hybrid composites. Further research on the economic feasibility of these composites will be made so as to create a comprehensive picture of the acceptability of their use.

Author Contributions: Conceptualization, O.Z., M.H.M. and C.C.; methodology, O.Z. and C.C.; software, A.S.; validation, O.Z. and C.C.; formal analysis, M.H.M.; investigation, O.Z., C.C. and A.S.; resources, M.H.M. and A.S.; data curation, O.Z., C.C. and M.H.M.; writing—original draft preparation, O.Z. and A.S.; writing—O.Z. and M.H.M.; visualization, C.C.; supervision, O.Z. and C.C.; project administration, C.C.; funding acquisition, A.S. and C.C. All authors have read and agreed to the published version of the manuscript.

Funding: This research received no external funding.

Institutional Review Board Statement: Not applicable.

Informed Consent Statement: Not applicable.

Data Availability Statement: Data are contained within the article.

Acknowledgments: We hereby acknowledge the structural funds project PRO-DD (POS-CCE, O.2.2.1., ID 123, SMIS 2637, No. 11/2009) for providing the infrastructure used in this work (https://icdt.unitbv.ro/en/research-and-development-projects/the-r-d-institute-project.html, accessed on 21 September 2023).

Conflicts of Interest: The authors declare no conflicts of interest.

References

1. UNECE/FAO. Forest Products Annual Market Review, 2020–2021, United Nations, Geneva. 2021. Available online: https://unece.org/sites/default/files/2021-11/2114516E_Inside_Final_web.pdf (accessed on 29 November 2023).
2. Emin, A. How Composite Materials Influence Sustainable Development. *Entrep. Perspect.* **2021**, *40*, 7–16.
3. Espinach, F.X. Advances in Natural Fibers and Polymers. *Materials* **2021**, *14*, 2607. [CrossRef]
4. Kamarudin, S.H.; Abdullah, L.C.; Aung, M.M.; Ratnam, C.T. A study of mechanical and morphological properties of PLA based biocomposites prepared with EJO vegetable oil based plasticiser and kenaf fibres. *IOP Conf. Ser. Mater. Sci. Eng.* **2018**, *368*, 85314. [CrossRef]
5. Kumar, S.S.; Muthalagu, R.; Chakravarthy, C.N. Effects of fiber loading on mechanical characterization of pineapple leaf and sisal fibers reinforced polyester composites for various applications. *Mater. Today Proc.* **2021**, *44*, 546–553. [CrossRef]
6. Huner, U. Effect of Water Absorption on The Mechanical Properties of Flax Fiber Reinforced Epoxy Composites. *Adv. Sci. Technol. Res. J.* **2015**, *9*, 1–6. [CrossRef]
7. Hojo, T.; Xu, Z.; Yang, Y.; Hamada, H. Tensile Properties of Bamboo, Jute and Kenaf Mat-Reinforced Composite. *Energy Procedia* **2014**, *56*, 72–79. [CrossRef]
8. Neves, A.C.C.; Rohena, L.A.; Mantovani, D.P.; Carvalho, J.P.R.G.; Vieira, C.M.F.; Lopes, F.P.D.; Simonassi, N.T.; Luz, F.S.; Monteiro, S.N. Comparative mechanical properties between biocomposites of epoxy and polyester matrices reinforced by hemp fiber. *J. Mater. Res.Technol.* **2020**, *9*, 1296–1304. [CrossRef]
9. Siregar, J.P.; Zalinawati, M.; Cionita, T.; Rejab, M.R.M.; Mawarnie, I.; Jaafar, J.; Hamdan, M.H.M. Mechanical properties of hybrid sugar palm/ramie fibre reinforced epoxy composites. *Mater. Today Proc.* **2021**, *46*, 1729–1734. [CrossRef]
10. Baley, C.; Bourmaud, A.; Davies, P. Eighty years of composites reinforced by flax fibres: A historical review. *Compos. Part A Appl. Sci. Manuf.* **2021**, *144*, 106333. [CrossRef]
11. Rasoul, Z.S.; Juoi, J.M.; Mohamad, M.; Fawzi, N.M. Date Palm Fiber(DPF) and its Composites: A Comprehensive Survey. *Int. J. Adv. Sci. Technol.* **2020**, *29*, 1776–1788.
12. Ghori, W.; Saba, N.; Jawaid, M.; Asim, M. A review on date palm (*Phoenix dactylifera*) fibers and its polymer composites. *IOP Conf. Ser. Mater. Sci. Eng.* **2018**, *368*, 012009. [CrossRef]

13. Hanieh, A.A.; Hasan, A.; Assi, M. Date palm trees supply chain and sustainable model. *J. Clean. Prod.* **2020**, *258*, 120951. [CrossRef]
14. Mahdavi, S.; Kermanian, H.; Varshoei, A. Comparison of mechanical properties of date palm fiber-polyethylene composite. *BioResources* **2010**, *5*, 2391–2403. [CrossRef]
15. Chandrasekaran, M.; Bahkali, A.H. Valorization of date palm (*Phoenix dactylifera*) fruit processing by-products and wastes using bioprocess technology–Review. *Saudi J. Biol. Sci.* **2013**, *20*, 105–120. [CrossRef]
16. Bamaga, S.O. Review on the Utilization of Date Palm Fibers as Inclusion in Concrete and Mortar. *Fibers* **2022**, *10*, 35. [CrossRef]
17. Al-Oqla, F.M.; Alothman, O.Y.; Jawaid, M.; Sapuan, S.M.; Es-Saheb, M.H. Processing and properties of date palm fibers and its composites. In *Biomass Bioenergy*; Springer: Cham, Switzerland, 2014; pp. 1–25. [CrossRef]
18. Faiad, A.; Alsmari, M.; Ahmed, M.M.Z.; Bouazizi, M.L.; Alzahrani, B.; Alrobei, H. Date Palm Tree Waste Recycling: Treatment and Processing for Potential Engineering Applications. *Sustainability* **2022**, *14*, 1134. [CrossRef]
19. Elseify, L.A.; Midani, M.; Shihata, L.A.; El-Mously, H. Review on cellulosic fibers extracted from date palms (*Phoenix dactylifera* L.) and their applications. *Cellulose* **2019**, *26*, 2209–2232. [CrossRef]
20. Amirou, S.; Zerizer, A.; Pizzi, A.; Haddadou, I.; Zhou, X. Particleboards production from date palm biomass. *Eur. J. Wood Prod.* **2013**, *71*, 717–723. [CrossRef]
21. Haseli, M.; Layeghi, M.; Hosseinabadi, H.Z. Characterization of blockboard and battenboard sandwich panels from date palm waste trunks. *Measurement* **2018**, *124*, 329–337. [CrossRef]
22. Hosseinkhani, H.; Euring, M.; Kharazipour, A. Utilization of date palm (*Phoenix dactylifera* L.) pruning residues as raw material for MDF manufacturing. *J. Mater. Sci. Res.* **2015**, *4*, 46–62. [CrossRef]
23. Sadik, T.; Muthuraman, S.; Sivaraj, M.; Negash, K.; Balamurugan, R.; Bakthavatsalam, S. Mechanical Behavior of Polymer Composites Reinforced with Coir and Date Palm Frond Fibers. *Adv. Mater. Sci. Eng.* **2022**, *2022*, 9882769. [CrossRef]
24. Aly, N.M.; ElNashar, D. Polyester/flax biocomposites reinforced using date palm leaves and wood flour as fillers. *Int. J. Eng. Technol.* **2016**, *8*, 1579–1588.
25. Hammood, A.S. Effect of erosion on water absorption and morphology for treated date palm fiber-reinforced polyester composites. *Int. J. Mech. Mechatron. Eng.* **2015**, *15*, 108.
26. Elmously, H.; Elerian, K.M. Determination of Mechanical Properties for Date Palm Midribs and Leaflets. *Int. J. Curr. Res.* **2017**, *9*, 61251–61258.
27. Elseify, L.A.; Midani, M.; Hassanin, A.H.; Hamouda, T.; Khiari, R. Long textile fibres from the midrib of date palm: Physiochemical, morphological, and mechanical properties. *Ind. Crops Prod.* **2020**, *151*, 112466. [CrossRef]
28. Mazaherifar, M.H.; Hosseinabadi, H.Z.; Coşereanu, C.; Cerbu, C.; Timar, M.C.; Georgescu, S.V. Investigation on *Phoenix dactylifera/Calotropis procera* Fibre-Reinforced Epoxy Hybrid Composites. *Forests* **2022**, *13*, 2098. [CrossRef]
29. Wankhede, B.; Bisaria, H.; Ojha, S.; Dakre, V.S. A review on cotton fibre-reinforced polymer composites and their applications. *Proc. Inst. Mech. Eng. Part L J. Mater. Des. Appl.* **2023**, *237*, 1347–1362. [CrossRef]
30. Yahaya, R.; Sapuan, S.; Jawaid, M.; Leman, Z.; Zainudin, E. Effect of layering sequence and chemical treatment on the mechanical properties of woven kenaf–aramid hybrid laminated composites. *Mater. Des.* **2015**, *67*, 173–179. [CrossRef]
31. Elfaleh, I. A comprehensive review of natural fibers and their composites: An eco-friendly alternative to conventional materials. *Results Eng.* **2023**, *19*, 101271. [CrossRef]
32. Naveen, J.; Jawaid, M.; Zainudin, E.S.; Sultan, M.T.H.; Yahaya, R.B. Selection of Natural Fiber for Hybrid Kevlar/Natural Fiber Reinforced Polymer Composites for Personal Body Armor by Using Analytical Hierarchy Process. *Front. Mater.* **2018**, *5*, 52. [CrossRef]
33. Goud, B.N.; Sura, S.; Aravind, P.; Jawahar, B.L.; Sanskruti, K.; Pavan, C. An experimental study on mechanical properties of Kevlar composite for aircraft structural applications. *Mater. Today Proc.* **2022**, *64*, 909–916. [CrossRef]
34. Sekhar, V.C.; Pandurangadu, V.; Rao, T.S. TGA, DSC, DTG properties of epoxy composites reinforced with feather fibers of 'Emu'bird. *Int. J. Innov. Res. Sci. Eng. Technol.* **2014**, *3*, 13017–13023.
35. Khan, M.U.; Abas, M.; Noor, S.; Salah, B.; Saleem, W.; Khan, R. Experimental and Statistical Analysis of Saw Mill Wood Waste Composite Properties for Practical Applications. *Polymers* **2021**, *13*, 4038. [CrossRef] [PubMed]
36. Gheith, M.H.; Aziz, M.A.; Ghori, W.; Saba, N.; Asim, M.; Jawaid, M.; Alothman, O.Y. Flexural, thermal and dynamic mechanical properties of date palm fibres reinforced epoxy composites. *J. Mater. Res. Technol.* **2019**, *8*, 853–860. [CrossRef]
37. Ahmadi, M.S.; Gholami, M.; Tavanaie, M.A.; Mehrizi, M.K. Tensile and flexural properties of epoxy-date palm fiber composites. *J. Sci. Technol. Compos.* **2018**, *5*, 69–78.
38. Hussein, S.M. Incorporation of palm fiber to enhance the mechanical properties of epoxy. *Iraqi J. Sci.* **2020**, *61*, 1960–1970. [CrossRef]
39. *EN 323:1996*; Wood-Based Panels-Determination of Density. European Committee for Standardization: Brussels, Belgium, 1996.
40. *EN 322:1996*; Wood-Based Panels-Determination of Moisture Content. European Committee for Standardization: Brussels, Belgium, 1996.
41. *EN 317:1996*; Particleboards and Fibreboards: Determination of Swelling in Thickness after Immersion in Water. European Committee for Standardization: Brussels, Belgium, 1996.
42. *EN 310:1993*; Wood-Based Panels-Determination of Modulus of Elasticity in Bending and of Bending Strength. European Committee for Standardization: Brussels, Belgium, 1993.

43. *EN 321:2003*; Wood-Based Panels-Determination of Moisture Resistance under Cyclic Test Conditions. European Committee for Standardization: Brussels, Belgium, 2003.
44. Alshammari, B.A.; Saba, N.; Alotaibi, M.D.; Alotibi, M.F.; Jawaid, M.; Alothman, O.Y. Evaluation of Mechanical, Physical, and Morphological Properties of Epoxy Composites Reinforced with Different Date Palm Fillers. *Materials* **2019**, *12*, 2145. [CrossRef]
45. Munawar, S.S.; Kusumah, S.S.; Subiyanto, B.; Purnomo, D.; Akbar, F.; Syahrir, A. Physical and Mechanical Properties of Natural Fiber Polyester Laminate Composites. *IOP Conf. Ser. Earth Environ. Sci.* **2021**, *891*, 012002. [CrossRef]
46. Mohammad Khanlou, H.; Hall, W.; Woodfield, P.; Summerscales, J.; Francucci, G. The mechanical properties of flax fibre reinforced poly (lactic acid) bio-composites exposed to wet, freezing and humid environments. *J. Compos. Mater.* **2018**, *52*, 835–850. [CrossRef]
47. Belgacem, C.; Serra-Parareda, F.; Tarrés, Q.; Mutjé, P.; Delgado-Aguilar, M.; Boufi, S. The Integral Utilization of Date Palm Waste to Produce Plastic Composites. *Polymers* **2021**, *13*, 2335. [CrossRef]
48. Supian, A.B.M.; Jawaid, M.; Rashid, B.; Fouad, H.; Saba, N.; Dhakal, H.N.; Khiari, R. Mechanical and physical performance of date palm/bamboo fibre reinforced epoxy hybrid composites. *J. Mater. Res. Technol.* **2021**, *15*, 1330–1341. [CrossRef]
49. Kabir, M.; Wang, H.; Lau, K.; Cardona, F. Chemical treatments on plant-based natural fibre reinforced polymer composites: An overview. *Compos. Part B Eng.* **2012**, *43*, 2883–2892. [CrossRef]
50. Oladele, I.O.; Okoro, A.M.; Omotoyinbo, J.A.; Khoathane, M.C. Evaluation of the mechanical properties of chemically modified chicken feather fibres reinforced high density polyethylene composites. *J. Taibah Univ. Sci.* **2018**, *12*, 56–63. [CrossRef]
51. Tamrakar, S.; Kiziltas, A.; Mielewski, D.; Zander, R. Characterization of kenaf and glass fiber reinforced hybrid composites for underbody shield applications. *Compos. Part B Eng.* **2021**, *216*, 108805. [CrossRef]
52. Azwa, Z.N.; Yousif, B.F.; Manalo, A.C.; Karunasena, W. A review on the degradability of polymeric composites based on natural fibres. *Mater. Des.* **2013**, *47*, 424–442. [CrossRef]
53. Padmaraj, N.H.; Vijaya, K.M.; Dayananda, P. Experimental investigation on fatigue behavior of glass/epoxy quasi-isotropic laminate composites under different ageing conditions. *Int. J. Fatigue* **2021**, *143*, 105992. [CrossRef]
54. Böer, P.; Holliday, L.; Kang, T.H.-K. Independent environmental effects on durability of fiber-reinforced polymer wraps in civil applications: A review. *Constr. Build. Mater.* **2013**, *48*, 360–370. [CrossRef]
55. Cormier, L.; Joncas, S. Effects of cold temperature, moisture and freeze-thaw cycles on the mechanical properties of unidirectional glass fiber-epoxy composites. In Proceedings of the 51st AIAA/ASME/ASCE/AHS/ASC Structures, Structural Dynamics, and Materials Conference 18th AIAA/ASME/AHS Adaptive Structures Conference, Orlando, FL, USA, 12–15 April 2010.
56. Noonan, K.; Hess, K.; Srubar, W. Moisture-and freeze-thaw-induced deterioration of natural fiber composites with low fiber contents. *Acad. J. Civ. Eng.* **2019**, *37*, 398–404.
57. Karthi, N.; Sathish, S.; Gokulkumar, S.; Gokulnath, R.; Aravinth, T.; Govindaraj, G. Synthesis and Characterization of Jute and Banana Fiber Reinforced Polyester Composites. *Nano World J.* **2023**, *9*, S672–S678. [CrossRef]

applied sciences

MDPI

Article

Applying the Artificial Neural Network and Response Surface Methodology to Optimize the Drilling Process of Plywood

Bogdan Bedelean *, Mihai Ispas and Sergiu Răcășan

Faculty of Furniture Design and Wood Engineering, Transilvania University of Brasov, Bd-ul Eroilor nr. 29, 500036 Brasov, Romania; ispas.m@unitbv.ro (M.I.); sergiu.racasan@unitbv.ro (S.R.)
* Correspondence: bedelean@unitbv.ro

Abstract: Plywood is a wood-based composite with many applications in construction, shipbuilding, and furniture production. One of the basic plywood processing and mandatory operations is drilling. Up to now, considerable and very diverse thematic research has been recently carried out on drilling, but little of that deals with modeling of the drilling process of plywood. Therefore, in this work, the artificial neural network modeling technique and response surface methodology were applied to model and optimize the drilling process of plywood. Two artificial neural network models were developed to predict the thrust force and the drilling torque based on drill tip angle, tooth bite, and drill type. The developed ANN models were used to complete the value of responses in the experimental design, which was requested by the response surface methodology. The trust force during the drilling of plywood is significantly influenced by the drill type (helical or flat). The most significant factor that affects the drilling torque during the drilling of plywood is the tooth bite. A helical drill assures a lower minimum thrust force and drilling torque than a flat drill. The proposed method could be used as an optimization tool during the design phase of the furniture manufacturing process.

Keywords: plywood drilling; modeling; optimization; neural networks; response surface methodology; thrust force; drilling torque

Citation: Bedelean, B.; Ispas, M.; Răcășan, S. Applying the Artificial Neural Network and Response Surface Methodology to Optimize the Drilling Process of Plywood. *Appl. Sci.* **2023**, *13*, 11343. https://doi.org/10.3390/app132011343

Academic Editor: Krzysztof Koszela

Received: 23 September 2023
Revised: 10 October 2023
Accepted: 13 October 2023
Published: 16 October 2023

1. Introduction

Plywood is an engineered product made from wood and it is a part of the large family of wood-based composites, a family that also includes other materials widely used in various economic sectors: particleboards (chipboards), Medium Density Fiberboards (MDFs), Oriented Strand Boards (OSBs), and others. Plywood is a wood-based composite with many uses. Its applications include construction, shipbuilding, furniture production, etc. In 2020, the world's total plywood production was over 118 billion cubic meters, 1.97% more than in 2019 and 15.1% more than 10 years before [1].

Along with circular blade cutting, drilling is one of the basic plywood processing operations. Unlike these two, milling and sanding are much less used when processing plywood.

Plywood drilling is carried out on the same general principles, with the same tools and machines, and with roughly the same cutting parameters as drilling solid wood or other wood-based materials (chipboard, MDF, OSB etc.). Over time, the drilling of wood composites has been the focus of various research work to study and optimize the process. Various factors such as feed rate, diameter of drill, spindle speed, tip angle, type of drill etc. affects the efficiency of drilling operation that is often quantified through tool durability, cost, and quality of holes. An extensive review regarding the drilling of wood and wood-based materials is presented in the study performed by Gorski [2]. However, regarding the drilling of plywood, there is limited information in the literature.

Podziewski and Górski [3] showed that the feed rate has a significant linear influence on the axial force and on the cutting torque during the drilling of plywood. Moreover, the same scholars, in other work [4], obtained that the effect of spindle speed on the drill advance force is negligible and the relation between feed per revolution and feeding force is proportional.

Rogoziński et al. [5] reveal that plywood produces a smaller amount of fine dust than drilling in high-density fiberboard.

Podziewski and Górski [4] studied how different machining conditions affect the drilling of wood-based materials such as plywood, MDF, and particleboard. They analyzed the effect of the cutting speed and the feed speed on the feed force. The results revealed that the effect of the cutting speed on the feed force is almost negligible in comparison to the other factors such as feed speed and the type of material.

Podziewski et al. [6] studied the machinability of plywood and other types wood-based materials by drilling. The aim of the experiments was to evaluate the quality of the edges of the holes, the cutting force, and torque. The obtained data were used to establish a quality index and cutting force index defined by referencing them to one selected material (MDF). The results revealed that the quality index was not correlated to the cutting force index due to the fact that it is influenced by the internal structure of the type of material used.

Agarwal and Mishra [7] used wooden fibers and epoxy resin to manufacture laminated plywood and medium density fiberboard. The purpose of the study was to examine the effect of drilling parameters on the manufactured materials. Four types of drill bits, with a point angle of 30°, 60°, 90°, and 120°, different cutting speeds and feed rates were used for drilling in order to examine the effect of the delamination factor. The obtained results shows that the delamination factor increases with the increase in the point angle of the drill bit, and also with the increase in the feed speed.

Sydor et al. [8] analyzed the accuracy of holes drilled in the side surface of plywood. The results showed the adhesive layer negatively influences the accuracy of the position and the angle of the holes. In addition to, the study recommends not to use a drill diameter smaller than 3.8 mm. In addition, in order to maintain the hole accuracy, a feed per revolution lower than 0.15 mm/rev. in the second phase of drilling is highly recommended to maintain hole accuracy.

Modeling of the wood drilling process represents a new management principle that leads to high-quality wooden products [9]. Various approaches could be used to model the drilling process of wood and wood-based materials. The most applied method to model and optimize the drilling process is Response Surface Methodology (RSM). In addition, the artificial neural network (ANN) modeling technique was, also, applied to predict the various outputs of the drilling process. Nowadays, ANN models are included in various decision support systems that are developed for the forest industry [10].

Davim et al. [11] studied how to control the delamination factor in the drilling of MDF by testing two types of MDF panels with different coating layers, using cemented carbide drills. They analyzed the main interaction effects of the drilling parameters on the delamination factor. Feed rate and cutting speed were used as independent variables and delamination was measured at the entry and exit of the holes. The authors used response surface methodology to design and model the experiments. ANOVA, also, was used to check the validity of the models.

The article published by Prakash and Palanikumar [12] explores the prediction of surface roughness during the drilling of MDF using response surface methodologies with TiN-coated carbide step drill bits. The experiments, guided by Taguchi's design technique, involve spindle speed, feed rate, and drill diameter as factors. The study establishes a second-order model through response surface methodology, revealing that the feed rate is the most influential parameter on surface roughness, followed by the cutting speed. The developed model shows a promising predictive accuracy, validated through additional experiments.

The study published by Wang et al. [13] examines delamination's impact on drilling quality in MDF, specifically focusing on variable feed pressure in multi-spindle drilling machines. Second-order polynomial models are established using response surface methodology to analyze the relationship between delamination and machining parameters. The analysis of variance (ANOVA) validates these models, revealing that delamination increases with decreased drill hardness and increased feed pressure at the MDF entry and exit. At low drill hardness and high feed pressure, the delamination factor remains relatively insensitive to variations in these factors.

The paper of Valarmathi et al. [14] explores how to achieve good surface quality in drilling particleboard by changing the input parameters. It measures the surface roughness (Ra) and finds that it is lowest with a high cutting speed, low feed speed, and small point angle. It uses ANOVA to show that feed speed has the most effect on Ra, followed by point angle and cutting speed. It also uses RSM and adaptive neuro-fuzzy inference system (ANFIS) to make mathematical models to predict Ra and compares their accuracy. The R^2 coefficient values of 98.5% (RSM) and 99.9% (ANFIS) indicates that the models are useful to predict the surface roughness of particleboards.

Ayyildiz et al. [15] used a parallel robot to drill MDF and examined the effect of cutting parameters on surface roughness. They used the Taguchi technique and ANOVA to find the best drilling parameters and fit a quadratic regression model. They report that the surface roughness is optimal when the point angle is 118°, the cutting speed is 47.1 m/min, and the feed rate is 0.01 mm/rev.

Bedelean et al. [16] studied how to optimize the drilling process of wood particleboards by adjusting selected factors. They used ANN models and RSM to predict and analyze the effects of drill tip angle, tooth bite, and drill type on delamination factor, thrust force, and drilling torque. The authors found that a helical drill assures a lower delamination factor, a lower thrust force, and a lower drilling torque than a flat drill. The designed ANN models were reliable.

Zbieć [17] developed a neural network to monitor tool wear in MDF milling based on machined surface temperature, cutting force, thrust force, and power consumption. The article of Jegorowa et al. [18] proposes an automatic system to detect the drill bit condition by measuring various signals during machining. It uses k-NN (k-nearest neighbor algorithm) classifier in MATLAB to classify the drill bit as Green (tool can still be used), Red (tool change is necessary), or Yellow (intermediate, warning state). It reports a 76% accuracy of classification.

Based on the information presented above, one could observe that there is a lack of studies regarding the application of ANN and RSM in modeling of the drilling process of plywood. Therefore, the objective of this research is to apply the ANN modeling technique and RSM to model and optimize the drilling process of plywood. The analyzed factors (independent variables) that are considered in this work are drill tip angle, tooth bite, and drill type (flat or helical). The responses (dependent variables) are the thrust force and drilling torque—two dynamic parameters that affect both quality and the energy consumption during drilling of wood and wood-based materials.

2. Materials and Methods

2.1. Experimental Part

The experiments carried out to determine the drilling torque and thrust force used the methodology presented by Ispas and Răcăşan [19]. Two sets of drills were used: four flat drills and four helical drills (Figure 1). All eight drills had a drilling diameter of 10 mm. The drills were sharpened with four values of tip angle ($2\kappa r$ = 30°, 60°, 90°, 120°) and with the same clearance angle α = 20°.

Eighty samples of beech plywood (Figure 2), with a thickness of 18 mm, were used for processing. The rotation speed of the drills was n = 3000 rpm. Four feed speeds were used, vf = 0.6, 1.8, 3.0 and 4.2 m/min. Forty samples were processed with the flat drills, and the other forty with the helical ones. Each group of forty samples was divided into four groups of ten samples, each such group being processed with one of the four feed speeds.

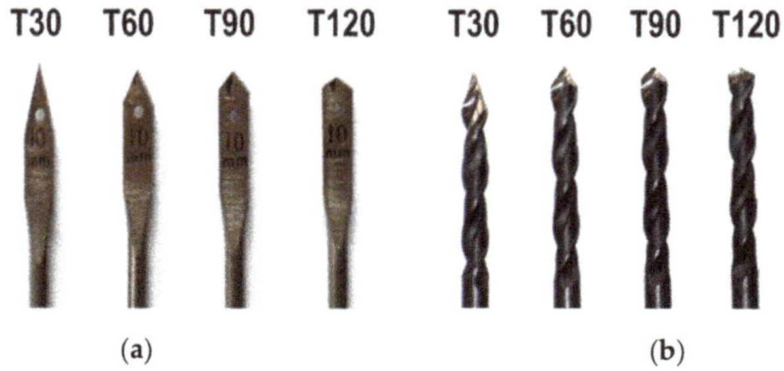

Figure 1. The drill bits used in the experiments: (**a**) flat drills; (**b**) helical drills.

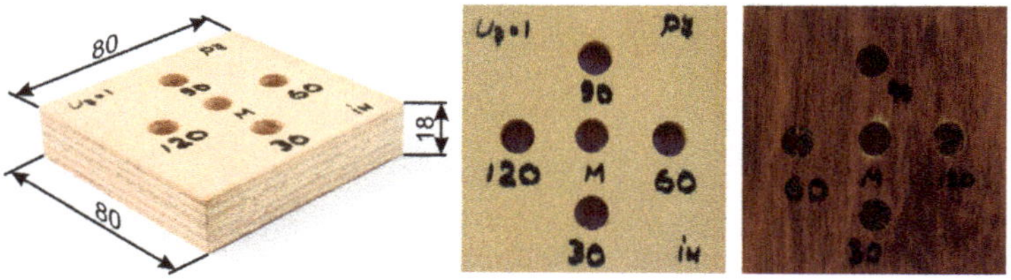

Figure 2. The plywood samples used during drilling experiments.

Drilling was performed on an ISEL GFV/GFY CNC machining center. The measured parameters were the thrust force and the active power consumed by the spindle motor during drilling. The thrust force was measured with a device consisting of three HBM force transducers, type S2 (nominal force: 500 N). The measurement of the active power consumed by the spindle motor was performed by a Camille Bauer Sineax P530/Q531 transducer for active and reactive power. The equipment connection scheme is shown in Figure 3.

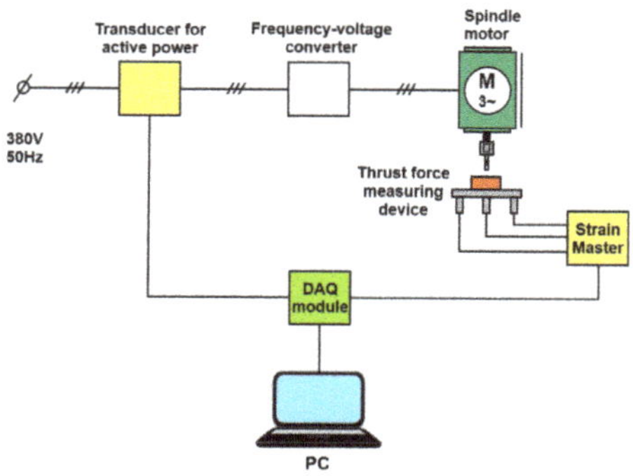

Figure 3. Connection diagram of the equipment for measuring the dynamic parameters of the drill.

For data recording, a DAQ Board Keithley Model KUSB-3108 was used with the help of the Keithley KUSB QuickDataAcq software v.1.5.5.0. Data recording was carried out with a frequency of 100 values/s simultaneously on all four channels: three for force sensors and one for the active power.

The processing of the experimental data was carried out with the help of Microsoft Excel under Windows 10. To begin with, a filtering of the data was performed with a fourth-order Butterworth digital filter, in order to remove the parasitic components of the signal.

To determine the thrust force F_T, the data from the three force sensors were summed 3 by 3.

To determine the drilling torque T_D, the data from the measurement of the active power consumed by the spindle motor were transformed using Formulas (1) and (2).

$$P_D = P_T - P_0 \tag{1}$$

where:

P_D is the power consumed only for the actual drilling, in W;

P_T is the total active power consumed by the spindle motor during drilling (and measured), in W;

P_0 is the power consumed during idle running, in W.

As known, in rotational motion the mechanical power P_D:

$$P_D = T_D \cdot n$$

where:

T_D is the torque, in Nm;

n is the rotational speed, in rad/s.

Therefore, the torque is: $T_D = \frac{P_D}{n}$ $[Nm]$ if P_D is given in [W] and n in [rad/s].

However, if the rotational speed n is given in [rpm], we must transform the equation according to the new measurement units.

So:

$$1\frac{rot}{min} = \frac{1rot}{60s} = \frac{2\pi rad}{60s} = \frac{2\pi}{60} \cdot \frac{rad}{s} \cong \frac{1}{9.55}\frac{rad}{s}$$

Therefore, if P_D is given in [W] and n in [rpm]:

$$T_D = 9.55\frac{P_D}{n} \ [Nm] \tag{2}$$

where:

n is the rotation speed of the drill, in rpm.

For further processing and analysis, the maximum values of both thrust force F_T and drilling torque T_D were taken into account.

2.2. Data Modeling

2.2.1. Modeling with Artificial Neural Network

The obtained data set was randomly divided in two subsets. One subset was used to train and test the neural networks that were designed to predict the thrust force (Y_1) and drilling torque (Y_2) based on drill tip angle (X_1), tooth bite (X_2), and drill type (X_3). The sample size for this subset was 107 in the case of flat drill and 112 in the case of helical drill. The other subsets were used to validate the developed models. In the case of flat drill, the sample size was equal to 15 and in the case of helical drill, the sample size was equal to 17. The main descriptive statistics of both subsets are presented in Table 1. One could observe that both subsets are roughly the same from the point of view of analyzed statistical indicators.

Table 1. Descriptive statistics of training and validation data sets.

Variables	Training Data Set (n_{Flat} = 107, n_{Helix} = 112)								Validation Data Set (n_{Flat} = 15, n_{Helix} = 17)							
	M		SD		Min		Max		M		SD		Min		Max	
	Flat	Helix	Flat	Helix	Flat	Helix	Flat	Helix	Flat	Helix	Flat	Helix	Flat	Helix	Flat	Helix
Drill tip angle (X_1)	75	76.82	33.46	33.26	30	30	120	120	78.00	77.64	33.64	35.27	30	30	120	120
Tooth bite (X_2)	0.40	0.40	0.22	0.22	0.1	0.1	0.7	0.7	0.42	0.40	0.22	0.21	0.10	0.10	0.7	0.70
Thrust force (Y_1)	309.79	69.92	138.42	23.42	166.16	16.05	505.49	116.36	318.04	74.56	92.12	32.88	180.55	31.44	454.68	170.67
Drilling torque (Y_2)	1.55	0.95	0.63	0.43	0.47	0.28	2.91	1.97	1.59	0.94	0.72	0.44	0.50	0.29	2.76	1.76

M—mean; SD—standard deviation; Min—minimum value; Max—maximum value; n—data set size.

The ANN models were developed by involving the Neural Ware Predict Software (Predict), NeuralWare Inc., Carnegie, 239, PA, USA [20]. This software integrates all the facilities needed to apply neural modeling techniques, namely, data analysis, variable transformation, and reveal the network architecture. Predict is built based on cascade correlation learning algorithm. This algorithm starts with a minimum architecture of the multi-layer perceptron (MLP), which is one of the most popular architectures, and new hidden units are added step by step till the optimum configuration of MLP is achieved. The main structure of MLP architecture is depicted in Figure 4.

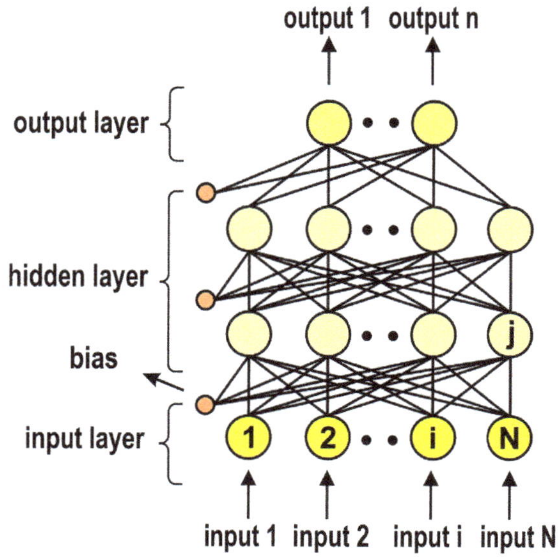

Figure 4. The multi-layer perceptron architecture.

The MLP architecture is formed by a number of layers, namely, the input layer, the hidden layer, and the output layer. The input layer receives the input signals and transmits the signals to the hidden layer to be processed. The output layer shows the results of the network. An MLP architecture may contain one or more hidden layers in function to the complexity of the problem that is going to be addressed through ANN modeling technique. Each layer contains various number of neurons or processing elements, which are not connected in the same layer.

The process of developing ANN models begins with training the neural network with a known input and output data set. During the training process, the values of weights and biases are adjusted till the best performance is obtained. The value of weights and biases are saved and could be used to predict outputs with new input values, if the designed neural networks performed well during the testing phase [21].

According to Özşahin [22], the job of one neuron is to receive signals (x_i) from the neurons that are located in the previous layer together with its corresponding weight factor (w_{ij}), to sum the weighted signals (net_j) and the bias of the layer (θ_j) and to communicate the output (y_j) to the next layer. A mathematical (transfer) function (f (.)) is applied to compute the output value (Equations (3) and (4)).

$$net_j = \sum_{i=1}^{n} x_i w_{ij} - \theta_j \quad (3)$$

$$y_j = f(net_j) \quad (4)$$

The number of neurons in the input layers is equal to the number of inputs of the network.

The number of processing elements or neurons in the hidden layer is typically found by applying the trial–error approach or various optimization algorithms. The number of neurons in the output layers is equal to the number of desired outputs.

$$Y_j = W_0 + \sum_{j=1}^{n} x_i w_{ij} \tag{5}$$

Since most of the default settings of Predict are useful in solving useful problems, these settings, also, were applied in this work [20]. The models were designed to have a hyperbolic–tangent transfer function in the hidden layer and a sigmoid transfer function in the output layer. The adaptive gradient learning rule and a learning rate equal to 0.01 was used in this study. The adaptive gradient learning is based on back-propagation technique designed to train the MLP network. This technique employs an objective function to figure out how closely the outputs of the network match the target outputs in the training set of data. The result of objective function is improved by adjusting the weights of each individual processing element (neuron).

The performance of ANN models developed to predict the thrust force and drilling torque was measured using the most applied indicators in literature [20–22], namely, coefficient of correlation (R), coefficient of determination (R^2), root mean square error (RMSE), and the mean absolute percentage error (MAPE). The corresponding equations are:

$$\mathrm{R} = \frac{\sum_{i=1}^{n} (p_i - \overline{p})(a_i - \overline{a})}{\sqrt{\sum_{i=1}^{n} (p_i - \overline{p})^2} \sqrt{\sum_{i=1}^{n} (a_i - \overline{a})^2}} \tag{6}$$

$$\mathrm{R}^2 = 1 - \frac{\sum_{i=1}^{n} (a_i - p_i)^2}{\sum_{i=1}^{n} (a_i - \overline{p}_i)^2} \tag{7}$$

$$\mathrm{RMSE} = \sqrt{\frac{1}{n} \sum_{i=1}^{n} (a_i - p_i)^2} \tag{8}$$

$$\mathrm{MAPE} = \frac{1}{n} \left(\sum_{i=1}^{n} \left[\left| \frac{a_i - p_i}{a_i} \right| \right] \right) \times 100 \tag{9}$$

where: a_i is the experimental value, p_i is the predicted value of ANN model, $\overline{a}$ is the mean of experimental values; $\overline{p}$ is the mean of predicted values, n is the total number of values.

Coefficient of correlation and coefficient of determination measure the strength of the relationship between experimental and predicted values. When the value of R or R^2 is close to one, there could be concluded that there is a strong relationship between experimental and predicted values. On the other hand, a value of R or R^2 close to zero indicates that between the experimental and predicted values is a lower relationship. In addition, the value of R^2 explains what percentage of variation in the dependent variable could be explained by the analyzed independent variables.

The root mean square error measures the average difference between the experimental and predicted values by the developed models. A value of RMSE close to zero tells us that the predicted values are similar with those that were obtained during the experiments.

The mean absolute percentage error, which represents most performance indicator of ANN models, describes how well the developed models predict the desired outputs. According to Lewis (1982) cited by Özşahin and Singer [21], a value of MAPE lower than 10% indicates that the developed models have a high precision. On the other hand, when the MAPE is higher than 50%, the models are inaccurate. When the MAPE is between 10 and 20%, the models have a good accuracy. A reasonable model is obtained when the MAPE is greater than 20% and lower than 50%.

2.2.2. Modeling with Response Surface Methodology

The first step to applying the Response Surface Methodology is to generate a central composite design. In this design, various combinations among the values of factors are generated. The corresponding response of these combinations could be revealed by means of experiments or simulations. Once the experimental design is completed, the next step consists of figuring out the regression equation that best describes the relationship between inputs and output. Usually, the ANOVA is applied to reveal which factors and coefficients are significant from a statistical point of view. The equation in coded form could be used to analyze the influence of factors (inputs) on the analyzed response (output). The equation in real form could be used to make predictions for various scenarios. Last but not least is the step that consists of using the regression equation to reveal the optimum combination of factors that corresponds to the desired output. These steps were performed by means of the Design Expert software (Version 9, Stat-Ease Inc., Minneapolis, MN, US). The experimental design generated by the software is a two-level face centered design (FCCD), which is pictured in Figure 5. The FCCD generates four combinations for the flat drill (#1, #16, #23, and #24) and four combinations for helical drill (#6, #8, #13, and #25) among the low (−1) and high (+1) levels of analyzed factors, namely, drill tip angle (X_1) and tooth bite (X_2) (Tables 2 and 3). The two-level factorial points are needed to estimate the linear effect of factors on the analyzed responses. On the other hand, to reveal whether there is a non-linear effect of factors, the FCD design requires taking into account the axial points (α). In Table 4, the axial points are combinations #3, #10, #17, and #18 for the flat drill and combination #4, #14, #19, and #22 in the case of the helical drill. This design also includes combinations among the center level (0) of analyzed factors, which are used during the analysis to estimate the experimental error. These combinations are #2, #11, #15, #20, and #21 in the case of the flat drill and combinations #5, #7, #9, #12, and #26 for the helical drill (Table 3). Summing up, 26 combinations were analyzed. The corresponding value of analyzed responses (thrust force and drilling torque) were revealed by means of designed ANN models.

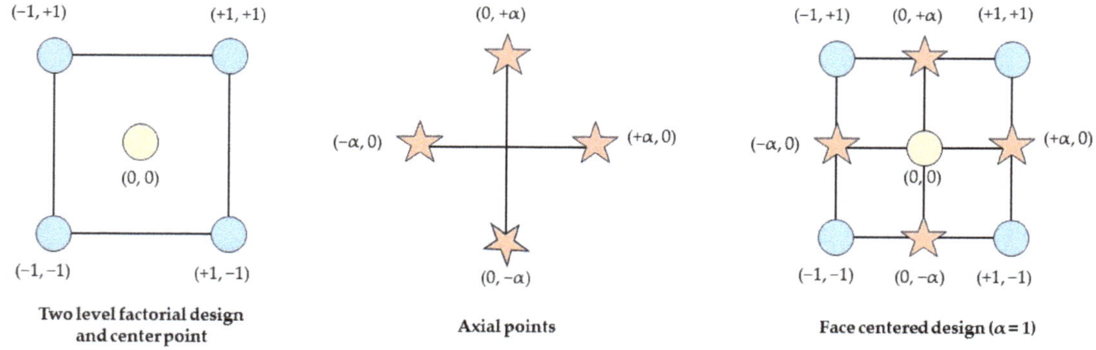

Figure 5. The structure of a face centered composite design.

Table 2. The name of factors and their analyzed levels.

Numeric Factor	Level				
	$-\alpha$ *	−1	0	+1	$+\alpha$ *
Drill tip angle (X_1), °	30	30	75	120	120
Tooth bite (X_2), mm	0.1	0.1	0.4	0.7	0.7
Categoric factor	**Level 1**			**Level 2**	
Drill type (X_3)	Flat (−1)			Helical (+1)	

* axial points.

Table 3. The experimental plan that was used for the optimization study.

Combination #	Factors (Inputs)			Responses (Outputs)	
	Drill Tip Angle (X_1), °	Tooth Bite (X_2), mm	Drill Type (X_3)	Thrust Force (Y_1), N	Drilling Torque (Y_2), Nm
1	30 (−1)	0.1 (−1)	Flat (−1)	173.87	1.14
2	75 (0)	0.4 (0)	Flat (−1)	313.98	1.55
3	120 (+α)	0.4 (0)	Flat (−1)	333.09	1.17
4	120 (+α)	0.4 (0)	Helical (1)	91.91	0.73
5	75 (0)	0.4 (0)	Helical (1)	71.08	0.96
6	120 (1)	0.7 (1)	Helical (1)	114.54	1.11
7	75 (0)	0.4 (0)	Helical (1)	71.08	0.96
8	120 (1)	0.1 (−1)	Helical (1)	72.29	0.37
9	75 (0)	0.4 (0)	Helical (1)	71.08	0.96
10	75 (0)	0.7 (+α)	Flat (−1)	329.95	2.29
11	75 (0)	0.4 (0)	Flat (−1)	313.98	1.55
12	75 (0)	0.4 (0)	Helical (1)	71.08	0.96
13	30 (−1)	0.7 (1)	Helical (1)	42.46	1.89
14	75 (0)	0.1 (-α)	Helical (1)	76.53	0.42
15	75 (0)	0.4 (0)	Flat (−1)	313.98	1.55
16	30 (−1)	0.7 (1)	Flat (−1)	387.75	2.58
17	75 (0)	0.1 (-α)	Flat (−1)	193.95	0.77
18	30 (-α)	0.4 (0)	Flat (−1)	365.28	2.01
19	75 (0)	0.7 (+α)	Helical (1)	84.72	1.46
20	75 (0)	0.4 (0)	Flat (−1)	313.98	1.55
21	75 (0)	0.4 (0)	Flat (−1)	313.98	1.55
22	30 (-α)	0.4 (0)	Helical (1)	33.88	1.24
23	120 (1)	0.1 (−1)	Flat (−1)	190.58	0.51
24	120 (1)	0.7 (1)	Flat (−1)	459.36	1.82
25	30 (−1)	0.1 (−1)	Helical (1)	36.40	0.53
26	75 (0)	0.4 (0)	Helical (1)	71.08	0.96

Table 4. Results obtained during the influence of independent variable on the performance of ANN models.

Independent Variable	Number of Neurons in the Hidden Layer		Testing Performance Indicator, R	
	ANN Model for Thrust Force	ANN Model for Drilling Torque	ANN Model for Thrust Force	ANN Model for Drilling Torque
Drill tip angle (X_1)	1	3	0.074	0.389
Tooth bite (X_2)	0	3	0.377	0.768
Drill type (X_3)	0	0	0.866	0.436
Drill tip angle (X_1) × Tooth bite (X_2)	4	2	0.372	0.839
Drill tip angle (X_1) × Drill type (X_3)	2	6	0.872	0.627
Tooth bite (X_2) × Drill type (X_3)	2	7	0.977	0.890
Drill tip angle (X_1) × Tooth bite (X_2) × Drill type (X_3)	14	4	0.986	0.988

Once the values of analyzed responses were figured out, a quadratic model could be obtained. The general form is presented in Equation (10) [23]:

$$Y = \beta_0 + \sum_{i=1}^{k} \beta_i X_i + \sum_{i=1}^{k} \beta_{ii} X_i^2 + \sum_{i=1}^{k} \sum_{j=1}^{k} \beta_{ij} X_i X_j + \varepsilon \quad (10)$$

where Y is the dependent variable or the response (thrust force or drilling torque), X_i and X_j are the analyzed factors, namely, drill tip angle (X_1) and tooth bite (X_2), β_0 is the constant, β_i, β_{ii}, and β_{ij} are the coefficients of the equation, ε is the error term, and k represents the number of analyzed factors.

The coefficient of mathematical models were developed through regression analysis. In order to figure out if the model parameters of each factor are significant, the analysis of variance (ANOVA) was involved. ANOVA was performed at a 1% and 5% level of significance. The coefficient of determination (R^2) was used to establish the model accuracy.

For the optimization study, a various combinations of analyzed factors are analyzed to fulfill the optimization criteria, namely, to minimize or maximize the investigated response or responses. In this study, the optimization criteria was to minimize both the thrust force and drilling torque. To determine the best combination among factors, the Design Expert software uses the desirability function approach, which is one of the most widely used methods in the industry for the optimization of multiple response processes. The method finds operating conditions that provide the most desirable response values. According to literature [24,25], for each response *Yi(X)*, the desirability function *di(Yi)* assigns numbers between 0 and 1 to the possible values of *Yi*. When the *di(Yi)* is close to zero (0) there is an undesirable value of *Yi*. On the other hand, when the *di(Yi)* is close or equal to one (1), there is a desirable value of analyzed dependent variable (response). All individual desirabilities are combined by means of geometric mean that describes the overall desirability, D (Equation (11)):

$$D = (d_1(Y_1) \times d_2(Y_2) \times \ldots \times d_n(Y_n))^{1/n} \quad (11)$$

where n represents the number of responses being optimized.

3. Results and Discussion

3.1. ANN Modeling

In Table 4, various architectures of designed ANN models that are able to predict the thrust force and drilling torque during drilling of plywood can be observed. Based on the coefficient of correlation (R), the following findings were revealed. The most important input that affects the performance of a designed ANN model to predict the thrust force (Y_1) is the drill type (X_1), followed by the tooth bite (X_2) and drill tip angle (X_1). In the case of the model developed to predict the drilling torque (Y_2), the tooth bite has a greater influence than the other two analyzed variables, namely, drill tip angle (X_1) and drill type (X_3). The drill type has a more important role in the performance of model than the drill tip angle (X_1). However, the best performance of both designed ANN models is obtained for the case when all three input variables were considered in the model. The coefficient of correlation was equal to 0.98 for both ANN models.

The performance indicators of ANN developed models are presented in Table 5. As can be observed, the developed models have a good accuracy, namely, a high R or R^2, and a low RMSE or MAPE. This affirmation is also supported by Figures 6 and 7.

The architecture of designed ANN models is different in the hidden layer both in the case of thrust force and drilling torque when it is compared with the architecture developed to predict the same outputs. However, in the case of prelaminated wood particle panels [16], the number of neurons in the hidden layer was equal to four in the case of the ANN model developed to predict the thrust force and nine for the model designed to predict the drilling torque. Since the same inputs and outputs were analyzed, the difference in the number of neurons could be attributed to the analyzed material in this work, namely, plywood, which has a different structure than prelaminated wood particle panels. Thus, it could be stated that the type of material influences the architecture and performance of artificial neural networks. Consequently, in a further study, the type of material should be included as one of the input of neural networks among other factors.

Table 5. The performance indicators of ANN models designed to predict the thrust force and drilling torque.

Performance Indicator	ANN Model for Thrust Force	ANN Model for Drilling Torque
R	0.986	0.988
R^2	0.972	0.976
RMSE	16.59	0.10
MAPE	6.80	7.39

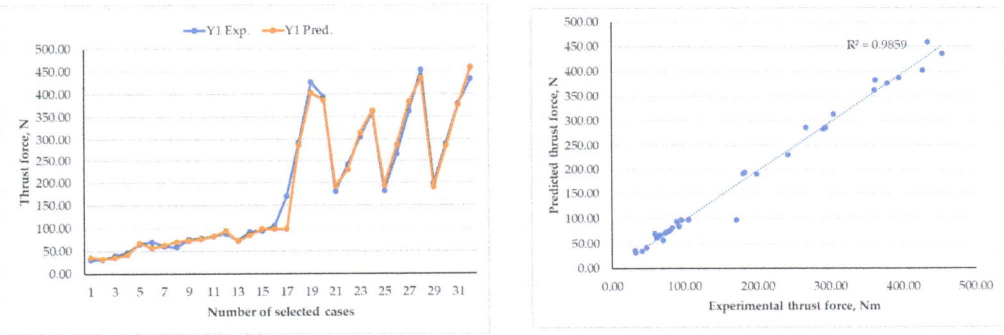

(**a**) (**b**)

Figure 6. The graphical comparison between the experimental and predicted values of thrust force: linechart (**a**) and scatter plot (**b**).

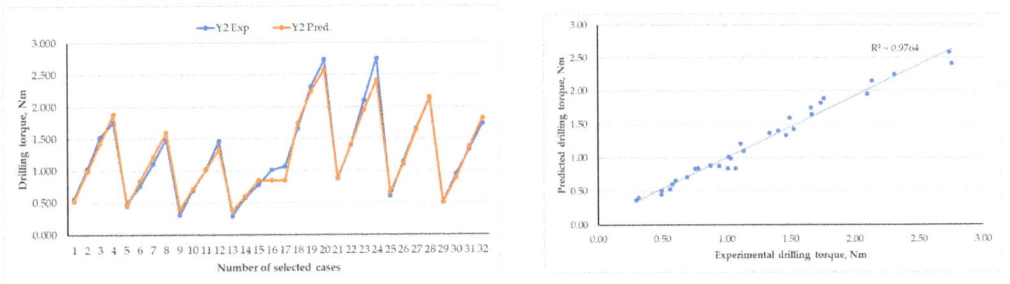

(**a**) (**b**)

Figure 7. The graphical comparison between the experimental and predicted values of drilling torque: line chart (**a**) and scatter plot (**b**).

To summarize, both developed ANN models have a high coefficient of determination ($R^2 \approx 0.97$). Consequently, the designed models explain at least 97% of experimental data used during the modeling process. By comparing the value of coefficient of determination with the results from other studies, one could observed that the value is close to the value obtained by Tiryaki et al. [26] to predict power consumption during wood planning, Özşahin and Singer [21] to predict noise emissions during wood machining, Tiryaki et al. [27] to predict surface roughness and power consumption during the abrasive machining of wood, and Bedelean et al. [16] to predict the thrust force and drilling torque during the drilling of prelaminated wood particle panels.

3.2. Response Surface Methodology

3.2.1. Thrust Force

The Design Expert software suggests that the relation between the analyzed factors (drill tip angle, tooth bite, and drill type) and thrust force could be best described by Equation (12), which is in coded form. The model is significant at 1% (Table 6). Therefore, one could observe that the trust force during the drilling of plywood is significantly influenced by the drill type (X_3). In addition, the drill tip angle (X_1) and tooth bite (X_2) influence the thrust force (Y_1). The tooth bite has a bigger influence than the drill tip angle ($X_2 > X_1$) in the case of thrust force during the drilling of plywood. This result is the same as that obtained in our previous work wherein the drilling of prelaminated wood-particle boards was analyzed and optimized [16]. In addition, this result is correlated with the data reported in the literature for the plywood and prelaminated particle board panels, namely, it is stated that the feed rate affects the thrust force [3,28]. Consequently, a low feed rate leads to a low thrust force, and a high feed rate implies a high thrust force. The tooth bite has a non-linear effect on the thrust force. This finding tells us that the optimum value of the tooth bite stands inside the analyzed range (0.1–0.7 mm). Moreover, the statistical analysis reveals the fact that there is an interaction effect of analyzed factors on thrust force ($X_2X_3 > X_1X_2 > X_1X_3$). The most significant interaction is between the tooth bite (X_2) and drill type (X_3). This result is the same with that obtained in the case of prelaminated wood-particle boards [16]. The influence of the drill tip angle (X_1) and tooth bite (X_2) on thrust force (Y_1) is plotted in Figure 8. Equations (13) and (14) could be used to predict the thrust force (Y_1) based on the drill tip angle (X_1) and tooth bite (X_2) during drilling of plywood. Equation (13) is for the flat drill and Equation (14) is for the helix drill.

$$\hat{Y}_{1coded} = 192.55 + 18.51X_1 + 56.27X_2 - 120.64X_3 + 11.39X_1X_2 + 9.16X_1X_3 - 46.85X_2X_3 + 13.44X_1^2 \\ -21.31X_2^2 - 2.34X_1X_2X_3 - 24.53X_1^2X_3 + 27.95X_2^2X_3 \quad (12)$$

$$\hat{Y}_{1flat} = 208.52 - 3.01X_1 + 705.33X_2 + 1.016X_1X_2 + 0.018X_1^2 - 547.35X_2^2 \quad (13)$$

$$\hat{Y}_{1helix} = 14.34 + 1.16X_1 - 77.87X_2 + 0.67X_1X_2 - 0.0055X_1^2 + 73.75X_2^2 \quad (14)$$

Table 6. Results obtained during ANOVA analysis of the model selected to predict the trust force.

Source	Sum of Squares	df	Mean Square	F-Value	*p*-Value Prob > F	Observation
Model	447,346.21	11	40,667.83	98.51	<0.0001	Significant
Drill tip angle (X_1)	4110.97	1	4110.97	9.96	0.0070	Significant
Tooth bite (X_2)	37,989.69	1	37,989.69	92.02	<0.0001	Significant
Drill type (X_3)	168,825.11	1	168,825.11	408.95	<0.0001	Significant
X_1X_2	1037.29	1	1037.29	2.51	0.135	Not Significant
X_1X_3	1005.84	1	1005.84	2.44	0.140	Not Significant
X_2X_3	26,335.85	1	26,335.85	63.79	<0.0001	Significant
X_1^2	997.94	1	997.94	2.42	0.142	Not Significant
X_2^2	2508.90	1	2508.90	6.08	0.02	Significant
$X_1X_2X_3$	43.78	1	43.78	0.11	0.749	Not Significant
$X_1^2X_3$	3324.29	1	3324.29	8.05	0.0132	Significant
$X_2^2X_3$	4315.11	1	4315.11	10.45	0.006	Significant
R^2				0.98		

The ANOVA results are significant at 5% ($\alpha = 0.05$) and are presented in Table 6.

When the *p*-value was less than 0.05, the analyzed factors were significant at 95% confidence level. Therefore, it could be concluded that the next model terms are significant: X_1, X_2, X_3, X_2X_3, X_2^2, $X_1^2X_3$, $X_2^2X_3$.

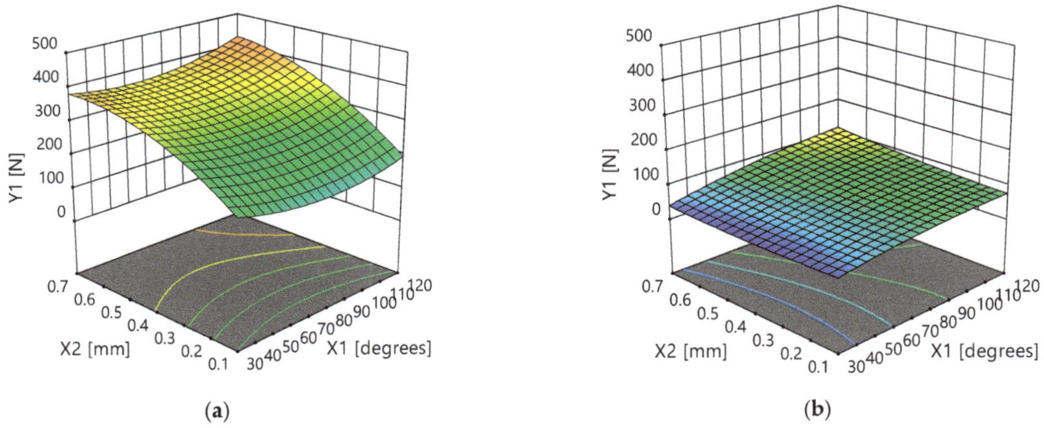

Figure 8. The influence of the drill tip angle and tooth bite on the thrust force: flat drill (**a**) and helical drill (**b**).

3.2.2. Drilling Torque

The relation between the analyzed factors, drill bit angle (X_1), tooth bite (X_2) and drill type (X_3), and drilling torque (Y_2) is described by a quadratic regression equation, which was also revealed by the Design Expert software. The coded form of the suggested mathematical model, which is significant at 1% (Table 7), is presented in Equation (15). This regression equation (in coded form) tells us that the most significant factor (from the analyzed ones) that affects the drilling torque during the drilling of plywood is the tooth bite (X_2). The next ones are the drill tip angle (X_1) and drill type (X_3) with almost the same influence (since both factors have a coefficient close to 0.30). The obtained results are almost identical with those obtained in the case of drilling prelaminated wood particleboard panels [16]. There is a significant interaction between the analyzed factors (X_1X_2, X_1X_3, X_2X_3, $X_1X_2X_3$) on the drilling torque (Y_2). The most important interaction is between tooth bite (X_2) and drill type (X_3). The influence of drill tip angle (X_1) and tooth bite (X_2) on the drilling torque (Y_2) is presented in Figure 9. Equations (16) (in the case of a flat drill) and 17 (in case of a helical drill) could be used to predict the drilling torque during the drilling of plywood (Y_2) based on the drill tip angle (X_1) and tooth bite (X_2). The ANOVA results at 5% significance level are presented in Table 7.

$$\hat{Y}_{2coded} = 1.26 - 0.31X_1 + 0.62X_2 - 0.30X_3 - 0.092X_1X_2 + 0.067X_1X_3 - 0.094X_2X_3 + 0.020X_1^2 - 0.032X_2^2 - 0.062X_1X_2X_3 + 0.0077X_1^2X_3 + 0.017X_2^2X_3 \quad (15)$$

$$\hat{Y}_{2flat} = 1.11 - 0.008X_1 + 2.97X_2 - 0.022X_1X_2 + 6.058e - 006X_1^2 - 0.54X_2^2 \quad (16)$$

$$\hat{Y}_{2helix} = 0.37 - 0.0028X_1 + 2.73X_2 - 0.0113X_1X_2 + 1.373e - 005X_1^2 - 0.164X_2^2 \quad (17)$$

Table 7. Results obtained during ANOVA analysis of the model selected to predict the drilling torque.

Source	Sum of Squares	df	Mean Square	F-Value	p-Value Prob > F	Observation
Model	8.15	11	0.74	613.64	<0.0001	Significant
Drill tip angle (X_1)	1.13	1	1.13	935.99	<0.0001	Significant
Tooth bite (X_2)	4.59	1	4.59	3803.62	<0.0001	Significant
Drill type (X_3)	1.04	1	1.04	862.460	<0.0001	Significant
X_1X_2	0.067	1	0.067	55.79	<0.0001	Significant
X_1X_3	0.054	1	0.054	44.52	<0.0001	Significant
X_2X_3	0.11	1	0.11	88.57	<0.0001	Significant

Table 7. *Cont.*

Source	Sum of Squares	df	Mean Square	F-Value	*p*-Value Prob > F	Observation
X_1^2	0.0022	1	0.0022	1.84	0.196	Not Significant
X_2^2	0.0055	1	0.0055	4.60	0.05	Not Significant
$X_1X_2X_3$	0.030	1	0.031	25.33	0.0002	Significant
$X_1^2X_3$	0.00033	1	0.0003	0.28	0.60	Not Significant
$X_2^2X_3$	0.00157	1	0.0015	1.31	0.27	Not Significant
R^2			0.99			

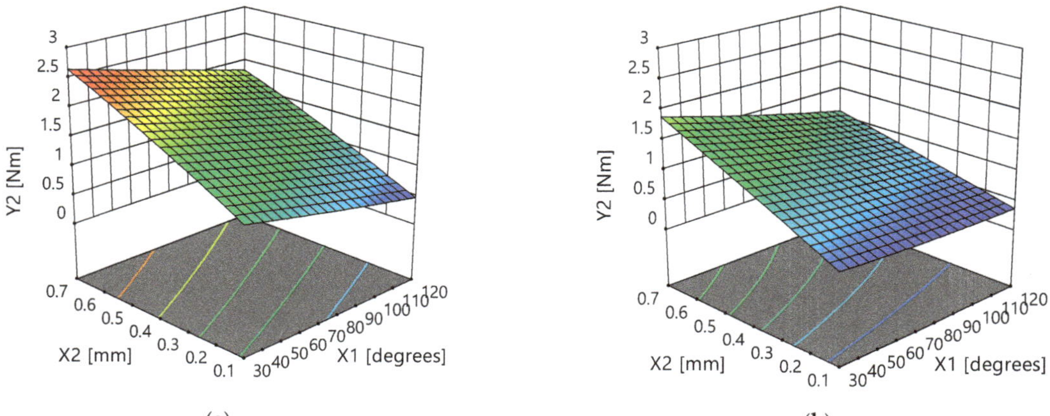

Figure 9. The influence of the drill tip angle and tooth bite on the thrust force: (**a**) flat drill (**a**,**b**) helical drill.

The optimum combinations of independent variables during the drilling of plywood in order to achieve both a lower thrust force and a lower drilling torque are presented in Table 8. In the case of the helical tool, the drill tip angle (X_1) should be equal to 30° and the tooth bite (X_2) equal to 0.1. The relative error, which was calculated using Equation (18), between predicted and experimental values is around 25% in the case of thrust force (Y_1) and around 5% in the case of drilling torque (Y_2). Regarding the flat tool, the minimum thrust force (Y_1) and drilling torque (Y_2) is obtained when the drill tip angle (X_1) is equal to 118° and the tooth bite (X_2) is adjusted at the minimum analyzed level (0.1 mm). The relative error of regression equations that were obtained in the case of the flat drill is lower (between 2.14% and 2.65%) than the error obtained for the helical drill (Table 8). Since the desirability coefficient has a high value in the case of the helical drill, it is obviously that this kind of drill performs better in terms of energy consumption during drilling of plywood, than a flat tool. The obtained results are correlated with our previous study regarding the drilling of prelaminated wood particle boards [16].

$$E_R = \frac{\left| Y - \hat{Y} \right|}{Y} \times 100 \quad (18)$$

where E_R is the relative error (%), Y is the experimental value, and $\hat{Y}$ is the predicted value.

Table 8. Optimization criteria and optimal solutions obtained in the case of plywood samples drilling.

Independent Variables	Goal Settings	Minimum Value	Maximum Value	Level of Factor Importance
Drill tip angle (X_1)	In range	30	120	3
Tooth bite (X_2)		0.1	0.7	3
Drill type (X_3)		Flat	Helical	3
Dependent variables				
Thrust force (Y_1)	Minimize	38.88	459.36	3
Drilling torque (Y_2)		0.36	2.58	3
Optimal solutions				

X_1	X_2	X_3	Trust Force (N)			Drilling Torque (Nm)			D
			$\hat{Y}_1$	Y_1	E_{R1}	$\hat{Y}_2$	Y_2	E_{R2}	
30	0.1	Helical	39.42	31.44	25%	0.535	0.56	4.46%	0.955
118	0.1	Flat	191.80	196 [a]	2.14%	0.477	0.49 [a]	2.65%	0.773

a—drill tip angle was considered equal to 120°.

4. Conclusions

In this study, it was shown that the artificial neural network modeling technique could be successfully applied to predict the trust force and drilling torque during the drilling of plywood. The inputs of ANN models were drill tip angle, tooth bite, and drill type. To reveal the optimum combination of factors during the drilling of plywood, the response surface methodology was also applied in this study, together with a face centered composited design, which was completed by using the designed ANN models. Based on the applied methodology, it was revealed that the most important factor that affects the thrust force during the drilling of plywood is drill type. On the other hand, the drilling torque is most influenced by the tooth bite. A helical drill assures a lower thrust force and drilling torque than a flat drill during drilling of plywood. To increase the performance and applicability of designed ANN models, other factors should be taken into account such as the diameter of the drill bit, tool wear, properties of material, rotation speed, and direction of drilling (perpendicular or parallel to veneer layers). Moreover, a probabilistic modeling approach could represent a tool that could improve the predictability of ANN models. In addition, these kinds of models could be integrated with sensor–fusion technique to assure the transition towards the fourth phase of Industrial Revolution (Industry 4.0 concept) that is based on interconnectivity, machine learning, automation, and real-time data processing.

Author Contributions: Conceptualization, B.B. and M.I.; methodology, B.B., M.I. and S.R.; software, B.B. and S.R.; validation, B.B.; resources, S.R.; writing—original draft preparation, B.B. and M.I.; writing—review and editing, M.I. and B.B.; visualization, S.R. All authors have read and agreed to the published version of the manuscript.

Funding: This research received no external funding.

Institutional Review Board Statement: Not applicable.

Informed Consent Statement: Not applicable.

Data Availability Statement: Not Applicable.

Conflicts of Interest: The authors declare no conflict of interest.

References

1. Food and Agriculture Organization Corporate Statistical Database (FAOSTAT). Available online: https://www.fao.org/faostat/en/#home (accessed on 30 August 2023).
2. Górski, J. The Review of New Scientific Developments in Drilling in Wood-Based Panels with Particular Emphasis on the Latest Research Trends in Drill Condition Monitoring. *Forests* **2022**, *13*, 242. [CrossRef]

3. Podziewski, P.; Górski, J. Effect of the feed rate on the cutting forces during drilling in a standard hardwood plywood. *Ann. WULS–SGGW. For. Wood Technol.* **2012**, *79*, 170–172.
4. Podziewski, P.; Górski, J. Relationship between machining conditions and feed force during drilling in some wood-based materials. *Ann. Wars. Univ. Life Sci.-SGGW For. Wood Technol.* **2011**, *75*, 216–219.
5. Rogoziński, T.; Wilkowski, J.; Górski, J.; Czarniak, P.; Podziewski, P.; Szymanowski, K. Dust creation in CNC drilling of wood composites. *BioResources* **2015**, *10*, 3657–3665. [CrossRef]
6. Podziewski, P.; Szymanowski, K.; Górski, J.; Czarniak, P. Relative machinability of wood-based boards in the case of drilling–Experimental study. *BioResources* **2018**, *13*, 1761–1772. [CrossRef]
7. Agarwal, G.; Mishra, A. Fatigue Behavior of Wooden Fiber Reinforced Epoxy Composites. *AIP Conf. Proceeding* **2021**, *2317*, 020024. [CrossRef]
8. Sydor, M.; Rogoziński, T.; Stuper-Szablewska, K.; Starczewski, K. The accuracy of holes drilled in the side surface of plywood. *BioResources* **2020**, *15*, 117–129. [CrossRef]
9. Hosseini, S.M.; Peer, A. Wood Products Manufacturing Optimization: A Survey. *IEEE Access* **2022**, *10*, 121653–121683. [CrossRef]
10. Lapshin, V.P.; Turkin, I.A.; Omelechko, V.U. Automation the Processes of Wood Processing by Drilling, Due to the Development a Mathematical Apparatus for Accounting the Interrelated Feed Drive and Drive of Cutting. In *Advances in Automation III; RusAutoCon* 2021. Lecture Notes in Electrical Engineering 2021; Radionov, A.A., Gasiyarov, V.R., Eds.; Springer: Cham, Switzerland, 2021; Volume 857, pp. 14–24. [CrossRef]
11. Davim, J.P.; Gaitonde, V.N.; Karnik, S.R. An investigative study of delamination in drilling of medium density fibreboard (MDF) using response surface models. *Int. J. Adv. Manuf. Technol.* **2008**, *37*, 49–57. [CrossRef]
12. Prakash, S.; Palanikumar, K. Modeling for prediction of surface roughness in drilling MDF panels using response surface methodology. *J. Compos. Mater.* **2011**, *45*, 1639–1646. [CrossRef]
13. Wang, K.; Wang, C.; Gao, M. Investigation on Delamination Factor in Drilling Medium Density Fiberboards with Variable Feed Pressure using Multi-Spindle Drill. *Adv. Mater. Res.* **2013**, *690–693*, 2529–2534. [CrossRef]
14. Valarmathi, T.N.; Palanikumar, K.; Sekar, S.; Latha, B. Investigation of the effect of process parameters on surface roughness in drilling of particleboard composite panels using adaptive neuro fuzzy inference system. *Mater. Manuf. Process.* **2020**, *35*, 469–477. [CrossRef]
15. Ayyildiz, E.A.; Ayyildiz, M.; Kara, F. Optimization of Surface Roughness in Drilling Medium-Density Fiberboard with a Parallel Robot. *Adv. Mater. Sci. Eng.* **2021**, *2021*, 1–8. [CrossRef]
16. Bedelean, B.; Ispas, M.; Răcășan, S.; Baba, M.N. Optimization of Wood Particleboard Drilling Operating Parameters by Means of the Artificial Neural Network Modeling Technique and Response Surface Methodology. *Forests* **2022**, *13*, 1045. [CrossRef]
17. Zbieć, M. Application of neural network in simple tool wear monitoring and identification system in MDF milling. *Drv. Ind.* **2011**, *62*, 43–54. [CrossRef]
18. Jegorowa, A.; Górski, J.; Kurek, J.; Kruk, M. Use of nearest neighbors (k-NN) algorithm in tool condition identification in the case of drilling in melamine faced particleboard. *Maderas. Cienc. Tecnol.* **2020**, *22*, 18–196. [CrossRef]
19. Ispas, M.; Răcășan, S. The influence of the tool point angle and feed rate on the dynamic parameters at drilling coated particleboard. *Pro Ligno* **2015**, *11*, 457–463.
20. NeuralWare. NeuralWorks Predict. In *User Guide 2009*; Carnagie: New York, NY, USA, 2009.
21. Özşahin, S.; Singer, H. Prediction of noise emission in the machining of wood materials by means of an artificial neural network. *N. Z. J. For. Sci.* **2022**, *52*, 1–10. [CrossRef]
22. Ozsahin, S. Optimization of process parameters in oriented strand board manufacturing with artificial neural network analysis. *Eur. J. Wood Prod.* **2013**, *71*, 769–777. [CrossRef]
23. Hazir, E.; Koc, K.H.; Baray, S.A.; Esnaf, S. Improvement of adhesion strength for wood-based material coating process using design of experiment methodology. *Eur. J. Wood Prod.* **2020**, *78*, 301–312. [CrossRef]
24. Ebadi, S.E.; Ashaari, Z.; Jawaid, M. Optimization and empirical modelling of physical properties of hydrothermally treated oil palm wood in different buffered media using response surface methodology. *BioResources* **2021**, *16*, 2385–2405. [CrossRef]
25. NIST/SEMATECH. e-Handbook of Statistical Methods. Available online: http://www.itl.nist.gov/div898/handbook/ (accessed on 28 August 2023).
26. Tiryaki, S.; Malkoçoğlu, A.; Özşahin, Ş. Artificial neural network modeling to predict optimum power consumption in wood machining. *Drewno* **2016**, *59*, 109–125. [CrossRef]
27. Tiryaki, S.; Özşahin, Ş.; Aydin, A. Employing artificial neural networks for minimizing surface roughness and power consumption in abrasive machining of wood. *Eur. J. Wood Prod.* **2017**, *75*, 347–358. [CrossRef]
28. Valarmathi, T.N.; Palanikumar, K.; Latha, B. Measurement and analysis of thrust force in drilling of particle board (PB) composite panels. *Measurement* **2013**, *46*, 1220–1230. [CrossRef]

applied sciences

MDPI

Article

Handcrafted Reproduction of a 17th Century Bema Door Supported by 3D Digitization and CNC Machining

Ioannis Barboutis [1], Vasiliki Kamperidou [1,*] and George Economidis [2]

1 Laboratory of Wood Technology, Department of Forestry and Natural Environment, Aristotle University of Thessaloniki, 54635 Thessaloniki, Greece; jbarb@for.auth.gr
2 "Xilogliptiki-Wood Carving-Emilios Ekonomidis & Sons" Company, N. Efkarpia, 57013 Thessaloniki, Greece; info@xilogliptiki.gr
* Correspondence: vkamperi@for.auth.gr

Abstract: During the Ottoman period, wood had replaced marble in Orthodox churches. Between the 16th and 19th century, common, unknown craftsmen created wonderful wood-carved ecclesiastical works by hand, though many of them have been partially or completely destroyed, usually without any of their images being saved. Advances in technology now make it possible to develop a detailed digital copy of an object of our cultural heritage, creating an unaltered prototype material to be studied by all, while offering in parallel the ease of reproducing precise copies, ensuring even museum quality standards. In the present study, a corresponding attempt to record, scan and then reconstruct a 17th century "Bema Door" (located in the Holy Monastery of Hilandar in Mount Athos) is implemented, presented step-by-step and discussed, in order to highlight the challenges and potential of the applied methods. The process included on-site recording of a 3D digital scan, and then the creation of the Bema door copy, first applying Computerized Numerical Control (CNC) machining and then, with elaborate handwork by experienced wood-carvers who successfully completed the process, imparting the necessary and unique human artistic character. The current work transfers the experience on this reproduction technique applied in a Bema door element, shares significant information and details on tools, materials, process steps, etc., in order to be an inspiration and stimulus for the preservation or reproduction of other cultural heritage monuments of high importance.

Keywords: ancient; artifact; carving; culture; digital; gild; heritage; reconstruction; reproduction; structure; wood

Citation: Barboutis, I.; Kamperidou, V.; Economidis, G. Handcrafted Reproduction of a 17th Century Bema Door Supported by 3D Digitization and CNC Machining. *Appl. Sci.* **2023**, *13*, 11690. https://doi.org/10.3390/app132111690

Academic Editors: Lidia Gurau, Mihaela Câmpean and Emilia-Adela Salca

Received: 3 September 2023
Revised: 22 October 2023
Accepted: 23 October 2023
Published: 25 October 2023

1. Introduction

Wood, among its several advantages compared to other materials, is an aesthetically incomparable material, presenting a wide variety of colors, textures and designs, and in addition, it is easy to join and accept varnishes or paints, while its processing can be implemented even using only basic tools. It is certainly reasonable to say that all of these advantageous characteristics have contributed to its widespread application since prehistoric times and it has served as the fundamental material in a wide range of applications since then [1].

During the Byzantine era, the use of wood in Orthodox churches was typically found only in their different portable objects. However, during the Ottoman period wood replaced marble and became the main material used in the interior decoration of Orthodox churches. The economic power that began to appear in the enslaved Greek population from the 17th century on, with the development of trade and the production of certain special products, was reflected, better than in any other expression of social life, in the construction of new houses, mansions and especially new churches, when the Ottomans allowed them to be built [2]. Since the conquerors imposed the external appearance of these churches to be more humble and poorer compared to the mosques, the believers put emphasis on enriching the interior decoration of their churches with magnificent wood-carved structures. Therefore, the artisans had the chance to display all their artistic skills in the interior decoration and to provide them with a distinctly imposing and reverential atmosphere inside the churches, with the wooden iconostases being the most important jewels of these churches.

Monastic complexes and villages that had acquired privileges from the Ottoman power became important centers of artistic wood-carving activity since the 16th century [3]. Residents of certain villages of Epirus and West Macedonia had been specialized in wood-carving and they had undertaken the decoration of churches in Greece as well as in many Balkan countries [4]. These ordinary, unknown carpenters, usually working under very tough conditions though with passion and a high sense of aesthetics, created wonderful wood-carved iconostases and other ecclesiastical objects by hand from the 16th to the 19th century. These authentic works of folk art made of wood nowadays constitute unique cultural heritage monuments of the post-Byzantine period.

However, in addition to its many benefits, wood also presents a number of drawbacks. As a lignocellulosic, biological material, it can burn or rot, and when it absorbs moisture it becomes more vulnerable to attack of bacteria, fungi and insects [5]. Taking into account the aforementioned disadvantages of wood, mainly resulting from its biological nature, combined with the environmental conditions prevailing in Greece that are favorable to its biological degradation, as well as the high number of wars and lootings that the country has suffered from in the past, it is not surprising that the archaeological findings concerning wooden objects and structures in Greece are really few [6]. Even the magnificent iconostases and other wooden artifacts of the post-Byzantine period did not remain unharmed by the impact of gradual deterioration. Many of them have been completely destroyed, with the loss of some valuable works of art of the past, usually without even a single image of them being saved, resulting in a great loss to civilization. Additionally, those that are still preserved have suffered varying degrees of damage, and there is always the risk of their partial or total destruction and loss [7]. The access to and study of these valuable wooden objects of cultural heritage is, most of the times, difficult or even impossible for ordinary citizens, as well as for scientists who are not involved in the field of archaeology. This difficulty applies in cases when they remain in their original places of installation, but also when they were transferred to museum spaces.

Nevertheless, even humble works of art reflect the culture of people, and humanity should consider them as a common heritage of all. Therefore, we have a duty to preserve them and pass them on to future generations, preserving all their richness and authenticity. Therefore, effective and active measures should be taken to protect them, since cultural heritage is increasingly threatened by destruction, unfortunately not only by the common causes of deterioration [8].

Nowadays, the development of technology makes it possible to create a detailed digital copy of a monument of our cultural heritage, thus generating an unaltered study material for all, but also offering ease in the production of precise copies of them [9,10]. Currently, the E.U., in the Horizon Europe program, funds the Cultural Heritage Cloud initiative which aims to add a new digital dimension to the preservation, conservation, restoration and enhancement of cultural heritage by providing cutting-edge technologies for the digitization of artefacts and the research of works of art [11].

Digital technologies have been adopted in various fields related to the museum experience and digital preservation, as well as the digitization and study of archaeological artifacts [12]. In fact, over the past decade technological advances in the fields of portable electronics have revolutionized the use and range of applications of tablet computers, smart phones and mobile devices [12]. With these opportunities provided by the ongoing technological advancement, academics have considered several techniques to create more accurate and convenient 3D digital copies of cultural heritage items of value.

For the continuous 3D scanning of small- and medium-sized heritage objects, Kesik et al. [13] proposed the "3DScaMOTO" methodology that achieved scanning results of good quality. A handheld Artec EVA 3D scanner (Senningerberg, Luxembourg) with structured light was used for the scanning, which offered precision of 50–100 microns and preserved photorealistic texturing. This methodology would contribute to successfully overcome scanning challenges such as an object being movable but with a glossy surface, an object being not movable but of good accessibility, and at last, an object being not movable combined with bad accessibility. Kochov and Stoleska [14] described the methodology they applied to the woodcarving digitization and reproduction process through practical examples. The creation of the digital copy of one wood-carving was implemented using NextEngine 3D Scanner (NextEngine, Santa Monica CA, USA), while the other two were created using the principle of photogrammetry together with the AutoDesk ReCap program v.23.0.0.258. The reproduction of the wood-carvings was carried out using Computerized Numerical Control (CNC) machines ("Uniteam professional 5 axis", Uniteam, Grgar, Slovenia), as well as 3D printing, while mentioning the cost of equipment and materials. Specifically, they developed a methodology that constitutes an example of reversible engineering for wood-carvings: the digitalization is implemented with one of the available tools, the carving is being "cleaned" in one of the programs found easily on the market, then the section is transferred to appropriate CAD/CAM programs and the 3D printer and/or CNC machine are allowed to do the "magic" work, which constitutes a fusion of engineering with art that is of great historical and cultural importance. Učakar et al. [15] created 3D graphical representations of wooden sculptures with photogrammetry (converting photos into 3D digital models) that were used for 3D printing reproduction, jewelry production, etc., while stating that the new technologies cannot replace real works of art, but that 3D representations offer significant help to conservators and, in addition, allow us to preserve at least the cultural and historical values of those art objects that cannot be preserved for various reasons. Neamțu et al. [16] presented the methodology applied in the digitization and digital restoration of imperial gates from wooden churches in Transylvania, to show the degradation of these cultural heritage objects over time. The procedure for digitally restoring the object to its pristine state included X-ray fluorescence (non-destructive elemental analysis (XRF) with a handheld Bruker spectrometer), FTIR spectroscopy measurements (with a Jasco 6100 spectrometer, Easton, MD, USA) and scanning and digitization using a Creaform Go!SCAN 50 structured light scanner (Go!SCAN, AMETEK, Bangkok, Thailand). Carvajal et al. [17] used the Mu3D digitization system based on digital photogrammetry to create virtual museums, applying digital photogrammetry and computer 3D modeling. They concluded that the same virtual environment can be arranged either for different collections or the same collection in a different layout that fits different groupings of objects that respond to different contexts, uses and users.

For the copyright protection of 3D digitized works of art, it is proposed, by Vasiljević et al. [18], to add some unique local indistinguishable errors, so that in case of any misuse of the 3D model with digital error, such as if someone would try to print a 3D model of this sculpture, the authors who made the sculpture available to the public could easily determine the originality of any copy of the sculpture that appears anywhere in the world.

Ramm et al. [19] presented a method they had developed for in situ high-resolution 3D digitization of cultural heritage objects, with a strong focus on portability. The digitization system consists of a 3D structured light sensor module, a connected DSLR camera and a data processing software on a separate workstation, with the workstation being recommended to be a laptop so that the entire system is portable.

Adamopoulos et al. [20] compared the quality of 3D digitization of cultural heritage objects from a variety of photogrammetric software, as well as photo-capturing equipment of various characteristics and prices and scanners of various technologies.

The digitization of exhibited objects, together with the digitization and information enrichment of the museum building, provides the possibility to improve the conditions of management and enjoyment of the museum, such as making a part of the heritage that is not accessible so far accessible to the public through the virtualization of the exhibits stored in the repositories [21].

The aim of the current work is to investigate the possibility of reproducing a precise copy of a wood-carved object of medium size and considerable cultural value, which may continue to be in its place of use or be exhibited in a museum space, taking advantage of the possibilities offered by the development of digital technology through 3D scanning, combined with traditional wood-carving handwork. The applied methodology is discussed in terms of advantages, challenges and future potential utilization.

2. Materials and Methods

For the purposes of this study towards the creation of a precise copy, a unique Bema door characterized by rich wood-carved decoration, was used as the model object (Figure 1).

Figure 1. View of the Bema door dated from 1615 AD found in the museum area of the Hilandar Monastery of Mount Athos.

Specifically, the particular Bema door constitutes a part of an iconostasis, constructed in the year 1615 and belongs to the Hilandar Monastery of Mount Athos (Figure 2), which only a few people have access to, as there are several restrictions on the access to Mount Athos (northern Greece) and the respective monasteries, whereby women are not allowed access at all. The main aim of this project was the reproduction and creation of a precise copy of this original Bema door, which at the moment is exhibited in a museum of the Hilandar Monastery.

The Bema door consists of two door leaves, which, in addition to the impressive and dense carving they contain, are also decorated with additional high-relief decorative parts, also of similar rich carving, which protrude considerably from the surface of the door leaves and are attached to them using adhesive, as well as the semi-cylindrical "astragal" placed at the junction of these two door leaves. The Bema door also includes flat sections covered by painted decoration. Its dimensions are approximately 150 $\times$ 105 cm, with its upper side having an arched shape [22]. The wood-carving of a Bema door leaf is considered shallow by wood-carvers when the depth of the carving does not exceed 2–2.5 cm, as is the case for this specific Bema door. An important feature of the Bema door is the gilding that has been applied to its entire surface. Gilding is a treatment applied, mainly for aesthetic reasons, to wood-carved objects after the completion of their construction, offering to them an impressive appearance, though it can alter the original details of the wood-carving to a small or large extent due to the several coating layers applied. In particular, the gilding, as described in detail in an 18th century manuscript, included the following individual stages [10,23]: (a) the production of animal adhesive, using this for coating of the wooden structure (mixed with water) and rubbing it with a brush for better penetration into the deeper parts of the carvings; (b) mixing gypsum plaster with water and adhesive, followed by coating using a brush (five–six layers) and sanding to smooth any irregularities; (c) the construction and coating of two or three layers of "bole", i.e., a mixture of clay and other materials (ochre or constantinopolitan ochre, mercury, soap, egg white, etc.), different depending on the case and the desired color; (d) spreading the surface with raki (using a sponge, soft brush or cotton) and, finally, carefully placing the layers of gold. It is logical that these multiple coating layers on the wood-carved Bema door adversely affect the process of producing a precise copy of it.

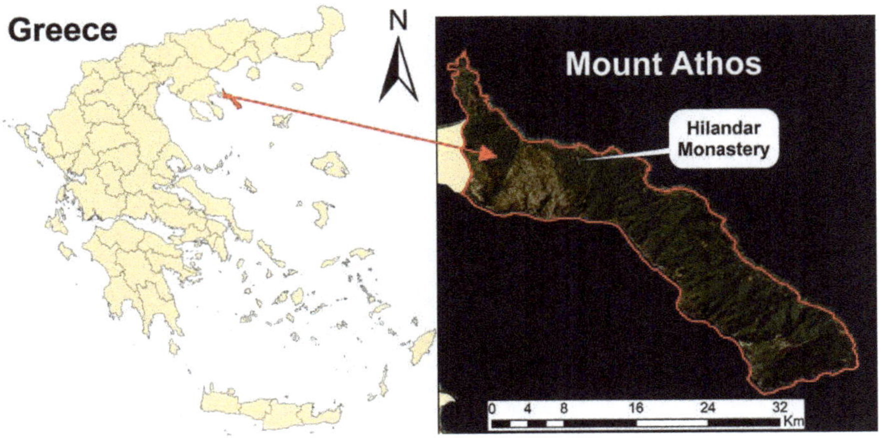

Figure 2. Location of Hilandar Monastery on the Holy Mount Athos Peninsula.

Identification of the wood species that the specific Bema door is originally made of has not been implemented so far. Verbal estimations indicate the possibility of it being made of walnut wood. However, it should be mentioned that in Greece there is no wooden object of archaeological and cultural value, exhibited in museums, with any relevant reference to the wood species.

For the scanning process, a portable Artec Space Spider 3D Scanner (Figure 3A) was used, which is based on blue light technology and is ideal for capturing small objects or complex details in high resolution. All the work of scanning and final sculpting was carried out by a family wood-carving business (Thessaloniki, Greece). Multiple scans were obtained from different directions in order to capture all the features, and then a point cloud was created, which was processed using a special CAD (Rhinoceros 5) and CAM (Rhinocam 2015) software to produce a precise 3D digital copy (Figure 3B), in order to be used in the machining in an automated way, by means of a CNC machine, a key part in the process of reproduction of a wood-carved copy of the original Bema door that followed. The final digitization processing was implemented using the Uniterm 3D modeling Rhinoceros 5 commercial software. The 3D model generated by the scanner initially corresponded to a 2 GB file (which was decreased to 180 MB in the following, to be more easily processed) with 0.015 mm initial accuracy (in the following the accuracy was decreased to 0.05 mm for the same reason). Smoothing and abnormal triangles filter options were applied during the process. The reverse engineering software "Geomagic Design X" (version 2015) was used for the filling of gaps and holes in the scanned 3D model. There were indeed specific gaps in the scanning results, attributed to the given conditions of the scanning process (limited time; location, since it is permanently exhibited in a monastery museum room; and other difficulties and limitations in the scanning process primarily brought about by the monastery). A resampling process was not applied given the circumstances of limited time and access to the exhibition place of the Bema door.

(A) (B)

Figure 3. Scanning process of the original Bema door (**A**) and digital scan result (**B**).

It should be mentioned that since the current stage of copy production of the Bema door involved the wood-carving of only the parts of it that are on the surface of its main wooden body, all the additional decorative wood-carved elements attached to it were removed from the digital copy. The scanning duration for the whole Bema door was about 8 h (1 day), while the processing time of all these scan data (until the reproduction phase in CNC was ready) was about 1 week.

The CNC machine used for mechanical carving was a ROVER K type (Biesse, Pesaro, Italy) with the following characteristics: the number of controlled axes was 3, the total number of drilling spindles 16 and the total number of tool changer positions was 16. For the machining of the specific Bema door, 2 machining drill bits were used, 12 mm in diameter, with a step of 3 mm for the initial rough machining and 1.6 mm in diameter with a step of 0.4 mm for the elaborate work of the details.

In order to reproduce the Bema door, lime wood (*Tilia cordata*) was initially used, which is light in color, has a fine and even texture and is soft enough to carve while being resistant to tearing [24]. Its ease of processing, which derives from these favorable characteristics, contributes to its classification as one of the most popular wood species intended for carving, as noted already during the Middle Ages [25]. However, the Greek philosopher Theophrastus (371–287 BC) mentioned in ancient times that it was known that the wood of lime trees is the most refined and easily processed of all wood species (*Historia plantarum*, V, 5-1). For this reason, lime constitutes the most popular wood species chosen by beginners in wood-carving. It should be noted, however, that lime wood is not resistant to biological attacks [24]. More specifically, according to the EN 350 standard [26], it is classified in class DC 5 as non-durable to fungal attacks and in class S as of low natural durability and being susceptible to termite attacks and action. Therefore, this wood species is not recommended to be used in high-value constructions such as iconostases. Nevertheless, in this particular case, the reproduction was first carried out with the soft lime wood in order to render the carving of details more accurate and easy, while in the second stage, a wood species of higher density and hardness was used. In both cases of wood species, for the construction of each Bema door leaf, wood boards of 10–12 cm width were used and were glued together using D3 PVA adhesive (AKFIX, Riga, Latvia).

In particular, the process of reproduction began with carving on the CNC machine and then the "rough" result of the mechanical carving was corrected and further improved by hand-made work carried out by experienced wood-carvers. The resulting new lime wood-carving was then scanned again, following the same process as the scanning of the original object. By scanning the copied object in a laboratory infrastructure this time, the opportunity for a more detailed and cleaner recording of its sculptural details was provided. Since the new 3D digital copy resulting from the scan required additional processing and shaping because of its adjustment to a new iconostasis, it was decided to make the wood-carved copy of a different, smaller width (by 10 cm) compared to the original one. Additionally, since the original Bema door was gilded and, thus, had various layers of materials on its surface, the digital copy was cleaned and corrected thoroughly in its various decorative patterns, a step which requires experience and deep knowledge of the use of designing and reproduction programs.

Oak constitutes the oldest wood species that is widely known to be used in wood-carving, especially the species *Quercus robur*. Oak wood is hard and has an attractive brown color with a smooth and shiny surface and very beautiful grain pattern known as "silver grain". It is a good choice of wood species for use by experienced wood-carvers, though, because of its hardness, it is not that suitable for beginners. Its sapwood part is very susceptible to insect attacks, while its heartwood is considered very durable under any conditions [27]. According to the EN 350 standard [25], it is classified in class DC 2 as "durable to fungal attack" and in class M as "moderately durable" concerning the natural durability against termites' action; therefore, it is widely used in wood-carving elements of high value, such as construction of iconostases.

At the end of the reproduction process, the Bema door copy made of oak wood was cautiously sanded using No 120 and 150 grain sandpapers to smooth the surfaces, and then the whole copy of the Bema door was disinfested by being exposed to phosphine. The next step to follow was the process of gilding, leaving only the flat areas intended to be painted vacant (iconography creation).

3. Results and Discussion

The CNC machining of the lime wood resulted in the overall appearance of the wood-carving, it bearing all the embossed features of the original Bema door, but with the surface quality being quite rough and without the final depth of carving achieved, details which are necessary since they enable the wood-carver to process it (Figure 4).

Figure 4. Detailed view of the CNC machining process of the lime wood.

Moreover, due to the gilding that exists on the original Bema door, many of its wood-carved details had been blunted to some extent. Since gilding was to be applied to the final copy of the Bema door in one of the next stages, it is evident that, by preserving these faded features of the original gilded Bema door in the final gilded copy as they were, there was a possibility that many of its wood-carved details would no longer be apparent. In addition, in an old wood-carving over 400 years old, such as the studied one, there will certainly be some micro-cracks or other alterations/deteriorations of the material surface and probably of the design/shape. The handmade work that followed resulted in the highlighting of the details of the wood-carving and refining of its surface configuration (Figure 5).

Figure 5. Detailed view of the finishing work on the lime wood implemented by wood-carvers' handiwork.

It is worth mentioning that the processing of the high-relief decorative parts of the Bema door was separated from the shaping work of its superficial wood-carved parts, which preceded it (Figure 6). Moreover, in the original Bema door all its high-relief parts are added and glued on the surface of the Bema door leaves.

Figure 6. Refinement of the wood-carving in the lime wood after applying handiwork by wood-carvers.

The completed Bema door that resulted from the processing of lime wood was a precise copy of the original one (Figure 7) and constituted the new model item for the construction of the final copy in oak wood. This new final model, now given the convenience of time, laboratory infrastructures and controllability potential of the result, was scanned in detail in order to be able to use the new digital copy for the CNC machine to produce one or more copies using any wood species desired. In this particular case, as mentioned, oak wood was chosen since the copy is to be used in an iconostasis, in a similar place to that of the original Bema door. The lime wood copy could be used as an exhibition piece, though, due to its low biological resistance and hardness, it is not recommended for any functional use.

The new digital copy could also be used for the production of copies using a 3D printer, especially in smaller dimensions or the reproduction of individual parts of it, which could be made available for sale for the financial support of the monastery.

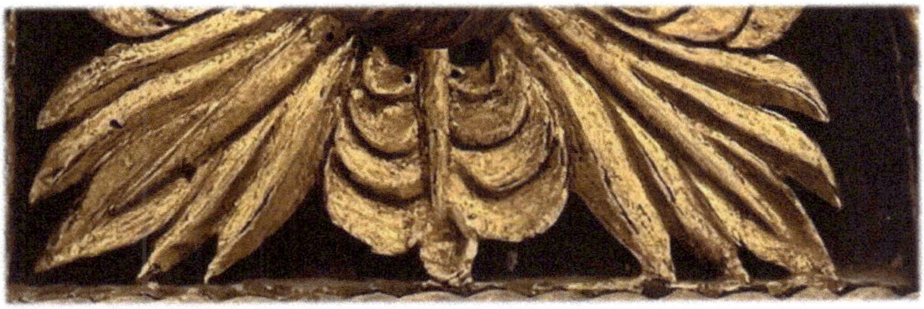

Figure 7. Details of the original gilded Bema door.

The CNC-machined oak wood-carving (Figure 8) was improved and refined in its carving details by hand-crafting by an experienced wood-carver (Figures 9–11). Human contribution is particularly important, since it is the part that provides the final details' configuration to a wood-carving and, moreover, the appearance of a hand-made wood-carving piece similar to the original one. During the reproduction process, an attempt has been implemented, in order any potential signs of destruction, wear or tear caused to the original Bema door during its 408-year life (since the construction year) were not reproduced.

Figure 8. View of the CNC-machined wood-carving details in the oak wood.

Figure 9. View of the refined by wood-carvers' wood-carving details in the oak wood.

Figure 10. Copy of the whole Bema door leaf (**left**) made of oak wood and the whole image of the final version of the Bema door copy in oak wood bearing all the attached decorative parts (**right**).

Figure 11. A pattern of the original gilded Bema door (**left**) and its corresponding copy in oak wood (**right**).

The refining of the piece is always conducted by removing material from the main wooden substrate by wood-carvers. If there are several missing parts in the original artwork (here only few missing parts were detected), they are usually digitally corrected during the 3D model creation process, prior to the manufacturing. Sometimes this process is implemented while also taking into account previous evidence, such as images or drawings of the initial design, the logical continuity of the design or similar patterns found replicated in different places on the same art-piece, etc. The wood-carver contributes significantly to this work as well, and especially, does not leave any destroyed or missing points/parts without the necessary correction, based again on the initial design of the original work.

The wood-carving tool traces do not constitute a problematic issue, rather a desirable one that, of course, cannot be precisely replicated (even by the same wood-carver), although the tools used in wood-carving have been more or less the same for thousands of years. In general, the wood-carver in the last step of reproduction process bestows it with this special unique appearance of the handmade work. The contribution of the human factor (the wood-carver's view and artistic skills) may have slightly varied some curves or other points, always respecting the initial design and detailed images of the original Bema door elements.

The reproduction of precise copies of old wood-carved icons or parts of them, exclusively in the traditional way, is a very difficult process, as it requires many hours of laborious specialized work of an experienced and talented wood-carver and correspondingly, very high cost [28]. The machining, combined with the digitization of the original object and the correct reproduction of the wood-carving patterns in the produced copy, contributes significantly to the reduction in the wood-carver's working time and the corresponding total cost of reproduction [28]. Of course, the level of resolution and choice of material determine the final cost of the copy as well [29].

Particularly, for each of the two leaves of the Bema door, the machining process on the CNC machine lasted about 7 h (keeping in mind the step of 3D digitization that lasted 4 h for each leaf and the scanning data processing that lasted one week for the whole Bema door), followed by four more days of manual work by the wood-carvers to refine it. Without the CNC machine, applying only the manual wood-carving, it would take twice as long.

In the present project, the enormous importance of the coexistence of human contribution and CNC machine involvement towards a successful, easy, quick and of low-cost reproduction of high cultural value and heritage wood-carvings was highlighted.

4. Conclusions

From the 16th to the 19th century, unknown folk craftsmen without special education and working under very difficult conditions, created, by hand, amazingly designed and beautiful wooden church objects, which are unique cultural heritage monuments of the post-Byzantine period. The chemical composition and biological resistance of wood, combined with many other unfavorable factors led to the destruction or significant degradation of many of them. Additionally, the access to these valuable wooden objects of cultural heritage is often difficult or even impossible for the general public.

The development and application of 3D imaging in many fields has motivated many researchers to study the utilization of the potential of this technology in the field of cultural heritage. Three-dimensional (3D) digitization can be used in a variety of ways, such as in the production of digital files, remote study, but also to create precise copies of various constructions, contactless and with great precision.

Digital files combined with mechanical processing now offer a unique opportunity to reproduce precise copies of old wood-carved elements or parts of them faster and more economically than the traditional way which is a very difficult, laborious process, as it requires many hours of specialized work and consequently, a very high cost. The location or size of the original wood-carving were not proven to be a restriction.

The reproduction, in slightly smaller dimensions, as required of an exhibition gilded, wood-carved Bema door from the 17th century carried out in the context of this work, was based on the modern technology of digital copying and machining, while it was completed with hand-made processing by an experienced wood-carver, thereby providing an excellent detailed result in copying its wood-carved designs, as well as the appearance of a hand-made result, such as the appearance of the prototype, while significantly reducing the total time of wood-carving work.

For the successful completion of such a project, the choice of suitable wood species for each step during the reproduction of copies (lime and oak) was of high importance, and, therefore, the in-depth knowledge of each wood species' properties and performance, as well as the excellent knowledge of the digitization, use of scanning design programs and the mechanical equipment for the wood processing. Finally, it seems to be a matter of utmost importance to support and preserve the "woodcarver" profession, as well as provide proper training to future wood-carvers, during which the thorough knowledge of wood structures and properties should play a prominent role, in order to contribute to the preservation of such high-value monuments and artifacts, through reproduction of high accuracy.

Author Contributions: Conceptualization, I.B. and G.E.; methodology, G.E.; software, I.B. and G.E.; validation, I.B. and G.E.; formal analysis, I.B., V.K. and G.E.; investigation, G.E.; resources, G.E.; data curation, I.B., V.K. and G.E.; writing—original draft preparation, I.B; writing—review and editing, V.K.; visualization, I.B. and G.E.; supervision, I.B.; project administration, I.B.; funding acquisition, G.E. All authors have read and agreed to the published version of the manuscript.

Funding: This research received no external funding.

Institutional Review Board Statement: Not applicable.

Informed Consent Statement: Not applicable.

Data Availability Statement: Not applicable.

Acknowledgments: We express sincere gratitude to the Abbot of Hilandar Monastery on Mount Athos, Archimandrite Methodius, for the permission to digitally copy the Bema door that is exhibited in the museum of the monastery. Additionally, we express our thanks to the group of wood-carvers of the "Xilopliptiki" company, who worked on the refinement and finishing steps, and their contribution to the whole process of the Bema door reproduction in the frame of this project.

Conflicts of Interest: Author George Economidis is employed by the company "Xilogliptiki-Wood Carving-Emilios Ekonomidis & Sons". The remaining authors declare that the research was conducted in the absence of any commercial or financial relationships that could be construed as a potential conflict of interest.

References

1. Agnantopoulou, E.; Barboutis, I.; Kamperidou, V. Wood Utilization in Windmill Mechanisms on Sikinos Island (Greece). *Appl. Sci.* **2023**, *13*, 9216. [CrossRef]
2. Moutsopoulos, N. *Macedonia Thessaloniki: The Preservation Problem of the Traditional Settlements of Macedonia (Greece)*; Etaireia Makedonikon Spoudon: Thessaloniki, Greece, 1980.
3. Makris, K. *Ecclesiastical Wood Carvings*; Apostolic Ministry of the Church of Greece: Athens, Greece, 1982.
4. Moutsopoulos, N. *Churches in the Prefecture of Pella*; Institute for Balkan Studies: Thessaloniki, Greece, 1989.
5. Shumka, L.; Marku, M.; Shumka, S. The study of biological decay with church wooden materials in Prespa lake area, Albania. *Nat. Monten.* **2010**, *9*, 827–831.
6. Barboutis, I.; Koutis, P.; Kamperidou, V.; Chryssoulaki, S. Knowledge and quality of woodworking in ancient Greece revealed from a simple finding. *ProLigno* **2019**, *15*, 144–151.
7. Barboutis, I. The wooden econostases of the post-Byzantine churches of Thessaloniki. In Proceedings of the 7th Scientific Symposium on Modern Greek Ecclesiastical Art Modern Greek Ecclesiastical Art, Athens, Greece, 31 March–2 April 2023.
8. UNESCO, World Heritage Convention 1972. Available online: https://whc.unesco.org/en/conventiontext/ (accessed on 1 January 2022).
9. Pieraccini, M.; Guidi, G.; Atzeni, C. 3D digitizing of cultural heritage. *J. Cult. Herit.* **2001**, *2*, 63–70.
10. Agnantopoulou, E.D. Wood in the Cultural Heritage of the Cyclades: Types, Uses, Protection. Doctoral Dissertation, Aristotle University of Thessaloniki, Thessaloniki, Greece, 2022.
11. European Commission, Directorate-General for Research and Innovation, European Collaborative Cloud for Cultural Heritage—Publications Office of the European Union. 2022. Available online: https://research-and-innovation.ec.europa.eu/research-area/social-sciences-and-humanities/cultural-heritage-and-cultural-and-creative-industries-ccis/cultural-heritage-cloud_en (accessed on 1 February 2020).
12. Barmpoutis, A.; Bozia, E.; Fortuna, D. Interactive 3D digitization, retrieval, and analysis of ancient sculptures, using infrared depth sensors for mobile devices. In *Universal Access in Human-Computer Interaction. Access to the Human Environment and Culture, Proceedings of the 9th International Conference, UAHCI 2015, Held as Part of HCI International 2015, Los Angeles, CA, USA, 2–7 August 2015*; Springer International Publishing: New York, NY, USA, 2015; Part IV, Number 9; pp. 3–11. Available online: https://link.springer.com/chapter/10.1007/978-3-319-20687-5_1 (accessed on 1 January 2023).

13. Kęsik, J.; Żyła, K.; Montusiewicz, J.; Miłosz, M.; Neamtu, C.; Juszczyk, M. A Methodical Approach to 3D Scanning of Heritage Objects Being under Continuous Display. *Appl. Sci.* **2023**, *13*, 441. [CrossRef]
14. Kochov, A.; Stoleska, L. Methodology of Reverse Engineering Implemented in the Process of Digitalization and Conservation of Wooden Carvings. *Teh. Glas.* **2022**, *16*, 121–128. [CrossRef]
15. Učakar, A.; Sterle, A.; Vuga, M.; Trček Pečak, T.; Trček, D.; Ahtik, J.; Košak, K.; Muck, D.; Tomc, H.G.; Kočevar, T.N. 3D Digital Preservation, Presentation, and Interpretation of Wooden Cultural Heritage on the Example of Sculptures of the FormaViva Kostanjevica Na Krki Collection. *Appl. Sci.* **2022**, *12*, 8445. [CrossRef]
16. Neamțu, C.; Bratu, I.; Măruțoiu, C.; Măruțoiu, V.C.; Nemeș, O.F.; Comes, R.; Bodi, Ș.; Buna, Z.; Popescu, D. Component materials, 3D digital restoration, and documentation of the imperial gates from the Wooden Church of Voivodeni, Sălaj county, Romania. *Appl. Sci.* **2021**, *11*, 3422. [CrossRef]
17. Carvajal, D.A.L.; Morita, M.M.; Bilmes, G.M. Virtual museums. Captured reality and 3D modeling. *J. Cult. Herit.* **2020**, *45*, 234–239. [CrossRef]
18. Vasiljević, I.; Obradović, R.; Đurić, I.; Popkonstantinović, B.; Budak, I.; Kulić, L.; Milojević, Z. Copyright protection of 3D digitized artistic sculptures by adding unique local inconspicuous errors by sculptors. *Appl. Sci.* **2021**, *11*, 7481. [CrossRef]
19. Ramm, R.; Heinze, M.; Kühmstedt, P.; Christoph, A.; Heist, S.; Notni, G. Portable solution for high-resolution 3D and color texture on-site digitization of cultural heritage objects. *J. Cult. Herit.* **2022**, *53*, 165–175. [CrossRef]
20. Adamopoulos, E.; Rinaudo, F.; Ardissono, L. A critical comparison of 3D digitization techniques for heritage objects. *Int. J. Geo-Inf.* **2021**, *10*, 10. [CrossRef]
21. Lo Turco, M.; Calvano, M. Digital museums, digitized museums. In Proceedings of the 1st International and Interdisciplinary Conference on Digital Environments for Education, Arts and Heritage: EARTH, Brixen, Italy, 5–6 July 2018; Springer International Publishing: Cham, Switzerland, 2019; Number 1; pp. 387–389. Available online: https://link.springer.com/chapter/10.1007/978-3-030-12240-9_41 (accessed on 1 September 2020).
22. Nikonanos, N. Bema doors, Chelandari Monastery. In *Treasures of Mount Athos*, 2nd ed.; Holy Community of Mount Athos, Ministry of Culture Museum of Byzantine Culture, Organization for the Cultural Capital of Europe Thessaloniki: Thessaloniki, Greece, 1997; pp. 270–272. Available online: https://www.elpenor.org/athos/en/e218bh5.asp (accessed on 15 October 2020).
23. *Dionysios of Fournas. Interpretation of Painting Art*; Papadopoulos-Kerameus, Athanasios: Petroupoli, Greece, 1900. Available online: https://anemi.lib.uoc.gr/metadata/2/c/a/metadata-141-0000170.tkl (accessed on 1 January 2023).
24. Barboutis, I.; Vasileiou, V.; Mitani, A.; Kamperidou, V. Effects of short time thermal treatment on some properties of lime wood. *ProLigno* **2011**, *7*, 39–49.
25. Eaton, E.; Caudullo, G.; De Rigo, D. Tilia cordata, *Tilia platyphyllos* and other limes in Europe: Distribution, habitat, usage and threats. In *European Atlas of Forest Tree Species*; Publications Office of the EU: Luxembourg, 2016; pp. 184–185.
26. *EN 350*; Durability of Wood and Wood-Based Products—Testing and Classification of the Durability to Biological Agents of Wood and Wood-Based Materials. CEN: Brussels, Belgium, 2016.
27. Hasluck, P.N. *Manual of Traditional Wood Carving*; Dover Publications, Inc.: New York, NY, USA, 1977. Available online: https://archive.org/details/Manual_of_Traditional_Wood_Carving_by_Paul_N._Hasluck (accessed on 1 September 2022).
28. Santoš, I.Z.; Šokac, M.; Korolija-Crkvenjakov, D.; Kosec, B.; Soković, M.; Budak, I. Reconstruction of 3d models of castsculptures using close-range photogrammetry. *Metalurgija* **2015**, *54*, 695–698.
29. Wachowiak, M.; Karas, B.V. 3D Scanning and Replication for Museum and Cultural Heritage Applications. *J. Am. Inst. Conserv.* **2009**, *48*, 141–158. [CrossRef]

applied sciences

MDPI

Article

Application of Additive Manufacturing Technology for Chair Parts Connections

Antoniu Nicolau [1], Mihai Alin Pop [2], Sergiu Valeriu Georgescu [1] and Camelia Coșereanu [1,*]

1 Faculty of Furniture Design and Wood Engineering, Transilvania University of Brasov, B-dul Eroilor, nr. 29, 500036 Brasov, Romania; antoniu.nicolau@unitbv.ro (A.N.); sergiu.georgescu@unitbv.ro (S.V.G.)
2 Faculty of Materials Science and Engineering, Transilvania University of Brasov, B-dul Eroilor, nr. 29, 500036 Brasov, Romania; mihai.pop@unitbv.ro
* Correspondence: cboieriu@unitbv.ro

Abstract: Additive manufacturing (AM) is a potential technology to be applied for replacing wood joints in the furniture industry. Overall, 3D-printed connectors have a lot of advantages in chair construction: the possibility of improving the design by using colored materials and new concepts with no limits of shape and size, avoiding the use of synthetic materials such as adhesives and finishing materials. The present research shows the results of the research conducted on 3D-printed connectors for a chair leg-stretchers joint. Larch (*Larix decidua* Mill.) wood was used for chair components and selective laser sintering (SLS) technology was employed for the 3D-printing connector designed to assemble the chair elements. Diagonal tensile and compression tests were performed on the L-type joints with 3D-printed connectors and the bending moments were determined and compared to the results obtained after testing the leg-stretchers mortise–tenon joints made of larch wood, taken as a reference. Micrograph investigation was finally performed on the broken areas of the 3D-printed connectors as a result of mechanical testing failures. The results show that the SLS printing technology applied to build the connector does not ensure enough strength to the assembly, and changes in printing parameters are needed for further research, in terms of laser power and scan speed.

Keywords: additive manufacturing; 3D printing; chair parts; connector

Citation: Nicolau, A.; Pop, M.A.; Georgescu, S.V.; Coșereanu, C. Application of Additive Manufacturing Technology for Chair Parts Connections. *Appl. Sci.* **2023**, *13*, 12044. https://doi.org/10.3390/app132112044

Academic Editor: Manoj Gupta

Received: 25 September 2023
Revised: 23 October 2023
Accepted: 30 October 2023
Published: 4 November 2023

1. Introduction

Nowadays, additive manufacturing (AM) is gaining more and more attention in the industrial field for rapid prototyping, rapid manufacturing, mass customization and mass production [1]. Based on CAD modelling, this technology is supported by a diversity of professional software able to create bold shapes, with no limits of creativity and has the potential to transform new ideas and methods outside of 'conventional' industry norms [2].

One of the methods applied in additive manufacturing is selective laser sintering (SLS). Powder raw material, such as nylon, bio-plastics, ceramics, wax, bronze, stainless steel, cobalt chrome and titanium, is sintered or fused by the application of a laser beam for each layer specified by the 3D model [1,3,4]. The minimum layer thickness for this technology is 0.1 mm and the advantage of applying this method is the fast build speed, but the quality of the surface is appreciated as an average one [4]. According to the findings in the literature, the process input parameters, such as scanning rate, laser power, hatching distance or scanning pattern, can cause defects in the structure of the printed part or an insufficient density of the sintered material [5]. During the SLS processing, sintered, semi-sintered/semi-melted or completely melted surfaces can be formed [6]. The most important factor affecting the quality of the SLS printed surfaces is the scanning pattern, varying from the raster in X or Y direction, alternating in the X and Y direction, spiral, fractal, or scanning of successive sectors [5]. Applying a raster in alternating in the X and Y direction resulted in a higher fractional density comparing to the use of the raster strategy in the X or Y direction [7]. On the other hand, the particle fusion can be improved by high energy density, with a benefit on the mechanical properties of the final part. The energy radiated into the powder bed produces the melting and the layers adhesion, so the higher the energy, the better the connection between the two successive sintered layers, and the sintered part can be generated without layer delamination [8].

Despite its advantages, the ability of SLS to process polymers is affected by some defects, such as the porous structure and balling effect [6,9]. The causes of these defects were found to be the high laser power and the fast scan speed and interruptions near fused powders. The balling defect affects the uniformity of the powder layer deposited on the previously sintered one and often leads to cracks, porosity and a rough surface [9]. Another particularity of this technology is that the final product needs a long cooling time and also, the non-fused powder remaining on the part exterior surface has to be removed.

Accompanying the Industry 4.0 digital revolution, additive manufacturing (AM) is considered a key technology [10]. At first being applied for fast prototyping in medicine, aerospace and automotive industry, it became rapidly widespread in other sectors, including the furniture industry. There are three types of applications where 3D-printing technology is used in furniture manufacturing, as follows: the entire piece of furniture is manufactured in 3D printing, some parts of the furniture are 3D printed, or only connection parts or fasteners are produced using this technology.

The variety of materials available today for AM and the corresponding 3D-printing technologies and printers are the source of inspiration for freedom design with complex geometries. In the furniture industry, the most used materials are thermoplastic materials in the form of filaments, used in the case of applying fused filament fabrication (FFF), or fused deposition modeling (FDM) [10–14], and thermoplastic powders, used when applying selective laser sintering (SLS) technology [10]. PLA and ABS are the most commonly used materials to manufacture the filaments for AM in furniture manufacturing, but they are confronted with several drawbacks: PLA has a poor toughness, whilst the ABS is damaged by sunlight [15], for which reasons various composite materials are being developed to replace them [16,17]. In addition to the above remarks, another drawback of FFF technology is that for highly complex parts, printing supports are necessary, which in certain cases are either difficult or even impossible to remove. Instead, SLS is an expensive technology but has the advantage of using a large variety of materials comprised in thermoplastic, ceramic, glass or metal powders, and has no limitations in producing geometrically complex parts for artistic purposes and pieces of furniture without using supports [18]. The powder that is not melted by the laser beam behaves like the supports made with FFF technology, only it is easily removed by brushing or blowing with air. Depending on the needs of the product designer, if you only want a quick check of the shape and the physical-mechanical properties are not very important, then FFF technology is chosen, and if you want a prototype with good properties, high precision, etc., and even if the time to obtain it is longer (compared to FFF technology—but lower than by other classic methods) then SLS technology can be chosen [19]. Using SLS technology, it is possible to realize microstructures at the scale of tens of microns and based on the elastic behavior of the aperiodic microstructures inspired by Voronoi open-cell foams, large objects can be generated with specific elasticity [20–22], including furniture parts or connectors.

L-type corner joints for furniture are tested for bending strain, in the case of chairs [23], or case furniture [24–26]. The bending moment is calculated, depending on the test type, namely a tensile or compression test. Tests were performed on chair solid wood parts [23,27], or wood-based panels [24,25]. Wood dowels [24,27], dowels and innovative fasteners made from polyamide and obtained by 3D printing using the SLS technology [24,26] were used for jointing the panels. The strength of the corner joints connected with the wood dowels was proved to be significantly higher than those connected with the 3D-printed dowels [24]. In order to test the strength and stiffness of the L-type corner joints, both for solid wood and wood-based panels, two methods are applied in the literature, namely the diagonal compression and the diagonal tensile test, respectively [11,23–25]. A chair was built with 3D-printed connectors designed for chair leg and side rails jointing, made from ABS filament by using the FDM-printing technology [28]. The wooden parts were made of beech wood, and connectors with wall thicknesses of 3 mm and 5 mm, respectively, were bonded with the wooden parts with a polyurethane adhesive. Average failure forces for the connectors (in the range of 790 N–905 N) and for the wood mortise-and-tenon joints (1769 N) were recorded, showing the mechanical performance of the connectors at the level of at most 51% of the mechanical performance of the traditional wooden jointing.

The present research investigates the quality and the strength and stiffness of the 3D-printed connectors designed to assemble the leg and two stretchers of a chair. The proposed 3D-printing method in the research is SLS technology and the material used for 3D printing is Polyamide 12 (PA12). The assembly composed of connector and chair components forms an L-type corner joint. Both a diagonal compression test and a diagonal tensile test were performed on the assembled parts and the bending moment was calculated. The results were compared to those of the reference assembly, made of larch solid wood components assembled by a glued mortise and tenon joint.

The use of a 3D-printed connector instead of a classical wood joint brings the advantage of removing from the technological process several operations, such as machining and gluing the mortise–tenon joint, conditioning time for bonded parts, and also contributes to the reduction in solid wood consumption needed for the chair components. All these processing operations are time and cost consuming. In addition to this, jointing with 3D connectors provides a one-time ready-to-assemble (RTA) construction of the chairs and implicitly the ease of assembly. They also allow free design with complex geometrical structures, a multitude of colors and a high surface roughness when using the SLS technology.

The literature in the field provides various models of 3D-printed fasteners and connectors made of different materials; for each strength and stiffness, corner joints were studied and resulted in a low performance compared to the traditional wooden joints. The challenges in the use of SLS technology for connectors and fasteners need more study to optimize the printing parameters and to predict the defects, such as the porous structure. The results obtained thus far in this field represent a starting point for modeling the shape and dimensions of the connectors and fasteners, so as to improve their mechanical properties.

The present research is a contribution to the use of PA12 material and SLS technology as a future perspective for jointing the wooden components of chairs, and bringing novelty in their design and assembly method. Previous research employed FDM and FFF technology for the 3D-printed connectors used for jointing the chair elements, but the SLS technology is preferred because the quality of the parts is clearly superior, no supports are required for complex parts and any geometric shape can be obtained without restrictions. By assessing the results of bending moments under the tensile and compression loads and the defects occurring after testing, the present study aims to offer input data to build a safe and solid chair structure using 3D-printed connectors.

2. Materials and Methods

2.1. Materials

The wooden components of the experimental samples consisted of one leg part and two stretcher parts were made from larch wood (*Larix decidua* Mill.), the material used to manufacture a set of chairs in the frame of Faculty of Furniture Design and Wood Engineering at Transylvania University of Brasov. The objective of the research is to replace the classical mortise–tenon joint between the leg and stretcher with the 3D-printed connector designed for this purpose. The classical mortise–tenon joint and the dimensions of the sample used in the experiment as a reference one are presented in Figure 1a, and the chair assembled parts using the designed connector are presented in Figure 1b.

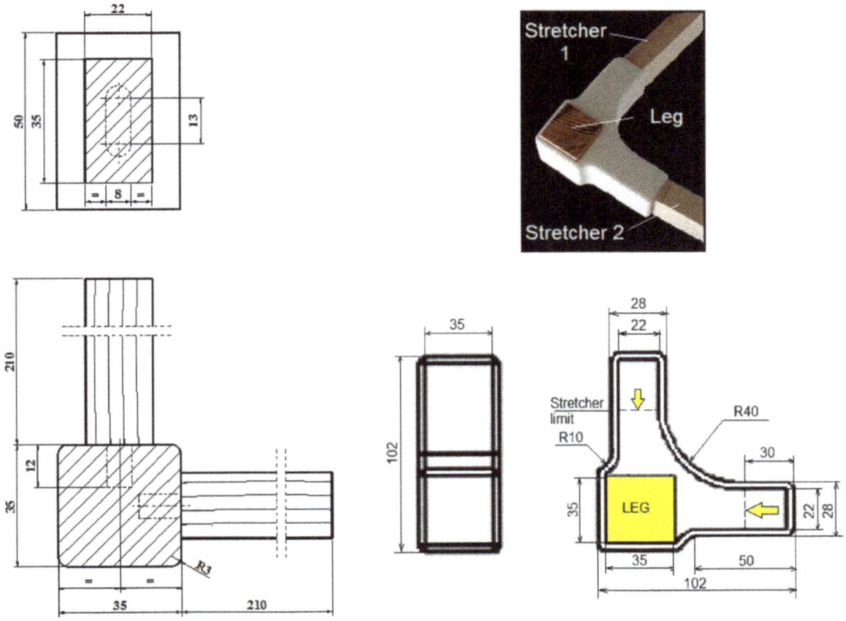

(**a**) (**b**)

Figure 1. Dimensions of samples prepared for the experimental research: (**a**) Reference sample with stretchers and legs mortise–tenon jointed; (**b**) Assembled leg and stretchers with the 3D-printed connector (up) and connector's dimensions (the yellow arrows represent the limit of introducing the wooden part into the connector).

As seen in Figure 1a, the sizes of the piece of the leg in the assembly was 50 mm × 35 mm × 35 mm, and the sizes of the stretcher including the tenon were 222 mm × 35 mm × 22 mm, and without tenon were 210 mm × 35 mm × 22 mm. The larch wood had a moisture content of 9.2%. A tight fit was used between the wooden parts and connector. Mortise and tenon were glued with commercial polyvinyl acetate Novobond D2.

The material selected to be used for 3D printing with SLS technology is DuraForm PA Plastic powder manufactured by 3D Systems, 2nd Floor, West and South Wings, The Maylands Building, 200 Maylands Avenue, Hemel Hempstead, Hertfordshire. HP2 7TG, United Kingdom. The characteristics of this material, according to the technical sheet of the material, are presented in Table 1.

Table 1. Characteristics of DuraForm PA Plastic.

Property	Value
Specific Gravity, g/cm^3	1.00
Moisture Absorption—24 h	0.07%
Tensile Strength Ultimate, MPa	43
Tensile Modulus, MPa	1586
Elongation at break, %	14
Charpy Impact *Unnoched*, J/m^2	336
Charpy Impact *Noched*, J/m^2	32
Flexural Strength, Ultimate MPa	48
Flexural Modulus, MPa	1387
Shore Hardness	73

2.2. 3D Printing

The software used to design the 3D model is SolidWorks 3D CAD, version 2016 developed by Dassault Systèmes, France. The file .STL type was imported in the software of SPRO 60 SD printer. The installation assists the manufacturing of products made of ceramic, metallic and polymeric or composite powders, regardless of their geometrical complexity. The sintering furnace attached to the SLS installation sinters the ceramic and metallic powders, and according to the desired specifications, their infiltration with different metals. The maximum dimensions for the specimen can be 315 × 315 × 415 mm.

The SLS technology used for 3D printing the connectors uses a high-power laser beam, which, by melting (sintering) powder in alternating layers, leads to obtaining the desired 3D model. The 3D model is initially converted into cross-sections and then sent to the printer. Based on the information received, the moving laser beam selectively melts (sinters) the powder layer on the build platform, according to each cross-section. The following parameters were used for 3D printing the experimental connectors: fill laser power of 16 W, outline laser power of 12 W, 1 for outline scan count, slicer fill scan spacing of 0.15 mm, and 1 for sorted fill enabled. It was applied to the raster alternating in the X and Y direction in order to obtain a high density, as recommended in the literature [7]. The position of printing the connector is shown in Figure 2.

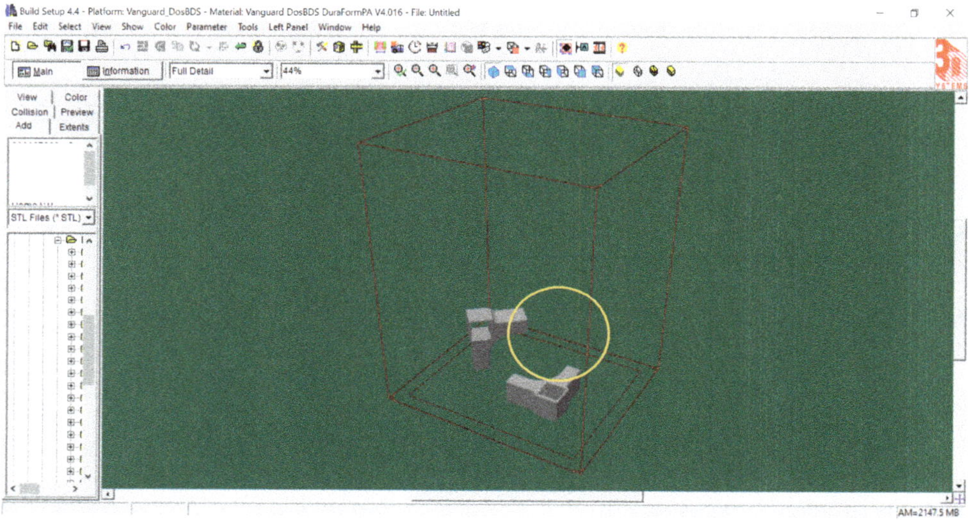

Figure 2. Position of printing the connector (yellow circled one).

The deposition of the successive powder layers measures between 20 μm and 150 μm [5]. The microscopic images taken with 200× magnification show the sizes of the particles of the new powder (Figure 3), measuring up to 100 μm.

In the present research, new powder has been used, and according to the image in Figure 3, the sizes of the particles varied from about 20 μm to 100 μm, and the height of each sintered layer by the laser beam is 100 μm (Figure 4). In case of the used powder, the non-fused powder is removed after the specimen is 3D printed. In Figure 4, there is a schematic representation of the sintering process by laser beam. The red area is the scanned area by the laser beam. A part of the particle, as seen in the detail (left), was sintered, whilst the rest remained unfused and was removed and reused.

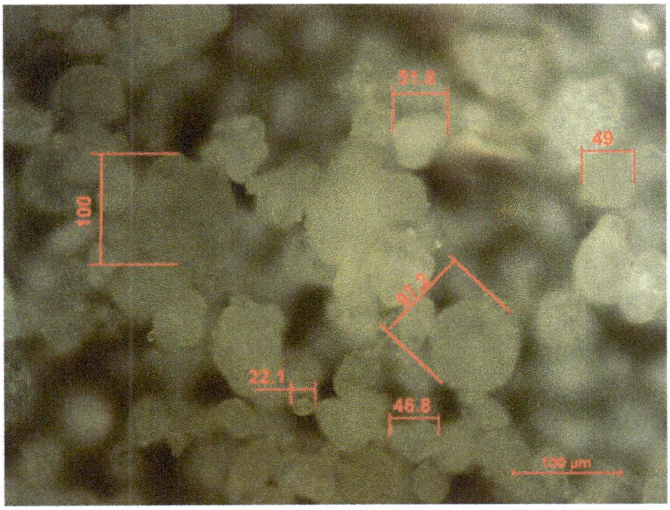

Figure 3. Sizes of the particles of new PA12 powder.

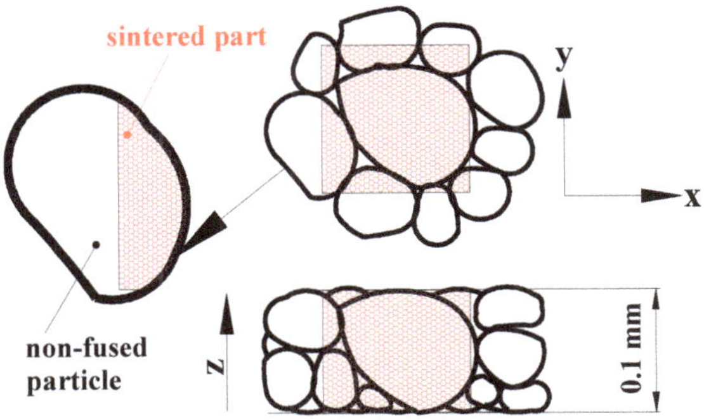

Figure 4. Schematic representation of the laser beam radiation.

2.3. Microscopic Investigation

Microscopic investigation of the PA12 powder and of the 3D-printed connectors before and after mechanical testing was conducted using the optical microscope from Nikon OmniMet-Buehler (Tokyo, Japan). The magnification power used to investigate the powder and the defects occurring in the connectors after their mechanical testing was of 100×, and 200×, respectively. Longitudinal and crosscut sections of the connectors were investigated, in order to evaluate the cracks and ruptures occurring due to the tensile and compression loads applied during the tests on the leg-stretches assembly.

2.4. Mechanical Tests

The L-type corner joint specimens were tested under both diagonal compression and tensile tests, using the models described in the literature [23–25]. The tests were performed on a Zwick/Roell Z010 universal testing machine (Ulm, Germany) for six samples of each category. The compressive forces tend to open the joint, while the tensile forces tend to close the joint and bending moments occur at the corner joints under these loads [24]. The model and the equations used to calculate the bending moment under the tensile load (M_t) and under the compression load (M_c) are all presented in Figure 5. In these equations, F

represents the maximum failure load, in N, and Lt and Lc are the length of the arm of the force F under tensile and compression loads, respectively.

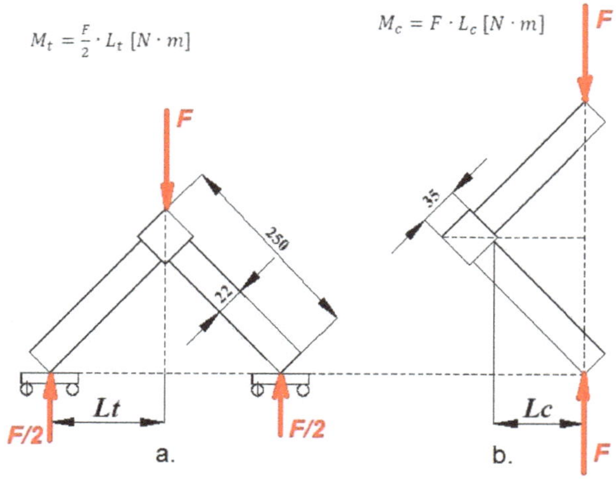

Figure 5. Loading of the L-type corner joint specimens: (**a**) under tensile testing, (**b**) under compression.

The loadings of the L-type corner joint specimens are presented in Figure 6 for diagonal tensile testing (a) and diagonal compression testing (b), both for the assembly with the connector and for the reference.

(**a**) (**b**)

Figure 6. L-type corner joint test specimens: (**a**) diagonal tensile testing for mortise–tenon jointed parts (**top**) and specimen with a 3D-printed connector (**bottom**); (**b**) diagonal compression testing for mortise–tenon jointed parts (**left**) and specimen with a 3D-printed connector (**right**).

3. Results and Discussion

3.1. Mechanical Properties

The bending moment capacities under tension (Mt) and compression (Mc) of the joints were calculated (in Nm) based on the maximum failure load. The results are presented in the diagram in Figure 7.

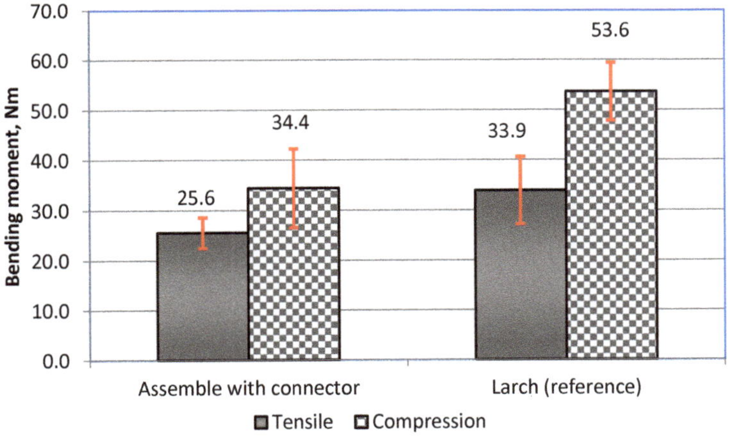

Figure 7. Tensile and compression bending moments of the joints.

The results from Figure 7 show that the mechanical performance of the chair parts assembled with the 3D-printed connector using the sintering technology and PA12 powder is lower than that of the classical mortise–tenon joints made of larch wood. Unsatisfactory results regarding the strength of the L-corner joints for case furniture were obtained by other researchers [24], who replaced the traditional wooden dowels with 3D-printed ones, made of PA12 using the SLS technology.

First of all, it has to be noticed that the bending moment under compression (*Mc*) is higher than the bending moment under tensile testing (*Mt*) both for the investigated assembly and the reference sample. Whilst the bending moment at compression for the chair parts assembled with the 3D-printed connector is 24.5% lower than the one of the reference, in the case of the bending moment at tensile testing, that difference is higher, with a percentage of 35.8%. The maximum forces recorded by the L-corner joints were higher in the case of diagonal tensile testing than in the case of diagonal compression testing both for the reference assembly and for the assembly where the 3D-printed connector was used. The average values of the maximum forces recorded when using the 3D-printed connector were of 350 N for the tensile load and 209 N for the compression load, representing 65% and 58%, respectively, of the values recorded by the reference assembly. Comparatively, when testing the chairs [28], the results showed that joints with 3D-printed connectors had a 42–51% lower failure load than that achieved with traditional wooden mortise–tenon joints with similar dimensions. Even so, the chair had enough strength for a person to sit on it [28]. It comes to the conclusion that an excessive strength of the products could bring unnecessary costs, which can be reduced by finding cheaper solutions with a lower but sufficient strength of the product [24]. Based on the conclusions of the researchers mentioned before [24,28], by using the 3D-printed connectors proposed in the present paper for jointing the chair elements, it is possible to achieve enough strength of the final product and this will be the subject of further research.

An explanation of the unsatisfactory results obtained by the structures with 3D-printed connectors could be the fact that the adjacent powder layers are not sintered and bonded together in the same way. A cause can be an insufficient density of the sintered material determined by the scanning rate, laser power or hatching distance [5], or due to the semi-sintered/semi-melted surfaces that can be formed during the SLS processing [6]. In an ideal case, when using optimized laser energy, the laser sinters the new powder layer and bonds it to the adjacent one. In this case, the 3D-printed objects have a high density (728 kg/m^3) and good mechanical properties in terms of the flexural modulus with a value of 337.1 MPa [9,29]. The defects in the structure of 3D-printing parts with SLS technology occur because the top layer might be removed during the recoating of new layers, or because of warping which stops the deposition and sintering of new powder layer, or due to the balling effect which hinders the uniform deposition of the new powder on the previous one, favoring the occurrence of cracks and porosity [9].

According to the results obtained by other researchers [13,24], the strength and stiffness of the corner joints connected with auxetic dowels made from Polyamide 12 powder by using the 3D-printing SLS technology were significantly lower than those connected with the reference traditional wooden dowels used for case furniture joints. As in the case of case-type furniture [13,24], the scope of the present study is to provide a one-time ready-to-assemble (RTA) construction of the chairs. As remarked by the authors of the previous mentioned research works, the aim of the engineering design of the products is a compromise between technical and economic considerations, in terms of strength–costs appreciation. Over the traditional glued wooden joints used in the furniture industry, the 3D-printed fasteners or connectors are easy to assemble and have the advantages of shortening the production time by reducing the mounting operations and implicitly the final cost of the product.

Figure 8 presents the typical defects occurring after applying tensile (a) and compression (b) loads on the mortise–tenon jointed chair elements. Both compression and tensile testing affect the joint and destroy the adherence between the mortise and tenon and in other cases lead to the occurrence of deep cracks in the leg element. The failure of these assembled parts happens for breaking forces in the range 333 N–573 N for the tensile test and 329 N–415 N for the compression test. Under these loads, the assembled parts were destroyed so that the jointing became nonfunctional.

(a) (b)

Figure 8. Typical defects occurred in the reference sample after mechanical testing: (**a**) at tensile; (**b**) at compression.

Contrarily, the use of 3D-printed connector in the assembled parts did not affect the integrity of the wood stretchers and leg after mechanical testing, all stresses being supported by the connector in this case, and the joint failing in the 3D-printed materials. The same remark was made by other researchers [28] for the connectors used between the leg and side rails of a chair made of beech wood, for which an ABS filament was used and the FDM-printing technology was employed. In this case, the FEM analyses have shown that the maximal values of stresses occurred at the upper edges of the connectors, in the area of the connection between the side rail and the leg, showing that the end part of the rail presses from inside to the wall of the connector, as noticed in the present research.

Figure 9 shows the typical defects occurring on the connector after conducting the mechanical tests under tensile (a) and compression (b) loads. For both types of loads, cracks occurred on the corner of the connector part where the stretchers were mounted, on the longitudinal direction of the stretcher. Only one connector was totally broken perpendicular to the stretcher direction, because of the force exerted by the end of the stretcher to the connector wall (Figure 9b right). In this case, the stress was applied along with the layers and the fracture occurred at the weak area, probably between two adjacent layers sintered by partially melting. This phenomenon was explained by other researchers [9], and it is related to the sizes of the powder particles, reaching a maximum of 100 μm, as presented in Figure 3. The small powder particles are melted during the sintering process and work as a functional binder, whilst the large particles remain solid. Other researchers noticed that the size distribution of the particles has an influence on the porosity formation of the sintered layers [30]. Thus, the smallest particles are molten even with a low energy density, and the molten material is welding the layers. If large particles are more present in the powder, the layers could be partially melted and the porosity occurs, influencing the mechanical properties of the printed object, or, in other words, if a high fraction of small particles is neglected, the SLS process with the powder may fail [31].

(**a**) (**b**)

Figure 9. Typical defects occurred in the 3D-printed connector after mechanical testing: (**a**) at tensile; (**b**) at compression. The red frames highlight the cracks occurring in the 3D-printed connector.

In order to visualize better the defects occurring on the connectors, microscope investigation with a 100× magnification was further conducted.

3.2. Micrography of the Connectors after Mechanical Testing

The image in Figure 10a shows the detachment of the powder granules due to the tensile loads applied to the chair elements assembled with a 3D-printed connector. The crack was placed longitudinally on the corner of the connector. As can be seen in this image, there is a random distribution of the detached powder granules, indicating bonding damage of the different deposited powder layers using the SLS technology. The particle fusion can be explained by several factors that can affect the uniform deposition of the new powder layer on the previous one, such as the energy density and scan speed. Low energy density creates a porous structure and layer delamination with a negative impact on the mechanical properties of the printed object [8]. Contrarily, the high laser power and the fast scan speed can produce the balling defect [6,9], which often leads to cracks.

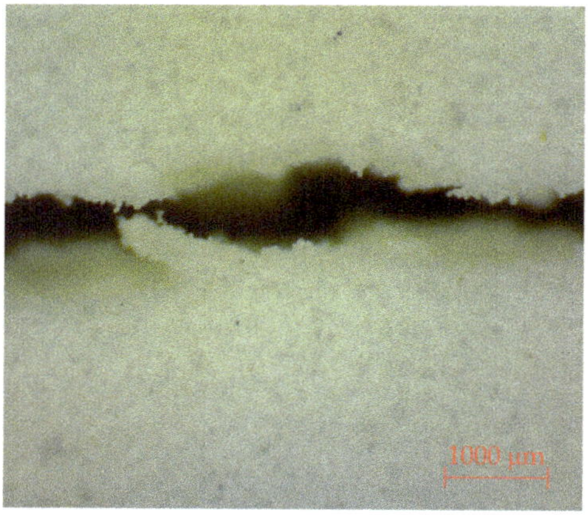

(a)

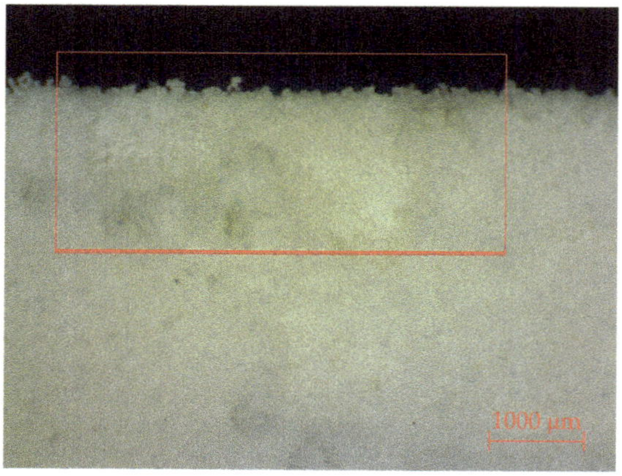

(b)

Figure 10. Defects of the connector after mechanical testing: (**a**) A crack occurred after tensile testing on the longitudinal direction at the corner of the part; (**b**) A rupture of the connector on the transversal direction, after the compression testing. The red frame indicates the area where printing defects were noticed.

The image in Figure 10b shows the linear detachment of the powder granules in the case of breaking the connector in the transversal direction (as seen in Figure 9b right), as a result of compression loading. This can be explained by the low bonding between two adjacent sintered layers of powder. In the marked area in Figure 10b, the balling defect is highlighted, which could cause the breaking of the part. According to the observations of researchers [9,26], the orientation of the part during the SLS printing process has an influence on the nature of the fracture. Thus, on the 0° orientated parts, a serrated fracture occurs after mechanical testing, because the force is applied along the layers, and the weak areas are fractured, a phenomenon which explains the fracture presented in Figure 10a. In the case of the 90° orientation of the part, the force is applied perpendicular to the layers, causing a straight fracture between two subsequent layers, a fact that may explain the occurrence of the fracture presented in Figure 10b.

As noticed by other researchers [24,26] who studied the behavior of 3D-printed dowels made from PA12 by applying SLS technology and which were used for L-type corner joints designed for case-type furniture, during the diagonal tensile and compression tests, the normal and shear stresses occurring in the cross-section of the dowels exceeded the stress limits of the PA12 material. As a result, the failures of the dowels occurred on their entire cross-section, as for the connector presented in Figure 9b.

Reducing the unnecessary excessive strength of the classical wooden joints by replacing them with joints with 3D-printed connectors can bring important advantages: no limitation in design and build complex structures, rapid execution, saving money, time and materials [18].

4. Conclusions

The present paper investigated the possibility of replacing the classical mortise–tenon joint between the leg and two stretchers of a chair with a 3D-printed connector using SLS technology. The L-type corner joint models built with the two variants were subjected to diagonal compression and diagonal tensile testing, and the results of the bending moment capacities calculated based on the maximum breaking load were compared. The 3D-printed connector did not enrich the mechanical performance of the glued mortise–tenon joint used to assemble the chair leg with the stretchers made from larch wood. Whilst the maximum failure load damaged the joint and produced deep cracks in the leg element of the wooden

L-corner type mortise–tenon joint, in the case of using the 3D-printed connector, cracks occurred at the corners where the stretcher edges pressed the connector, favoring the stretcher to come out from the assembly, but the wood elements were not affected. This may be an advantage of using the connectors instead of wooden joints, because they can be replaced and the chair becomes functional again. The microscopy of the printed parts shows the defects in the structure of the connector, which can be eliminated in further research by modifying the printing parameters such as the laser power and the scan speed. The change in the method of 3D printing and also the use of other materials are the next steps of the research in this field. Extending the testing methods to the final product is also a challenge. On the other hand, the change in the shape of the connector for the round section of the stretcher is another subject of further research.

Author Contributions: Conceptualization, A.N. and C.C.; methodology, M.A.P.; software, M.A.P.; validation, C.C., M.A.P. and A.N.; formal analysis, C.C.; investigation, S.V.G.; resources, A.N., C.C., M.A.P. and S.V.G.; data curation, A.N., C.C. and M.A.P.; writing—original draft preparation, C.C.; writing—review and editing, C.C. and M.A.P.; visualization, A.N. and S.V.G.; supervision, C.C.; project administration, A.N.; funding acquisition, A.N. All authors have read and agreed to the published version of the manuscript.

Funding: We hereby acknowledge the structural funds project PRO-DD (POS-CCE, O.2.2.1., ID 123, SMIS 2637, No. 11/2009) for providing the infrastructure used in this work and the Contract No. 7/9.01.2014.

Institutional Review Board Statement: Not applicable.

Informed Consent Statement: Not applicable.

Data Availability Statement: Data are contained within the article.

Conflicts of Interest: The authors declare no conflict of interest.

References

1. Ayidin, M. Additive Manufacturing: Is It a New Era for Furniture Production? *J. Mech. Eng. Autom.* **2015**, *5*, 338–347.
2. Reddy, K.S.; Dufera, S. Additive Manufacturing Technologies. *Int. J. Manag. Inf. Technol. Eng.* **2016**, *4*, 89–112.
3. Wong, K.V.; Hernandez, A. A Review of Additive Manufacturing. *Int. Sch. Res. Not.* **2012**, *2012*, 208760. [CrossRef]
4. Ramya, A.; Vanapalli, S. 3D Printing Technologies in Various Applications. *IJMET* **2016**, *7*, 396–409.
5. Przestacki, D.; Sieniawski, J.; Stambolov, G.; Lisiak, P. An Overview of Selective Laser Sintering Technologies. *Arch. Mech. Technol. Mater.* **2015**, *16*, 23–32.
6. Tolochko, N.K.; Mozzharov, S.E.; Yadroitsev, I.A.; Laoui, T.; Froyen, L.; Titov, V.I.; Ignatiev, M.B. Balling processes during selective laser treatment of powders. *Rapid Prototyp. J.* **2004**, *10*, 78–87. [CrossRef]
7. Simchi, A.; Pohl, H. Effects of laser sintering processing parameters on the microstructure and densification of iron powder. *Mater. Eng.* **2003**, *A359*, 119–128. [CrossRef]
8. Korycki, A. Study of the Selective Laser Sintering Process: Materials Properties and EFFECT of Process Parameters. Ph.D. Thesis, Institut National Polytechnique de Toulouse (Toulouse INP), Labège, France, 2020. Available online: https://oatao.univ-toulouse.fr/27651/1/Korycki_Adrian.pdf (accessed on 15 September 2023).
9. Han, W.; Kong, L.; Xu, M. Advances in selective laser sintering of polymers. *Int. J. Extrem. Manuf.* **2022**, *4*, 042002.
10. Felek, S.Ö. A new Era in Furniture Production: 3D Printer. International Conference on Knowledge & Innovation in Engineering, Science &Technology, March 2020, Budapest, Hungary. Available online: www.kiconf.org (accessed on 15 September 2023).
11. Smardzewski, J.; Rzepa, B.; Kılıç, H. Mechanical Properties of Externally Invisible Furniture Joints Made of Wood-Based Composites. *Bioresources* **2016**, *11*, 1224–1239. [CrossRef]
12. Nicolau, A.; Pop, M.A.; Coșereanu, C. 3D Printing Application in Wood Furniture Components Assembling. *Materials* **2022**, *15*, 2907. [CrossRef]
13. Kasal, A.; Kuşkun, T.; Smardzewski, J. Experimental and Numerical Study on Withdrawal Strength of Different Types of Auxetic Dowels for Furniture Joints. *Materials* **2020**, *13*, 4252. [CrossRef] [PubMed]
14. van Wassenhove, R.; De Laet, L.; Vassilopoulos, A.P. A 3D printed bio-composite removable connection system for bamboo spatial structures. *Compos. Struct.* **2021**, *269*, 114047. [CrossRef]
15. Swetham, T.; Reddy, K.M.M.; Huggi, A.; Kumar, M.N. A Critical Review on of 3D Printing Materials and Details of Materials used in FDM. *Int. J. Sci. Res. Sci. Eng. Technol.* **2017**, *3*, 353–361.
16. Li, X.; Ni, Z.; Bai, S.; Lou, B. Preparation and Mechanical Properties of Fiber Reinforced PLA for 3D Printing Materials. *IOP Conf. Ser. Mater. Sci. Eng.* **2018**, *322*, 022012. [CrossRef]

17. Liang, L.; Huang, T.; Yu, S.; Cao, W.; Xu, T. Study on 3D printed graphene/carbon fiber multi-scale reinforced PLA composites. *Mater. Lett.* **2021**, *300*, 130173. [CrossRef]
18. Valiyousefi, M.; Alihedarloo, A. A Study the impact of 3D-printed joints on the complex wooden structures. In Proceedings of the International Congress on Science & Engineering University of Tokio, Tokyo, Japan, 14–15 October 2019.
19. Ngo, T.D.; Kashani, A.; Imbalzano, G.; Nguyen, K.T.Q.; Hui, D. Additive manufacturing (3D printing). A review of materials, methods, application and challenges. *Compos. Part B* **2018**, *143*, 172–196.
20. Martínez, J.; Dumas, J.; Lefebvre, S. Procedural Voronoi Foams for Additive Manufacturing. *ACM Trans. Graph.* **2016**, *35*, 1–12. [CrossRef]
21. Martínez, J.; Hornus, S.; Song, H.; Lefebvre, S. Polyhedral Voronoi diagrams for additive manufacturing. *ACM Trans. Graph.* **2018**, *37*, 1–15. [CrossRef]
22. Capone, M.; Lanzara, E.; Portioli, F.P.A.; Flore, F. Digital Form Finding Using Voronoi Pattern. *Nexus Netw. J.* **2021**, *23*, 959–975. [CrossRef]
23. Ayrilmis, N.; As, N.; Dündar, T.; Şendağ, A. Determination of Bending Moment of L-Type Corner Joints Used in Chair Production and Their Effects on Mechanical Performance of Chairs. *Mater. Int.* **2020**, *2*, 0318–0323.
24. Kasal, A.; Smardzewski, J.; Kuşkun, T.; Güray, E. Analyses of L-Type Corner Joints Connected with Auxetic Dowels for Case Furniture. *Materials* **2023**, *16*, 4547. [CrossRef] [PubMed]
25. Yerlikaya, N.C. Failure load of corner joints, which are reinforced with glass-fiber fabric in case-type furniture. *Sci. Res. Essays* **2013**, *8*, 325–339.
26. Krzyżaniak, Ł.; Smardzewski, J. Impact damage response of L-type corner joints connected with new innovative furniture fasteners in wood-based composites panels. *Compos. Struct.* **2021**, *255*, 113008. [CrossRef]
27. Ke, Q.; Lin, L.; Chen, S.; Zhang, F.; Zhang, Y. Optimization of l-shaped corner dowel joint in pine using finite element analysis with Taguchi method. *Wood Res.* **2016**, *61*, 243–254.
28. Hajdarevic, S.; Kuzman, M.K.; Obucina, M.; Vratuša, S.; Kušar, T.; Kariž, M. Strength and stiffness of 3D-printed connectors compared with the wooden mortise and tenon joints for chairs. *Wood Mater. Sci. Eng.* **2023**, *18*, 870–883. [CrossRef]
29. Dupin, S.; Lame, O.; Barrès, C.; Charmeau, J.-Y. Microstructural origin of physical and mechanical properties of polyamide 12 processed by laser sintering. *Eur. Polym. J.* **2012**, *48*, 1611–1621. [CrossRef]
30. Schmid, M.; Wegener, K. Additive manufacturing: Polymers applicable for laser sintering (LS). *Proc. Eng.* **2016**, *149*, 457–464. [CrossRef]
31. Beal, V.; Paggi, R.A.; Salmoria, G.; Lago, A. Statistical Evaluation of Laser Energy Density Effect on Mechanical Properties of Polyamide Parts Manufactured by Selective Laser Sintering. *J. Appl. Polym. Sci.* **2009**, *113*, 2910–2919. [CrossRef]

applied sciences

MDPI

Article

A Stratified Characterization of Surface Quality of Beech Processed by Profile Milling

Luminița-Maria Brenci and Lidia Gurău *

Faculty of Furniture Design and Wood Engineering, Transilvania University of Brasov, B-dul Eroilor 29, 500036 Brasov, Romania; brenlu@unitbv.ro
* Correspondence: lidiagurau@unitbv.ro

Abstract: The evaluation of surface quality after milling still represents a challenge due to the stratified structure of wood. In this paper, the surface quality of beech wood obtained by profile milling was analyzed differentiating between tool marks, fuzzy grain, and accidental surface gaps overlapped onto the deep anatomical cavities. The samples were milled at two rotation speeds, 3308 rpm (n_1) and 6594 rpm (n_2), and two feed speeds of 6.53 m/min (v_{f1}) and 23.74 m/min (v_{f2}). After the samples were cut, approximately 600 m of beech wood were further processed before measuring the surface quality again. The surface waviness measured by Wa increased more than double when the feed speed increased from v_{f1} to v_{f2}. Rk increased with the feed speed (with the feed per tooth, f_z, for a constant n, respectively) for n_2 by a significant 57%. An increase in the rotation speed from n_1 to n_2 has reduced Rk, but the effect was not significant. The processing quality, Rk, measured at different depths did not differ statistically, in spite of the local differences in the cutting speed. The moment of measuring the surface roughness, such as immediately after sharpening or after a working period, influenced the surface quality. After the tool processed 600 m of beech material, the surface quality improved by 30%. The statistical analysis showed that the most important factor affecting the surface quality was the feed speed (implicitly f_z, for constant n and number of cutters).

Keywords: beech wood; profiled milling; surface quality; roughness; waviness; tool marks; raised fibers; anatomical cavities

Citation: Brenci, L.-M.; Gurău, L. A Stratified Characterization of Surface Quality of Beech Processed by Profile Milling. *Appl. Sci.* **2024**, *14*, 129. https://doi.org/10.3390/app14010129

Academic Editor: Giuseppe Lazzara

Received: 25 November 2023
Revised: 19 December 2023
Accepted: 20 December 2023
Published: 22 December 2023

1. Introduction

The assessment of the surface quality following wood processing is very important in order to establish the impact of various processing parameters combinations [1–3]. The optimization of milling operations by considering the surface quality will reduce the amount of further sanding and can contribute to a reduction in the finishing costs. Therefore, it is important to find a combination of processing parameters suitable for a minimum surface roughness. Furthermore, the surface roughness of wood is very complex and represents a superposition of morphological features, such as waviness, raised fibers, tool marks, wood anatomical voids, as well as possible processing surface defects, such as material detached from the surface [4–9]. In this respect, the most used roughness parameters such as Ra or Rz are not enough to give a complete picture of the origin of the different irregularities present on a processed surface.

According to the studies performed by Gawroński [10], Sutcu [11], and Koc et al. [12], the cutting parameters affect the surface roughness of beech wood. The study performed by Gürgen et al. [13] on *Pinus sylvestris* after CNC milling showed that the surface roughness decreased with increasing rotation speed and decreased with decreasing the feed speed for all machining parameters. The evaluation of surface roughness for different depths of cut revealed that the maximum surface roughness value was registered for the highest depth of cut. Jiang et al. [14] have found that the cutting speed has not significantly influenced the surface roughness of walnut processed by straight-tooth milling, but the depth of cut was significant. Costes and Larricq [15] stated that the quality of beech, when routed on a CNC machine, improves with an increase in the cutting speed. Piernik et al. [16] concluded that an increase in the feed speed and the chip thickness increases the surface roughness of beech milled by CNC, while the cutting speed had no effect on surface roughness, but just on the cutting power.

Most researchers have used Ra and Rz [11–13,17–19] to assess the quality of wood after milling, but these parameters do not consider the nature of wood irregularities and include the inherent wood anatomical gaps in the evaluation. The roughness parameters Ra and Rz decreased with a lower feed rate and a higher spindle speed in a study by Kok et al. [12] on beech (*Fagus orientalis*), Ayous (*Triplochiton Scleroxylon*), and MDF processed by CNC. Similarly, in another study of Rabiei and Yaghoubi [18], the surface roughness evaluated by Ra decreased with increasing the spindle speed and decreasing the feed rate as well as the depth of cut when beech wood was milled on a CNC machine. However, these two parameters employed by the researchers are not sufficient to correctly interpret the quality of wood and wood-based surfaces, which means that they must be evaluated together with other roughness parameters [7].

Kilic et al. [20] evaluated the effect of milling on the surface roughness of beech and aspen wood. On the basis of statistical analysis, they found no significant difference between the surface roughness determined on the tangential and radial surfaces. They observed that a number of roughness parameters, such as Ra, Rz, Rk, Rpk, and Rvk roughness are useful, to be used together, to evaluate the surface quality of wood.

Magoss [21] studied the influence of the wear of the cutting edges on the surface roughness for several species (oak, larch, beech, and pine wood) processed on CNC milling machines. The experimental results showed that the Rk parameter, which measures the core roughness and is the closest approximation of the tool marks, may be an important indicator of the tool edge wear. In the case of beech wood, the Rk value after milling with a blunt tool was approximately twice the value of milling with a sharp tool.

Gurău et al. [22] investigated the surface roughness of larch wood (*Larix decidua* Mill.) cut through by laser and CNC routing at different feed speeds, by parameters Ra, Rp, Rv, Rsk, Wa, Pa, Pt and the Abbot-curve parameters, Rk, Rpk, and Rvk, in order to find a more comprehensive understanding of the effect of processing on wood surface topography.

Learning from the above studies, it can be concluded that a complex surface such as that of wood requires a larger set of roughness parameters capable of differentiating between the various layers of irregularities present on a processed surface.

Therefore, this study uses a comprehensive set of standard parameters for surface characterization, such as Ra, Rv, Rsk, and Wa [23], and the Abbot curve parameters, Rk, Rpk, Rvk, A1, and A2 [24]. These parameters allow the analysis of surface quality on stratified levels of irregularities differentiating between tool marks, fuzzy grain, and accidental surface gaps overlapping onto the deep anatomical cavities of the wood, such as pores from earlywood. Such analysis will provide a more in-depth understanding and quantification of the tool interaction with the wood material and will help for a reliable selection of the most appropriate processing parameters. Furthermore, in a calculation of those parameters, a robust Gaussian regression filter tested and found suitable for wood was used in order to avoid the bias caused by the simple Gaussian filter contained in the majority of measuring instruments [9,25].

2. Materials and Methods

2.1. Wood Samples

Beech wood (*Fagus sylvatica* L.), with a mean density of 745 kg/m^3 and 7% moisture content, was processed by profiled milling in two processing passes in order to obtain profiles with 16 mm maximum depth (Figure 1).

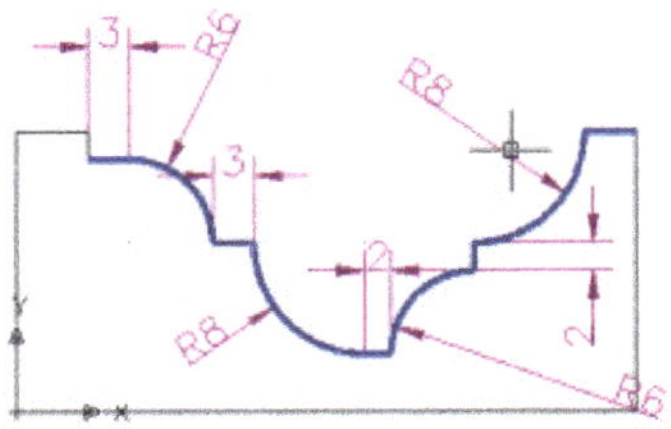

Figure 1. The samples' profile.

The milling was carried out on 1000 mm × 50 mm × 20 mm strips, along the grain direction, by using various cutting parameter combinations, which will be further detailed.

2.2. Processing by Profiled Milling

A vertical milling machine was used to process the beech strips with a profile cutterhead made of C45 steel [26], with 4 new and sharp cutters, made of C120 steel [27] (Figure 2). The dimensions and geometry of the cutting tool and the cutters are presented in Tables 1 and 2, respectively.

Figure 2. The profile cutterhead used in the milling process.

Table 1. Geometric parameters of the cutterhead.

D (mm)	B (mm)	d (mm)	Clearance Angle α (°)	Rake Angle γ (°)
178	40	30	25	18

Table 2. The dimensions of the cutters.

Length (mm)	Width (mm)	Thickness (mm)	Height (mm)
58	40	7.2	16

The profiled milling was performed with two rotation speeds, such as 3308 rpm (n_1) and 6594 rpm (n_2), and two feed speeds, 6.53 m/min (v_{f1}) and 23.74 m/min (v_{f2}), respectively. The cutting parameters combinations were n_1-v_{f1}, n_1-v_{f2}, n_2-v_{f1}, and n_2-v_{f2}. This provided a first set of samples. A second set of samples was prepared by using the same cutting parameter combinations, but after the tool milled 600 m of beech wood. The intention was to observe the impact of the cutting distance, as it further shows the influence of the cutting tool wear [28] on the surface quality, when the same working variables are used.

The strips of each cutting case combination were further shortened to 60 mm × 50 mm × 20 mm specimens available for inspection. The samples were labeled to identify their cutting parameters combination. Three samples were selected from the first set of data (sharpened tool) and two samples from the second set of data (after tool has cut 600 m), for each rotation speed and feed speed combination (n_1-v_{f1}, n_1-v_{f2}, n_2-v_{f1}, and n_2-v_{f2}), for both tool conditions: sharpened and after 600 m cutting distance. This meant 20 processed samples were available for the surface quality measurement.

Table 3 summarizes the machining parameters used during the milling experiments with indication of the local cutting speed and feed per tooth values in 4 regions/profile depths, marked in Figure 3, and in which the surface quality was further checked.

Table 3. The machining parameters.

Tool Rotation Speed n (rpm)		Feed Rate v_f (m/min)		Measured Point	Profile Depth (mm)	Cutting Speed v_c (m/s)	Feed per Tooth f_z (mm/Tooth)	
n_1	3308	v_{f1}	6.53	Point 1	2	28.39	f_{z11}	0.49
		v_{f2}	23.74				f_{z12}	1.79
n_2	6594	v_{f1}	6.53			56.59	f_{z21}	0.25
		v_{f2}	23.74				f_{z22}	0.90
n_1	3308	v_{f1}	6.53	Point 2	8	29.43	f_{z11}	0.49
		v_{f2}	23.74				f_{z12}	1.79
n_2	6594	v_{f1}	6.53			58.66	f_{z21}	0.25
		v_{f2}	23.74				f_{z22}	0.90
n_1	3308	v_{f1}	6.53	Point 3	16	30.82	f_{z11}	0.49
		v_{f2}	23.74				f_{z12}	1.79
n_2	6594	v_{f1}	6.53			61.43	f_{z21}	0.25
		v_{f2}	23.74				f_{z22}	0.90
n_1	3308	v_{f1}	6.53	Point 4	10	29.78	f_{z11}	0.49
		v_{f2}	23.74				f_{z12}	1.79
n_2	6594	v_{f1}	6.53			59.35	f_{z21}	0.25
		v_{f2}	23.74				f_{z22}	0.90

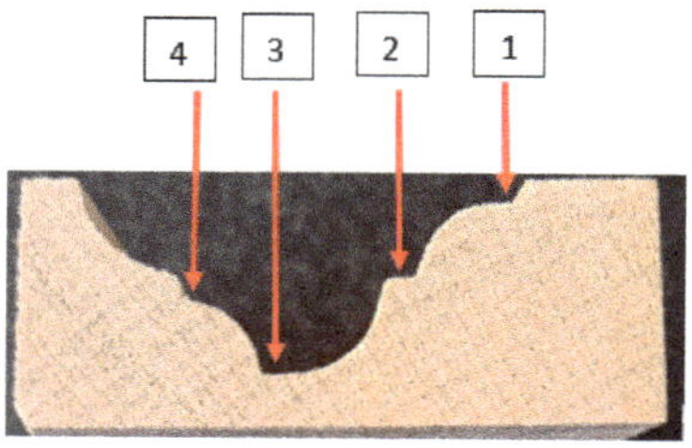

Figure 3. Sample cross-section with four zones of linear measurement.

From Table 3, note that the f_z is directly influenced by the feed speed v_f for a constant n and for the same number of cutters (a single type of tool was used in this research). Another notable observation is that the cutting speed, v_c, differs very slightly with the local tool diameter (depth of cut, respectively), for a given rotation speed, n. Therefore, the interpretation of the results will be mainly focused on variations in v_f, respectively, n.

2.3. Roughness Measurement

The surface morphology of the beech samples was measured with MarSurf XT20 equipment manufactured by MAHR Gottingen GMBH, Göttingen, Germany (Figure 4a). It used an MFW 250 scanning head with a tracing arm in the range of ±750 µm, and a stylus with 2 µm tip radius and 90° tip angle, as shown in Figure 4b.

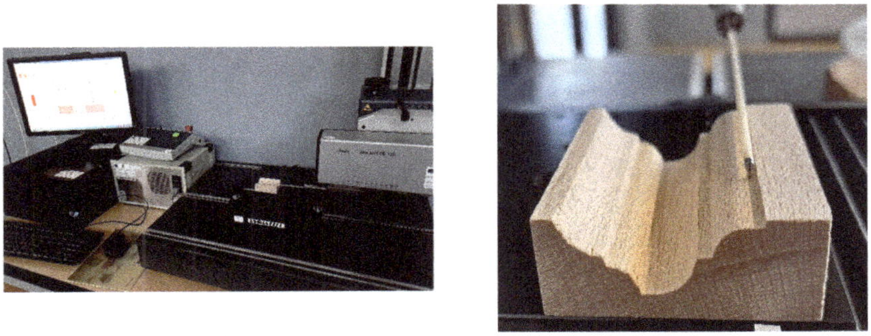

(a) (b)

Figure 4. Measurements of surface morphology. (**a**) MarSurf XT20 profilometer and data acquisition; (**b**) linear surface scanning by stylus (zone 1 in Figure 3).

For measuring the beech milled profiles, only the straight zones were considered, as marked by red arrows in Figure 3. The depth for the marked points were 2 mm for point 1, 8 mm for point 2, 16 mm for point 3, and 10 mm for point 4. This decision was drawn from the requirement that the stylus remains normal in relation to the processed surface. Further studies will take into consideration the possibility of measuring round profiles as well, by using adequate supports. Four profiles, 30 mm long, were measured from the numbered zones (Figure 3) along the milling direction with a speed of 0.5 mm/s, a scanning force of 0.7 mN, and a resolution of 5 µm. This meant 4 profiles were taken from each sample. In case of the first set of samples, there were 12 profiles for one parameters combination. For all 4 parameter combinations (ex. n_1-v_{f1}; n_1-v_{f2}; etc.), the total number of measurements reached 48. Similarly, for the second set of samples, there were 32 measurements.

Data acquired were further processed with MARWIN XR20 software provided together with the measuring equipment. Data evaluation began with removing any unwanted form errors, followed by a procedure of filtering with a robust Gaussian regression filter, contained in ISO/TS 16610-31 [25]. This filter was tested and it was recommended for wood surfaces because it reduces the distortions associated with deep grouped valleys in the profiles and it is more reliable in comparison with a simple Gaussian filter, common in most measuring instruments [9,29–32]. In this way, the measured data were separated by filtering the wavelengths of irregularities in waviness profiles and roughness profiles, corresponding to W parameters for the longest wavelengths, while R parameters evaluated the shortest wavelengths. The cut-off length used for filtering was 2.5 mm, recommended for wood surfaces [30,32]. After filtering, a number of useful parameters were calculated, such as Ra (arithmetic mean deviation of the roughness profile), Rv (the largest absolute profile valley depth), Rsk (skewness of the profile), and Wa (arithmetic mean deviation of the waviness profile) according to ISO 4287 [23]. In addition, Rk (the core roughness depth), Rpk (the reduced peak height), Rvk (the reduced valley depth), A1 (peak area in μm^2/mm), and A2 (valley area in μm^2/mm) were employed according to ISO 13565-2 [24].

The parameters from ISO 13565-2 [24] are particularly useful to characterize the stratified structure of a processed surface. Compared with other parameters, Rk depends the least on the wood anatomical structure and measures the height of the core roughness, the region defined by the highest concentration of data points in a measured profile, which excludes isolated peaks and valleys. For a milling process, it can be an approximation of the tool marks left on wood. Instead, Rpk is defined by the isolated peaks, which for a wood species, according to Westkämper and Riegel [4], can be considered an evaluation of wood raised grain and fuzziness of the surface. The fuzziness is caused by groups of fibers that are attached to the surface at only one end. Rvk is defined by isolated valleys, which in the case of wood can be caused by accidental material pull-out during processing and/or deep anatomical cell lumens.

Another test was conducted to monitor the surface quality after the tool has milled 600 m of the same profile and species, in order to see which is the effect of the tool condition.

Surface profiles were measured in the same way as in the case when the tool was sharpened, but on two specimens. This meant there were 8 measured profiles for analysis of each of the 4 parameters combination, making a total of 32 measurements. The surface morphology measurement and evaluation followed the same steps as for the sharpened tool.

2.4. Stereo-Microscopy Analysis

A stereo-microscope NIKON SMZ 18-LOT2 (Nikon Instruments, Melville, NY, USA) was used for the microscopic investigations of the profile milled beech samples. The straight zone 1 from Figure 2 was in focus. In order to better observe the occurrence of the raised fibers and possible surface defects such as material detachment, the samples were tilted by approximately 45°.

2.5. Statistical Analysis

The surface quality parameters were tested for their significance for all processing combinations (rotation speed–feed speed). The software used was an IBM SPSS Statistics version 23.

3. Results and Discussion

The mean values and standard deviation for parameters evaluated in this analysis are contained in Table 4.

Table 4. The mean values roughness parameters (in μm). Standard deviation is presented in parentheses.

Tool Condition	Processing Parameters	Ra	Rk	Rpk	Rvk	Rv	Rsk	Rpk/Rvk	Rk+Rpk+Rvk	A1/A2	Wa
sharpened	3308−6.53 (n_1-v_{f1})	5.74 (0.94)	12.10 (1.70)	6.81 (2.19)	18.43 (5.30)	46.02 (13.25)	−2.02 (0.58)	0.37 (0.41)	37.34 (6.62)	0.18 (0.22)	6.79 (1.18)
	3308−23.74 (n_1-v_{f2})	7.4 (0.88)	14.99 (3.41)	16.51 (6.21)	20.77 (3.24)	46.98 (5.12)	−0.71 (0.91)	0.79 (1.92)	52.27 (5.48)	0.54 (1.11)	14.81 (2.20)
	6594−6.53 (n_2-v_{f1})	4.83 (0.67)	10.19 (1.94)	7.54 (2.30)	11.64 (2.52)	30,40 (7.52)	−1.08 (1.00)	0.64 (0.91)	29.37 (4.49)	0.48 (0.79)	5.38 (0.90)
	6594−23.74 (n_2-v_{f2})	6.59 (0.82)	16.02 (2.38)	10,68 (3.35)	16.39 (3.29)	42.29 (10.06)	−1.01 (0.89)	0.65 (1.02)	43.09 (5.32)	0.39 (0.50)	11.83 (2.99)
after 600 m milling length	3308−6.53 (n_1-v_{f1})	4.46 (0.63)	10.15 (1.38)	6.24 (1.83)	12.27 (3.17)	30.43 (8.02)	−1.34 (0.97)	0.51 (0.61)	28.66 (3.71)	0.28 (0.34)	10.28 (1.85)
	3308−23.74 (n_1-v_{f2})	4.93 (0.63)	12.26 (2.09)	8.67 (4.00)	13.54 (2.71)	35.37 (4.93)	−1.15 (0.87)	0.64 (1.48)	34.46 (5.33)	0.42 (1.58)	12.35 (5.65)
	6594−6.53 (n_2-v_{f1})	5.03 (1.62)	10.93 (2.11)	9.71 (4.72)	14.24 (8.02)	36.09 (14.57)	−0.99 (1.49)	0.68 (0.59)	34.89 (11.90)	0.47 (0.45)	10.44 (2.31)
	6594−23.74 (n_2-v_{f2})	4.97 (0.70)	11.43 (1.15)	7.12 (3.60)	14.42 (3.67)	36.38 (9.79)	−1.50 (0.88)	0.49 (0.98)	32.96 (4.66)	0.29 (0.59)	10.47 (4.20)

The surface morphology was first studied by considering the largest wavelengths irregularities produced by milling on the surface, namely the surface waviness. This can be observed directly by comparing primary profiles, which contain both roughness and waviness (Figure 5), selected for the four groups/combinations of rotation speeds and feed speeds.

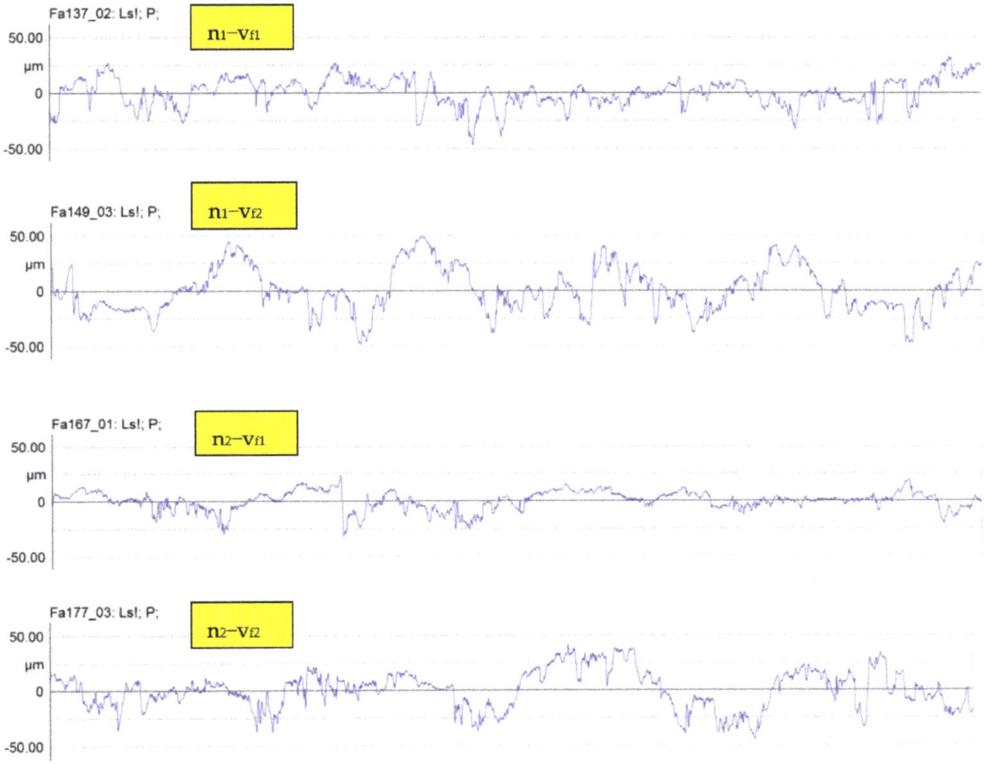

Figure 5. Primary profiles containing waviness and roughness together, for the four combinations of rotation speeds and feed speeds, when the tool was sharpened.

It can be seen from Figure 5 that an increase in the feed speed from v_{f1} to v_{f2} (3.6 times) has increased surface waviness, in the primary profiles, for both rotation speeds. The profiles are considerably wavier, containing large undulations, which can be spotted as patterns repeated at almost regular distance (approximately 6 mm), in Figure 5 and by a waviness profile overlapped on the ridged beech surface, in Figure 6. This observation is in agreement with the values of Wa parameter, which doubled in the case of both rotation speeds (Figure 7).

This result was statistically significant for $p < 0.05$ (test ANOVA, followed by Dunnett T3 multiple comparisons) (Tables 5 and 6). An explanation may be that an increase in the feed speed increases tool vibration, which leaves deeper waves in the surface. When the rotation speed almost doubled from n_1 to n_2, a decrease of 20% in surface waviness was measured, but this reduction had no statistical support (Table 6).

Table 5. Dependent variable Wa—Test between groups' effects.

Source	Sum of Squares	df	Mean Square	F	Sig
Groups	693.397	3	231.132	51.894	0.000

Figure 6. Beech surface processed by n_1-v_{f2} parameters combination with a waviness profile overlapped.

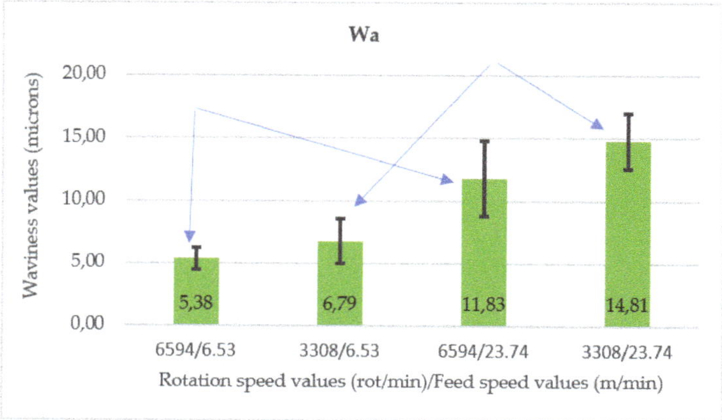

Figure 7. The variation of waviness parameter, Wa, with combinations of the rotation speed with the feed speed. With blue arrows, pair means that differ statistically are marked.

Table 6. Dependent variable Wa− Dunnet T3 Post Hoc Test for multiple comparisons between groups rotation speed–feed speed.

Groups		Mean Difference	Std. Error	Sig
n_1-v_{f1}	n_1-v_{f2}	−8.017 [1]	0.82	0.000
	n_2-v_{f1}	1.415	0.58	0.141
n_2-v_{f2}	n_1-v_{f2}	−2.98	1.07	0.063
	n_2-v_{f1}	6.45 [1]	0.90	0.000

[1] The error term is Mean Square (Error) = 4.454. The mean difference is significant at the 0.05 level.

Figure 8 displays the roughness profiles measured for each pair rotation speed–feed speed. Isolated raised fibers are marked with red ellipses, while the yellow ellipses indicate isolated deep valleys. From definition, the core roughness of the roughness profiles does not contain the isolated deep valleys associated with the deep pores from earlywood or isolated detached wood material, nor the isolated raised fibers or surface fuzziness. Instead, it can be viewed as the density of data above and below the zero line. Therefore, it appears that an increase in the feed speed (respectively, in f_z, for a constant n and z) is causing more dispersion above and below the zero line (Figure 6), which translated in an increase in Rk. These visual observations are confirmed by the measured Rk values, which increased, when the feed speed (respectively, f_z) increased 3.6 times, by 24% for n_1 and by a significant

57% for n_2 (from 10.19 μm to 16.02 μm) (Figure 9). The conclusion is that an increase in the feed speed (respectively, in f_z, for a constant n and z) increases the core roughness Rk and is more pronounced when the rotation speed increases as well (Figure 9).

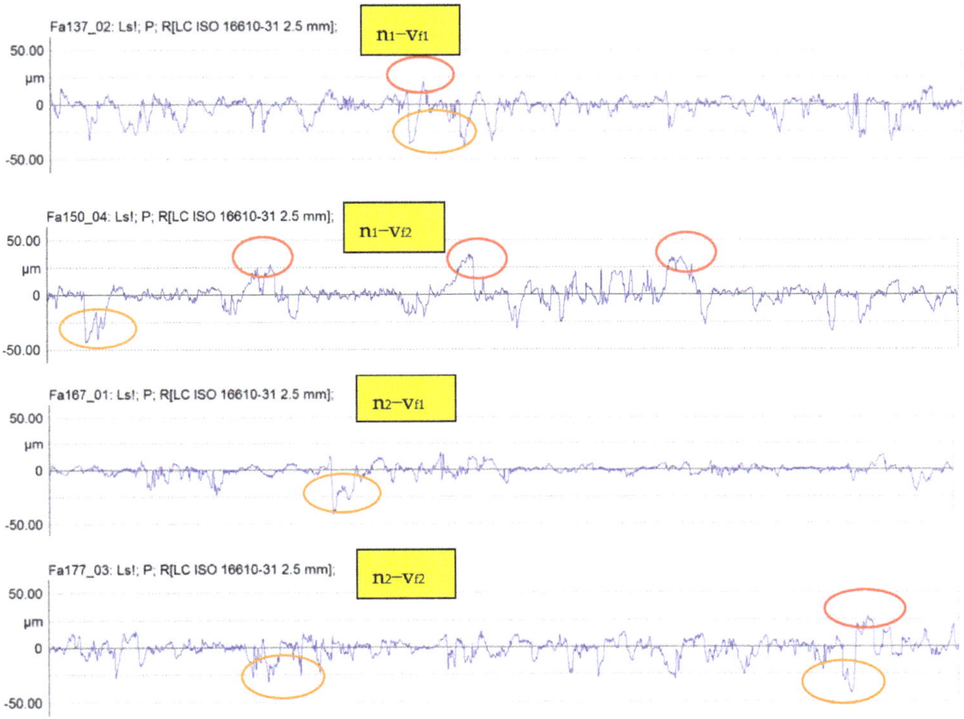

Figure 8. Roughness profiles of each group combination, rotation speed–feed speed. Raised fibers are marked with red; deep isolated valleys are marked with yellow ellipses.

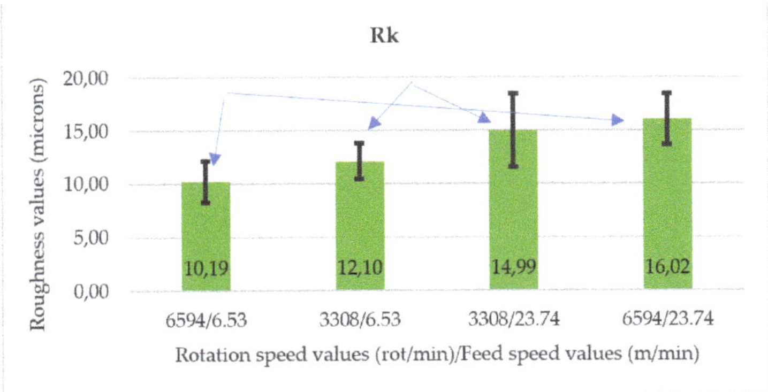

Figure 9. The variation of the core roughness parameter, Rk, with combinations of the rotation speed with the feed speed. With blue arrows, pair means that differ statistically are marked.

Since Rk is the parameter that approximates the best the tool marks (processing roughness), a supplementary statistical test was performed with the direct cutting parameters v_c–f_z on local profile depths D_1–D_4 (as in Table 3), when the rotation speed n was kept constant and the variable was the feed speed v_f (consequently, f_z). The conclusion from

Table 7 was the same, respectively, for a constant n; the increase in the feed speed vf (respectively, f_z) was statistically significant for Rk, even at different cutting levels (D_1-D_4) and mostly for the n_2 rotation speed.

Table 7. Pairwise comparison of parameters combinations (v_c/f_z) for Rk and the level of significance for an increase in v_f (f_z, respectively).

Parameters Combination 1 (v_c/f_z)	Parameters Combination 2 (v_c/f_z)	*p*-Value	Significance of an Increase in v_f, and in f_z, Respectively, at Different Profile Depths (Having Very Slight Changes in v_c in between Them)
$n_2-v_{f1}-D_1$ 56.59/0.25	$n_2-v_{f2}-D_1$ 56.59/0.90	0.00	Increase of v_f significant for n_2
$n_2-v_{f1}-D_2$ 58.66/0.25	$n_2-v_{f2}-D_2$ 58.66/0.90	0.015	Increase of v_f significant for n_2
$n_2-v_{f1}-D_3$ 61.43/0.25	$n_2-v_{f2}-D_4$ 59.35/0.90	0.019	Increase of v_f significant for n_2
$n_2-v_{f1}-D_4$ 59.35/0.25	$n_2-vf_2-D_4$ 59.35/0.90	0.002	Increase of v_f significant for n_2
$n_1-v_{f1}-D_1$ 28.39/0.49	$n_1-v_{f2}-D_1$ 28.39/1.79	0.229	Increase of v_f not significant for n_1
$n_1-v_{f1}-D_2$ 29.43/0.49	$n_1-v_{f2}-D_2$ 29.43/1.79	0.275	Increase of v_f not significant for n_1
n_1-v_{f1}-D_3 30.82/0.49	$n_1-v_{f2}-D_3$ 30.82/1.79	0.02	Increase of v_f significant for n_1
$n_1-v_{f1}-D_4$ 29.78/0.49	$n_1-v_{f2}-D_4$ 29.78/1.79	0.03	Increase of v_f significant for n_1

Also, a visual examination of the roughness profiles in Figure 8 shows a reduction in Rk when the feed speed was kept constant while the rotation speed increased. For example, an increase in the rotation speed from n_1 to n_2 decreased Rk by approximately 16% for the vf1 feed speed. However, the effect on Rk of changing the rotation speed was not overall significant. This result was also confirmed by testing the local significance for the local profile depths (D_1–D_4) in Table 8.

Similar trends to Rk were observed in the values of Ra parameter (Figure 10), for which an increase in the feed speed has caused a significant increase no matter which rotation speed was used.

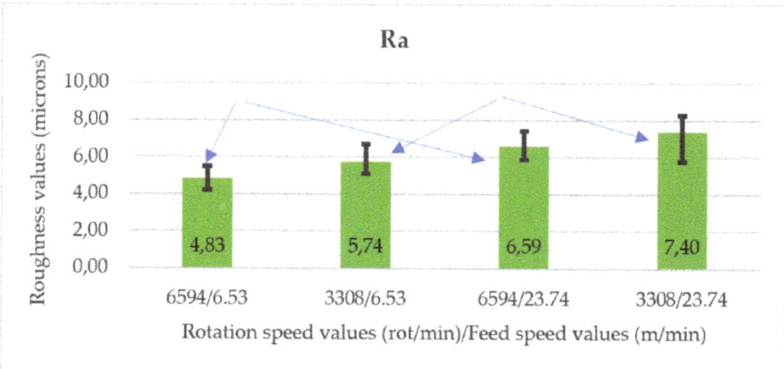

Figure 10. The variation of the mean roughness parameter, Ra, with combinations of the rotation speed with the feed speed. With blue arrows, pair means that differ statistically are marked.

Table 8. Pairwise comparison of parameters combinations (vc/fz) for Rk and the level of significance for an increase in n (and local v_c, respectively).

Parameters Combination 1 (v_c/f_z)	Parameters Combination 2 (v_c/f_z)	*p*-Value	Significance of an Increase in n, When v_f (f_z) is Kept the Same
$n_1-v_{f1}-D_1$ 28.39/0.49	$n_2-v_{f1}-D_1$ 56.59/0.25	0.03	Increase of n significant for v_{f1}
$n_1-v_{f1}-D_2$ 29.43/0.49	$n_2-v_{f1}-D_2$ 58.66/0.25	0.05	Increase of n significant for v_{f1}
$n_1-v_{f1}-D_3$ 30.82/0.49	$n_2-v_{f1}-D_3$ 61.43/0.25	0.752	Increase of n not significant for v_{f1}
$n_1-v_{f1}-D_4$ 29.78/0.49	$n_2-v_{f1}-D_4$ 59.35/0.25	0.792	Increase of n not significant for v_{f1}
$n_1-v_{f2}-D_1$ 28.39/1.79	$n_2-v_{f2}-D_1$ 56.59/0.90	0.001	Increase of n significant for v_{f2}
$n_1-v_{f2}-D_2$ 29.43/1.79	$n_2-v_{f2}-D_2$ 58.66/0.90	0.559	Increase of n not significant for v_{f2}
$n_1-v_{f2}-D_3$ 30.82/1.79	$n_2-v_{f2}-D_3$ 61.43/0.90	0.598	Increase of n not significant for v_{f2}
$n_1-v_{f2}-D_4$ 29.78/1.79	$n_2-v_{f2}-D_4$ 59.35/0.90	0.934	Increase of n not significant for v_{f2}

Another morphological aspect studied in relation to the processing parameters was the raised fiber. For the rotation speed n_1 = 3308 rot/min, the increase more than 3.6 times in the feed speed from v_{f1} = 6.53 m/min to v_{f2} = 23.74 m/min has caused Rpk to increase 2.4 times (from 6.81 µm to 16.51 µm) which was statistically significant (ANOVA followed by Dunnett T3). This means, that an increase in the feed speed is raising the wood fiber (Figure 11). This increase can also be observed from the microscopical images in Figure 12c,d, illustrating a detail of zone 1, zone depicted with red arrows in Figure 12a,b. The raised fibers also appear in the roughness profiles, where they are encircled in red (Figure 8). For the rotation speed n_2, although Rpk increased by 42% when the high feed speed was used and some raised fibers are visible in the roughness profiles, this had no statistical support. The increase in the rotation speed does not produce any significant effect regarding fuzziness or raised fiber.

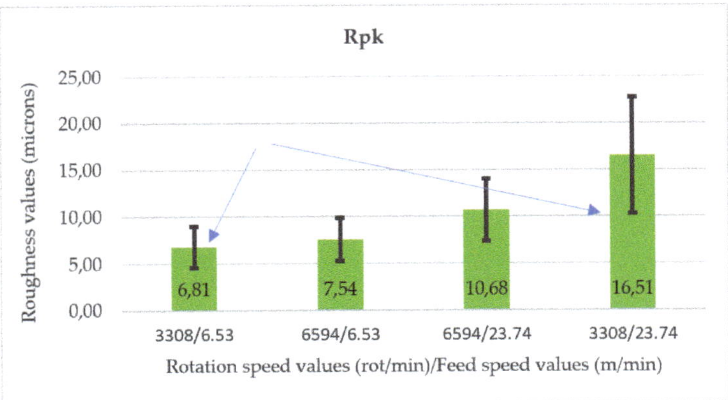

Figure 11. The variation of the reduced peak heights parameter, Rpk, with combinations of the rotation speed with the feed speed. With blue arrows, pair means that differ statistically are marked.

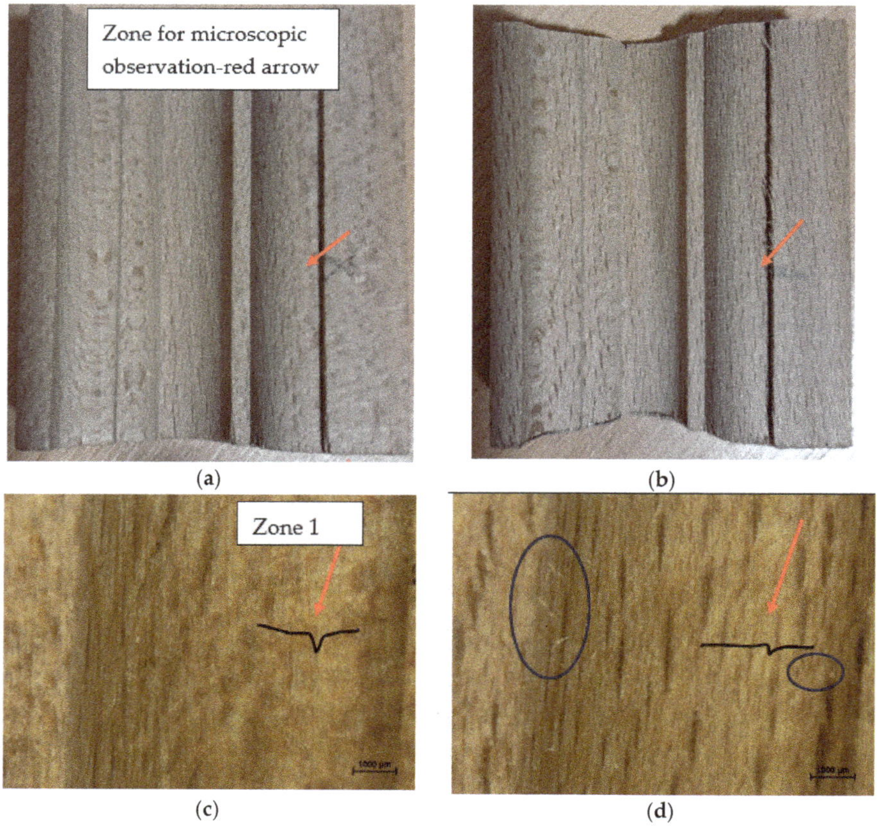

Figure 12. Images of beech milled profiles (**a**,**b**) and microscopic details (**c**,**d**) at magnification 22.5×. (**a**,**c**) Processing by $n_1 - v_{f1}$; (**b**,**d**) processing by $n_1 - v_{f2}$. Examples of raised fibers are marked with blue ellipses.

Wood is a stratified heterogeneous material and apart from outliers located above the core roughness it also has outliers located below the core roughness. They can be seen as surface gaps, exemplified by yellow ellipses in Figure 8. Their origin can be deep anatomical cavities going below the marks caused by the cutting tool, but can also include accidental surface gaps when wood material is detached leaving unwanted voids in the material. Both of them will depend on the type and proportion of anatomical cells encountered by the cutting plane. The weakest zones should be those from the earlywood and those coinciding with the wood rays (radial surfaces). The processed samples did not consider a certain top wood surface orientation, which varied from radial to tangential and combinations like semi-radial and semi-tangential. For the above-mentioned reasons, the parameter Rv, which measures the deepest valley in the profile, had a high standard deviation (Table 4) in comparison with the core roughness Rk, depending on the local wood anatomy. Furthermore, Rv had values in the range of earlywood pores cavities reported for beech (8–45–85 μm) [33] (p. 107) with the majority of them around the mean values given in the literature (45 μm), while some of the deepest valleys were below 70 μm. These observations can lead to the assumption that the isolated deep valleys seen in the roughness profiles, in most cases, belong to inherent wood anatomical cavities and are not processing defects. An exception was made in the situations in which the cutter encountered radial surfaces and this caused detachment of some ray tissues, as can be seen in Figure 13b, which significantly increased Rv (Figure 14), with approximately 40%, for the feed speed

v_{f2} in comparison with v_{f1} (Table 4, Figure 14). Wood pores went beyond the marks left by the tool on the wood, and this is observed by a negative Rsk and a subunitary ratio Rpk/Rvk. The same is true for the ratio A1/A2, which indicates a larger share of the area of the valleys compared to the area of the peaks (Table 4). An increase in the rotation speed from n1 to n2 has significantly reduced Rv (Figure 14) with approximately 33% for the feed speed v_{f1} (ANOVA and Dunnett T3).

(a) (b)

Figure 13. Samples processed by milling with n_2 rotation speed. (**a**) Feed speed v_{f1}; (**b**) feed speed v_{f2}. The red rectangle exemplifies a radially cut surface which contains detached ray tissues.

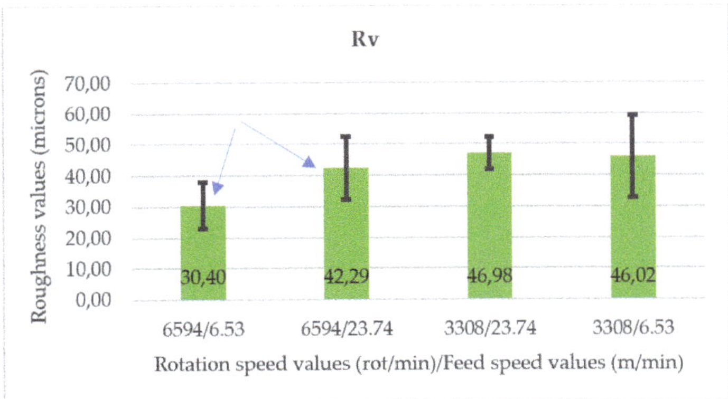

Figure 14. The variation of the deepest valley, Rv, with combinations of the rotation speed with the feed speed. With blue arrows, pair means that differ statistically are marked.

Interpretation of surface quality in the case of wood is complex. For optimization of the cutting quality and selection of the best working parameters, the tool marks are of interest. From this perspective, Rk parameter is representative and has shown it is significantly influenced by the feed speed and was less affected by the selection of the rotation speed. The surface fuzziness and accidental material detachment will be species-dependent and of wood local anatomy. However, their effect on the overall surface quality cannot be neglected. A statistical analysis ANOVA followed by Dunnett T3 test performed on the cumulative effect of processing plus the raised fiber (Rk+Rpk) has led to the same conclusions as in the case of Rk alone: the feed speed was the most important factor affecting the surface quality by milling. Although the surface quality has improved by increasing the rotation speed, its effect was not significant (Tables 9 and 10). The best surface quality was obtained when combining n_2 (6594 rot/min) with v_{f1} (6.53 m/min) and the worst when the rotation speed n_1 (3308 rot/min) was combined with the high feed speed v_{f2} (23.74 m/min), as shown in Figure 15.

Table 9. Dependent variable Rk+Rpk—Test between groups' effects.

Source	Sum of Squares	df	Mean Square	F	Sig
Groups	1541.92	3	513.97	30.7	0.000

Table 10. Dependent variable Rk+Rpk—Dunnet T3 Post Hoc Test for multiple comparisons between groups' rotation speed–feed speed.

Group		Mean Difference	Std. Error	Sig
n_1-v_{f1}	n_1-v_{f2}	-12.59 [1]	513.97	0.000
	n_2-v_{f1}	1.17	1.26	0.921
n_2-v_{f2}	n_1-v_{f2}	-4.80	1.99	0.134
	n_2-v_{f1}	8.96 [1]	1.55	0.000

[1] The error term is Mean Square (Error) = 16.738. The mean difference is significant at the 0.05 level.

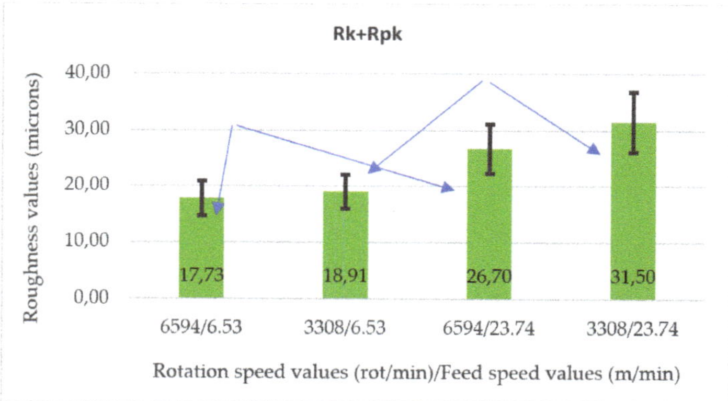

Figure 15. The variation of the cumulative parameter Rk+Rpk with combinations of the rotation speed with the feed speed. With blue arrows, pair means that differ statistically are marked.

When the surface isolated gaps were included, together with the tool marks and the raised fiber, not only the feed speed was a significant factor, but also the rotation speed, the surface quality improving for the highest rotation speed n_2 in comparison with n_1 (Figure 16, Tables 11 and 12).

Another test was conducted to monitor the surface quality after the tool has milled 600 m of the same profile and species, in order to see which is the effect of the tool condition.

Comparative profiles when the tool was sharpened and profiles after the tool milled 600 m are displayed in Figure 17.

Table 11. Dependent variable Rk+Rpk+Rvk—Test between groups' effects.

Source	Sum of Squares	df	Mean Square	F	Sig
Groups	3350.23	3	1116.74	36.49	0.000

Table 12. Dependent variable Rk+Rpk+Rvk—Dunnet T3 Post Hoc Test for multiple comparisons between groups' rotation speed–feed speed.

Group		Mean Difference	Std. Error	Sig
n_1-v_{f1}	n_1-v_{f2}	−14.93 [1]	2.28	0.000
	n_2-v_{f1}	7.97 [1]	2.30	0.015
n_2-v_{f2}	n_1-v_{f2}	−9.18 [1]	2.20	0.002
	n_2-v_{f1}	13.72 [1]	2.00	0.000

[1] The error term is Mean Square (Error) = 30.6. The mean difference is significant at the 0.05 level.

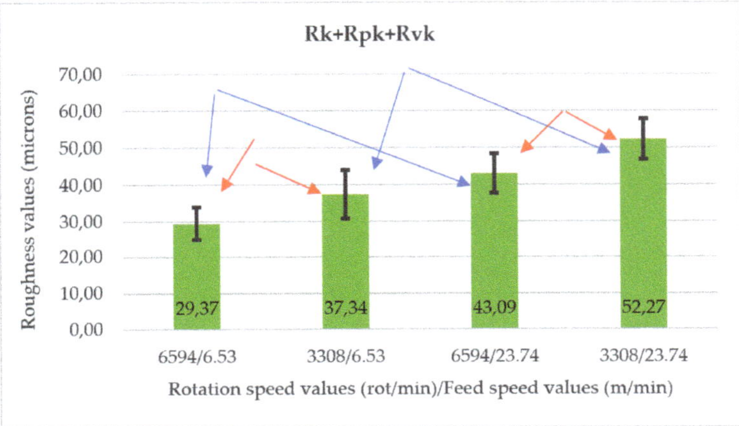

Figure 16. The variation of the cumulative parameter Rk+Rpk+Rvk with combinations of the rotation speed with the feed speed. With blue and red arrows, pair means that differ statistically are marked.

From the profiles presented in Figure 17, it can be observed that after using the tool by continuous processing for 600 m cutting distance, the surface quality improved (Figure 18). The surface improvement was also sensed by touch, but it is visible also in Figure 18c,d taken as example.

A statistical comparison of the composed parameter Rk+Rpk+Rvk was performed in order to obtain a global evaluation of the surface improvement. The ANOVA test followed by a Post Hoc test Games–Howell (groups had unequal number of measurements) indicated a surface improvement after the tool was working 600 m, which was significant when the high feed speed was used (v_{f2} = 23.74 m/min). The improvement was quantified to 30% roughness decrease for the rotation speed n_2 and to 34% when the rotation speed n1 was used.

This result is an indication that the surface quality is not only dependent on the wood material and the processing parameters, but also on the tool working load, which is expected to vary with the milling parameters as well as with the species. It will be interesting, in further tests, to explore the surface quality from the moment the tool is sharpened to the moment when it is worn out, together with microscopic observations regarding the state of the tool cutting edge. This may help establish an ideal working interval for the tool before it needs resharpening or replacement. A worn tool will increase the surface roughness and the occurrence of surface defects, such as pull-out material and fiber detachment.

Finally, another analysis addressed the surface quality at the four profile depths (from Figure 3), in order to see if the variation of the local cutting speed, v_c, implicitly of the feed per tooth, f_z, has a significant impact on the processing roughness quantified by the Rk parameter. The analysis was made for all combinations of rotation speed–feed speed. An

ANOVA single factor test was performed, with a level of significance $p < 0.05$. The results are presented in Table 13.

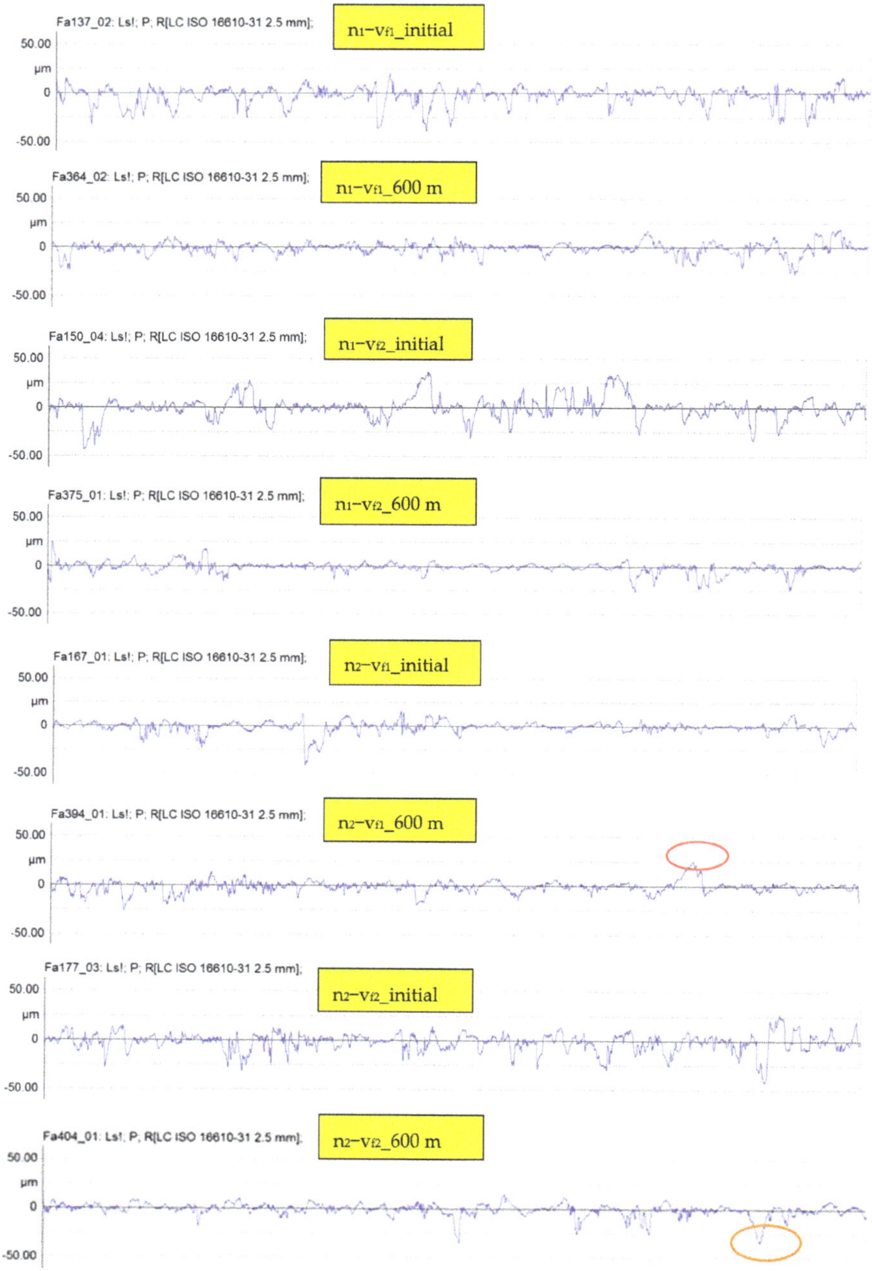

Figure 17. Roughness profiles of each group combination, rotation speed–feed speed comparing the effect of tool condition (when new and sharpened and after 600 m cutting distance). Raised fibers are marked with red; deep isolated valleys are marked with yellow ellipse.

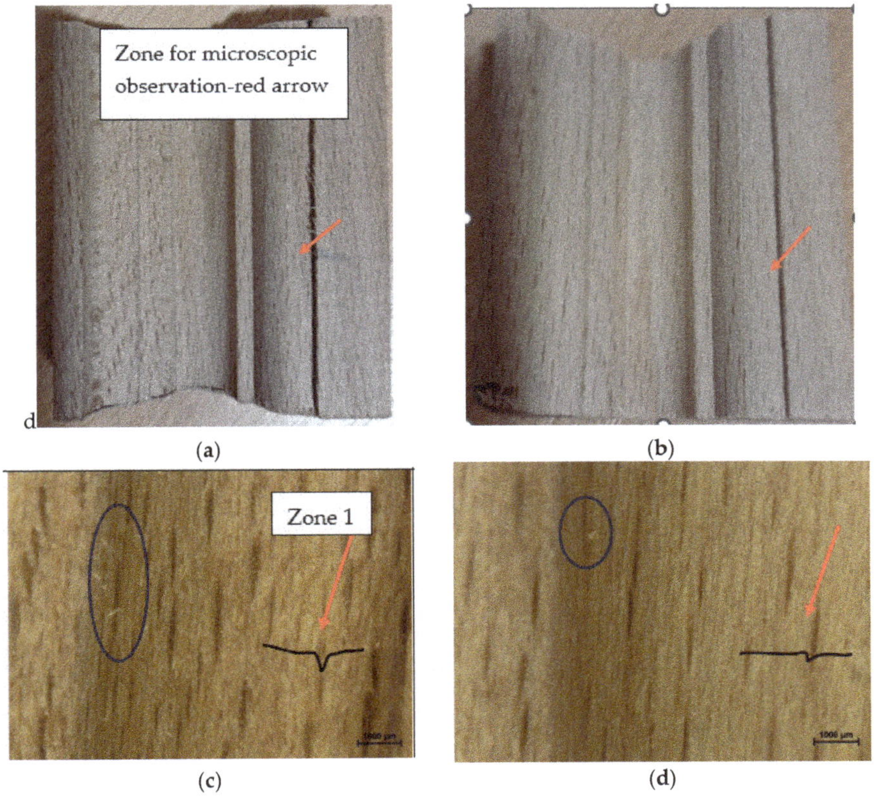

Figure 18. Images of beech milled profiles (**a**,**b**) and microscopic details (**c**,**d**) at magnification 22.5×. (**a**,**c**) Processing by n_1-v_{f2}, sharpened tool; (**b**,**d**) processing by n_1-v_{f2} after 600 m profile milling. Raised fibers are encircled with blue.

Table 13. Rk mean values at the four measured profile depths and the analysis of their statistical difference. Standard deviation in parentheses.

Parameters Combination	*p*-Value	Depth 1	Depth 2	Depth 3	Depth 4	Overall Mean
n_1-v_{f1}	0.04	13.84 (0.05) A	12.83 (1.67) AB	11.05 (1.61) B	10.69 (0.77) B	12.10 (1.70)
n_1-v_{f2}	0.21	11.64 (0.48) A	14.82 (1.52) A	17.02 (4.70) A	16.49 (3.59) A	14.99 (3.41)
n_2-v_{f1}	0.51	9.77 (3.12) A	9.17 (1.21) A	11.62 (0.94) A	10.21 (1.93) A	10.19 (1.94)
n_2-v_{f2}	0.20	17.91 (1.43) A	13.76 (1.66) A	16.06 (2.29) A	16.34 (2.84) A	16.02 (2.38)

Note: Values on the same row with the same letter do not differ statistically.

From Table 13, it can be concluded that the processing quality measured at different depths does not differ statistically, in spite of the local differences in the cutting speed.

There are different opinions in the literature regarding the effect of the cutting speed during machining processes. Some researchers found that the cutting speed has no effect on the surface quality, while others found that the cutting speed has a slight effect depending on process conditions [15].

The results regarding the effect of milling parameters on the wood surface quality are not surprising and are expected, but in comparison with other studies from the literature bring a novel approach to the evaluation of the surface quality, regarding wood as a stratified material. A stratified analysis of surface roughness parameters can not only clarify the surface complex morphology, but it is helpful to determine the contribution of processing parameters to the overall surface quality and their significance.

4. Conclusions

This study has shown that the surface quality should be analyzed on stratified levels of irregularities differentiating between, waviness, tool marks, fuzzy grain, and accidental surface gaps overlapped onto the deep anatomical cavities of the wood, such as pores from earlywood. This is because the milling quality will be dependent not only on the processing parameters and tool characteristics, but also on the wood species and its anatomical structure.

The interpretation of the results was focused on variations in v_f, respectively n, since f_z is directly influenced by the feed speed v_f, for a constant n and for the same number of cutters (a single type of tool was used in this research), while the cutting speed, v_c, differs very slightly with the local tool diameter (depth of cut, respectively), for a given rotation speed, n.

An increase in the feed speed, v_f, increases tool vibration, which leaves deeper waves in the surface. The surface waviness measured by Wa increased more than double when the feed speed increased approximately 3.6 times, from v_{f1} = 6.53 m/min, to v_{f2} = 23.74 m/min. Increasing the rotation speed from n_1 to n_2 (double) has decreased Wa with 20%, but without statistical support.

An increase in the feed speed increases the toolmarks approximated by Rk which are more pronounced as the rotation speed becomes higher. Rk increased with the feed speed (respectively, f_z for a constant n and number of cutters) for n_2 = 6594 rot/min by a significant 57%. An increase in the rotation speed from n_1 to n_2 has reduced Rk, but the effect was not significant. For example, an increase in the rotation speed from n_1 to n_2 decreased Rk by approximately 16% for the v_{f1} feed speed. Similar behavior was noticed for the Ra parameter. The processing quality (Rk) measured at different depths did not differ statistically, in spite of the local differences in the cutting speed.

An increase in the feed speed raises the wood fiber. Therefore, the Rpk parameter increased 2.4 times when the feed speed increased 3.6 times. The increase in the rotation speed does not produce any significant effect regarding fuzziness or raised fiber.

Wood pores were deeper than the marks left by the tool on the wood and their effect on surface quality was disregarded in the above conclusions. However, some gaps deeper than the core roughness and caused by material detachment were noticed when the milling encountered radial surfaces, thus affecting the ray tissues. This did not change the conclusion that the smoothest beech surface is obtained when the smallest feed speed (6.53 m/min) is used. However, by including the surface gaps in the analysis by referring to the parameter Rk+Rpk+Rvk, it attributed a significance to the rotation speed as well, such as the higher rotation speed (6594 rot/min) also contributing to the surface quality improvement.

The processing quality measured at different depths does not differ statistically, in spite of the local differences in the cutting speed.

The statistical analysis helped us to understand the hierarchy of influence factors and their significance. In this respect, the most important factor affecting the surface quality was the feed speed.

The moment of measuring the surface roughness, such as immediately after sharpening or after a working period, influences the result of the surface quality. Measurements of surface quality after the tool processed 600 m of beech material improved the surface quality by 30%.

In further work, such stratified analysis will provide a more in-depth understanding and quantification of the tool interaction with the wood material and will help the selection of the most appropriate processing parameters as a function of the tool and species variables.

Author Contributions: Conceptualization, L.-M.B. and L.G.; data curation, L.-M.B. and L.G.; formal analysis, L.-M.B.; funding acquisition, L.-M.B.; investigation, L.-M.B. and L.G.; methodology, L.-M.B.; project administration, L.-M.B.; resources, L.-M.B. and L.G.; software, L.-M.B. and L.G.; supervision, L.-M.B. and L.G.; validation, L.-M.B. and L.G.; writing—original draft, L.G.; writing—review and editing, L.-M.B. and L.G. All authors have read and agreed to the published version of the manuscript.

Funding: This research received no external funding.

Institutional Review Board Statement: Not applicable.

Informed Consent Statement: Not applicable.

Data Availability Statement: Data are contained within the article.

Acknowledgments: We hereby acknowledge the structural funds project PRO-DD (POS-CCE, O.2.2.1., ID 123, SMIS 2637, No. 11/2009) for providing the infrastructure used in this work and the Contract No. 7/9.01.2014.

Conflicts of Interest: The authors declare no conflict of interest.

References

1. Thoma, H.; Peri, L.; Lato, E. Evaluation of Wood Surface Roughness Depending on Species Characteristics. Maderas. *Cienc. Tecnol.* **2015**, *17*, 285–292. [CrossRef]
2. Hazir, E.; Koc, K.H. Optimization of wood surface machining parameters in CNC routers: Response surface methodology (RSM) approach. *Int. J. Sci. Res. Eng. Technol.* **2016**, *5*, 494–501.
3. Hazir, E.; Koc, K.H. Optimization of wood machining parameters in CNC routers: Taguchi orthogonal array based simulated angling algorithm. *Maderas Cienc. Tecnol.* **2019**, *21*, 493–510. [CrossRef]
4. Westkämper, E.; Riegel, A. Qualittskriterien fur Geschlieffene Massivholzoberflchen. *Holz Roh Werkst.* **1993**, *51*, 121–125. [CrossRef]
5. Krisch, J.; Csiha, C. Analysing wood surface roughness using an S3P Perthometer and computer based data processing. In Proceedings of the XIII Sesja Naukowa Badania dla Meblarstwa, Poznan, Poland, 26 November 1999.
6. Magoss, E.; Sitkei, G. Influence of wood structure on the surface roughness at milling operations. In Proceedings of the 4th ICWSF, Missenden Abbey, UK, 14–16 July 1999.
7. Gurău, L. The Roughness of Sanded Wood Surfaces. Ph.D. Thesis, Forest Products Research Centre, Buckinghamshire Chilterns University College, Brunel University, London, UK, 2004.
8. Thibaut, B.; Denaud, L.; Collet, R.; Marchal, R.; Beauchêne, J.; Mothe, F.; Méausoone, P.J.; Martin, P.; Larricq, P.; Eyma, F. Wood machining with a focus on French research in the last 50 years. *Ann. For. Sci.* **2016**, *73*, 163–184. [CrossRef]
9. Gurău, L.; Irle, M. Surface roughness evaluation methods for wood products: A review. *Curr. For. Rep.* **2017**, *3*, 119–131. [CrossRef]
10. Gawronski, T. Optimisation of CNC routing operations of wooden furniture parts. *Int. J. Adv. Manuf. Technol.* **2013**, *67*, 2259–2267. [CrossRef]
11. Sutcu, A. Investigation of parameters affecting surfaces roughness in CNC Routing operation on Wooden EGP. *Bioresources* **2013**, *8*, 795–805.
12. Koc, K.H.; Erdinler, E.S.; Hazir, E.; Öztürk, E. Effect of CNC application parameters on wooden surface quality. *Measurement* **2017**, *107*, 12–18. [CrossRef]
13. Gürgen, A.; Çakmak, A.; Yldiz, S.; Malkoçoğlu, A. Optimization of CNC operating parameters to minimize surface roughness of pinus sylvestris using integrated artificial neural network and genetic algorithm. *Maderas Cienc. Tecnol.* **2022**, *24*, 5. [CrossRef]
14. Jiang, S.; Buck, D.; Tang, Q.; Guan, J.; Wu, Z.; Guo, X.; Zhu, Z.; Wang, X. Cutting Force and Surface Roughness during Straight-Tooth Milling of Walnut Wood. *Forests* **2022**, *13*, 2126. [CrossRef]
15. Costes, J.P.; Larricq, P. Towards high cutting speed in wood milling. *Ann. For. Sci.* **2002**, *59*, 857–865. [CrossRef]
16. Piernik, M.; Pinkowski, G.; Krauss, A. Effect of chip thickness, wood cross-sections, and cutting speed on surface roughness and cutting power during up-milling of beech wood. *BioResources* **2023**, *18*, 6784–6801. [CrossRef]
17. Pakzad, S.; Pedrammehr, S.; Hejazian, M. A Study on the Beech Wood Machining Parameters Optimization Using Response Surface Methodology. *Axioms* **2023**, *12*, 39. [CrossRef]

18. Rabiei, F.; Yaghoubi, S. A comprehensive investigation on the influences of optimal CNC wood machining variables on surface quality and process time using GMDH neural network and beest optimization algorithm. *Mater. Today Commun.* **2023**, *36*, 106482. [CrossRef]
19. Zhu, Z.; Jin, D.; Wu, Z.; Xu, W.; Yu, Y.; Guo, X.; Wang, X. Assessment of Surface Roughness in Milling of Beech Using a Response Surface Methodology and an Adaptive Network-Based Fuzzy Inference System. *Machines* **2022**, *10*, 567. [CrossRef]
20. Kilic, M.; Hiziroglu, S.; Burdurlu, E. Effect of machining on surface roughness of wood. *Build. Environ.* **2006**, *41*, 1074–1078. [CrossRef]
21. Magoss, E. General regularities of wood surfaces roughness. *Acta Silv. Lignaria Hung.* **2008**, *4*, 81–93.
22. Gurău, L.; Coşereanu, C.; Paiu, L. Comparative Surface Quality of Larch (*Larix decidua* Mill.) Fretwork Patterns Cut through by CNC Routing and by Laser. *Appl. Sci.* **2021**, *11*, 6875. [CrossRef]
23. *ISO 4287*; Geometrical Product Specifications (GPS)—Surface Texture—Profile Method—Terms, Definitions and Surface Texture Parameters. International Organization for Standardization: Geneva, Switzerland, 2009.
24. *ISO 13565-2*; Geometrical Product Specifications (GPS)—Surface Texture: Profile Method; Surfaces Having Stratified Functional Properties. Part 2: Height Characterisation Using the Linear Material Ratio Curve. International Organization for Standardization: Geneva, Switzerland, 1998.
25. *ISO/TS 16610-31*; Geometrical Product Specification (GPS)—Filtration—Part 31: Robust Profile Filters: Gaussian Regression Filters. International Standards Organisation: Geneva, Switzerland, 2010.
26. *EN ISO 683-1*; Heat-Treatable Steels, Alloy Steels and Free-Cutting Steels. Part 1: Non-Alloy Steels for Quenching and Tempering. International Organization for Standardization: Geneva, Switzerland, 2016.
27. *EN ISO 4957*; Tool Steel. International Organization for Standardization: Geneva, Switzerland, 2018.
28. Rolleri, A.; Burgos, F.; Aguilera, A. Surface Roughness and Wettability Variation: The effect of Cutting Distance during Milling of Pinus Radiata Wood. *Drv. Ind.* **2016**, *67*, 223–228. [CrossRef]
29. Fujiwara, Y.; Fujii, Y.; Sawada, Y.; Okumura, S. Assessment of wood surface roughness: A comparison between tactile roughness and three-dimensional parameters derived using a robust gaussian regression filter. *J. Wood Sci.* **2004**, *50*, 35–40. [CrossRef]
30. Gurău, L.; Mansfield-Williams, H.; Irle, M. Filtering the Roughness of a Sanded Wood Surface. *Holz Roh Werkst.* **2006**, *64*, 363–371. [CrossRef]
31. Piratelli-Filho, A.; Sternadt, G.H.; Arencibia, R.V. Removing deep valleys in roughness measurement of soft and natural materials with mathematical filtering. *Ciência Eng.* **2012**, *21*, 29–34. [CrossRef]
32. Tan, P.L.; Sharif, S.; Sudin, I. Roughness models for sanded wood surfaces. *Wood Sci. Technol.* **2012**, *46*, 129–142. [CrossRef]
33. Wagenführ, R. *Holzatlas*, 5th ed.; Fachbuchverlag: Leipzig, Germany, 2000; p. 707.

applied sciences

Article

Adhesion of Varnish Coatings as a Background for Analogue and Digital Printing Technologies

Maciej Tokarczyk [1,2], Barbara Lis [1], Emilia Adela Salca [3,*] and Tomasz Krystofiak [1,*]

1 Department of Wood Science and Thermal Techniques, Faculty of Forestry and Wood Technology, Poznan University of Life Sciences, Wojska Polskiego 28, 60-627 Poznan, Poland; maciej.tokarczyk@up.poznan.pl (M.T.); barbara.lis@up.poznan.pl (B.L.)
2 BORNE-FURNITURE Company, Złotego Smoka St. 23, 66-400 Gorzów Wielkopolski, Poland
3 Faculty of Furniture Design and Wood Engineering, Transilvania University of Brasov, Universitatii 1, 500068 Brasov, Romania
* Correspondence: emilia.salca@unitbv.ro (E.A.S.); tomasz.krystofiak@up.poznan.pl (T.K.)

Abstract: In analogue and digital printing technologies, from 3 up to 12 layers of lacquer products are applied. Technological parameters significantly influence the adhesion in the coating system. This article refers to the analysis of the influence of selected technological parameters, such as the number of layers, energy doses distributed by the radiators, and line speed, on the topography and adhesion of varnish coatings formed in the process of varnishing with rollers and UV-curing systems. The appropriately prepared surface can be used as a background layer for the analogue and digital printing technology. Manufacturers must adapt the production process to the particular varnish to obtain finished products with the best possible performance properties. The state of surface free energy and finally adhesion can be assessed by theoretically determining the possibility of an adhesive bonding between the product and the substrate, taking into account the assumptions of the adsorption theory of adhesion and measurement of the contact angle (Θ). An experimental confirmation of adhesion measurements included removing the coatings from the substrate via stamps glued to the coating.

Keywords: adhesion; analogue technology; printing technology; roughness; surface; topography; UV lacquer system; wettability

Citation: Tokarczyk, M.; Lis, B.; Salca, E.A.; Krystofiak, T. Adhesion of Varnish Coatings as a Background for Analogue and Digital Printing Technologies. *Appl. Sci.* **2024**, *14*, 304. https://doi.org/10.3390/app14010304

Academic Editor: Giuseppe Lazzara

Received: 30 November 2023
Revised: 24 December 2023
Accepted: 26 December 2023
Published: 29 December 2023

1. Introduction

The increasing shortage of wood promotes the development of technologies that apply other materials. Wood-based boards provide an excellent alternative to wood in the furniture industry, the advantage of which is a lower price and wide physical-mechanical parameters [1,2]. The properties of boards shaped during the technological process determine their application in various wood-processing industries. In the furniture and furnishing industry, MDF and HDF boards are widely used [3]. One of the many possibilities of using them in the production of furniture or interior doors made on cellular wood boards are the outer layers. The boards in their raw state are considered unattractive for various projects, so they are subjected to finishing via various methods. Analogue and digital printing techniques offer interesting solutions for increasing product attractiveness. In the former, the pattern carriers are engraved rollers, while in the latter, data in the digital form are transferred to the printhead [4–6]. UV varnish products, which are considered environmentally friendly, have been widely used in these technologies [5–10]. Their crosslinking occurs via polymerization based on a mechanism of free-radical reaction, which is initiated by UV radiation. Manufacturers need to adapt the production process to the particular varnish to obtain finished products with the best possible performance properties. Adhesion is one of the most important parameters determining the functional properties and durability of a finish [11–13]. It is the result of the wetting and curing processes of varnish products on the substrate. Its value is influenced by a number of factors that can include the type and preparation of the substrate, the type of varnish, application, and curing parameters [14,15]. Adhesion can be assessed by theoretically determining the possibility of an adhesive bonding between the product and the substrate, taking into account the assumptions of the adsorption theory of adhesion and experimental studies. In theoretical terms, adhesion requires the determination of the work of adhesion (Wa) and the surface free energy of the solid (γ_S), taking into account their dispersive and polar components. To determine them, measurement of the contact angle (Θ) is used.

The contact angle depends on a number of factors, which include the type of film-forming polymer and its degree of cross-linking, the content of auxiliary agents, the type of substrate, and the surface roughness [16]. On the other hand, experimental measurements include removing the coatings from the substrate via stamps glued to the coating. For this purpose, manual, hydraulic, or pneumatic instruments are used. The results are given in numerical form (MPa), which expresses the minimum tensile stress required for detachment [14]. In the case of inadequate adhesion and cohesive strength, the coating will exhibit defects in the form of protruding from the substrate, contributing to the formation of incomplete products. Therefore, process parameters must be developed to ensure maximum adhesion, both between the substrate and the varnish coating and between the coatings of a given finish. The most important parameters include adjusting the thickness of the applied product, conveyor speed, radiator type, minimum radiation dose, and distance between the lamp and the substrate.

In addition, using very high line speeds poses quite a challenge in determining the above data to obtain the optimum degree of curing and coating properties. Many studies can be found in the references on this important technological issue of adhesion. They have focused, among other things, on the influence of surface preparation method, surface roughness [13,14,17–20], application technology, type of varnish product [11,13], the composition of varnish products [13,20–25], layer thickness [11,13], number of varnish coatings, and aging [20,26–32]. The wide range of experiments conducted previously proves the importance of this topic. The experiments illustrate what factors need to be considered when specifying process parameters for a quality product.

No results of studies devoted to this subject have been encountered in the literature for HDF surfaces finished with UV LED varnishes. It seems fully justified as work constantly is carried out on developing, improving, and implementing various UV varnish products in industrial practice and technology, and the scope of the undertaken projects is subject to the highest confidentiality clause. This constitutes a competitive advantage for the company. In this context, there is a legitimate need to research and develop an understanding of the phenomena of wetting and final adhesion.

This article is devoted to the analysis of the influence of selected technological parameters, such as the number of layers, energy doses distributed by the radiators, and line speed, on the adhesion of varnish coatings formed in the process of varnishing with a roller. The objective of this paper was implemented by evaluating the phenomenon of wetting on coatings formed on HDF board, determined based on the measurements of angle Θ, the γ_S, and Wa, together with their dispersion and polar share, and experimentally using the pull-off method, taking into account the impact of the surface roughness on this process.

2. Materials and Methods

The surface finishing of a board with 100 $\times$ 60 cm HDF outer cladding was carried out with the roller method under the technological conditions of Borne Furniture. The application process of multilayer systems of varnish products was carried out at line feed rates of 25 and 40 m/min according to the general idea presented in Figure 1.

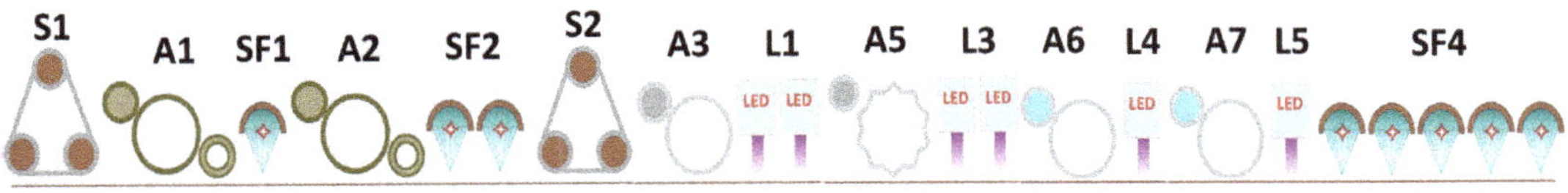

Figure 1. Selected schematic diagram for finishing the board furniture components with a roller method. Process description: (**A1**) putty 1–30 g/m^2; (**A2**) putty 2–20 g/m^2; (**A3**) basecoat 1–10 g/m^2; (**A5**) basecoat 2–25 g/m^2; (**A6**) topcoat 1–12 g/m^2; (**A7**) topcoat 2–6 g/m^2; (**S1**, **S2**) sanding; (**L1**, **L3**, **L4**, **L5**) LED lamp 396 nm 12 W/cm^2; (**SF1**, **SF2**, **SF4**) super focus mercury lamp 120 W/cm.

Table 1 shows the basic properties of UV varnish products based on the manufacturer's technical sheets.

Table 1. Properties of UV varnish products.

	Name of UV Varnish Products		
Properties	**Q-UV 03040 Putty**	**IQ-UVC03284 Basecoat**	**IQ-UVC03392 Topcoat**
Polymer base Color Solid content [%]	acrylic colorless 95.3 ± 3	acrylic white 97.6 ± 3	acrylic white 97.6 ± 3
Viscosity of delivery	65–85 (flow cup 8 mm)	90–140 (flow cup 6 mm)	35–50 (flow cup 6 mm)
Brookfield CAP 2000+/cone 2 [mPa·s]	1100–1500	-	-
Processing temperature [°C]		Between 20–50	

The templates prepared at different stages were transported to the laboratory, cut into samples, and left in an air-conditioned laboratory at 23 ± 2 °C, RH 50 ± 5% (Borne Furniture Company in Gorzow Wlkp. Poland). A summary of the samples with the accepted labeling is given in Table 2.

Table 2. Preparation of the test substrate.

Labeling of Samples (Systems)	Operations Performed in Preparation of the Test Substrate	Speed of Line [m/min]
1-3d	The surface of a cellular board with 1 layer 30 g/m^2 of a putty at the maximum setting of lamps focusing a beam of electromagnetic radiation	40
1-3a	The surface of a cellular board with 1 layer 30 g/m^2 of a putty at the maximum setting of lamps focusing a beam of electromagnetic radiation	25
1-7d	The surface of a cellular board with 1 layer of a filler with increased input, 1 layer of a filler with decreased input, and 1 primer at the maximum setting of lamps focusing a beam of electromagnetic radiation behind the 2 filler and maximum setting of LED modules behind the 1 basecoat 20 g/m^2	40
1-7a	The surface of a cellular board with 1 layer of a filler with increased input, 1 layer of a filler with decreased input, and 1 primer at the maximum setting of lamps focusing a beam of electromagnetic radiation behind the 2 filler and maximum setting of LED modules behind the 1 basecoating 20 g/m^2	25
1-8d	The surface of a cellular board with 1 layer of a putty with increased input, 1 layer of a filler with decreased input, and 1 primer at 50% setting of lamps focusing a beam of electromagnetic radiation behind the 2 filler and maximum setting of LED modules behind the 1 basecoating 10 g/m^2	40

Table 2. *Cont.*

Labeling of Samples (Systems)	Operations Performed in Preparation of the Test Substrate	Speed of Line [m/min]
1-8a	The surface of a cellular board with 1 layer of a putty with increased input, 1 layer of a putty with decreased input, and 1 primer at 50% setting of lamps focusing a beam of electromagnetic radiation behind the 2 filler and maximum setting of LED modules behind the 1 basecoating 10 g/m^2	25
1-14d	The surface of a cellular board with 2 layers of a putty, 2 layers of a primer using OPTI, and 2 layers of topcoat 12 g/m^2 at the maximum setting of lamps focusing a beam of electromagnetic radiation and maximum for each LED module.	40
1-14a	The surface of a cellular board with 2 layers of a putty, 2 layers of a primer using OPTI, and 2 layers 12 g/m^2 of topcoat at the maximum setting of lamps focusing a beam of electromagnetic radiation and maximum for each LED module.	25
1-15d	The surface of a cellular board with 2 layers of a putty, 2 layers of a primer using OPTI, and 1 layer 6 g/m^2 of topcoat at the maximum setting of lamps focusing a beam of electromagnetic radiation and maximum for each LED module.	40
1-15a	The surface of a cellular board with 2 layers of a putty, 2 layers of a primer using OPTI and 1 layer of topcoat 6 g/m^2 at the maximum setting of lamps focusing a beam of electromagnetic radiation and maximum for each LED module.	25

After that, the visual assessment was carried out and the following measurements were taken: topographic, surface roughness, contact angle, and adhesion.

2.1. Visual Assessment

Visual assessment was conducted with an unaided eye at a distance of 250 mm, incident at an acute angle to highlight the presence of potential surface defects.

2.2. Surface Topography

Surface topography was measured with a OneAttension Theta optical tensiometer (Biolin Scientific AB, Västra Frölunda, Sweden). The parameters of the measurement system are given in Table 3.

Table 3. Measurement system's parameters [33].

Method	Fringe Projection Phase-Shifting
XY pixel size	1.1 μm × 1.1 μm
Measured range in Z direction	1–60 μm
Lateral sampling (XY)	1.41 mm × 1.06 mm
Measurement speed	5–30 s (1280 × 960 measurement points)
Imaging options	Optical image, 2D and 3D roughness graphs
Method	Fringe projection phase-shifting
XY pixel size	1.1 μm × 1.1 μm
Measured range in Z direction	1–60 μm
Lateral sampling (XY)	1.41 mm × 1.06 mm

2.3. Roughness

A SJ-210 profilometer was applied to test the geometric structure of the surface. Specialized software (Version; SJ-210 V.1.000) was also used, and it aimed to control the profilometer and determine the numerical values of roughness parameters. The element that represented the surface was a diamond blade with an apex angle of 60° and a radius of 2 μm. Prior to the experiment, the profilometer was calibrated according to the roughness standard ISO 4287-1998 [34]. After that, the measurement conditions were entered, and the roughness parameters were specified (speed 0.5 mm/s, ln = 12.5 mm (length of the measuring section), and λc = 0.25 mm (cut-off)). Three indices were determined: Ra—asymmetric mean of the profile ordinates, Rz—roughness height according to 10 points, R max—highest profile height.

2.4. Contact Angle

Measurements were carried out with a PG3 goniometer using distilled water (at a temperature of 20 °C) as the wetting liquid according to the PN-EN 828 standard [35]. A water droplet with a volume of 3.5 μL was applied to the surface with an integrated micro dosing pump and the camera recorded its behavior in contact with the substrate for 60 s. Based on the data obtained, specialized software analyzed the droplet's shape by determining the contact angle's value. Building upon the data of the contact angle obtained for a water droplet stabilized for 10 s and in compliance with the assumptions of the adsorption theory of adhesion, the interactions between the substrate–individual layers of coating system of the values of surface free energy (γ_S) and work of adhesion (Wa), together with the dispersive and polar components, were determined [16].

2.5. Adhesion of Coatings

The test was carried out in compliance with EN ISO 4624 [36]. Silane epoxy adhesive (viscosity 50,000 mPa·s, processing time 30 min, handling strength after 120 min) from Jowat Polska Company was used to glue stamps measuring 20 mm in diameter to the test surfaces. After a conditioning period of 168 h, the surfaces of the specimens were notched around the measuring stamps using a hob cutter. They were then peeled off with a hydraulically driven PostiTest test apparatus. Nine tests were carried out on each test system. The images of delamination under destructive loads were assessed visually, taking into account the rating scale in Table 4.

Table 4. Measurement system's parameters [36].

Detachment Type	Detachment Occurring in a Given System
A	Cohesive in a substrate
A/B	Adhesive between a substrate and the first coating
B	Cohesive in the first coating
B/C	Adhesive between the first and second coating
-/Y	Adhesive of the last coating and glue
Y	Cohesive in glue
Y/Z	Adhesive between glue and a measuring stamp

3. Results

The resulting systems were of high quality, with no defects registered, which indicates that the substrate finishing stages were carried out correctly. In addition, no defects were observed after their transport to the testing laboratory, which confirms the adequate delivery conditions and control over the mode of transport. When assessing the appearance of the surface, an increase in the homogeneity of the varnish coating was observed in the function of the applied coatings.

3.1. Visual Assessment

The visual assessment of the tested systems was complemented by quality verification based on tactile sensations. In contact with the hand, the final coatings evoked a pleasant tactile sensation and sensory satisfaction. It is notable that this effect is of great importance to furniture users. As an example, Table 5 shows a graphical representation of the surfaces of the tested systems.

Table 5. Image of surface topography with a travel rate of 25 m/min.

Sample Labelling	Surface Topography
1-3a	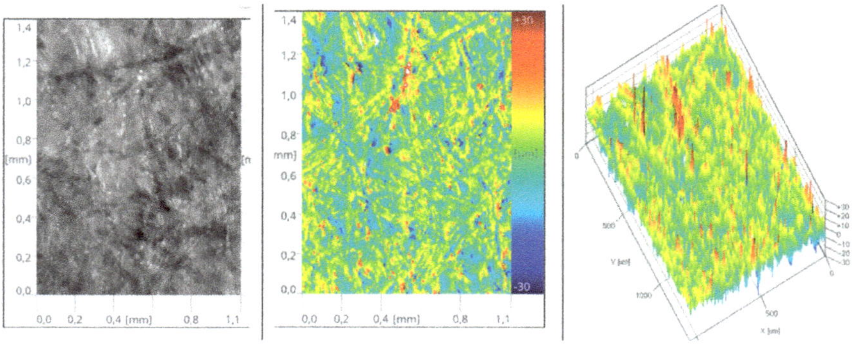
1-7a	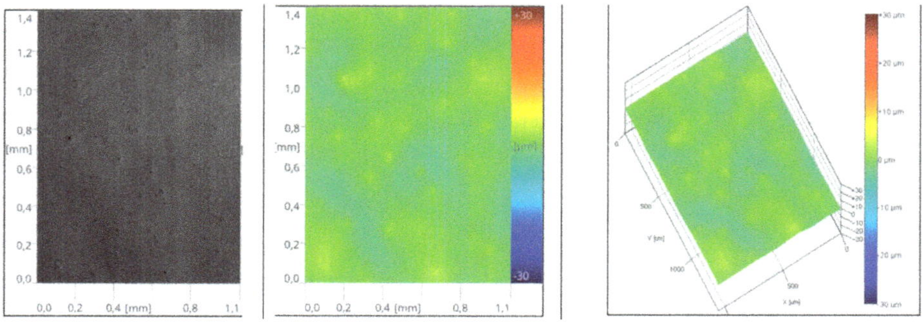
1-8a	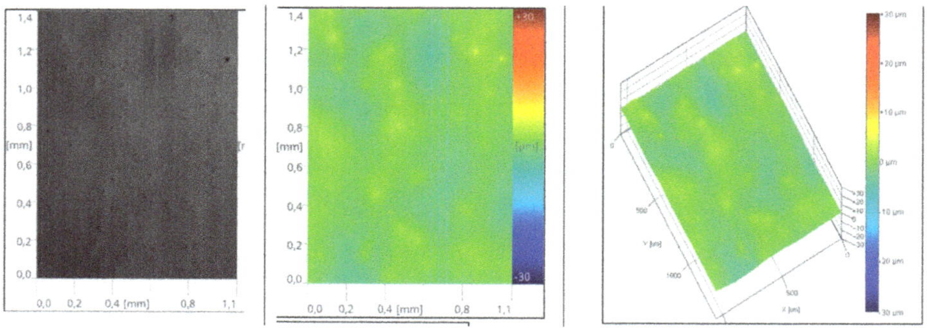
1-14a	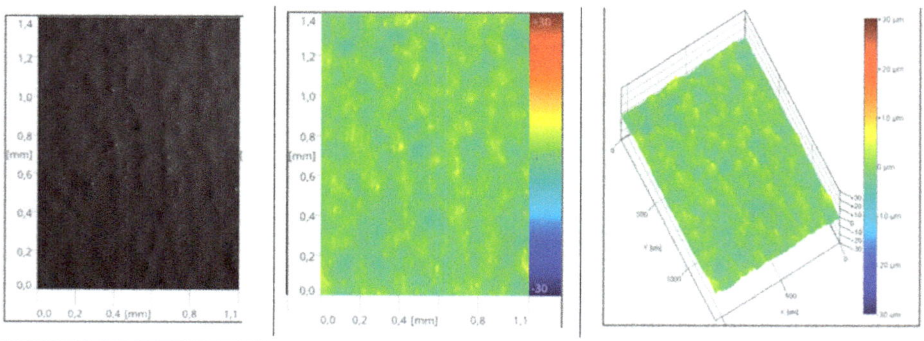

Table 5. *Cont.*

Sample Labelling	Surface Topography		
1-14a	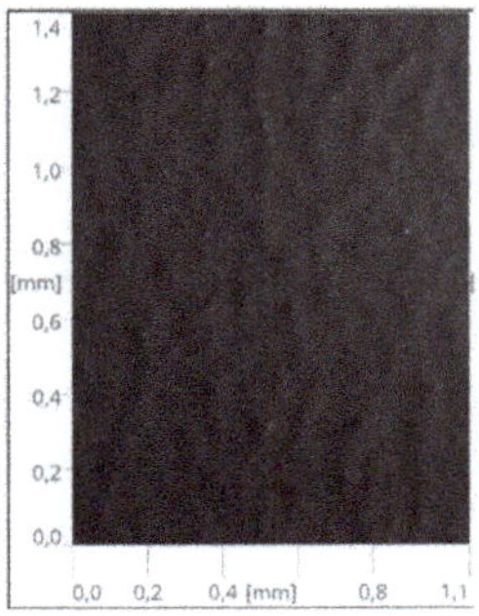	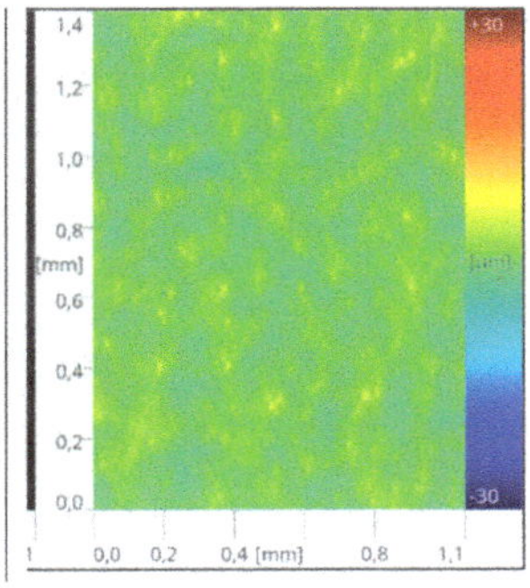	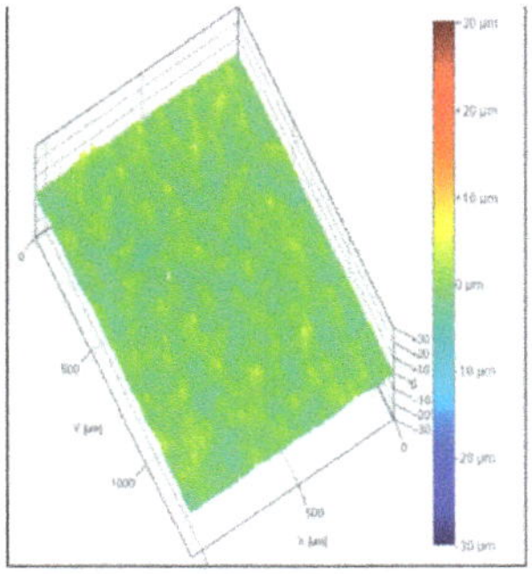

On the topographic images of the selected surfaces, it can be seen that both the applied layers and the applied doses of energy distributed by the radiators influenced the surface structure. The recorded changes were visible in the roughness measurements. However, the study showed values of final coatings that were unexpectedly higher than the values of primers. The average values of 10 measurements of the recorded roughness parameters are shown in Figure 2.

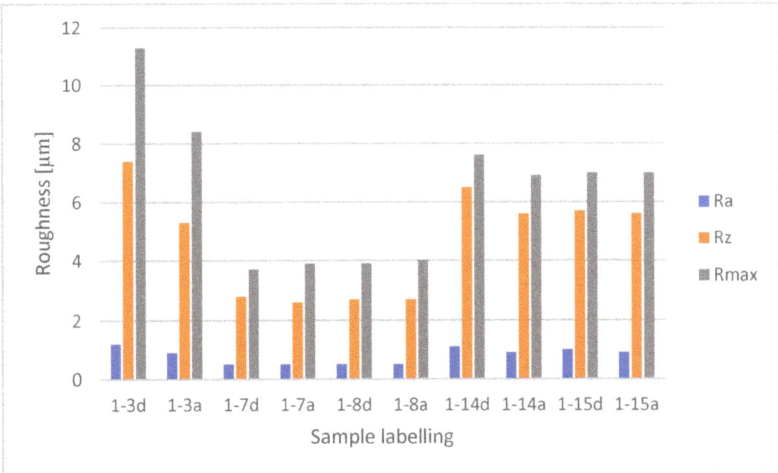

Figure 2. Values of Ra, Rz, and Rmax parameters for the tested systems.

From the results obtained, it can be observed that the determined average values of the Ra parameter, irrespective of the tested system, were approximately six times lower than Rz. In addition, lower values were found for the interlayer systems with the exception of the one layer of putty (systems 1-3). It is most likely that this layer, which is in direct contact with the HDF board despite the doubled input, penetrated the substrate, which influenced the recorded parameters. Studies conducted by Bialecki et al. in 2008 [37] proved the influence of the structure of the HDF board on the amount of varnish applied to achieve an adequate surface preparation. Parameters that were indicative of the above correlation were the selected roughness indices. This would confirm the penetration of the putty deep into the board while creating a rougher surface. The application of an increased putty layer, a second putty layer, sanding, and a primer layer reduced the roughness level by almost half regardless of the conveyor speed. For the final coatings, the roughness was almost an order of magnitude higher than the systems with putty and primer layers. However, with regard to the one putty layer, they showed similar values to the determined indices. This could be due to the different rheological properties of the applied products, different energy doses distributed by the radiators, the reduction in product volume, and their shrinking during the polymerization process [38–40]. The geometrical structure of the surface, shaped in the technological process, can influence the wetting processes. This relationship is indicated in the references [41,42].

3.2. Contact Angle

Figure 3 presents the average contact angle values as a function of the Ra parameter.

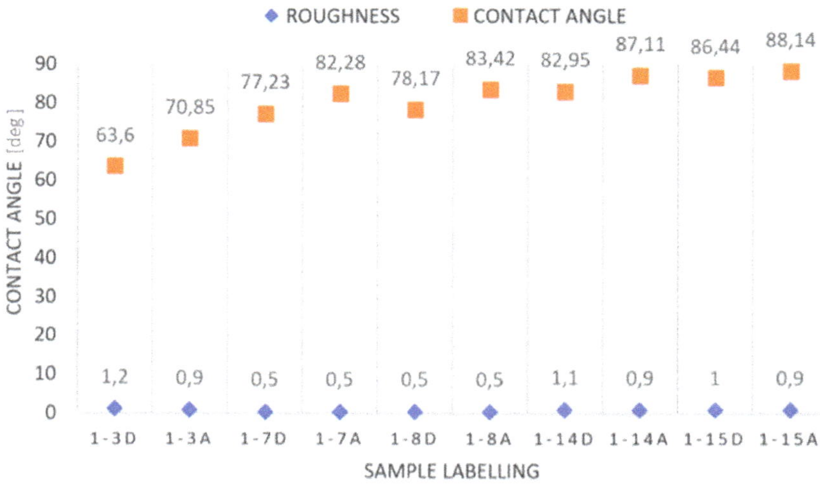

Figure 3. Formation of contact angle for the tested systems.

However, no unambiguous correlations between the change in contact angle and the recorded roughness index were found in the studies. The mechanisms of interaction between selected technological parameters do not enable us to determine the reasons for such discrepancies unambiguously. The coating formed from one putty layer at the assumed process parameters is characterized by the highest roughness. In this case, the hydrophobic character of the surface was weakened, which ensured good adhesion properties for the subsequent layers. Further layer application resulted in a decrease in wettability, which contributed to increased contact angle. Although final coatings exhibited a similar roughness to the system with one layer of putty, the contact angle was not reduced. In addition, it was observed that the systems prepared at varying conveyor speeds showed slight differences in the formation of the recorded values. These differences are due to the doses of energy distributed as a result of the longer residence of the samples under the radiator. Different levels for different speeds were also recorded between primer systems. The smallest impact of speed was observed in the case of topcoat systems. This is undoubtedly related to the type of products and their composition. In general, systems with undercoats only had lower values than final coatings with topcoats. From the point of view of the technological process, this is very advantageous. The contact area between the liquid and the substrate is then greater. Therefore, it can be indirectly concluded that the varnish products will have the ability to wet the substrate and achieve a favorable adhesive bond between the wetted material and the varnish coatings [43,44]. On the other hand, the recorded higher contact angle contributes to the formation of hydrophobic properties of the topcoats, which is desirable during the use of the final products. When considering the evaluated parameter as a function of the number of coatings applied, the lowest values were found for one putty coating at the maximum lamp setting regardless of the conveyor speed. An increase in this parameter was recorded with the application of subsequent layers. This is due to an increase in the homogeneity of the varnish coating and the deliberate action of changing the energy density supplied during the refinement process. Coatings that are less cross-linked are more wettable and, at the same time, more susceptible to sanding. The sanding process is an advisable treatment after the application of putty layers. With regard to the systems considered, the highest angle value was achieved by system 1-15a consisting of five coatings comprising two layers of putty and primer using the OPTI and one layer of topcoating 6 g/m^2 at the maximum settings of the electromagnetic beam focusing lamps and for each LED module (Table 6). The observed differences for each system are the result of different energy doses of absorbed electromagnetic radiation. In addition, it was noted that the use of higher conveyor speeds contributes to lower values of the contact angle, which is undoubtedly related to the absorption of different energy doses (Figure 4).

Table 6. UV curing system measurement.

Samples	Speed [m/min]	Amount for Last Layer [g/m^2]	Type	Wavelength Range	Power Peak [mW/cm^2]	Energy Density [mJ/cm^2]
1-3d	40	30	putty	UVV UVA	1050 1155	85 90
1-3a	25	30	putty	UVV UVA	1065 1170	135 150
1-7d	40	20	basecoat	UVV	11,550	575
1-7a	25	20	basecoat	UVV	11,567	922
1-8d	40	10	basecoat	UVV	11,550	575
1-8a	25	10	basecoat	UVV	11,567	922
1-14d	40	12	topcoat	UVV UVA	7050 2860	505 305
1-14a	25	12	topcoat	UVV UVA	7050 2860	800 490
1-15d	40	6	topcoat	UVV UVA	7050 2860	505 305
1-15a	25	6	topcoat	UVV UVA	7050 2860	810 485

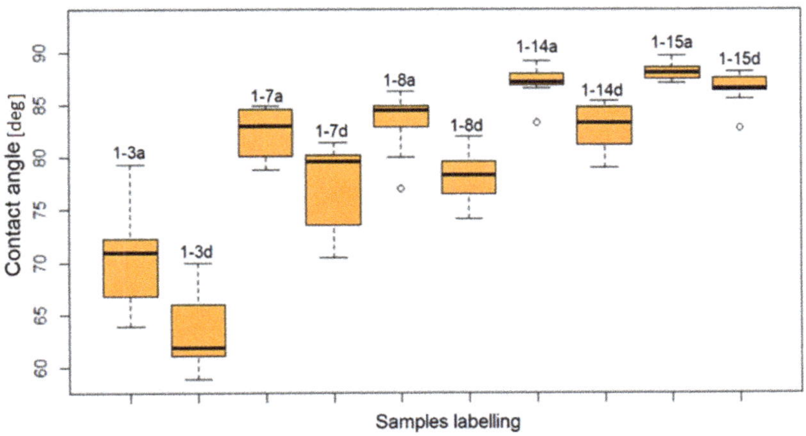

Figure 4. Contact angle's result.

A correlation analysis based on Tukey's criterion multiple comparisons of means with a 95% confidence level shows that the energy density applied to the surface has an influence on the formation of the contact angle. In many cases, for the same type of applied materials, no effects of changing the application amount on the contact angle were observed. The small impact of the amount of liquid materials applied on the surface with the same energy density was observed between the topcoat, samples 1-14d (12 g/m^2), and 1-15d (6 g/m^2).

The type of varnish used clearly influences the formation of the contact angle between the basecoating with a mean of 82.28° (1-7a) and the topcoating with a mean of 87.11° (1-14a), (p-value 0.0103271).

Additionally, it was stated that line speed has equal importance on the formation of the contact angle with the same power of lamp. For example, the mean of 63.60° for 1-3d at 40 m/min in comparison to 70.85° for 1-3a at 25 m/min (p-value 0.0000077) or the mean of 77.23° for 1-7d at 40 m/min in comparison to 82.28° for 1-7a at 25 m/min (p-value 0.0058). The samples stay under the lamps longer, which affects the density of energy supplied to the process.

Based on the value of the contact angle, a parameter characterizing the wetting process of the substrate (γ_S) with the varnish products and the interlayer interaction was calculated. It indicates the possibility of systems with compatible relations between the individual layers.

Preliminary analysis of the values of surface free energy influencing the adsorption, wetting, and adhesion processes to the most significant extent, being at the same time the effect of intermolecular interactions, indicates that the tested systems exhibited γ_S values in the range of 32–51 mJ/m^2 (Figure 5). Regarding the theoretical assumptions of the adsorption theory of polymer adhesion to the substrate, the γ_S value of the refined surface should be higher than that of the consolidated varnish. In the evaluated relations, this requirement was fulfilled. Based on literature data, for the HDF board, the values for beech wood (*Fagus sylvatica* L.) of $\gamma_S = 67.6$ mJ/m^2, $\gamma_S{}^d = 24.7$ mJ/m^2, and $\gamma_S{}^P = 42.9$ mJ/m^2 [16], on which the HDF board was based, were adopted. This allowed an indirect inference about the γ_S value of the HDF board despite the use of adhesive and hydrophobic agents in its production. The obtained γ_S values exceeded the relevant data determined for the varnish coatings under consideration.

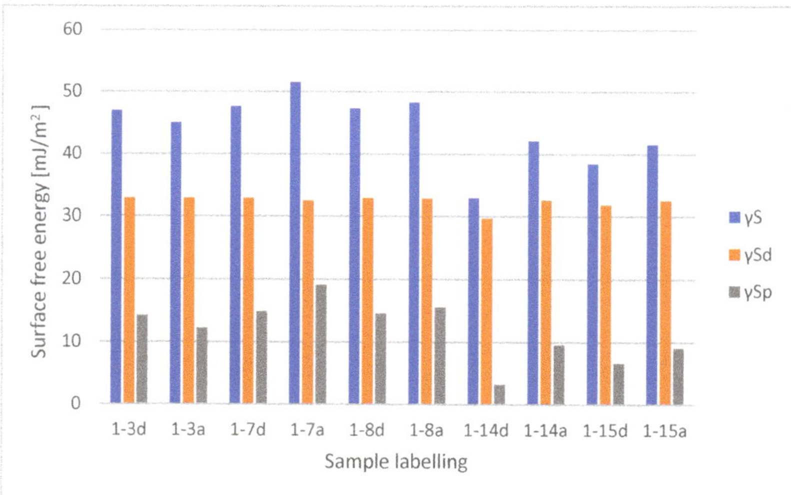

Figure 5. Results of the surface free energy.

When evaluating the relations between the applied products, interlayer coatings were characterized by higher γ_S values. The primer coatings can therefore be regarded as a typically intermediate factor in shaping the final adhesion. The application of topcoats resulted in a decreased surface energy. A clear influence of the dispersion component on the formation of γ_S values was found. After the phase transformation of the products into solidified layers, the resulting varnish coatings were characterized by favorable interlayer adhesion and substrate adhesion.

Unevenness and porosity increase the effective interaction surface between the surface forces of the substrate and varnish products and, therefore, increase the value of the work of adhesion of the varnish to the board compared to the work of adhesion and spreading of the varnish on a smooth surface. The work of adhesion (Wa) in the HDF–coat varnish system had a higher value (91.80–93.78 mJ/m^2) than the work of adhesion of the topcoats to the primers (75.79 to 87.82 mJ/m^2). The primer coatings can, therefore, be regarded as a factor typically facilitating the formation of the final adhesion. Considering the values of the dispersion component of the work of adhesion (Wad) for the systems under consideration, it was found that they were quite similar at 65 mJ/m^2, while the polar component of the work of adhesion (Wa^p) varied considerably. The results obtained for the variants under consideration, including both Wa and its components Wad and Wa^p, provide, as regards theoretical possibilities, favorable conditions to obtain varnish coatings exhibiting high adhesion in both interlayer and final systems in the finishing process. This was confirmed in pull-off adhesion tests.

Moreover, this parameter was found to be higher than the surface roughness. Furthermore, it was shown that the substrate, i.e., the cohesive forces of the HDF board, was predominantly the weakest link in the tested systems. This can be explained by the fact that the contact area between the substrate and the solidified putty layer increases with increasing roughness. In addition to this, it should be taken into account that on a rough surface, the putty layer may have a higher affinity with the surface. The observed relationships are confirmed by the mechanical theory of adhesion developed by Mc Bain and Hopkins in 1925 [45,46], according to which decohesion occurs within the bonded materials, among others, as a result of anchoring the applied product into the rough surface and not at the interface. The finishes' quality can be considered fully acceptable in light of the results obtained.

4. Conclusions

Research material has been previously presented on the patented technological line [47]. The obtained results in the current study are difficult to relate to the data of other authors since adhesion evaluation for varnish coatings of UV LED products formed on HDF boards under industrial conditions has not been carried out. In the surface interactions of the cured systems in the substrate–varnish products relationship, in terms of the adhesion phenomenon, the following was found:

1. Tested surfaces were characterized by Ra values in the range of 0.5–1.2 μm, and contact angle θ was in the range of 63.60 ÷ 88.14° for individual surfaces.
2. The power of the radiators and speed of the line have a greater influence on the formation of the contact angle than the amount of liquid materials applied on each roller machine.
3. The speed of the line is important for the formation of the contact angle at the same power of the lamp.
4. No correlation between the change in the contact angle and recorded roughness index was observed.
5. Final coatings exhibited lower γ_S values than interlayer coatings. A clear influence of the dispersion component on the formation of γ_S values was observed. No clear effect of the applied curing parameters on the value of the $\gamma_S{}^d$ component was noted.
6. The obtained coatings were characterized by high Wa values, ranging from 75.79–97.94 mJ/m^2. For the individual systems, the level of Wa values achieved depended primarily on the type of individual varnish product layers. The relationship

between Wa and its components, Wad and Wa^p, provided favorable conditions for obtaining coatings with high adhesion in the finishing process.

7. The adhesion of individual layers to the HDF board's surface and the interlayer system in the analyzed systems ranged from 1.25–1.56 MPa. However, cohesive-type delamination in the HDF board was clearly the dominant delamination mechanism under destructive loads.
8. The substrate refinement parameters adopted in the roller refinement provide favorable conditions for obtaining final finishes in terms of adequate adhesion. This applies to the doses of energy distributed by the radiators, in Ga, Hg, and LED versions, the module focusing the electromagnetic radiation beam, and the distance of the radiator from the cured surface layers and varnish coatings.

Author Contributions: Conceptualization, M.T. and T.K.; methodology, M.T., B.L. and T.K.; software, M.T.; validation, M.T., B.L. and T.K.; formal analysis, M.T.; investigation, M.T. and B.L.; resources, M.T., B.L. and T.K.; data curation, M.T.; writing—original draft preparation, M.T., B.L. and T.K.; writing—review and editing, M.T., B.L., E.A.S. and T.K.; visualization, M.T.; supervision, B.L. and T.K.; project administration, M.T., B.L. and T.K.; funding acquisition, M.T. All authors have read and agreed to the published version of the manuscript.

Funding: This research was funded by the Borne Furniture Company in Gorzow Wielkopolski Poland and the APC was funded by the same company.

Institutional Review Board Statement: Not applicable.

Informed Consent Statement: Not applicable.

Data Availability Statement: The data presented in this study is available upon request from the corresponding author. The data are not publicly available due to privacy.

Acknowledgments: The authors would like to gratefully acknowledge the National Center for Research and Development (Poland). All samples were prepared by using the finishing UV-curing line created as a result of research and development works of project POIR.01.01.01-00-1235/17, Development of technology for finishing the surface of wood materials using selected energy-saving sources electromagnetic radiation at advanced speeds and varnish systems characterized by increased resistance.

Conflicts of Interest: Author Maciej Tokarczyk was employed by the BORNE-FURNITURE Company. The remaining authors declare that the research was conducted in the absence of any commercial or financial relationships that could be considered as a potential conflict of interest. The authors declare that this study received funding from Borne Furniture Company. The funder was not involved in the study design, collection, analysis, interpretation of data, the writing of this article or the decision to submit it for publication.

References

1. Bartoszuk, K.; Kowaluk, G. The Influence of the Content of Recycled Natural Leather Residue Particles on the Properties of High-Density Fiberboards. *Materials* **2023**, *16*, 5340. [CrossRef] [PubMed]
2. Henke, M.; Lis, B.; Krystofiak, T. Mechanical and Chemical Resistance of UV Coating Systems Prepared under Industrial Conditions. *Materials* **2023**, *16*, 4468. [CrossRef] [PubMed]
3. Rahaman, M.; Hossain, S.; Islam, M.; Uddin, M. Suitability of Medium Density Fiber Board Made From Rubber Wood for Household and Industrial Use. *Bangladesh J. Agric. Res.* **2021**, *46*, 203–209. [CrossRef]
4. Pawlak, D.; Boruszewski, P. Digital printing in wood industry. *Ann. WULS For. Wood Technol.* **2020**, *109*, 109–115. [CrossRef]
5. Sang, R.; Manley, A.; Wu, Z.; Feng, X. Digital 3D Wood Texture: UV-Curable Inkjet Printing on Board Surface. *Coatings* **2020**, *10*, 1144. [CrossRef]
6. Mendes, C.; Oliveira, J.; Etxebarria, I.; Vilas, J.; Lanceros-Méndez, S. State-of-the-Art and Future Challenges of UV Curable Polymer-Based Smart Materials for Printing Technologies. *Adv. Mater. Technol.* **2019**, *4*, 1800618. [CrossRef]
7. Altun, S.; Köse, D. Some of the Physical Properties of UV Jet Printed Furniture Surfaces. *Drv. Ind.* **2013**, *64*, 39–43. [CrossRef]
8. Rawat, R.S.; Nidhi Chouhan, N.; Talwar, M.; Diwan, R.K.; Tyagi, A.K. UV coatings for wooden surfaces. In *Progress in Organic Coatings*; MDPI: Basel, Switzerland, 2019; Volume 135, pp. 490–495. ISSN 0300-9440. [CrossRef]

9. Lu, Q.; Zhang, C.; Huang, B.; Wei, X.; Wang, Y. Study on the Application of UV Ink in Printing Manufacturing. In *Advances in Graphic Communication, Printing and Packaging Technology and Materials*; Springer: Berlin/Heidelberg, Germany, 2021; pp. 629–636. [CrossRef]
10. Sang, R.; Yang, S.; Fan, Z. Effects of MDF Substrate Surface Coating Process on UV Inkjet Print Quality. *Coatings* **2023**, *13*, 970. [CrossRef]
11. Ozdemir, T.; Bozdoğan, Ö.; Mengeloglu, F. Effects of Varnish and Film Thickness on Adhesion Strength of Coated Wood. *Proligno* **2013**, *9*, 164–168.
12. Nkeuwa, W.N.; Riedl, B.; Landry, V. Wood surfaces protected with transparent multilayer UV-cured coatings reinforced with nanosilica and nanoclay. Part I: Morphological study and effect of relative humidity on adhesion strength. *J. Coat Technol. Res.* **2014**, *11*, 283–301. [CrossRef]
13. Kılıç, K.; Söğütlü, C. Varnish Adhesion Strength in Natural Aged Wood Material. *Politek. Derg. J. Polytech.* **2023**, *26*, 847–854. [CrossRef]
14. Vitosytė, J.; Ukvalbergienė, K.; Keturakis, G. The Effects of Surface Roughness on Adhesion Strength of Coated Ash (*Fraxinus excelsior* L.) and Birch (*Betula* L.) Wood. *Mater. Sci.* **2012**, *18*, 347–351. [CrossRef]
15. Iis, Y.; Istie, R.; Lumongga, D.; Darmawan, W. Wettability and adherence of acrylic paints on long and short rotation teaks. *Wood Mater. Sci. Eng.* **2019**, *15*, 229–236. [CrossRef]
16. Pecina, H.; Paprzycki, O. Lack auf Holz. In *Lacquers on the Wood*; Vincentz Verlag: Hannover, Germany, 1995; p. 175.
17. Ozdemir, T.; Hiziroglu, S. Evaluation of surface quality and adhesion strength of treated solid wood. *J. Mater. Process. Technol.* **2007**, *186*, 311–314. [CrossRef]
18. Ozdemir, T.; Hiziroglu, S. Influence of surface roughness and species on bond strength between the wood and the finish. *For. Prod. J.* **2009**, *59*, 90–94.
19. Ozdemir, T.; Hiziroglu, S.; Kocapınar, M. Adhesion Strength of Cellulosic Varnish Coated Wood Species as Function of Their Surface Roughness. *Adv. Mater. Sci. Eng.* **2015**, *2015*, 525496. [CrossRef]
20. Guo, X.; Li, R.; Teng, Y.; Cao, P.; Wang, X.; Ji, F.; Vohringer, S. Effects of surface treatment on the properties of UV coating. *Wood Res.* **2015**, *60*, 629–638.
21. Jung, S.J.; Lee, S.J.; Cho, W.J.; Ha, C.S. Synthesis and properties of UV-curable waterborne unsaturated polyester for wood coating. *J. Appl. Polym. Sci.* **1998**, *69*, 695–708. [CrossRef]
22. Mashouf, G.; Ebrahimi, M.; Bastani, S. UV curable urethane acrylate coatings formulation: Experimental design approach. *Pigment Resin Technol.* **2014**, *43*, 61–68. [CrossRef]
23. Dai, J.; Ma, S.; Wu, Y.; Liu, X. High bio-based content waterborne UV-curable coatings with excellent adhesion and flexibility. *Prog. Org. Coat.* **2015**, *87*, 197–203. [CrossRef]
24. Dai, J.; Ma, S.; Liu, X.; Han, L.; Wu, Y.; Dai, X.; Zhu, J. Synthesis of bio-based unsaturated polyester resins and their application in waterborne UV-curable coatings. *Prog. Org. Coat.* **2015**, *78*, 49–54. [CrossRef]
25. Yan, X.; Qian, X.; Lu, R.; Miyakoshi, T. Synergistic Effect of Addition of Fillers on Properties of Interior Waterborne UV-Curing Wood Coatings. *Coatings* **2018**, *8*, 9. [CrossRef]
26. Gurleyen, L.; Ayata, U.; Esteves, B.; Cakicier, N. Effects of heat treatment on the adhesion strength, pendulum hardness, surface roughness, color and glossiness of Scots pine laminated parquet with two different types of UV varnish application. *Maderas. Cienc. Y Tecnol.* **2017**, *19*, 213–224. [CrossRef]
27. Çavuş, V. Weathering Performance of Mulberry Wood with UV Varnish Applied and Its Mechanical Properties. *Bioresources* **2021**, *16*, 6791–6798. [CrossRef]
28. Bekhta, P.; Lis, B.; Krystofiak, T.; Tokarczyk, M.; Bekhta, N. Gloss of Varnished MDF Panels Veneered with Sanded and Thermally Compressed Veneer. *Coatings* **2022**, *12*, 913. [CrossRef]
29. Salca, E.A.; Krystofiak, T.; Lis, B.; Mazela, B.; Proszyk, S. Some Coating Properties of Black Alder Wood as a Function of Varnish Type and Application Method. *Bioresources* **2016**, *11*, 7580–7594. [CrossRef]
30. Hazir, E.; Koc, K.H. Evaluation of wood-based coating performance for ultraviolet roller and conventional air-atomization processes, Maderas. *Cienc. Y Tecnol.* **2021**, *23*, 1–16. [CrossRef]
31. Kaygin, B.; Akgun, E. Comparison of Conventional Varnishes with Nanolacke UV Varnish With Respect to Hardness and Adhesion Durability. *Int. J. Mol. Sci.* **2008**, *9*, 476–485. [CrossRef]
32. Gürleyen, L. Effects of Artificial Weathering on the Color, Gloss, Adhesion, and Pendulum Hardness of UV System Parquet Varnish Applied to Doussie (Afzelia africana) Wood. *BioResources* **2021**, *16*, 1616–1627. [CrossRef]
33. *Anonymous Theta Flex User Manual–OneAttension Original Instructions*; Biolin Scientific Oy: Espoo, Finland, 2019; p. 74.
34. *ISO 4287 (1998)*; Geometrical Product Specification (GPS)—Surface Texture: Profile Method—Terms, Definitions and Surface Texture Parameters. International Organization for Standardization: Geneva, Switzerland, 1998.
35. *PN-EN 828*; Adhesives—Determination of Wettability by Measuring the Contact Angle and Surface Free Energy of a Solid Surface. iTeh Standard: Etobicoke, ON, Canada, 2013.
36. *PN-EN 4624*; Paints and Varnishes—Pull-Off Test for Adhesion. iTeh Standard: Etobicoke, ON, Canada, 2023.
37. Białecki, F.; Pohl, P.; Sydor, M. Investigations on sealing material consumption in relation with the roughness parameters of HDF boards. *EJPAU* **2008**, *11*, 17.

38. Park, J.W.; Shim, G.S.; Lee, J.G.; Jang, S.W.; Kim, H.J.; Choi, J.N. Evaluation of UV Curing Properties of Mixture Systems with Differently Sized Monomers. *Materials* **2018**, *11*, 509. [CrossRef] [PubMed]
39. Park, J.W.; Shim, G.S.; Back, J.H.; Kim, H.J.; Shin, S.; Hwang, T.S. Characteristic shrinkage evaluation of photocurable materials. *Polym. Test.* **2016**, *56*, 344–353. [CrossRef]
40. Sanai, Y.; Ninomiya, T.; Arimitsu, K. Improvements in the physical properties of UV-curable coating by utilizing type II photoinitiator. *Prog. Org. Coat.* **2021**, *151*, 106038. [CrossRef]
41. Papp, E.; Csiha, C. Contact angle as function of surface roughness of different wood species. *Surf. Interfaces* **2017**, *8*, 54–59. [CrossRef]
42. Yu, Q.; Pan, X.; Yang, Z.; Zhang, L.; Cao, J. Effects of the Surface Roughness of Six Wood Species for Furniture Production on the Wettability and Bonding Quality of Coating. *Forests* **2023**, *14*, 996. [CrossRef]
43. Casilla, R.C.; Chow, S.; Steiner, P.R. An immersion technique for studying wood wettability. *Wood Sci. Technol.* **1981**, *15*, 31–43. [CrossRef]
44. Wang, X.F.; Wei, X.F.; Huang, P.Q. Study on Performance of Adherence of Water-Base UV Varnish. *Adv. Mater. Res.* **2010**, *174*, 409–413. [CrossRef]
45. Schultz, J.; Nardin, M. Theories and mechanisms of adhesion. In *Adhesion Promotion Techniques: Technological Applications*; Mittal, K.L., Pizzi, A., Eds.; Marcel Dekker: New York, NY, USA, 2002; pp. 1–26.
46. Packham, D.E. The mechanical theory of adhesion. In *Handbook of Adhesive Technology*; Pizzi, A., Mittal, K.L., Eds.; Marcel Dekker Inc.: New York, NY, USA, 2003; pp. 69–93.
47. PL234055B1. Method for Manufacturing a Cellular Plate and a Cellular Plate Manufactured by This Method. Author Maciej Tokarczyk/BORNE Furniture Comp/. Available online: https://www.patentguru.com/PL234055B1 (accessed on 2 December 2023).

applied sciences

Article

Gloss Level, Type of Paper and Colour of HDF Boards Finished with the Energy-Saving Acrylic Paint System

Milena Henke [1], Tomasz Krystofiak [1,*], Emilia Adela Salca [2,*] and Barbara Lis [1]

1 Department of Wood Science and Thermal Techniques, Faculty of Forestry and Wood Technology, Poznan University of Life Sciences, Wojska Polskiego 28, 60-627 Poznan, Poland; milena.henke@up.poznan.pl (M.H.); barbara.lis@up.poznan.pl (B.L.)

2 Faculty of Furniture Design and Wood Engineering, Transilvania University of Brasov, Universitatii 1, 500068 Brasov, Romania

* Correspondence: tomasz.krystofiak@up.poznan.pl (T.K.); emilia.salca@unitbv.ro (E.A.S.)

Abstract: The global drive to eliminate mercury is stimulating the development of LED lamps for UV curing of coatings. Despite its many advantages, the transition to this system in the furniture industry presents technological issues. This paper aimed to determine the impact of the paint line's individual technological and process parameters on the coating's optical properties, such as colour and gloss of furniture components made of white-coloured light-weight boards. Samples were prepared under the conditions of an industrial paint line using LED and traditional mercury lamps at the rate of 50 m/min. The colour was tested using the CIELAB system on a BYK GARDNER spectrophotometer, while the gloss was measured using the PICO GLOSS meter. Based on the data obtained, the optical parameters were found to be highly stable irrespective of the power drop in the LED lamps. The coordinates L*, b* and gloss G were affected by the amount of topcoat and basecoat, the number of coats, the mercury lamps' power and the panel grinding programs used. The results of the tests showed that high L* and b* values could be achieved by increasing the number of overlays. This solution also ensures less basecoat and topcoat consumption, contributing to large material savings.

Keywords: UV lacquer system; LED radiators; high-density fibreboard; colour measurements; gloss level

Citation: Henke, M.; Krystofiak, T.; Salca, E.A.; Lis, B. Gloss Level, Type of Paper and Colour of HDF Boards Finished with the Energy-Saving Acrylic Paint System. *Appl. Sci.* **2024**, *14*, 394. https://doi.org/10.3390/app14010394

Academic Editor: Young-Wook Chang

Received: 28 November 2023
Revised: 20 December 2023
Accepted: 30 December 2023
Published: 31 December 2023

1. Introduction

The global paint products market is estimated to be worth USD 248.87 billion in 2023. Over the next five years, the market is projected to grow to USD 326.09 billion at a 7% CAGR [1]. Such a thriving industry is stimulated by global challenges, especially in the area of ecology [2]. The demand for environmentally friendly coatings is driven by the increasing awareness of consumers who expect sustainable products. Regulations that have been introduced in the United States, China and European countries are intensifying the efforts of entrepreneurs to implement low-emission technologies using organic compounds [3,4]. The use of UV varnishes is a good alternative to solvent-based varnishes in the protection of wood and wood–plastic surfaces [5,6]. The wood industry is the largest market segment for UV technology. In 2019, it accounted for 43% of the total UV curing market, and China proved to be the largest consumer. Asian countries are cited as the region where UV technology is developing most rapidly [2,3,7]. Over the years, UV-curable coatings have evolved, allowing them to be used in a wide variety of applications, such as automotive, biomedical, electronics and adhesive processes [6,8–10]. UV lamp-curing technology is considered an efficient process due to shorter curing times [11–15].

The coating process involves the conversion of functional monomers into a cross-linked polymer using free radicals under the influence of UV radiation [11–13]. Mercury lamps constitute the basis of UV curing technology. Ultraviolet, visible and infrared wavelengths are emitted due to the evaporation of mercury into plasma at high temperatures [16].

Gallium-doped mercury lamps are used for thick, pigmented coatings [17]. The disadvantages of curing technology using mercury lamps should, however, be mentioned. Only 5% of the energy is used to cure the varnish, while the rest is converted into heat, which raises the surface temperature, risking the flow of resins onto the wood surface and posing a fire hazard [18–24]. Furthermore, depending on its length and design, the lamp may contain between 10 and 100 mg of mercury. When adequately transported and operated, it poses no risk. Their recycling must, therefore, be based on the safe extraction of mercury and its transfer to sustainable disposal, storage or further processing [16]. International policy is leaning towards reducing this element in the environment. Global mercury consumption is high, at around 3,600 tonnes per year, though there has been a slight decrease from previous decades [25]. There are many regulations, e.g., REACH and the TSCA/Lautenberg Act, as well as regulatory bodies, e.g., the United States Environmental Protection Agency and the United Nations Environment Programme, which aim to create consumer awareness and accelerate the changes taking place in the industry. The most frequently cited document on the circulation of mercury is the European Commission's RoHS2 Directive. Currently, the exemption for the use of Hg in the manufacturing of UV curing lamps has been extended from 2021 to 2027 [16,26–28]. In 2017, the Minamata Convention international treaty was ratified by 137 countries. This does not prohibit the use of mercury in the manufacturing of UV lamps but mandates all member states to limit the use of mercury [29,30].

An alternative to high-energy mercury lamps are LED lamps, known since the mid-2000s. No mercury is used in their manufacturing, and their operation does not generate ozone [26]. LED radiation sources are based on semi-conductor components that, when energized, emit a narrow band of UVA energy [31,32]. The highest recorded temperatures during curing are 60 °C, about six times lower than in the case of mercury lamps. Air cooling of the LED lamps is unnecessary, resulting in lower sound levels. With LED lamps, coatings can be cured on resinous (pine, fir, spruce) or oily (mahogany) wood substrates and glued veneers. High energy efficiency, instant lamp activation/de-activation and elimination of lamp warm-up times generate electricity savings of 45–60%. The lifetime of LED lamps is predicted to be 10,000–20,000 h, with a 5% drop in performance over this time. Compared with conventional mercury lamps, the operating time is around 500–2000 h, and the power reduction at the end of life is around 50% [17,21,33,34]. UV LED technology's undoubted disadvantage is the coating's lower degree of cross-linking [35]. Finished product manufacturers seeking to increase production efficiency increasingly reduce curing times by increasing line speeds. However, this raises some problems with the cross-linking of paint products. Therefore, combined technologies are used in combination with conventional lamps. Lower line speeds were confirmed to provide a higher degree of curing [36]. Due to the efforts to maximize line efficiency, researchers are looking for suitable acrylate amine oligomers that allow the desired coating cross-linking to be achieved [37]. Intensive work is also underway on LED lamps with higher intensity and in the UV-B and UV-C spectrum [38]. The use of LED lamps involves the risk of failure of individual diodes. This, in turn, carries the risk of insufficient curing control. If the issue is not identified during manufacturing, the customer is exposed to residual monomers, photoinitiators and other harmful ingredients. The development of Industry 4.0 requires the definition of precise process requirements [39]. Thanks to the integrated sensors in the illuminator (e.g., bulb temperature, air temperature, UV intensity, humidity, air pressure), the Advanced Intelligent Monitoring System (AIMS) can transfer data to a cloud-based environment to co-create Industry 4.0. An intelligent IoT-enabled UV system provides real-time process performance data and allows analysis for predictive diagnostics [40].

One important aspect that determines the attractiveness of the finished products is their optical properties, such as colour and gloss. Many researchers have described the issue of obtaining the right gloss. After many years of the popularity of high-gloss coatings, matt surfaces are attracting increasing interest [41]. UV-cured coatings exhibit higher gloss values than other technologies [42]. New varnish formulations are being developed to provide a higher matt finish. Achieving good matting is about finding the right amount of matting agent, reactive ingredients and application and curing parameters. It has been proven that a high photoinitiator level, a high line speed and a high dose of UV light reduce the matting effect [41]. Roller application also produces a higher gloss than spray application [43]. Low substrate roughness, coating thickness and number of layers are all factors that affect the coatings' optical properties, and UV coatings turned yellow in aging tests [44,45].

Repeated reports in research about the lack of adequate coating curing with LED lamps and issues with optical properties at high line speeds provided the impetus to address the topic. Despite increasing pressure to implement LED-based UV curing technology, it still lacks clear coating curing procedures [16,46]. The challenge is determining the impact of technological parameter combinations that ensure adequate coating cross-linking and high final quality.

The analysis of the available literature includes studies, only fragmentarily connected with the signalled subject matter. This is mainly due to the fact that achievements of this kind are covered by the highest confidentiality clause. Therefore, it was decided to study the optical properties of varnish coatings obtained in industrial conditions to control the parameters of the technological process of surface finishing.

This paper aims to determine the impact of fibreboard (HDF) density, grinding parameters, the amount of basecoat and topcoat applied, the number of coats and different lamp powers on a combined industrial line using LED and mercury lamps at 50 m/min. It was assumed that these factors could significantly impact the coating's resistance and that simply learning about them and identifying the relationship could support entrepreneurs and manufacturing technologists and become a starting point for further work on varnish formulations and LED lamps.

2. Materials and Methods

2.1. Materials

The board used for the study was a frame version with a honeycomb core of 700 × 390 × 22 mm. The frame was made of 17 mm thick particleboard with a density of 540 +/− 20 kg/m^3 (Kronospan, Szczecinek, Poland). The 2.5 mm thick HDF boards from two different manufacturers, known as A and B, were used as the cladding material (Table 1). The total number of samples used for the experiments was 198.

Table 1. Primary characteristics of HDF boards.

Parameter	Board Marking	
	A	B
Density [kg/m^3] acc. to DIN EN 323:1993	850	830
Modulus of elasticity [MPa] acc. to the DIN EN 310:1993 standard	4300	4500
Moisture content [%]	7	7
Swelling resistance [%] acc. to the DIN EN 317:1999 standard	45	45

A one-component PVAC adhesive with a viscosity of 14,000 mPa·s indicated with a METLER TOLEDO S210 meter at 40 ± 0.5 °C and a pH of 4.1 was used to adhere the chosen boards together. After a conditioning period of 96 hours, the structures with a 22 mm thickness were subjected to finishing. Putty, basecoat and topcoat paint were used in a multi-layer paint scheme (Table 2).

2.2. Surface Varnish Finishing Process

Before the application of varnish products, the HDF boards were sanded using 180–220 and 220–220 sanding paper, after which the boards were passed through the rollers. The resulting varnish coatings consisted of 5 and 6 applied layers. The individual coatings were cured with 120, 90 and 60 W/cm^2 mercury lamps and 12, 9 and 6 W/cm^2 LED lamps. Before basecoat application, intermediate sanding was performed with 220–220 and 400–400 sanding paper. Finally, nine variants in white were prepared for experiments (Table 3). The finishing process was done under technological conditions on a prototype paint line.

Table 2. Primary physicochemical characteristics of varnish products.

Parameter	Heavy Putty	Basecoat	Topcoat
Density [g/cm^3]	1.63 ± 0.15	1.73	1.30 ± 0.15
Solids content [%] acc. to the PN-EN ISO 3251:2019 standard	95.3 ± 0.5	98.3 ± 0.5	97.8 ± 0.5
Viscosity [mPa·s] (Brookfield, Thermosel 35 °C, 20 rpm, spindle 27)	7700	400	1475

Table 3. Amounts of varnish application.

Type of Varnish Product—UV Acrylic	Amount of Applied Varnish [g/m^2]								
heavy putty	50	50	50	50	50	50	50	50	50
basecoat	40	45	50	40	45	50	40	45	50
topcoat	3	3	3	6.5	6.5	6.5	10	10	10

Figure 1 shows a schematic line for finishing boards with UV varnish products. In the first step of the process, the parts were sanded and then two layers of putty were applied. The first was cured with 1 mercury (Hg) lamp, while two Hg lamps were used to crosslink the second layer. The next operation was to sand the surface of the putty and apply the basecoat as two layers. Each layer was cured using two LED lamps. Some variants were prepared with three coats of basecoat. In the final step, a topcoat was applied and cured using two LED lamps and five mercury lamps. All operations were performed on the production line at a speed of 50 m/min at ambient temperature.

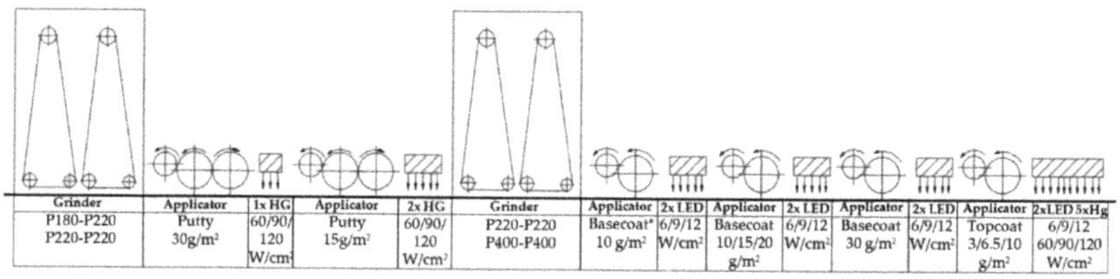

Figure 1. An overview of the varnish line on which the tests were conducted.

2.3. *Gloss Measurement*

Using a PICO GLOSS 503 photoelectric apparatus, the gloss was measured following the DIN 67530:1982 and ISO 2813:1994 standards [47,48]. Six measurements were taken for each sample in the direction opposite the varnish application using 60-degree geometries on each sample.

2.4. *Colour Measurement*

The colour was measured with the BYK GARDNER instrument. According to the ISO 7724-2 standard, six colour measurements were taken for each sample in the opposite direction to the varnish application [49]. The colour was assessed using the CIELab system. The following characteristics were used to measure colour: lightness (L*), degree of redness (a*) and degree of yellowness (b*). Because red and green are opposite colours and blue and yellow are complementary colours, the colour axes (a* and b*) are based on this idea.

The values on each axis range from positive to negative. Positive numbers on the a-a′ axis represent amounts of red, whereas negative values represent amounts of green. Yellow is positive and blue is negative on the b-b′ axis. Zero represents neutral grey for both axes. Lightness (denoted as L*) is represented by the central vertical axis, with values ranging from 0 (black) to 100 (white). A grid of lines parallel to the a* and b* axes surround planes of constant brightness L* in the graphic depiction of this colour space [50].

2.5. Processing Data

The test findings were analysed using the Minitab 19 and Microsoft Excel programs. The normality of the distribution was examined, and an analysis of variance (ANOVA) was carried out, to determine the impact of various parameters on the colour and gloss of the surface. The data were displayed using the main effects chart. A prediction and optimization report was produced.

3. Results

3.1. Initial Statistical Analysis

The normal distribution hypothesis was verified using the Ryan–Joiner test, similar to the popular Shapiro–Wilk test. At a significance level of $\alpha = 5\%$, the collected brightness L*, b* colour coordinate and gloss G data were normally distributed for the raw data. The null hypothesis that the error variance of the dependent variable was equal in all groups was assumed at the fixed confidence level (>0.05). The exception was the parameter a*, its data did not present a normal distribution and the null hypothesis was rejected [51].

ANOVA analysis was performed. The data were evaluated based on six variables: density of high-density fibreboard (HDF), sand belt used during sanding, amount of basecoat and topcoat applied, number of coats, different UV mercury lamps and different UV LED lamp powers used to cure the surface of coating (Table 4). It was found that the mean gloss level values differed at a significance level of $\alpha = 0.05$ for UV mercury lamp power W/cm, amount of topcoat, number of applicators and sandpaper. In the case of brightness L* and colour b*, the parameters related to the varnish application process (amount of topcoat and basecoat, number of applicators) were statistically different at $\alpha = 0.05$. The other factors examined had no statistically significant effect on the coating's optical parameters considered in the paper.

3.2. Colour Coordinates

The results obtained were further statistically analysed. Figures 2 and 3 show Pareto effect size plots in standard deviation for coordinates L* and b*. The most influential factors (larger than 2) were the number of applicators used and the amount of topcoat applied, followed by the amount of the basecoat applied. The sandpaper used was the factor in the intermediate importance group. Figure 2 also shows the main effect plot for the factors studied. Increasing the number of varnish applicators resulted in higher coordinates of L* and b*. An increase in brightness was noted during the increase in application when the topcoat and basecoat were applied. The finer the abrasive grit size (220–220 and 400–400), the higher the coating performance. The type of board was first in the group of factors with the least impact on L* and b*. Board A, with higher density, achieved higher brightness. For UV mercury and LED lamp power, the line is horizontal (parallel to the *x*-axis), so there was no main effect. Each level of the factor affected the response similarly, and the response mean was the same across all the factor levels.

Table 4. One-way ANOVA analysis of gloss as a function of the process variables.

One-Way ANOVA Response	Source	DF The Total Degrees of Freedom	Adj SS Adjusted Sums of Squares	Adj MS Adjusted Mean Squares	F-Value	*p*-Value
G.	UV LED lamp power W/cm	2	0.55	0.2756	0.01	0.991
	UV mercury lamp power W/cm	2	216.7	108.36	3.90	0.022
	Amount of topcoat	2	3426	1712.96	152.20	0.000
	Amount of basecoat	2	75.06	3753	1.32	0.271
	Number of applicators	1	504.0	504.04	19.25	0.000
	Type of board	1	9.69	9.689	0.34	0.562
	Sandpaper	1	646.6	646.63	25.41	0.000
L*	UV LED lamp power W/cm	2	0.0292	0.01459	0.05	0.947
	UV mercury lamp power W/cm	2	0.0677	0.03385	0.13	0.881
	Amount of topcoat	2	13.57	6.7839	34.50	0.000
	Amount of basecoat	2	7.974	3.9872	17.68	0.000
	Number of applicators	1	7.053	7.0528	30.79	0.000
	Type of board	1	0.3694	0.3694	1.40	0.238
	Sandpaper	1	0.6204	0.6204	2.37	0.125
b*	UV LED lamp power W/cm	2	0.0299	0.01497	0.04	0.963
	UV mercury lamp power W/cm	2	0.2822	0.1411	0.36	0.701
	Amount of topcoat	2	24.40	12.2012	44.88	0.000
	Amount of basecoat	2	6.835	3.4175	9.43	0.000
	Number of applicators	1	12.09	12.0884	36.23	0.000
	Type of board	1	0.4319	0.4319	1.10	0.296
	Sandpaper	1	1.759	1.7587	4.55	0.034

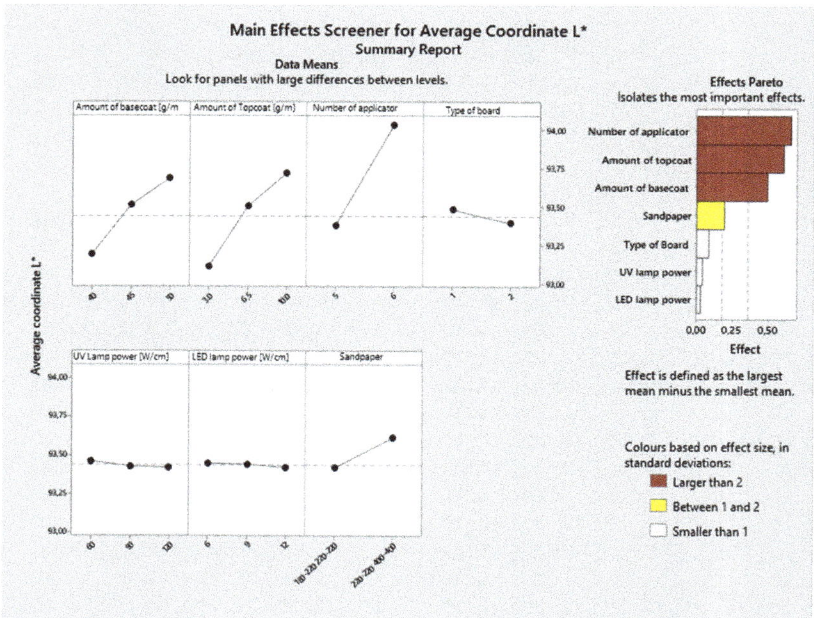

Figure 2. Main effects and Pareto effects plots for coordinate L*.

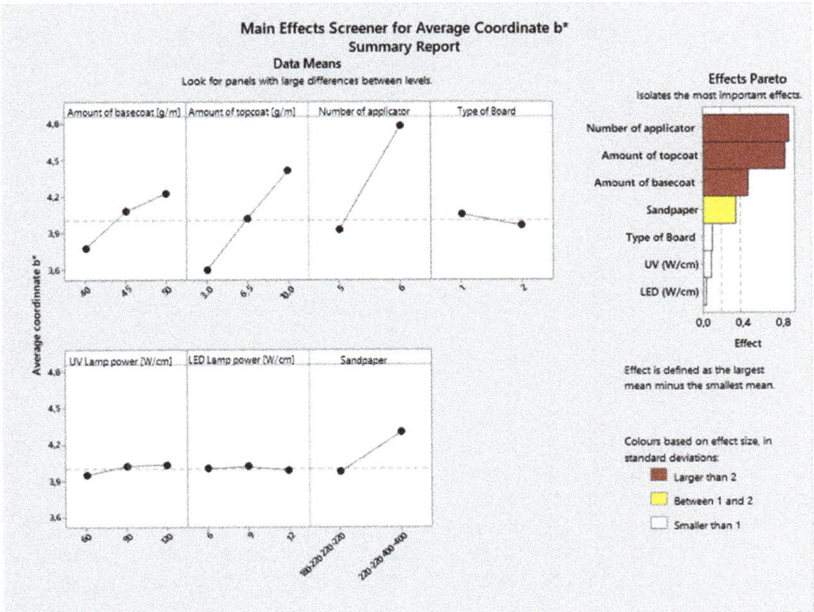

Figure 3. Main effects and Pareto effects plots for coordinate b*.

A Pareto analysis was carried out for L* and b* data to account for statistically significant interactions between factors. The results pointed out the following relations: the amount of basecoat–topcoat, basecoat–topcoat–UV mercury lamp power and basecoat–UV mercury lamp power. An interaction plot was then generated to visualize the interactions taking place (Figures 4 and 5). In the case of topcoat–basecoat interaction, the coordinates L* and b* increased when the amount of topcoat and basecoat increased. The largest spread occurred when the smallest assumed amounts of topcoat (3 g/m^2) and basecoat (40 g/m^2) were applied. Moreover, analysing the mercury lamps' power, it is possible to state that increasing the power resulted in a slight increase in L* and b*. The last significant interaction concerns the basecoat and mercury lamp power. The situation was analogous to that described previously, but it was surprising to see a decrease in both L* and b* when applying the largest amount of basecoat tested (50 g/m^2) when cured with a lamp power of 120 W/cm. The results of this variant equated to those obtained for an application of 45 g/m^2 and lamp power of 120 W/cm.

Factorial DOE analysis was carried out for L* and b* data, and optimization plots were generated (Figures 6 and 7). The impact of each factor (columns) on responses or composite efficacy (rows) is shown. The vertical red lines in the graph represent the current ratio settings. The numbers at the top of the column show the current factor level settings (in red). The horizontal blue lines and numbers show the responses for the current factor level. Each impact of the factors on responses or overall desirability is displayed in columns and rows. The graph's current factor settings are shown as vertical red lines. The current factor level settings are indicated (in red) by the numbers at the top of each column. The numbers and horizontal blue lines represent the replies for the current factor level. Based on the obtained data, it was estimated that by increasing the number of overlays in the process, a decrease of the basecoat spread rate from 50 g/m^2 to 40 g/m^2 and of the topcoat from 10 g/m^2 to 4.4 g/m^2 are allowed, when curing with the highest mercury lamp power of 120 W/cm and LED lamp power of 12 W/cm, while maintaining the same high level of coordinate L*. The same relationship applies to coordinate b*. To achieve a high coordinate b* = 4.7815, increasing the number of overlays from 5 to 6 is sufficient, thereby reducing the amount of basecoat applied by 10 g/m^2 and topcoat by 6.2 g/m^2.

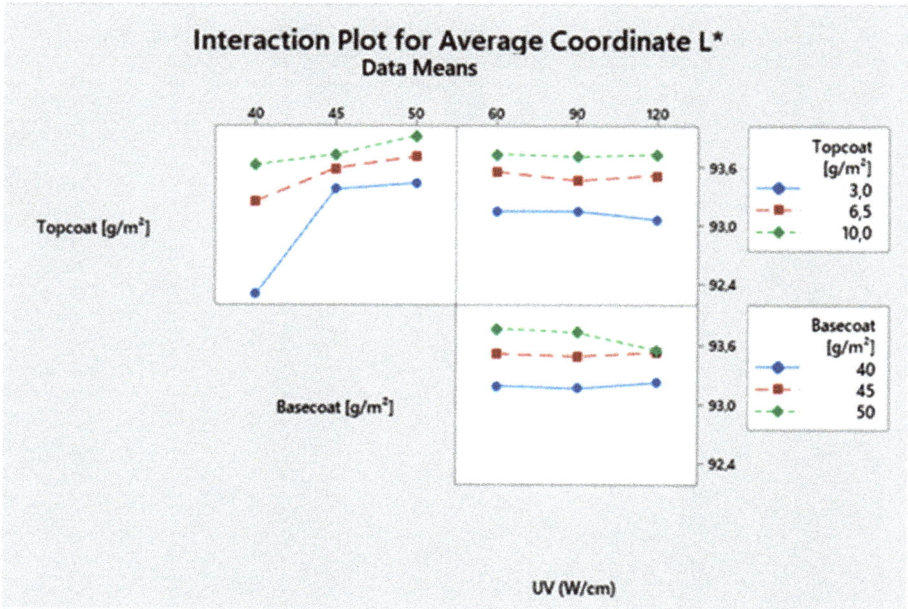

Figure 4. Interaction plot for coordinate L*.

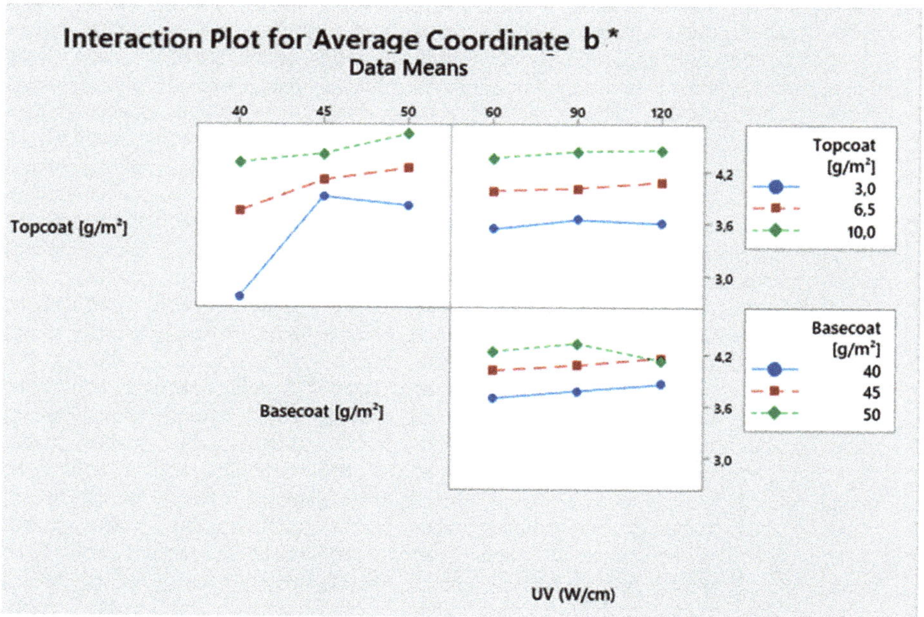

Figure 5. Interaction plot for coordinate b*.

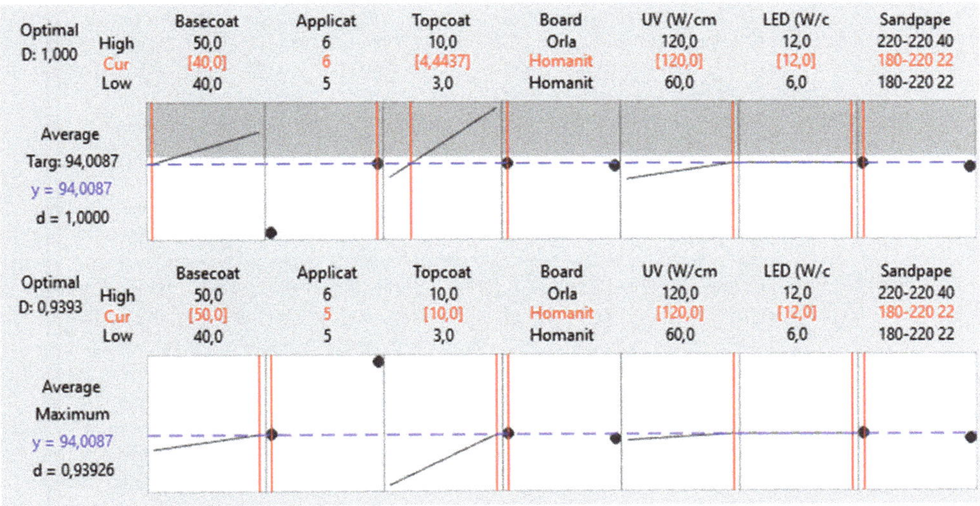

Figure 6. Optimal plot for L*=94,0087.

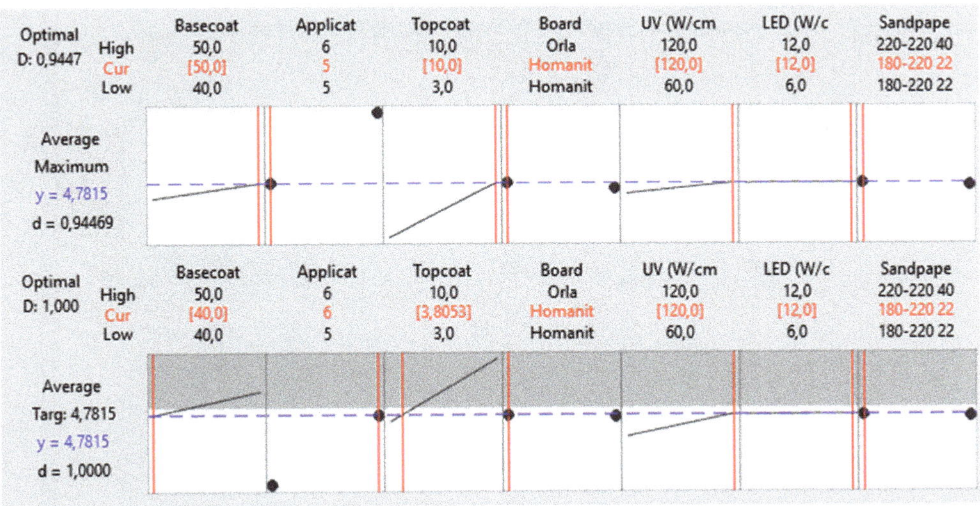

Figure 7. Optimal plot for b* = 4,7815.

The issue of curing coatings with LED lamps has been pointed out by several researchers, especially in the case of thick layers with a large number of pigments. One avenue has been to improve varnish formulations by changing the amount and type of photoinitiators [36] and developing LED lamps [38]. Another method is to reduce the line speed [36,37]. For floor painting lines, optimum curing is associated with reducing line speed from 9 m/min to 1.5 m/min. A better curing was also indicated for UV water-based formulations than for UV high-solid acrylates [36,37,52]. For thick coatings, hybrid curing is suggested [20]. In the present research, satisfactory coating curing was achieved by using LED and mercury lamps. The observed positive effect of increasing the number of applicators corresponds with the literature data. A better coating quality can be achieved by increasing the number of putty layers on the MDF and using a doctor blade on the application roller [53]. The resulting compromise not only allows a better coating curing, but also offers the opportunity to save considerable amounts of topcoat and basecoat, as shown in Figures 6 and 7. The increase in coordinates L* and b*, together with the increase in the number of varnishes applied, were related to the varnish properties that provide colour to the painted surface [54].

3.3. Gloss Level

A Pareto effect size plot in standard deviation was determined for the gloss results (Figure 8). It was found that the amount of the applied topcoat, the abrasives used and the number of applicators had the greatest effect on the achieved surface gloss (effect size larger than two). The next two factors are UV mercury lamp power and the amount of the basecoat (between 1 and 2). The type of board and the LED lamp power did not affect this parameter. The main effect plots describe the gloss change direction under individual factors' influence. Applying a topcoat at 3.0 and 6.5 g/m^2 had no significant effect on the analysed ratio. An increase of a further 3.5 g/m^2 resulted in a very sharp increase of almost 10 gloss level units. Using finer abrasive grit size and a greater number of applicators helped increase the recorded gloss values on the coating. Curing with mercury lamps in the intermediate power variant (90 W/cm) resulted in the lowest parameter, while the highest parameter was obtained for 120 W/cm. The gloss decreased as the amount of the basecoat increased. The higher-density HDF board achieved a slightly higher gloss level. The line is horizontal (parallel to the *x*-axis) for LED lamp power, so there was no main effect.

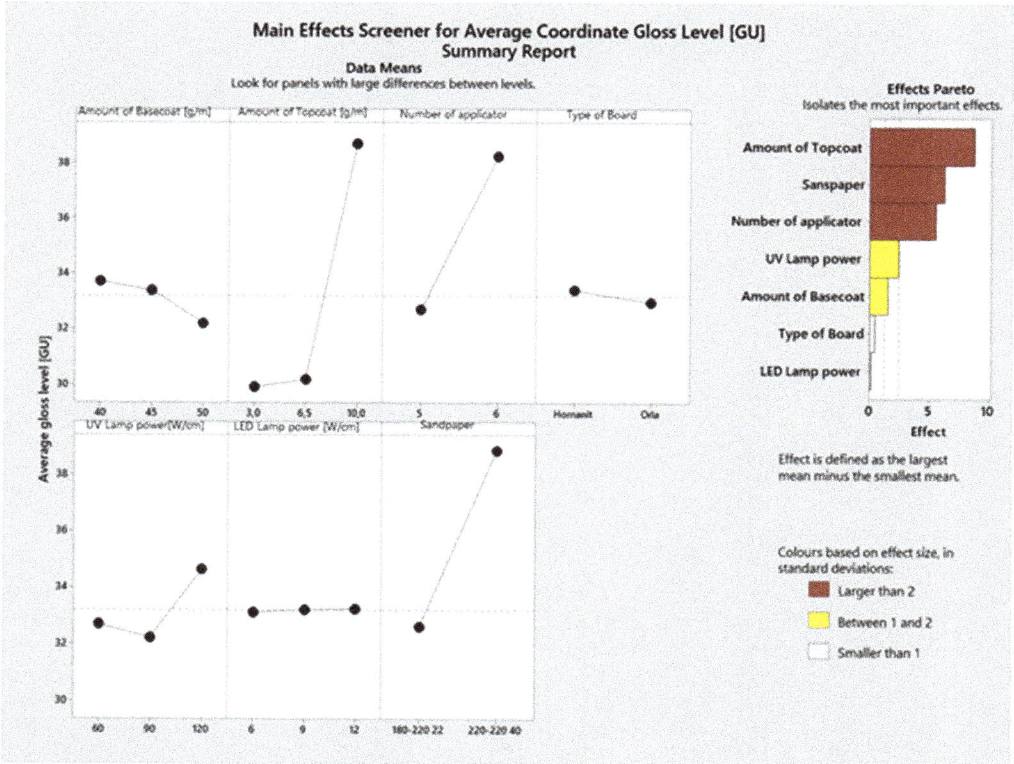

Figure 8. Main effects and Pareto effects plots for gloss level.

Analysing the Pareto chart of the standardized effects for interactions, the only statistically significant interaction was that between the amount of the topcoat applied and mercury lamp power (Figure 9). When 3 g/m^2 of topcoat was applied, the gloss level was similar for the surfaces cured with mercury lamps at 60 W/cm and 120 W/cm. The lowest values were achieved at 90 W/cm. The gloss level increased by applying 6.5 g/m^2, with the lowest increase recorded for curing with mercury lamps at 60 W/cm. The gloss increased as the lamps' power increased. A further increase for all curing variants proceeded at the same rate.

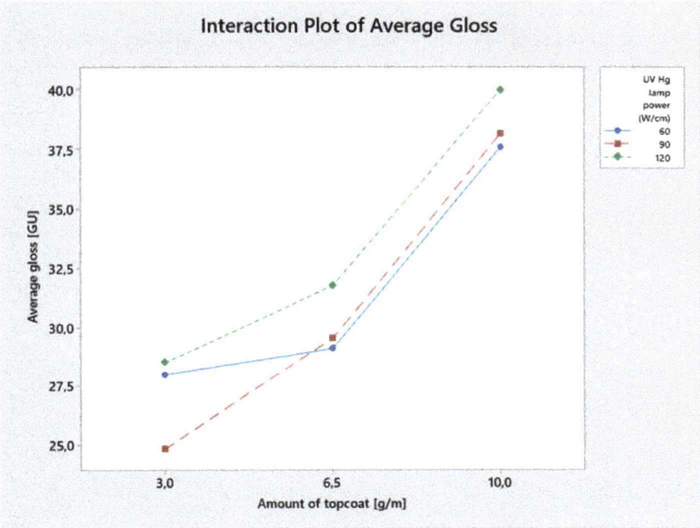

Figure 9. Interaction plot for average gloss level (topcoat–UV mercury lamp power).

Based on these, the author concluded that a thicker coating layer increases the coating resistance. This is consistent with the information provided on quality improvement when more topcoat is applied [54,55]. However, the same principle was not used with the priming varnish. The functions of varnishes need explanations. The basecoat varnish's chemical composition is designed to provide the varnished surface colour, guarantee evenness and provide initial resistance. The fundamental goal of a topcoat is to guarantee a quality finish. According to other reports, the ratios of the materials, the types of resin employed and other additives affect the scratch-resistant coatings [56].

There is a close relationship between the material after grinding and the coating [57], and that was confirmed by determining the effect of grinding on the surface gloss. It has been repeatedly proven that a gloss reduction occurs with an increase in roughness [58,59], which is provided by a smaller material grain size [43,60,61]. Overly aggressive grinding exposes the substrate structure, which can be filled with glue, thereby contributing to a reaction with the applied varnish layer [62]. The roughness of the first basecoat layer increased linearly, and the adhesion decreased for finer abrasives [13]. The cases described confirm the increase in gloss as the abrasive's grain size increased before and after grinding. In the literature, it is also possible to find publications on the increase in gloss with the application of successive varnish coats [45]. Lower gloss was observed with higher topcoat thickness when finishing using conventional UV lamps [55]. This may be due to differences in the composition of varnishes cured by mercury lamps and LED lamps. The researchers also conducted a comparative gloss study after scratch resistance tests of 200 cycles for coatings cured with LED and conventional UV lamps. No difference in gloss loss was noticed between the technologies [52].

The effect of conventional lamp power on colour and gloss of a coating is a common phenomenon known in manufacturing practice [55]. Painting on combined UV mercury and LED lines ensures stable colour and gloss performance. The smaller effect of mercury lamp power and the lack of effect of LED lamps on the test results obtained are a novelty.

This is an additional advantage of LED lamps. The present study will be extended to include testing of the coatings' mechanical resistance to obtain a better coating quality.

4. Conclusions

The scientific contribution of this paper is to present the impact of technological parameters in UV surface refinement using a paint line containing LED and mercury lamps.

In conclusion, this study indicates the intricate relationship between technological and process parameters in UV surface refinement, offering valuable insights into the optimization of paint line processes using LED and mercury lamps. The findings of this research not only contribute to the current body of knowledge, but also hold significant practical implications for the manufacturing and coating industries, particularly in the context of achieving desired colour coordinates and gloss levels. In essence, this study serves as a stepping stone toward a future where UV surface finishing becomes not only more technologically advanced but also environmentally conscious and economically viable.

Based on the results of the study, the following conclusions can be drawn:

- The number of varnish overlays used and the amount of the applied topcoat and basecoat had the greatest impact on colour coordinates L* and b*. Higher coordinates were obtained as the number of overlays and the number of varnishes increased. The interactions between the spread rates of varnish and the mercury lamp power were statistically significant. Increasing the lamp power contributed to the increase in colour coordinates.
- It was estimated that to obtain the highest possible parameter L* = 94., the number of overlays could be increased from 5 to 6, while reducing the amount of the applied basecoat by 10 g/m^2 (from 50 g/m^2 to 40 g/m^2).
- The statistically significant factors affecting the coatings' gloss were the spread rate of the applied topcoat, the abrasive used, the number of coats used and the mercury lamp power used in the process. An increase in gloss occurred with an increase in the spread rate of the applied topcoat, the number of coats and the abrasive grit size. The mercury lamps were also part of a significant interaction with the amount of the topcoat. The gloss increased as the lamps' power and the spread rate of the topcoat increased.
- LED lamp power did not affect the coatings' optical parameters. A high share of LED lamps in the painting line ensures a higher process stability. Moreover, it can be stated that LED lamps still do not affect the coatings' optical parameters, despite the passage of time and the power reduction, which may be a good prognosis for the extension of LED lamp life. It can also represent a large amount of savings due to the high price of LED modules.

Author Contributions: Conceptualisation, M.H. and T.K.; methodology, M.H., B.L. and T.K.; software, M.H.; validation, M.H., B.L. and T.K.; formal analysis, M.H.; investigation, M.H.; resources, M.H., B.L. and T.K.; data curation, M.H.; writing—original draft preparation, M.H., B.L. and T.K.; writing—review and editing, M.H., B.L., E.A.S. and T.K.; visualization, M.H.; supervision, B.L. and T.K.; project administration, B.L. and T.K.; funding acquisition, E.A.S. All authors have read and agreed to the published version of the manuscript.

Funding: This research was funded by the funding for PhD students under No. 953.224.012 of the Faculty of Forestry and Wood Technology, Poznań University of Life Sciences.

Institutional Review Board Statement: Not applicable.

Informed Consent Statement: Not applicable.

Data Availability Statement: The data presented in this study are available on request from the corresponding author.

Conflicts of Interest: The authors declare no conflict of interest.

References

1. The Business Research Company. Paints and Coatings Global Market Report 2023. 2023. Available online: https://www.thebusinessresearchcompany.com/report/paints-and-coatings-global-market-report (accessed on 2 November 2023).
2. Kelly, P. European Market Overview 2019. In Proceedings of the RadTech 2020 Conference, Orlando, FL, USA, 8–11 March 2020.
3. ReportLinker. Global Industrial Coatings Market Size, Share & Trends Analysis Report by Product, Technology, End Use, Region, and Segment Forecasts, 2021–2028. 2021. Available online: https://www.marketsandmarkets.com/Market-Reports/industrial-coatings-market-746.html (accessed on 2 November 2023).
4. Raport. Nowy raport przewiduje wzrost na globalnym rynku powłok przemysłowych [New report predicts growth in global industrial coatings market]. *Lakiernictwo Przemysłowe* **2022**, *1*, 135. Available online: https://www.lakiernictwo.net/dzial/142-aktualnosci-i-przeglad-rynku/artykuly/nowy-raport-przewiduje-wzrost-na-globalnym-rynku-p,4819 (accessed on 2 November 2023).
5. Jabłoński, M.; Świetliczny, M.; Ružińska, E. *Polimery Syntetyczne i Materiały Malarsko-Lakiernicze w Myślach Drzewnym [Synthetic Polymers and Paint and Varnish Materials in the Woodworking Industry]*; Wydawnictwo SGGW: Warsaw, Poland, 2009.
6. Marasinghe, L.; Croutxé-Barghorn, C.; Allonas, X.; Criqui, A. Effect of reactive monomers on polymer structure and abrasion resistance of UV cured thin films. *Prog. Org. Coat.* **2018**, *118*, 22–29. [CrossRef]
7. Weber, E. North American UV+EB Market Overview. In Proceedings of the RadTech 2020 Conference, Orlando, FL, USA, 8–11 March 2020; Available online: www.radtech.org (accessed on 2 November 2023).
8. Cano, M.; Núñez-Lozano, R.; Lumbreras, R.; González-Rodríguez, V.; Delgado-García, A.; Jiménez-Hoyuela, J.M.; de la Cueva-Méndez, G. Partial PEGylation of superparamagnetic iron oxide nanoparticles thinly coated with amine-silane as a source of ultrastable tunable nanosystems for biomedical applications. *Nanoscale* **2017**, *9*, 812–822. [CrossRef]
9. Jung, S.J.; Lee, S.J.; Cho, W.J.; Ha, C. Synthesis and properties of UV-curable waterborne unsaturated polyester for wood coating. *J. Appl. Polym. Sci.* **1998**, *69*, 695–708. [CrossRef]
10. Heuer, A.M.; Reinartz, R.; Skotarczak, A.; Presenter, M.N.; Lin, B. Novel approaches to improve scratch and abrasion resistance in UV coatings. In Proceedings of the RadTech 2020 Conference, Orlando, FL, USA, 8–11 March 2020.
11. Curatolo, B. UV technology for the protection of surfaces. *UV+ EB Technol.* **2015**, *2*, 22–29.
12. Li, Y.; Shao, H.; Lv, P.; Tang, C.; He, Z.; Zhou, Y.; Shuai, M.; Mei, J.; Lau, W.M. Fast preparation of mechanically stable superhydrophobic surface by UV cross-linking of coating onto an oxygen-inhibited layer of substrate. *Chem. Eng. J.* **2018**, *338*, 440–449. [CrossRef]
13. Guo, X.; Li, R.; Teng, Y.; Cao, P.; Wang, X.A.; Ji, F. Effects of surface treatment on the properties of UV coating. *Wood Res.* **2015**, *60*, 629–638.
14. Lustig, C.; Chakrapani, S. UV curable coatings-options for challenging substrates. In Proceedings of the RadTech 2020 Conference, Orlando, FL, USA, 8–11 March 2020.
15. Wnuk, R. Wodne materiały utwardzające promieniowaniem UV-przyszłość w lakierowanie drewna. *Uszlachetnianie Powierzchni Drzewnych cz. II.* 2011, pp. 16–18. Available online: https://woodchem.pl/uszlachetnianie-powierzchni-drewna/ (accessed on 2 November 2023).
16. Heathcote, J. Update on Mercury Regulation for UV Curing Lamps. In Proceedings of the RadTech 2020 Conference, Orlando, FL, USA, 8–11 March 2020.
17. Borysiuk, P.; Derda, M.; Auriga, R.; Boruszewski, P.; Monder, S. *1. Comparison of Selected Properties of Varnish Coatings Curing with the Use of UV and UV-LED Approach*; Annals of Warsaw University of Life Sciences—SGGW Forestry and Wood Technology: Warsaw, Poland, 2015; ISSN 1898-5912.
18. Studer, K.; Decker, C.; Beck, E.; Schwalm, R. Overcoming oxygen inhibition in UV-curing of acrylate coatings by carbon dioxide inerting, Part I. *Prog. Org. Coat.* **2003**, *48*, 92–100. [CrossRef]
19. Taki, K.; Yamada, R. Comparison of the degree of shrinkage under air and nitrogen atmospheres by laser displacement sensor. *J. Photopolym. Sci. Technol.* **2018**, *31*, 497–501. [CrossRef]
20. Park, J.W.; Sim, K.B.; Shim, G.S.; Back, J.H.; Baek, D.; Kim, H.J.; Shin, S. Depth profile of thin coating through surface and interfacial cutting analysis of UV curing system. *Mater. Des.* **2019**, *178*, 107855. [CrossRef]
21. Wang, T.; Shaw, J. Physical Properties of UV Curable Composites. In Proceedings of the RadTech 2020 Conference, Orlando, FL, USA, 8–11 March 2020; Available online: www.allnex.com (accessed on 2 November 2023).
22. Rudnicki, A. Lakiery utwardzane promieniami UV-wiele zalet [UV-cured varnishes—Many advantages]. *Lakiernictwo Przemysłowe. Uszlachetnianie Powierzchni Drzewnych cz. III.* 2011, pp. 16–17. Available online: https://woodchem.pl/uszlachetnianie-powierzchni-drewna/ (accessed on 2 November 2023).
23. Kien, W.; Osajda, M.; Boratyński, P. Linie lakiernicze z utwardzaniem naskórka promieniami UV-dobierz temperaturę urządzenia [Paint lines with UV surface curing—Adjust the device temperature]. *Lakiernictwo Przemysłowe. Uszlachetnianie Powierzchni Drzewnych cz. V.* 2013, pp. 40–47. Available online: https://woodchem.pl/uszlachetnianie-powierzchni-drewna/ (accessed on 2 November 2023).
24. Wnuk, R.; Utwardzanie powłoki lakierniczych przy użyciu LED. Utwardzanie powłoki lakierniczych przy użyciu LED. Oszczędne i „pożerające" mniej energii" [LED paint curing. Economical and consuming less energy]. *Lakiernictwo Przemysłowe. Uszlachetnianie Powierzchni Drzewnych cz. V.* 2013, pp. 34–35. Available online: https://woodchem.pl/uszlachetnianie-powierzchni-drewna/ (accessed on 2 November 2023).

25. Wang, X.; Feng, Y.; Zhang, L.; Protsak, I.; Jamali, R.; Shu, Y.; Pal, P.; Wang, Z.; Yang, J.; Zhang, D. Fast-cured UV-LED polymer materials filled with high mineral contents as wear-resistant, antibacterial coatings. *Chem. Eng. J.* **2020**, *382*, 122927. [CrossRef]
26. Heathcote, J. Update on Mercury Regulation for UV Curing Lamps. GEW (EC). 6 March 2023. Available online: https://www.pffc-online.com/news/17269-update-on-mercury-regulation-for-uv-curing-lamps (accessed on 1 May 2023).
27. European Commission. RoHS 2 FAQ. 2012. Available online: http://ec.europa.eu/enterprise/policies/single-market-goods/documents/blue-guide/ (accessed on 2 November 2023).
28. European Commission. *Fact Sheet Questions and Answers: EU Mercury Policy and the Ratification of the Minamata Convention*; European Commission: Brussels, Belgium, 2017.
29. United Nations Environment Programme. Minamata Convention on Mercury Text and Annexes. 2019. Available online: www.mercuryconvention.org (accessed on 2 November 2023).
30. Minamata Convention on Mercury. Strengthening the Specific International Programme to Support Capacity-Building and Technical Assistance of the Minamata Convention on Mercury: Enhancing the Effective Functioning of the Programme. 2021. Available online: https://minamataconvention.org/en/implementation/specific-international-programme (accessed on 2nd November 2023).
31. Beutler, P. 4. Mercury-Vapor Lamp or LED? Karlsruhe. 2022. Available online: http://assets.thermofisher.com/TFS-Assets/MSD/Product-Information/mercury-vapor-lamp-led-en-p072.pdf (accessed on 30 April 2023).
32. Kiyoi, E. 3. Keep It Cool to Keep It Clean Advantages of UV LEDs When Coating Heat-Sensitive Types of Wood. 2013. Available online: www.european-coatings.com (accessed on 2 November 2023).
33. Kiyoi, E. 5. The State of UV-LED Curing: An Investigation of Chemistry and Applications Characteristics of UV-LED Lamps. 2013. Available online: https://radtech.org (accessed on 2 November 2023).
34. Bekas, J. Skuteczniej utwardzają, a nie podgrzewają. *Gaz. Przemysłu Drzewnego* **2015**, *9*, 45.
35. Dawidowicz, A.L.; Nowakowski, P.; Dybowski, M.P.; Typek, R. Some physicochemical properties of polyacrylate varnish layers cured by classical and LED (395 nm) UV sources. *Food Packag. Shelf Life* **2020**, *24*, 100497. [CrossRef]
36. Europen Coatings. Radiation curing by LED technology for industrial wood flooring coatings. *Focus Powder Coat.* **2023**, *2023*, 6. [CrossRef]
37. Goodrich, J.E. Acrylated Amine Oligomers to Enhance Cure with UV LED Sources. In Proceedings of the RadTech 2020 Conference, Orlando, FL, USA, 8–11 March 2020.
38. Schmitz, C.; Poplata, T.; Feilen, A.; Strehmel, B. Radiation crosslinking of pigmented coating material by UV LEDs enabling depth curing and preventing oxygen inhibition. *Prog. Org. Coat.* **2020**, *144*, 105663. [CrossRef]
39. Ivarsson, D. The Impact of Different Intensities on the Performance of Led Cured Coatings. Available online: https://www.pcimag.com (accessed on 2 November 2023).
40. Gharagozloo, M.; Dombkowski, T.; Leonhardt, D.; Xu, D.; Macairan, J.; Skinner, B. IIoT Enabled Smart UV Systems. In Proceedings of the RadTech 2020 Conference, Orlando, FL, USA, 8–11 March 2020.
41. Ivarsson, D.; Bilinsky, H. Curing matt finishes into UV curable wood coatings by direct contactless microfabrication. In Proceedings of the RadTech 2020 Conference, Orlando, FL, USA, 8–11 March 2020.
42. Salca, E.A.; Krystofiak, T.; Lis, B.; Hiziroglu, S. Glossiness evaluation of coated wood surfaces as function of varnish type and exposure to different conditions. *Coatings* **2021**, *11*, 558. [CrossRef]
43. Salca, E.A.; Krystofiak, T.; Lis, B.; Mazela, B.; Proszyk, S. Some coating properties of black alder wood as a function of varnish type and applications method. *Bioresources* **2016**, *11*, 7580–7594. [CrossRef]
44. Jančovičová, V.; Kindernay, J.; Jakubíková, Z.; Mrlláková, I. Influence of photoinitiator and curing conditions on polymerization kinetics and gloss of UV-cured coatings. *Chem. Pap.* **2007**, *61*, 383–390. [CrossRef]
45. Gürleyen, L. Effects of artificial weathering on the colour, gloss, adhesion, and pendulum hardness of UV system parquet varnish applied to doussie (Afzelia africana) wood. *Bioresources* **2021**, *16*, 1616–1627. [CrossRef]
46. Okamura, H.; Niizeki, S.; Ochi, T.; Matsumoto, A. UV curable formulations for UV-C LED. *J. Photopolym. Sci. Technol.* **2016**, *29*, 99–104. [CrossRef]
47. *ISO 2813*; Paints and Varnishes—Determination of Specular Gloss of Non-Metallic Paint Films at 20 Degrees, 60 Degrees and 85 Degrees. International Organization for Standardization: Geneva, Switzerland, 1994.
48. *DIN 67530*; Reflectometer as a Means for Gloss Assessment of Plane Surfaces of Paint Coatings and Plastics. European Standards: Brussels, Belgium, 1982.
49. *ISO 7724*; Paints and Varnishes. Colourimetry. Part 2: Colour Measurement. International Organization for Standardization: Geneva, Switzerland, 1984.
50. Hill, B.; Roger, T.; Vorhagen, F.W. Comparative Analysis of the Quantization of Colour Spaces on the Basis of the CIELAB Colour-Difference Formula. *ACM Trans. Graph.* **1997**, *16*, 109–154. [CrossRef]
51. Wijekularathna, S.M.; Manage, D.K.; Scariano, A.B.W. Power analysis of several normality tests: A Monte Carlo simulation study. *Commun. Stat. Simul. Comput.* **2020**, *51*, 757–773. [CrossRef]
52. Landry, V.; Blanchet, P.; Boivin, G.; Bouffard, J.F.; Vlad, M. UV-LED curing efficiency of wood coatings. *Coatings* **2015**, *5*, 1019–1033. [CrossRef]
53. Evans, P.D.; Cullis, I. Effect of sanding and coating with UV-cured finishes on the surface roughness, dimensional stability and fire resistance of oriented strandboard. *Holz als Roh-und Werkstoff* **2008**, *66*, 191–199. [CrossRef]

54. Akkuş, M.; Akbulut, T.; Candan, Z. Application of electrostatic powder coating on wood composite panels using a cooling method. Part 1: Investigation of water intake, abrasion, scratch resistance, and adhesion strength. *BioResources* **2019**, *14*, 9557–9574. [CrossRef]
55. Henke, M.; Lis, B.; Krystofiak, T. Gloss Level of HDF Finished with Different Numbers of Layers and Hardened with UV Hg-Ga Lamps of Selected Power. *Coatings* **2022**, *12*, 533. [CrossRef]
56. Lis, B.; Krystofiak, T.; Proszyk, S. Studies of the resistance upon some factors of UV acrylic lacquer coatings on MDF boards. Part II. Mechanical factors. *Ann. Warsaw Univ. Life Sci. SGGW For. Wood Technol.* **2010**, *71*, 454–457.
57. Bekhta, P.; Lis, B.; Krystofiak, T.; Bekhta, N. Surface Roughness of Varnished Wood Pre-Treated Using Sanding and Thermal Compression. *Forests* **2022**, *13*, 777. [CrossRef]
58. Qi, L.; Chantler, M.J.; Siebert, P.; Dong, J. The joint effect of mesoscale and microscale roughness on perceived gloss. *Vis. Res.* **2015**, *115*, 209–217. [CrossRef] [PubMed]
59. Honson, V.; Huynh-Thu, Q.; Arnison, M.; Monaghan, D.; Isherwood, Z.J.; Kim, J. Effects of Shape, Roughness and Gloss on the Perceived Reflectance of Coloured Surfaces. *Front. Psychol.* **2020**, *11*, 485. [CrossRef] [PubMed]
60. de Moura, L.F.; Brito, J.O.; Nolasco, A.M.; Uliana, L.R. Evaluation of coating performance and colour stability on thermally rectified Eucalyptus grandis and Pinus caribaea var. Hondurensis woods. *Wood Res.* **2013**, *58*, 231–242.
61. Sinn, G.; Ginkl, M.; Reiterer, A.E.T.; Stanzl-Tschegg, S. Changes in the surface properties of wood due to sanding. *Holzforschung* **2004**, *58*, 246–251. [CrossRef]
62. Zeren, S.; Huguenard, S. UV-curable powder coatings: Formulation of thick white finish for MDF. *Surf. Coat. Int. Part B Coat. Trans.* **2004**, *87*, 71–148. [CrossRef]

applied sciences

MDPI

Article

Light-Induced Colour Changes in Wood Surfaces in Indoor Conditions Determined by an Artificial Accelerated Test: Influence of Wood Species and Coating Materials

Mihai-Junior Torcătoru * and Maria Cristina Timar *

Faculty of Furniture Design and Wood Engineering (FDMIL), Transilvania University of Brașov, 500036 Brasov, Romania
* Correspondence: mihai.torcatoru@unitbv.ro (M.-J.T.); cristinatimar@unitbv.ro (M.C.T.)

Abstract: Stabilising the natural colour of wood species exposed to light in indoor conditions is a challenge that could be better addressed based on a deeper understanding of the occurring phenomena and influencing factors. This paper investigates comparatively the light-induced colour changes for three hardwood species, namely, European maple (*Acer pseudoplatanus* L.), European ash (*Fraxinus excelsior* L.) and European walnut (*Juglans regia* L.), as well as the influence of finishing with three types of clear, colourless waterborne lacquers: acrylic-polyurethane (F1), acrylic (F2) and polyurethane (F3) on their colour stability. Colour measurements in the CIELab system in conjunction with an artificial accelerated ageing test under the action of UV-VIS radiation, simulating natural light passing through window glass, and two types of test samples were employed to highlight the influence and contributions of the wood substrate and of the coating films to the global colour modifications. Coating films applied on 1 mm clear glass slides were employed as a sort of "detachable" finish for this purpose. Direct exposure to UV-VIS light caused visible colour changes for both uncoated and coated wood surfaces, the values of the calculated colour differences (ΔE) after a 72 h exposure being dependant on both the wood species and the coating material. Excepting two situations for walnut, statistically significantly higher colour differences were obtained for the uncoated samples: maple (9.36 units), ash (8.39 units), walnut (6.20 units), compared to the coated ones: maple (4.92–5.71 units), ash (2.25–3.94 units), walnut (4.74–7.70 units). The wood substrates underlying the coating films were found to bring the maximum contribution to the overall colour changes in the clear coated surfaces, while the coating films employed in this research demonstrated a fairly good colour stability to UV-VIS light exposure, with maximum colour changes (ΔE) up to only 1.30 units. Overall, the wood species and the type of coating were found as influencing factors in interaction with the light-induced colour changes in wood surfaces in indoor conditions.

Keywords: wood; colour; clear waterborne coatings; UV-VIS radiation; accelerated artificial test; colour changes

Citation: Torcătoru, M.-J.; Timar, M.C. Light-Induced Colour Changes in Wood Surfaces in Indoor Conditions Determined by an Artificial Accelerated Test: Influence of Wood Species and Coating Materials. *Appl. Sci.* **2024**, *14*, 1226. https://doi.org/10.3390/app14031226

Academic Editor: Giuseppe Lazzara

Received: 8 December 2023
Revised: 24 January 2024
Accepted: 30 January 2024
Published: 1 February 2024

1. Introduction

Colour, alongside texture and grain, represent particular and valuable aesthetic features of the various wood species [1]. The coating of wood surfaces with transparent, colourless finishing materials is the usual treatment applied to highlight and preserve their beauty and natural aspect over time in various practical applications. Nevertheless, colour changes in both uncoated and coated wood surfaces occur over time when exposed to light/UV radiation, both indoors and outdoors. This practical issue has been only partly solved by the outcomes of extensive research worldwide, remaining an actual research topic [2–7].

Colour changes in wood surfaces exposed to light are primarily the result of a complex and dynamic photo-degradation process of the wood surface, having as main actors the UV radiation from the light source and the lignin in wood. The absorption of UV radiation by lignin results in chemical bond cleavage and the formation of reactive free radicals (e.g., phenoxyl), which are then transformed into carbonyl- and carboxyl-containing chromophores and quinoid structures, responsible for colour changes, especially yellowing of the exposed wood surfaces [8–13]. Not only UV light but also visible violet light and blue light up to 515 nm may cause surface and subsurface colour changes without affecting lignin [14,15]. Furthermore, a possible contribution of wavelengths in the visible-near infrared region was highlighted by hyperspectral imaging [16].

In addition to lignin, certain classes of wood extractives may play an important role in determining the natural colour of wood species and their behaviour when exposed to light. Poly-phenolic extractives, such as tannins, can absorb UV light, scavenge free radicals and undergo photo-oxidation, retarding the photo-degradation of lignin and wood [17–20]. On the other hand, extractives from dark coloured wood species are responsible for their sensitivity to visible light, resulting in a bleaching effect after longer exposure times [21].

The result is, therefore, that the light-induced colour changes in wood surfaces are significantly influenced by the wood species due to the differences in their chemical composition, as proven by comparative research on more than two different species [1,22–24]. Colour changes also depend on the type of light (natural/artificial), spectral range (wavelengths), type of light source, exposure conditions and duration [25–27]. Accordingly, data from different research publications are often difficult to compare.

The coating of wood can impart some protective effect against light-induced colour changes by reducing the amount of light/UV radiation reaching the underlying wood substrate [28], depending on the chemical structure of the film-forming resin and the specific additives in a certain formulation. The coating materials can also undergo photo-degradation processes (photolysis, photo-oxidation) under the action of UV light, resulting in property changes, including glass transition temperature, UV light transmittance, colour and their protective behaviour [29,30]. Different types of film-forming materials, such as: acrylic-polyurethanes [31], acrylics [28,29,32–34] and polyurethanes [35], as well as natural products [36], have been studied in terms of behaviour and protective effect against light/UV-induced colour changes in wood surfaces.

Due to their transparency for both visible and UV light, clear finishes offer generally less protection against the action of light/UV radiation, resulting in more pronounced colour changes [28,29,31,33,35], compared to semi-transparent and opaque finishes that contain colouring pigments with a UV protective effect.

The interaction of UV light with the underlying wood at the interface wood/coating material results in the photo-oxidation of lignin and associated colour changes, representing an important contribution to the overall colour changes in clear-coated wood surfaces [10,28,35]. Different principles for stabilising the colour of clear-coated wood have been in research focus: (i) pre-treatments of the wood substrate for an improved UV resistance, (ii) modification of the coating materials with additives for improved UV resistance and (iii) combined methods. Nanoparticles of TiO_2 [5], ZnO [37], CeO_2 [3], CH_2O [38] or α/γ Fe_2O_3 [6], UV absorbers (benzotriazole, [39], benzophenone [40]), HALS (hindered amine light stabiliser), (Tinuvin 292, Tinuvin 293 [41], Tinuvin 5151 [39]), or chemical modification treatments with succinic anhydride [42], maleic anhydride [43] or soybean oil modified epoxy resins [40] were employed as pre-treatments. Organic UV absorbers and HALS stabilisers [34,36] and inorganic UV absorbers and nanoparticles [6,44] were employed as additives for clear coatings. Approaches combining the two principles were also tested [4,13,33,39,41].

Most of the studied references reporting light-induced colour changes in the coated wood surfaces refer strictly to the changes observed and measured on the directly exposed coated wood surface [1,5,17,22,45], without any information on the possible photo-degradation of the wood substrate under the coating film. However, colour changes in clear-coated wood may be caused by colour changes in both the clear-coating film and underlying wood, or either of them [35], and their contribution should be acknowledged and evaluated to deeper understand the occurring phenomena. References addressing this important issue are rather rare [35,36,46,47]. Uncoated wood specimens and free coating films were employed to simulate clear-coated wood. A light reflection model was used to elucidate separately the colour changes caused by the clear-coating film and the underlying wood, showing a major contribution of the underlying wood compared to the coating film for two types of polyurethanes (approx. 60%/40%; 95%/5%) [35]. Different research concluded that in the case of spruce wood clear coated with a water-borne acrylic, the colour changes after 200 h UV irradiation were due almost entirely to the wood substrate, as colour changes in the coating film were negligible [47].

More studies have focused on the exposure of wood in outdoor conditions where UV radiation acts in conjunction with rainwater [2,32,45], while the topic of light-induced colour changes in indoor conditions by the natural light passing through the window glass [1,22,23,36] or light from artificial sources [26,27] has been less addressed, especially for clear-coated wood with natural aspect [46,48].

All these point to the idea that stabilising the natural colour of wood species exposed to light in indoors conditions remains a challenge that could be better addressed based on a deeper understanding of the occurring phenomena and influencing factors.

In this context, the objectives of the present study were: to evaluate comparatively the light-induced colour changes for uncoated and clear-coated wood surfaces with natural aspect in indoor conditions by accelerated testing; to determine the influence of the species and type of coating on the colour changes; and to highlight the specific contributions of the wooden support and of the coating films to the total colour changes in the finished surfaces. A dedicated laboratory method was employed for this purpose [46].

2. Materials and Methods

2.1. Wooden Material

Three hardwood species: European maple (*Acer pseudoplatanus* L.), European ash (*Fraxinus excelsior* L.) and European walnut (*Juglans regia* L.), were employed. The wood material was sourced from the company "Craiul Munților SRL", Stupărei village, Valcea county, Romania. These species were selected based on their aesthetic appearance and importance for furniture and other interior design element manufacturing, including wooden floors and panelling. These species differ in natural colour, anatomical structure, texture and the presence or not of colour-differentiated sapwood and heartwood areas, so that they can provide beautiful contrasts that can be valorised by design, while also representing a group with sufficient variability in structure and properties to provide useful research data. Moreover, the expected research results will add new information to a data base, including their ageing properties as bare substrates or coated surfaces with natural traditional finishing materials [24], allowing useful comparisons.

Test samples of a rectangular form of 120 × 80 × 8 (mm × mm × mm) with radial or semi-radial faces were prepared from air-dried wood material. All samples were sanded manually with abrasive paper (120 and 180 grit sizes).

2.2. Coating Materials

Three types of clear, waterborne, colourless lacquers, falling into the classes of acrylic-polyurethane F1 (YO 20-M702), acrylic F2 (YO 15-M864) and polyurethane F3 (YO 20-M838) coatings (Table 1), were employed for wood finishing with natural aspect. These are all products of Renner Italy (Minerbio, Italy) [49], commercialised in Romania by Kroncolor Brașov (Brașov, Romania) [50]. The selection of these materials took into account ecological aspects (preference for waterborne products instead of solvent-based), their market share as finishing materials, the quality of the cured coating films ensured by their chemistry, alongside the recommendation of the specialists from Kroncolor for transparent finishes with natural appearance.

Table 1. Waterborne transparent coating materials and additives (Renner–Italy) employed in the experimental research.

Lb. Code	Commercial Code	Film-Forming Resin	Solids Content, Csu, %	Density, g/cm^3	Formulation	Gardner Gloss Index
F1	YO 20-M702	Acrylic-polyurethane (non-yellowing)	25 ± 1	1.02	1k	20
F2	YO 15-M864	Acrylic (non-yellowing)	34 ± 1	1.03	1k	15
F3	YO 20-M838	Polyurethane	30 ± 1	1.03	1k	20

2.3. Test Sample Preparation

Wood specimens (uncoated and coated) and clear microscopic slides (1 mm thickglass lamellae) as such and coated with the experimental finishing materials (F1, F2, F3 as 1k formulations) were employed in this research. The finishing of the wood samples was carried out in laboratory conditions, applying two successive layers of approximately 90–100 g/m^2 (corresponding to an approximately 100 μm thickness as a liquid), at an interval of 2–4 h. The same finishing technology was applied for the glass lamellae, employing a special device for coating film application (2 layers of 100 μm), with the intention to ensure reproducible testing conditions such as film thicknesses for the test samples V1 (uncoated wood + coating film of glass lamella) and F2 (coated wood). It should be noted, however, that some differences resulted from the practical procedure and the variation in the solids content and viscosity of the three coating materials. Moreover, the differences in the anatomical features of the three wood species: maple—diffuse porous with small vessels not visible with the naked eye; walnut—diffuse porous with larger vessels visible with the naked eye; and ash—ring porous with distinctly larger vessels clearly visible in early-wood, resulted in different coating penetration and coverage of the wood grain. Accordingly, semi-filled/nearly filled grain finishes were obtained for maple, and open grain finishes for walnut and ash. The preliminary examination of the transversal cross-cut edges of the coated samples confirmed visual perception in terms of pores filling and highlighted the interaction between the coating and the wood substrate at the interface, resulting in different penetrations of the coatings into the wood substrates. It could be observed that strongly interconnected "composite coating layers", including coating films on top of wood surfaces and impregnated sub-surfaces, were obtained.

The actual spread rates applied and the (average values), determined by weighing each sample immediately after coating application, as well as the theoretical thickness of the resulting cured coating films, calculated considering the average spread rates and the solids content of each coating material but no wood penetration, are summarised in Table 2, The actual thicknesses of the cured coating films applied on glass lamellas, determined with a precision of 0.001 mm, are also included.

Two types of test samples for UV-VIS light exposure, coded V1 (uncoated wood support-Figure 1A) and V2 (coated wood support Figure 1B), resulted from assembling wood samples with glass lamellae, following an original concept [46]. This was developed to allow highlighting and measurement of the light-induced colour changes for the uncoated (V1/Z3) and coated wood surfaces (V2/Z1), the coating films (on glass lamellae) and the wood substrate under the coating film (simulated by V1/Z1, uncoated wood under coated glass lamella), in the various specific zones differentiated as exposure situations. Both types of test samples included a colour control area (V1/Z4, V2/Z2), tightly covered with black cardboard and aluminium foil to prevent the penetration and effect of light. The two types of test samples (V1, V2) were prepared in triplicate for each tested variant (wood species, type of coating), resulting in a number of 18 samples/species, respectively, 54 samples in total.

Table 2. Experimental values referring to the preparation of the coated wood samples and glass lamellae with the three types of coating materials: application rates, theoretical film thickness on wood samples, thickness film on glass lamellae, alongside the solids content and viscosity of the applied coating materials F1, F2, F3.

Wood Species Substrate	Experimental/Calculated Value	Type of Coating		
		F1	F2	F3
Maple	Application rate [g/m^2]	191.1 ± 16.2	196.3 ± 19.7	203.5 ± 20.2
	Film thickness [μm]	47.8	57.9	55.6
Ash	Application rate [g/m^2]	180.5 ± 25.8	203.3 ± 17.4	168.3 ± 4.0
	Film thickness [μm]	45.1	59.9	46.0
Walnut	Application rate [g/m^2]	161.6 ± 16.2	184.9 ± 28.6	189.4 ± 17.4
	Film thickness [μm]	40.4	54.5	51.8
Glass lamella	Film thickness [mm]	0.034 ± 0.004	0.043 ± 0.008	0.046 ± 0.007
Coating material	Solids content [%]	25.0	29.5	27.3
	Flow time, ϕ 6 mm, 20 °C [s]	12	49	6

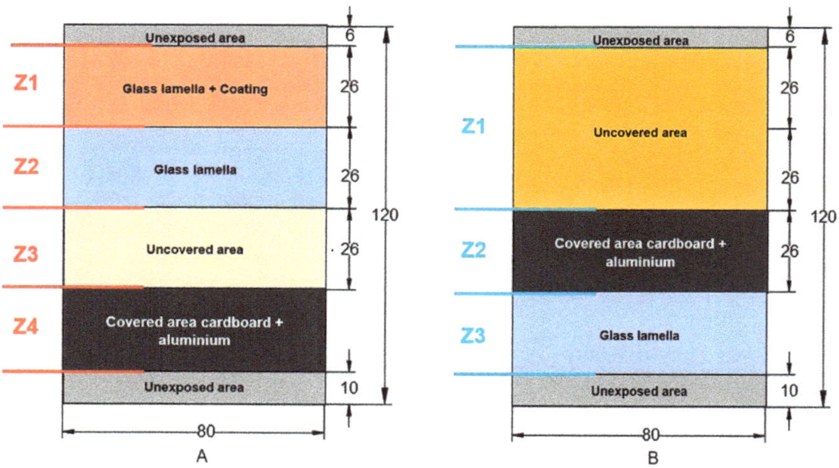

Figure 1. Test samples: (**A**). V1—Uncoated wood support, coating on glass lamella: dimensions and various specific zones/light exposure situations (Z1–Z4); (**B**). V2—Coated wood support: dimensions and various specific zones/light exposure situations (Z1–Z3). (reproduced from [46] with permission).

2.4. Colour Measurements

Colour measurements in the CIELab reference system were performed using an Avaspec-USB2 spectrometer from Avantes (Apeldoorn, The Netherlands) [51] equipped with an integration Ava Sphere with a diameter of 80 mm, with a circular measuring aperture of 8 mm in diameter. Measurements were carried out under a standard D65 illuminant and the measurement angle set at 10 degrees, according to the relevant standard methods [52]. The light connection to the sphere was via glass fibres and a COL-UV-VIS collimating lens connected to the light source. The colour data were processed employing Ava soft 7.0.

The measured colour parameters were: the lightness value (L*), varying from 0 for black to 100 for white; the degree of red (redness) on the green–red chromatic axis (−a*, +a*); and the degree of yellow (yellowness) on the blue–yellow colour axis (−b*, +b*). The colour measurements were performed on 12 fixed points (circular areas of 8 mm diameter) on each wood sample (uncoated or coated) and on 3 fixed points for the coated glass lamellae, prior to assembling the wood specimens with the glass lamellae. After each period of light exposure, the samples were disassembled and colour measurements repeated in the same areas on the wood substrate and the coated lamellae. The colour measuring points on the wood samples corresponded to 3 fixed locations on each zone of the test samples, excepting zones V2/Z1 with 6 measuring points. As the test samples were in triplicate, a minimum number of 9 and maximum 18 measurements resulted for each zone/exposure situation for each tested variant. All the colour data were collected and further processed in Excel to calculate the light-induced variation in colour coordinates (ΔL*, Δa* and Δb*) and the total (global) colour difference (ΔE) according to the relationships (1)–(4):

$$\Delta L^* = L_e^* - L_i^* \tag{1}$$

$$\Delta a^* = a_e^* - a_i^* \tag{2}$$

$$\Delta b^* = b_e^* - b_i^* \tag{3}$$

$$\Delta E = \sqrt{\left(\left(\Delta L^*\right)^2 + (\Delta a^*)^2 + \left(\Delta b^*\right)^2\right)} \tag{4}$$

where: e—exposed; i—initial.

A series of ΔL*, Δa*, Δb* and ΔE values calculated for each measuring point of a certain area/exposure situation (9 or 18 values) and their means were employed for statistical analysis of the colour changes after 72 h of light/UV exposure.

2.5. Accelerated UV–VIS Light-Induced Ageing Procedure

A Feutron 400 FKS environmental climatic chamber (Feutron Klima Simulation–GmbH-Greiz, Germany), equipped with a UVA Spot 400 T lamp (Hoenle UV Technology–Gilching, Germany), fitted with a glass H2 filter, was employed to expose the wood samples to UV-VIS light in the range of 295 to 600 nm, simulating natural light filtered by window glass, similarly to the previously reported research [23,53]. Considering the maximum emission of the lamp in the UV range (mainly UV-A, as indicated also by its name), this type of radiation with higher energy was considered to be the main ageing factor causing colour and chemical changes in our experiments. However, the expression UV lamp and UV exposure were still employed in this paper, including the graphs, for commodity reasons. It should be acknowledged, however, that visible light with wavelengths up to 515 nm may contribute to the surface and sub-surface wood colour changes, while visible violet radiation up to 430 nm may have a contribution to lignin degradation [14]. Accordingly, the visible light emitted by the source employed might have had a contribution to the colour changes measured in this research.

The samples were placed vertically on a rack at a distance of 60 cm from the UVA light source. The actual irradiation procedure, similar to the previously reported research [53], included an initial conditioning of 0.5 h (20 °C, 55% RH, no light), followed by 3 cycles of 24 h UV exposure separated by conditioning phases of 24 h (20 °C, 55% RH, no light), inserted to allow the removal of samples for measurements. Each cycle of 24 h UV exposure consisted of four steps of 6 h UV irradiation at 40 °C, alternated with dark periods of 0.5 h. Cycles of 24 h UV exposure are in accordance with a standardised procedure of exposure to daylight behind window glass (method B) [54]. This procedure was repeated three times so that the data refer to samples exposed for 24, 48 and 72 h to UV radiation. All the samples were in triplicate and exposure was carried out in two series. The position of the samples on the rack was changed after each period of 24 h exposure in order to ensure a similar degree of irradiation.

2.6. Data Processing and Statistical Analysis

Microsoft Excel version 2019 and the operating system Windows11 was employed for data collection and their basic processing by calculation of the average values and standard deviations for each group of 9 or 18 measurements. One-way Anova test in conjunction with Tukey's honestly significant difference (HSD) as post hoc tests were employed to check the significant differences between the means of different groups at a significance level $\alpha = 0.05$. Two-way Anova with replication was employed to evaluate the statistical significance of the wood species, type of coating material or exposure situation (specific areas of the test samples) as influencing factors on the global colour changes ΔE after 72 h light/UV exposure. The data analysis tool-pack in Excel 2019 and Statistica 8 software were used for statistics.

3. Results and Discussion

3.1. Light-Induced Colour Changes in the Uncoated Wood Surfaces

Light exposure of the uncoated wood surfaces (samples V1) resulted in progressive colour changes, visually perceived mostly as darkening, with some shade modifications, depending on the exposure situation and wood species. The evolution of these colour changes, expressed by the variation in the ΔE, ΔL*, Δa*and Δb* values as a function of the exposure time, is illustrated by the graphs in Figure 2, comparatively for the areas V1/Z3 directly exposed and the area V1/Z2 covered with a clear glass lamella.

The total colour changes (ΔE) in the unfinished wood support directly exposed to light/UV radiation (V1/Z3 areas) increased gradually during exposure, though these changes were more rapid in the first 24 h. The highest values after 72 h exposure were registered for maple (9.36 units), followed by ash (8.39 units) and then walnut (6.20 units). A similar order of light sensitivity for these three species was determined by [1], ΔE values of approximately 14.1 units, 12.5 units and 9.8 units resulting after longer exposure (600 h). Colour changes of 12.5 units for maple and 11.7 units for walnut were reported after 7 years exposure to natural light passing through the window glass [48].

The total colour changes resulted mostly from the decrease in lightness, highlighted by the negative ΔL* values evolving following a mirrored pattern when compared to ΔE and reaching the final values of (−5.07) units for maple, (−4.13) units for ash and (−3.25) units for walnut (Figure 2(A2)).

From a chromatic point of view, the positive Δa* values (up to 2.23 units) indicate only a slight increase in redness, while the positive Δb* values (up to 7.58 units) indicate a more important increase in the yellowness, specific to each wood species. The previous research and literature data indicated that the redness of wood, as well as UV/light-induced changes in redness, might be associated to the extractives [55], while the increase in yellowness can be correlated to UV-induced lignin degradation and formation of new quinones and stilbenes chromophores [8,9,11,12]. As lignin photo-degradation is the first and main photochemical process induced by UV light, higher changes in yellowness compared to redness are expected. In the present research, the highest increase in redness (Δa*) was determined in the case of walnut (2.23 units), followed by maple (1.48 units) and ash with quite similar values (1.43 units). Contrarily, the lowest change in yellowness was determined for walnut (Δb* = 4.42 units), compared to ash (7.05 units) and maple (7.58 units), with the highest value. This is in accordance with the previous research results and may be explained by the role of extractives [23,53]. Wood extractives may act as UV absorbers and free radical scavengers while themselves undergoing the photo-oxidation processes. All these effects result in retarding the photo-degradation of lignin in wood [18–20], which means less yellowing. A higher content of extractives in the walnut wood compared to maple and ash wood may explain both the darker natural colour of walnut and the reduced colour changes compared to maple following UV exposure, at least for the duration of the test carried out in this research (72 h).

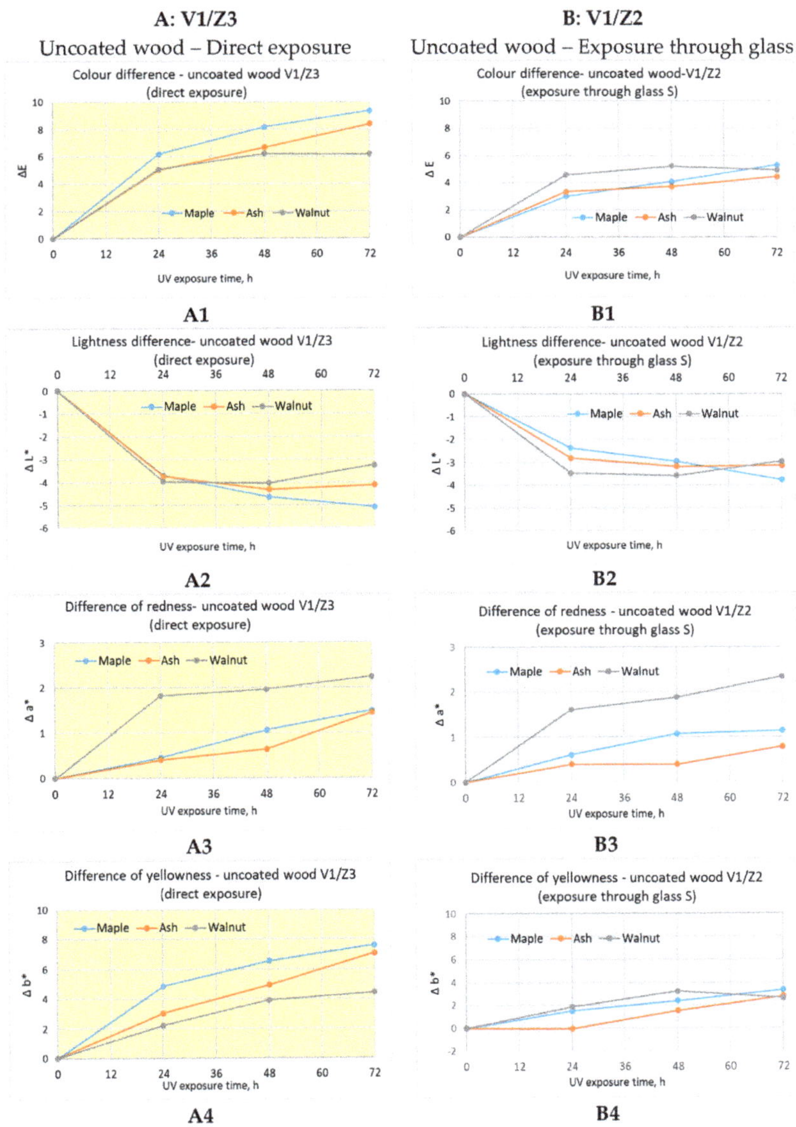

Figure 2. Comparative in-time evolution of light-induced colour changes (ΔE, ΔL*, Δa*, Δb*) for uncoated maple, ash and walnut wood surfaces (samples type V1) depending on the exposure situation: (**A**)—direct exposure to UV light (V1/Z3); (**B**)—exposure through a 1 mm thick clear glass lamella (V1/Z2).

The light-induced colour changes for the uncoated wood surfaces that were covered with a 1 mm thick clear glass slide (V1/Z2) are presented in Figure 2B. It can be observed that the total colour changes (ΔE) are smaller than those registered for surfaces directly exposed (V1/Z3) to UV light for all the three species, though the decrease in ΔE values varied among species, being higher for maple and ash (approximately −4 units) than for walnut (−1.24 units). This shows an effect of the glass by diminishing the intensity/energy and very likely the spectral distribution of light reaching the wood surface underneath. The literature data [56,57] and our own experience have shown reduced transmittance indexes of glass for UVB and UVA wavelength compared to visible radiation. The effect of glass is visible especially in the light-induced changes in lightness (ΔL) and yellowness (Δb*), while it is almost absent in the case of redness (Δa*). This may support the assumption of selective partial blocking of UV radiation from the spectral distribution of the light emitted by the source, so that wood under the glass is exposed not only to light of lower intensity/energy but also to light with a different spectral distribution, with less UV radiation and more VIS radiation. This may also explain why the colour changes were less affected by the cover glass slide in the case of walnut, with a darker, reddish natural colour due to the extractives, which may absorb more visible radiation. The previously published research has demonstrated that dark natural wood species undergo intensive photo-bleaching when exposed to daylight, the most detrimental effect being caused by wavelengths of the visible spectral range, which affects mostly the low molecular extractives [21].

The glass effect is clearly a limitation of the method employed in this research in the attempt of highlighting and quantifying the colour changes in the wood substrate occurring under the coating film by simulating this situation with coated glass lamella on top of uncoated wood (V1/Z1).

3.2. Light-Induced Colour Changes in the Coated Wood Surfaces

Exposure of the coated wood surfaces to light (V2 samples) resulted generally in smaller colour changes, dependant on the wood species and the type of finishing material.

The graphs in (Figure 3A) represent the colour changes in the wood samples coated with the acrylic-polyurethane lacquer, coded F1, in the areas directly exposed to light/UV radiation (F1/V2/Z1) as a function of the exposure duration. The evolution of the global colour changes, expressed by the ΔE values, show a quite similar pattern for maple and ash, with maximum changes occurring during the first 24 h of exposure (3.96 units for maple and 2.16 units for ash), followed then by a slower increase up to the 5.71 units for maple and only 3.37 units for ash. The evolution of colour changes for the walnut samples was totally different, almost linear in time, reaching the highest value of 7.70 units after 72 h of light exposure.

The changes in lightness varied depending on the wood species: an increase in lightness, expressed by positive ΔL* values, reaching (+3.40) units after 72 h exposure was registered for walnut, whilst for the maple and ash coated specimens the lightness decreased, as shown by the negative ΔL* values, reaching (−3.83) units for maple and only (−0.52) units for ash following a reversed evolution after 24 h exposure.

From a chromatic point of view, only slight variations in redness were registered as positive, with Δa* values up to (1.09) units for maple and values up to (1.14) units walnut, while those measured for ash were negative but very close to zero (−0.06 units). Positive Δb* values indicated for all the coated wood species a continuous increase in yellowness, up to 2.65 units for ash, 4.03 units for maple and 6.69 units for walnut. To conclude, the colour stability of the wood samples coated with the acrylic-polyurethane water-based lacquer decreased in the following order: F1-ash > F1-maple > F1-walnut, which is different from that observed for the uncoated wood samples: walnut > ash > maple.

The light-induced colour changes for the same coated wood surfaces covered with 1 mm thick clear glass (V2/Z3) are shown in (Figure 3B). It can be seen that the total colour changes ΔE are smaller than those recorded on the surfaces directly exposed (V2/Z1) to light for all the three species, but to a different extent. The effect of covering glass in reducing the global colour changes (ΔE) for the coated wood (Figure 3(B1)) was similar to uncoated wood, minimal for walnut and clearly more accentuated in the case of maple (maximum) and ash (medium). However, according to the graphs plotted in (Figure 3(B2–B4)), the effect of glass seemed to be more uniformly distributed on all the colour parameters L*, a*, b*, with some specificity related to the wood species. In the case of maple, for instance, it is clearly visible that the yellowness increase due to light exposure was considerably reduced (from approximately 4 units to approximately 1 unit) for the coated samples covered by glass. It is important to highlight that the order of colour stability among the F1 coated wood samples was not changed by the effect of glass cover.

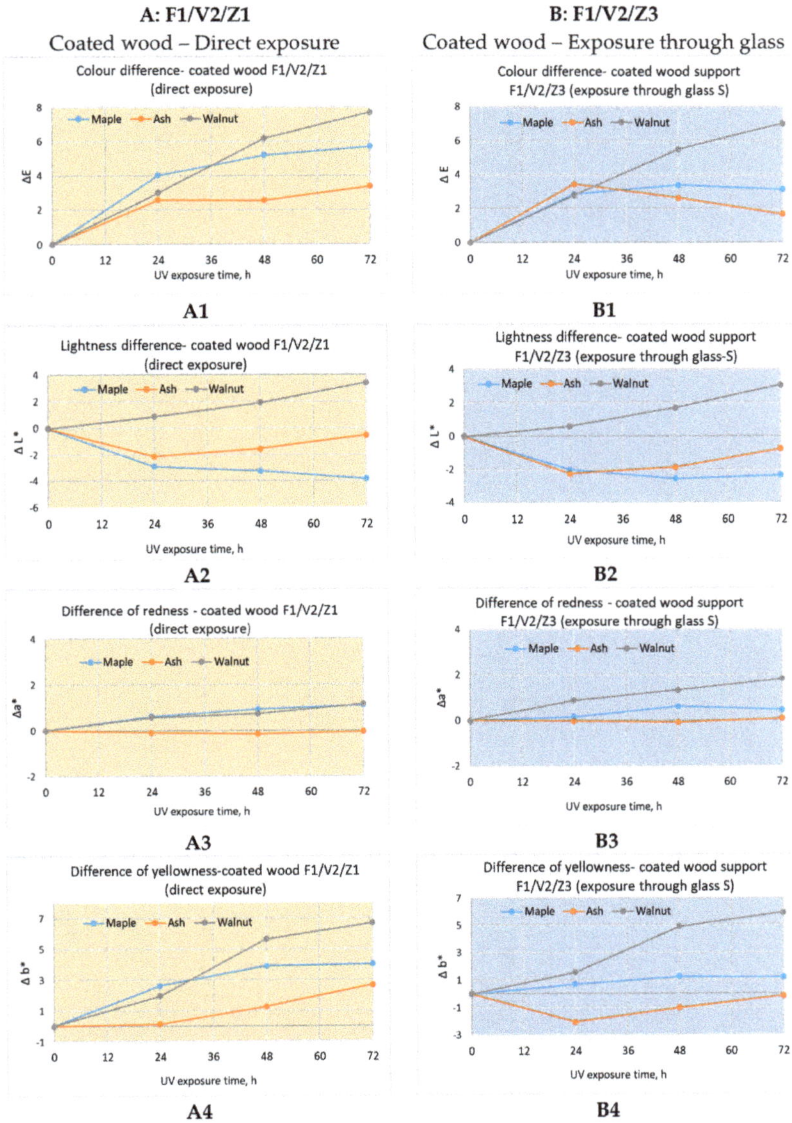

Figure 3. Comparative in-time evolution of light-induced colour changes (ΔE, ΔL*, Δa*, Δb*) for maple, ash and walnut wood surfaces coated with waterborne acrylic-polyurethane lacquer F1 (samples type V2), depending on the exposure situation: (**A**)—direct exposure to UV light (F1/V2/Z1); (**B**)—exposure through a 1 mm clear glass lamella (F1/V2/Z3).

A similar analysis of the light-induced colour changes for wood surfaces coated with the acrylic water-based lacquer, coded F2, is presented in (Figure 4), comparatively for the directly exposed areas (F2/V2/Z1) (Figure 4A) and areas exposed under a glass covering slide (F2/V2/Z3) (Figure 4B).

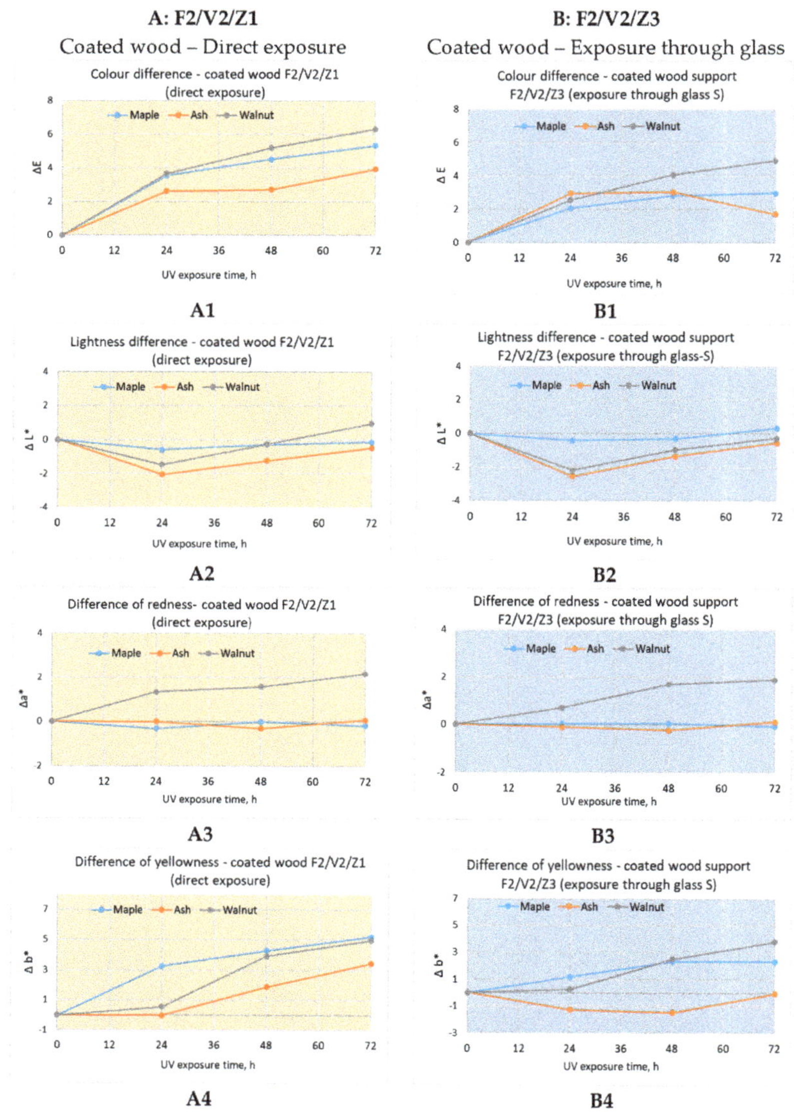

Figure 4. Comparative in-time evolution of the light-induced colour changes (ΔE, ΔL*, Δa*, Δb*) for maple, ash and walnut wood surfaces coated with waterborne acrylic lacquer F2 (samples type V2), depending on the exposure situation: (**A**)—direct exposure to UV light (F2/V2/Z1); (**B**)—exposure through a 1 mm clear glass lamella (F2/V2/Z3).

For the directly exposed areas (F2/V2/Z1) (Figure 4A), the evolution of the global colour changes, expressed by the ΔE values, showed a quite similar pattern for maple and ash, with changes occurring rapidly during the first 24 h of exposure (3.34 units for maple and about 2 units for ash), followed then by a slower increase up to the 5.23 units for maple and 3.49 units for ash. The colour changes for the walnut samples evolved somehow differently, almost linearly in time, reaching the highest value of 5.53 units after 72 h of light exposure. A trend of darkening expressed by negative ΔL* values was registered for all wood species after the first 24 h of light exposure, but this trend was reversed by prolonged exposure. After 72 h light exposure, a slight increase in lightness was registered for walnut

(ΔL^* = +0.96 units), while the ΔL^* values for ash (−0.12 units) and maple (−0.48 units) remained slightly negative but very close to zero.

From a chromatic point of view, the variation in redness Δa^* and yellowness Δb^* depended on the wood species. An increase in redness Δa^* was determined only for walnut (2.16 units), the values for ash and maple being close to zero. Positive Δb^* values indicated for all the coated wood species a continuous yellowing, following species dependant patterns, up to 3.45 units for ash, 4.96 units for walnut and 5.11 units for maple. To conclude, the colour stability of the wood samples coated with the acrylic water-based lacquer decreased in the following order: F2-ash > F2-maple ≅ F2-walnut. This is different from that observed for the uncoated wood samples (walnut > ash > maple) and those coated with the acrylic-polyurethane finish F1 (F1-ash > F1-maple > F1-walnut), pointing out that both the wood substrate and the coating film influence the colour light stability of the coated wood.

The light-induced colour changes for the same coated wood surfaces, covered with 1 mm thick transparent glass (F2/V2/Z3 areas), plotted in (Figure 4B), were again smaller than those determined for the directly exposed areas (Figure 4A), the influence of the covering glass being also dependant on the wood species. The highest decrease in the global colour changes (ΔE after 72 h of exposure) was observed again for maple (about −3 units), while the lowest was observed for walnut (about −0.5 units) (Figure 4(B1)). According to the graphs plotted in (Figure 4(B2–B4)), light exposure through the covering glass influenced mostly the evolution of the yellowness changes Δb^*, which were reduced for all species, with a more pronounced influence (about −3 units) for the maple and ash samples. It is important to highlight that the order of colour stability among the F2 coated wood samples was slightly changed by the effect of the glass cover (S), so that a clear difference could be observed between walnut and maple: S-F2-ash > S-F2-maple > S-F2-walnut.

Finally, the light-induced colour changes for wood surfaces coated with the polyurethane water-based lacquer (1 K), coded F3, are presented in Figure 5 comparatively for the directly exposed areas (F3/V2/Z1) (Figure 5A) and areas exposed under a glass covering slide (F3/V2/Z3) (Figure 5B).

For the directly exposed areas (F3/V2/Z1) (Figure 5A), the evolution of the global colour changes, expressed by the ΔE values, show more specific wood species patterns. For ash (coated F3), fast and important colour changes occurred during the first 24 h, a maximum ΔE value of approximately 4 units being registered after this period. The process was reversed with about the same speed by continuing the light exposure, the ΔE values determined after 48 h and 72 h being considerably lower. The individual patterns for maple and walnut remained very similar to those presented for the samples coated with the other two finishes in Figures 3(A1)(F1) and 4(A1)(F2), being characterised by: (i) changes occurring more rapidly during the first 24 h of exposure, followed then by a slower increase in the case of maple; (ii) changes increasing almost linearly in time in the case of walnut. After 72 h light exposure the measured ΔE values for the F3 coated samples, listed in increasing order, were: −2.25 for ash, 4.74 for walnut and 4.92 for maple.

A trend of darkening expressed by negative ΔL^* values was registered after the first 24 h of light exposure, especially for ash and maple. This trend was reversed by continuing the light exposure in the case of ash but continued at a lower slope in the case of maple. In the case of walnut, only a very slight darkening was registered after 24 h light exposure and the further reversing of the process resulted in a lightening effect after 72 h, contrarily to the other two species. The comparative ΔL^* values (glass/direct exposure) after 72 h were as follows: −3.17/−3.56 for maple, −0.17/−0.51 units for ash and +0.63/2.14 units for walnut.

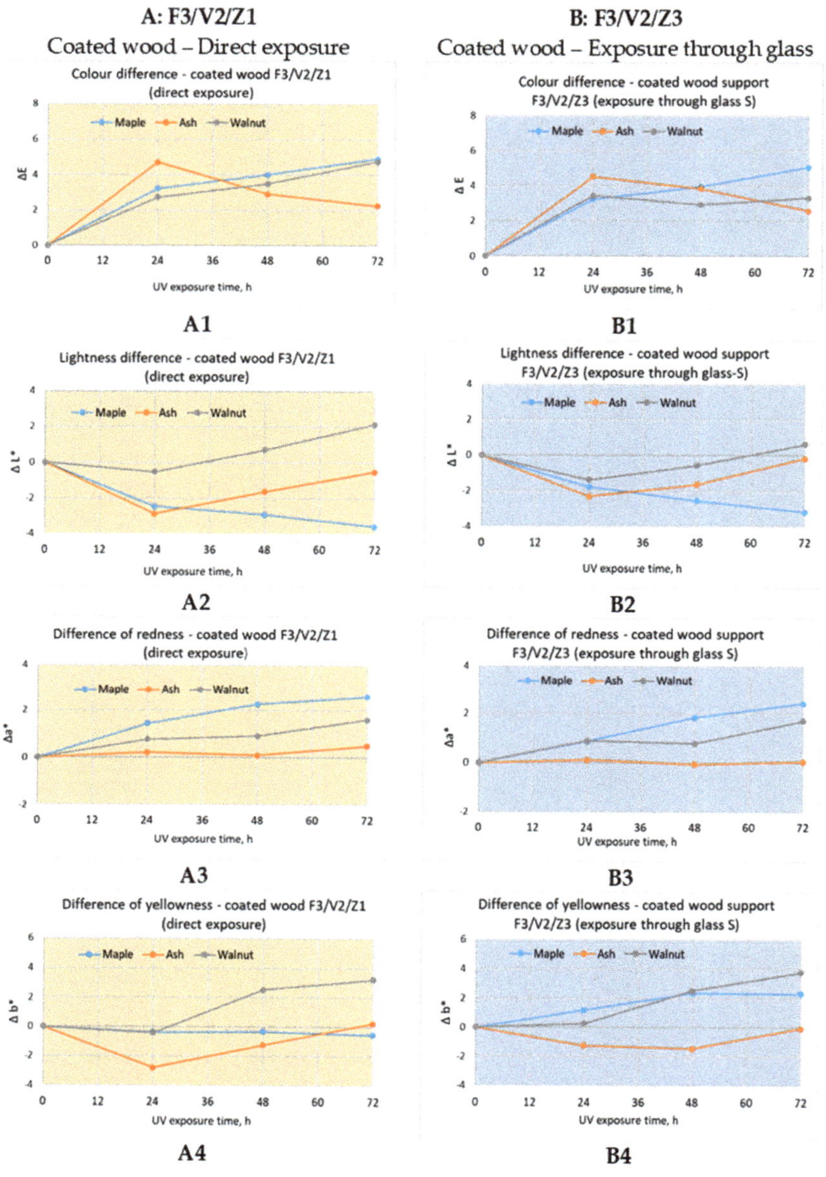

Figure 5. Comparative in-time evolution of light-induced colour changes (ΔE, ΔL*, Δa*, Δb*) for maple, ash and walnut wood surfaces coated with water-based polyurethane lacquer F3 (samples type V2), depending on the exposure situation: (**A**)—direct exposure to UV light (F3/V2/Z1); (**B**)—exposure through a 1 mm clear glass lamella (F3/V2/Z3).

From a chromatic point of view, the variation in redness Δa* and yellowness Δb* depended also on the wood species. A progressive increase in redness during light exposure was registered for maple (up to +2.60 units) and walnut (up to +1.61 units), while the values for ash had a slight wavy variation, resulting in only a very low increase (+0.49 units after 72 h).

The variation in the chromatic component b* was totally different. A trend of a very slight but continuous decrease in yellowness was registered for maple (up to about

−0.5 units). A more advanced decrease in yellowness (up to approximately −3 units) was registered for ash samples but only during the first 24 h of exposure; afterwards, this trend was reversed so that after 72 h light exposure, almost no variation in yellowness could be observed any longer (Δb^* = +0.24 units). An increase in yellowness after 72 h light exposure could be measured only for walnut (Δb^* = +3.24 units).

Based on the global colour differences ΔE after 72 h light exposure, the colour stability of the wood samples coated with the waterborne polyurethane (1 K) lacquer F3 decreased in the following order: F3-ash > F3-walnut ≅ F3-maple. This is different from those observed for the uncoated wood samples (walnut > ash > maple) and those coated with the acrylic-polyurethane finish F1 (F1-ash > F1-maple > F1-walnut) but closer to those resulting for the samples coated with the acrylic finish F2 (F2-ash > F2-maple ≅ F2-walnut).

The light-induced colour changes for the same coated wood surfaces, exposed behind 1 mm thick clear glass (F3/V2/Z3 areas), plotted in Figure 5B, were again different but not always smaller than those determined for the directly exposed areas. Looking comparatively at the corresponding evolution patterns of the colour differences in the Figure 5(A1–A4,B1–B4), it can be observed that the most obvious pattern change due to the effect of the covering glass appeared for the yellowness variation (Δb^*), especially for maple and ash. When exposed through the 1 mm clear cover glass, the F3 coated maple samples yellowed progressively, whilst when uncovered, there was a very slight trend of decrease. The pattern for ash was also different, with a less pronounced decrease in yellowness in the first 24 h, but finally after 72 h, there were no significant changes in yellowness, similarly to the uncovered samples. For walnut, the evolution pattern did not change shape, but the final increase in yellowness was slightly increased. Exposure through the covering glass reduced the variation in lightness to different extents depending on the wood species but did not change visibly the evolution patterns. It can be noted a smaller lightening for walnut, while the darkening effect for maple and ash seems less affected. Accordingly, there were some small differences in the evolution patterns of ΔE values for the three species and the final values recorded after 72 h light exposure. Based on these values, the colour stability of the wood samples coated with the waterborne polyurethane (1 K) lacquer F3 and covered by a 1 mm thick clear glass slide (S) decreased in the following order: S-F3-ash ≥ S-F3-walnut > S-F3-maple, which is slightly different from those observed earlier for direct exposure (F3-ash > F3-walnut ≅ F3-maple), especially as walnut is concerned. It has to be pointed out that for this type of finish (F3), the glass cover showed sort of a shielding effect only in the case of walnut, reducing the final ΔE value by 1.41 units, whilst the corresponding values for maple and ash substrates actually increased with 0.16, respectively, 0.35 units.

More research on an increased number of replicates and complementary investigation methods, including spectral analysis of the light emitted by the source and passing through the 1 mm thick clear glass cover, as well as the interaction of the respective light with the different wood substrates and the coating films in terms of absorption, reflection or transmittance are necessary to better understand and explain the experimental results obtained so far.

3.3. Light-Induced Colour Changes in the Wood Substrate under the Coating Film, Simulated by V1/Z1 Zones—Contribution of the Underlying Wood Substrate to Overall Colour Changes

The experimental method employed in this research envisaged a possibility to highlight, measure and compare the light-induced colour changes in the wood substrate under the coating film and in the coating film itself. For this purpose, sort of a "detachable" coating film was necessary, and this goal was achieved by applying the coating films on 1 mm thick clear glass slides. These were tightly attached to the uncoated wood substrate during light exposure and removed after each exposure period to allow evaluation of the colour changes in the coating film (on glass slide) and in the substrate under the protective coating film (areas V1/Z1). As was already mentioned, the influence of the 1 mm thick clear glass slide was a limitation of this method and was considered by comparative evaluations with areas V1/Z2 (exposure through clear glass). The graphs plotted in (Figure 6) illustrate the in-time evolution of the global colour changes, expressed as ΔE values, for wood substrates exposed to light under the three different types of coating films: acrylic-polyurethane (F1), acrylic (F2), polyurethane (F3) applied on glass slides (SF1, SF2, SF3), alongside the control curves highlighting the effect of the clear glass (S). It can be clearly observed that light penetrated through the transparent coating film and the clear glass slide, reaching the wood substrate underneath and causing colour changes that evolved with the duration of exposure. The evolution patterns in the colour changes were very similar for exposure under clear glass (S) and exposure under coated glass (SF1–SF3), with only some small differences in the actual ΔE values. Slightly lower values under the coated glass may suggest some protective effect of the respective coating films, whilst slightly higher values may indicate an opposite effect. However, when compared to the colour differences for wood exposed under clear glass the differences brought about by the coatings applied on glass were very low, as can be observed in Table 3, where the average data after 72 h light exposure are presented.

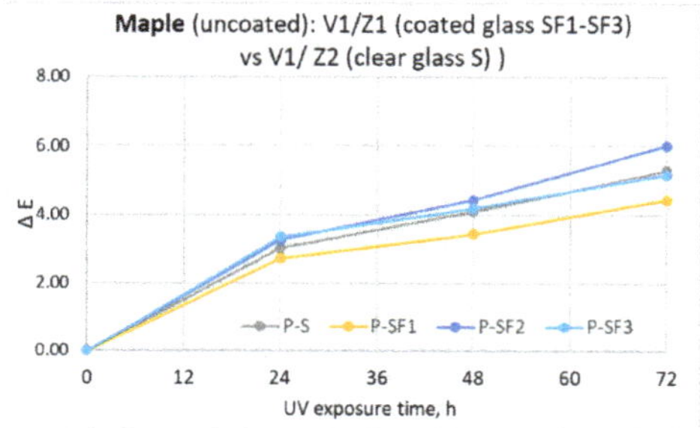

A

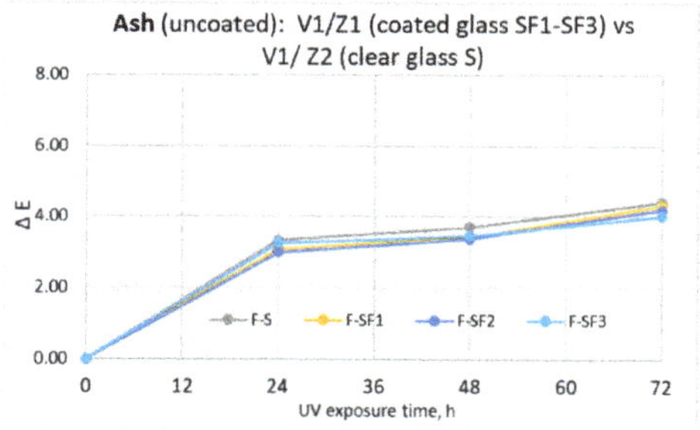

B

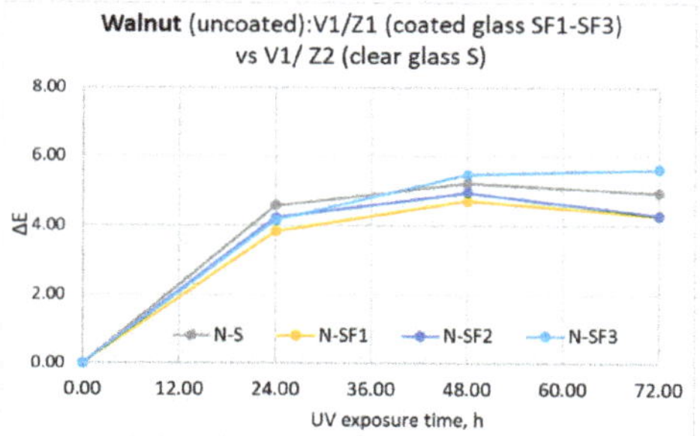

C

Figure 6. Light-induced colour changes in the wood substrate (uncoated) exposed under the protective effect of the three types of coatings (F1–F3) applied on clear glass slides (SF1, SF2, SF3), comparative to exposure through clear glass slides (S): (**A**)—maple; (**B**)—ash; (**C**)—walnut.

Table 3. Comparative data of colour changes after 72 h light/UV exposure for wood substrate (maple, ash and walnut) under the coating films (coating materials F1, F2, F3), simulated in V1/Z1 areas by uncoated wood exposed under coated glass slides; the data on colour changes in the same substrates under clear glass slides (areas V1/Z2) are presented as reference values to evaluate the effect of the coatings.

Wood Species		Colour Changes after 72 h Light /UV Exposure					
		V1/Z2 Uncoated Wood under Clear Glass (Reference Values)			**V1/Z1 Uncoated Wood under Coated Glass**		
		(F1)	**(F2)**	**(F3)**	**F1**	**F2**	**F3**
Maple	ΔE	4.90 *(1.39)*	5.80 *(1.13)*	5.20 *(1.86)*	4.44 *(1.36)*	6.02 *(1.55)*	5.17 *(1.73)*
	ΔL*	−3.56 *(1.42)*	−4.29 *(0.62)*	−3.43 *(1.46)*	−3.51 *(1.11)*	−4.37 *(1.16)*	−3.36 *(1.85)*
	Δa*	0.99 *(0.52)*	1.74 *(0.51)*	0.70 *(0.68)*	1.00 *(0.32)*	1.26 *(0.71)*	0.45 *(0.71)*
	Δb*	2.95 *(1.23)*	3.40 *(1.16)*	3.69 *(1.47)*	2.33 *(1.24)*	3.82 *(1.28)*	3.64 *(1.15)*
Ash	ΔE	4.84 *(0.76)*	4.23 *(0.82)*	4.23 *(1.34)*	4.32 *(1.15)*	4.20 *(0.77)*	4.03 *(0.96)*
	ΔL*	−3.25 *(0.55)*	−2.94 *(0.76)*	−3.23 *(1.30)*	−3.07 *(1.11)*	−3.16 *(0.59)*	−2.98 *(1.39)*
	Δa*	0.80 *(0.32)*	0.67 *(0.58)*	0.92 *(0.65)*	0.79 *(0.54)*	0.76 *(0.27)*	0.50 *(0.82)*
	Δb*	3.45 *(0.73)*	2.70 *(1.18)*	2.22 *(1.25)*	2.79 *(0.88)*	2.62 *(0.64)*	2.18 *(0.94)*
Walnut	ΔE	4.77 *(1.38)*	4.51 *(1.13)*	5.52 *(1.52)*	4.51 *(1.54)*	4.83 *(1.35)*	5.93 *(1.20)*
	ΔL*	−3.37 *(1.88)*	−2.07 *(0.74)*	−2.13 *(1.43)*	−3.18 *(1.60)*	−2.32 *(0.79)*	−2.22 *(1.38)*
	Δa*	2.26 *(0.66)*	2.55 *(1.10)*	2.51 *(0.74)*	2.19 *(0.61)*	2.08 *(1.12)*	2.41 *(0.72)*
	Δb*	2.99 *(1.03)*	2.99 *(1.34)*	4.15 *(1.55)*	1.83 *(1.33)*	2.94 *(2.35)*	4.56 *(1.77)*

Note: The data represent average values of colour changes in the 9 measuring points; standard deviations are presented in brackets in italics.

Analysis of paired data by using the one-way Anova test at significance level $\alpha = 0.05$ found no statistically significant differences between the mean ΔE values in the areas V1/Z1 and V1/Z2 (p values higher than 0.05), irrespective of the wood species and the type of coating material (F1, F2, F3). This means that no protective effect of the coating films on the colour changes in the wood substrate under the coating could be observed by the applied method. This statistics result was in good accordance with visual perception when analysing comparatively the light-induced colour changes in the areas V1/Z1 and V1/Z2 on the tested samples.

On the other hand, the fact that notable colour changes in the wood substrate occurring under the coated glass (V1/Z1) could be observed and quantified represents an important original contribution of this research. Moreover, the wood species and the type of coating were found as statistically significant influencing factors on the colour change values ΔE, determined for the V1Z1 areas after 72 h light/UV exposure, when the respective colour data were analysed by the Anova two-factor test with replication (summary in Table 4). The higher F value indicates wood species as a more important influencing factor than the coating film on the colour changes in the underlying substrate. The interaction of the two factors was not found statistically significant.

Table 4. Summary of two-factor Anova test results with replication of the colour change values ΔE, after 72 h light exposure, in the areas V1/Z1, considering the wood species (1) and the type of coating (2) as influencing factors. (α-0.05).

Anova: Two-Factor with Replication (α = 0.05)						
Source of Variation	**SS**	**df**	**MS**	**F**	***p*-Value**	**F Crit**
Wood species (1)	44.95563	2	22.47781	12.13096	2.88142×10^{-5}	3.123907
Type of coating (2)	13.78771	2	6.893853	3.720516	0.028997068	3.123907
Interaction (1)x(2)	18.01943	4	4.504858	2.431209	0.055229542	2.498919
Within	133.4109	72	1.852929			
Total	210.1737	80				

The colour changes in the wood substrate due to light exposure under coated glass slides (V1/Z1) were determined as a simulation of the behaviour (contribution) of wood substrate under the coating films (V2/Z1), situations where there is no effect of the glass. Accordingly, the real colour changes in the substrates underneath the coating films and the associated chemical changes should be higher than those determined by the method applied in this research. This major contribution of the wood substrate to the total colour changes in the clear-coated wood surfaces is an important key point in understanding the occurring phenomena with the view of their correction, as also pointed out by the previous research [35,47]. For equivalence as exposure situations, the colour and related chemical changes in the substrate in the areas V1/Z1 should be comparable with the colour changes in a wood substrate under the coating film after exposure to light passing through a covering clear glass slide, respectively, areas (V2/Z3) in this research.

3.4. Light-Induced Colour Changes in the Coating Films—Contribution of Finishing Film to the Overall Colour Changes

The colour changes (ΔE, ΔL*, Δa*, Δb*) in the finishing films applied on 1 mm clear glass slides as a result of light exposure for different periods of time are presented in Figure 7. It can be easily observed that for all the three types of coatings, there were registered similar wavy patterns in colour changes. These were maximal after the first 24 h of exposure and consisted mainly in slight darkening (ΔL* negative) and yellowing (Δb* positive). By further light exposure, the initial trend reversed twice. Redness evolved similarly but the changes were very low, close to zero. Consequently, the total colour changes (ΔE) after 72 h light exposure for the three types of finishing films were very low: 1.30 units for F1(acrylic-polyurethane), 1.25 units for F2 (acrylic) and 1.07 units for F3 (polyurethane). However, these differences in the means for the three types of coatings were not statistically significant when analysed by one-way Anova in conjunction with Tukey's HSD post hoc test (*p* values higher than 0.05 at a significance level $\alpha = 0.05$).

These low values demonstrate a fairly good colour stability to light exposure for all the tested coating materials. It would result, therefore, in a minimal contribution by the coating films to the overall colour changes in the wood coated surfaces, which strengthens the previous finding about the major contribution of the wood substrate to the total colour changes in the coated wood surfaces. However, the spectral properties of the coating films, meaning their specific capacity of absorbing, transmitting or reflecting/dispersing the UV-VIS radiation remain a key point in understanding and potentially controlling the colour changes in the substrate under the coating films and, finally, of the coated wood surfaces. These issues will be approached in future research.

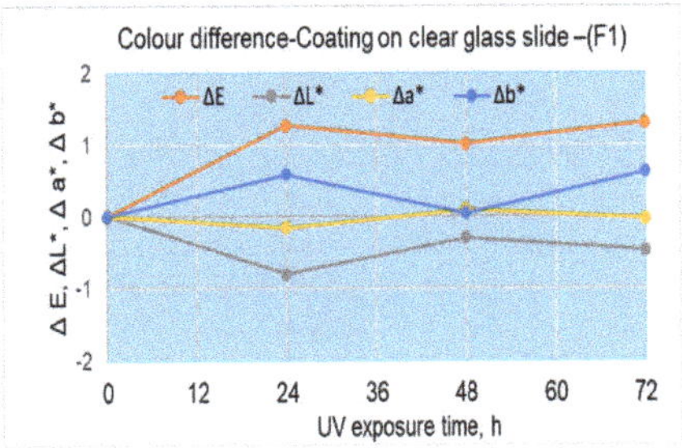

A

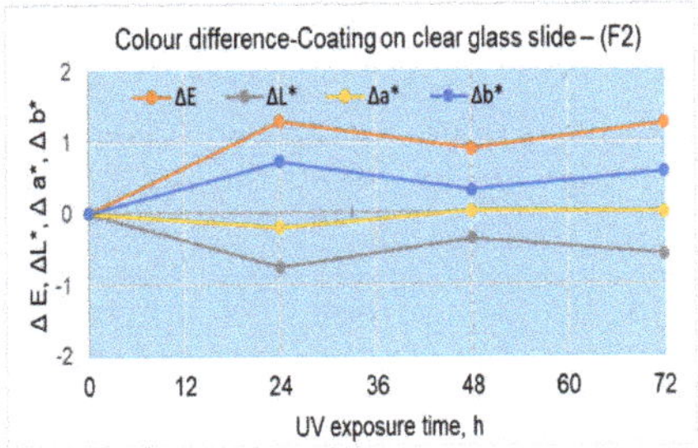

B

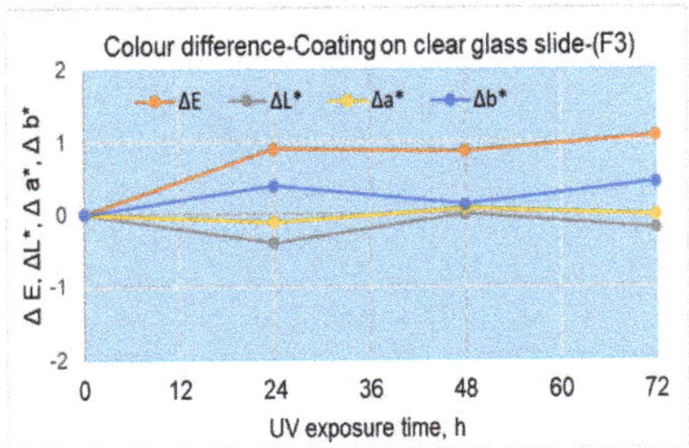

C

Figure 7. Light-induced colour changes in coating films applied on clear glass slides: (**A**). acrylic-polyurethane lacquer F1; (**B**). acrylic lacquer F2; (**C**). polyurethane lacquer F3.

3.5. Influence of Wood Species and Coating Materials on the Light-Induced Colour Changes in Wood Surfaces

The experimental results presented and discussed so far highlighted, in most cases, some protective effect of the three types of waterborne coatings (F1-acrylic-polyurethane, F2-acrylic, F3-polyurethane) against the UV-VIS light-induced colour changes in the wood substrates from the three wood species (maple, ash, walnut) under study. This protective effect was dependant on both the coating film and the wood species, which demonstrated different sensitivity to light-induced colour changes, in accordance with the previous research [22,24,53].

In order to have a cumulative and comparative image of the effect of clear coating on the colour stability of wood surfaces when exposed to light in indoor conditions, the data on the colour changes (ΔE, ΔL*, Δa*, Δb*) after 72 h light exposure in the conditions of this research are cumulated in Table 5. The data for uncoated and coated samples (F1, F2, F3) of the three species were also included to allow a final discussion on the influence of both wood species and type of coating. Lower values in colour changes (ΔE, ΔL*, Δa*, Δb*) mean better colour stability so that the effects of the above-mentioned influencing factors can be easily observed. The statistical significance of the differences between the mean values of the uncoated and coated specimens was tested by one-way Anova and Tukey's HSD as post hoc test, and this information is included as superscript codes.

Table 5. Comparative data of colour changes after 72 h light/UV exposure for uncoated and coated wood surfaces with the three finishing materials (F1, F2, F3).

Wood Species		Colour Changes after 72 h Light /UV Exposure			
		Uncoated V1/Z3	F1 V2/Z1	F2 V2/Z1	F3 V2/Z1
Maple	ΔE	9.36 ^{A} (1.71)	5.71 ^{B} (1.20)	5.35 ^{B} (2.11)	4.92 ^{B} (1.64)
	ΔL*	−5.07 ^{C} (1.61)	−3.83 ^{B} (0.84)	−0.11 ^{A} (1.18)	−3.56 ^{B} (1.76)
	Δa*	1.48 ^{B} (0.79)	1.09 ^{B} (0.58)	−0.14 ^{C} (0.66)	2.60 ^{A} (1.45)
	Δb*	7.58 ^{A} (1.38)	4.03 ^{B} (1.49)	5.11 ^{B} (2.30)	−0.52 ^{C} (1.48)
Ash	ΔE	8.39 ^{A} (1.28)	3.37 ^{B} (1.37)	3.94 ^{B} (1.24)	2.25 ^{C} (1.02)
	ΔL*	−4.13 ^{B} (1.13)	−0.52 ^{A} (1.74)	−0.48 ^{A} (1.50)	−0.51 ^{A} (1.57)
	Δa*	1.43 ^{A} (0.46)	−0.06 ^{B} (0.93)	0.06 ^{B} (0.86)	0.49 ^{B} (0.97)
	Δb*	7.05 ^{A} (1.38)	2.65 ^{B} (1.49)	3.45 ^{B} (1.45)	0.24 ^{C} (1.55)
Walnut	ΔE	6.20 ^{B} (1.34)	7.70 ^{A} (1.91)	6.31 ^{AB} (2.04)	4.74 ^{C} (1.40)
	ΔL*	−3.25 ^{C} (1.62)	3.40 ^{A} (1.36)	0.96 ^{B} (2.60)	2.14 ^{AB} (1.15)
	Δa*	2.23 ^{A} (0.85)	1.14 ^{B} (0.63)	2.16 ^{A} (0.71)	1.61 ^{AB} (1.02)
	Δb*	4.42 ^{B} (1.38)	6.69 ^{A} (1.76)	4.96 ^{AB} (2.67)	3.24 ^{B} (2.16)

Note: Values on the same line sharing a common ($^{A–C}$) code as superscript are not statistically different (one-way Anova, $\alpha = 0.05$, $p > 0.05$).

Based on the comparative ΔE values, it can be observed that all the three coating materials F1, F2, F3 protected to some extent maple and ash wood substrates against the light-induced colour changes. This protective effect was higher for ash than maple. In the case of walnut, some protective effect was registered only for the coating F3. No effect was registered for the coating F2, while the samples finished with the coating F1 suffered colour changes slightly more intense (statistically significant) than uncoated walnut wood.

Differences in the protective effects of various waterborne clear coatings have been reported before. For instance, an increased protective effect of a 2 k polyurethane topcoat (based on hydroxyl-bearing polyacrylate and aliphatic poly-isocyanate) compared to an acrylic topcoat was reported for Radiata pine boards primed with a moisture curing isocyanate [58]. Colour change values ΔE varied in the range of approximately 2–10 units after 700 h light exposure (accelerated aging test simulating natural light passing through window glass) for spruce samples coated with four different three-layer coating systems (primer, middle layer, top layer) with waterborne coatings based on acryl/alkyd or polyurethane resins and their mixture [59].

The coated maple and ash samples exposed to light darkened less compared to the uncoated ones, as reflected by the comparative negative ΔL* values for all the three finishing materials. In the case of maple, this effect appeared to be highest for the F2 coating material, while in the case of ash, all the three coatings seemed to have a similar effect. Contrarily, the coated walnut samples became lighter following 72 h light exposure, this effect being more pronounced for the coatings F1 and F3 and minimal for F2. This particular behaviour of walnut is due to its darker colour compared to the other two species, respectively, to the role of extractives and visible light in the occurring colour changes, as reported in the literature [18,20,21].

The statistical significance of wood species (1) and type of coating (2) as influencing factors on the light-induced colour change values ΔE was verified by an Anova two-factor with replication test, at a significance level $\alpha = 0.05$. According to the summary results presented in Table 6, respectively, the p values lower than 0.05, it can be stated that both the wood species (factor 1) and the type of coating (factor 2) are statistically significant influencing factors. Comparing the F values of the two factors, it results that wood species (F = 52.72) influence more the colour changes than the type of coating (F = 15.29). The test shows also that there is a significant two-way interaction between the two factors ($p < 0.05$; F = 3.31). This would mean a common and interconnected influence of the two factors for each wood species and type of coating. A significant impact of wood species and type of lacquer/coating system, alongside pre-treatments of wood surfaces and modification of coatings, on the light-induced colour changes in wood surfaces was previously reported [59].

Table 6. Summary of two-way Anova results with replication of the colour change (ΔE, after 72 h light exposure) in the coated wood surfaces (areas V2/Z1), considering the wood species (1) and the type of coating (2) as influencing factors.

	Anova: Two-Factor with Replication (α = 0.05)					
Source of Variation	**SS**	**df**	**MS**	**F**	***p*-Value**	**F Crit**
Wood species (1)	266.8111	2	133.4055	52.71715	3.83802×10^{-18}	3.055162
Type of coating (2)	77.40668	2	38.70334	15.29419	8.80515×10^{-7}	3.055162
Interaction (1)x(2)	33.53909	4	8.384773	3.313366	0.012333886	2.430772
Within	387.1804	153	2.530591			
Total	764.9372	161				

The interaction of the coating material with the wood substrate at the interface level, consisting in wetting and penetration, depending on both the structural features and surface properties of the substrate and the coating material, might be part of this significant two-way influence on the UV-induced colour changes determined in this research. According to the literature data, a combination of the wood anatomy and the ability of the coatings to flow into the wood capillaries, influenced by the viscosity, solids content and type of solvent, govern the degree of coating penetration [60]. Generally, waterborne coatings penetrate less than the solvent-borne ones, though large variations might be among different formulations. In this context, binder type, pigmentation, solids content and drying time were found to be important. Acrylic waterborne dispersions have been found to have a low penetration compared to alkyd emulsions [61,62].

Differences in the colour changes in clear-coated pine wood after an artificial weathering test were associated with different penetrations of an acrylic coating system (primer and topcoat), increased colour changes being explained as possibly due to a reduced coating penetration. Increased penetrations were obtained in the same research for beech wood, but no final conclusion was drawn on the influence of coating penetration on the colour changes caused by weathering [63].

An interesting study on the degradative behaviour of coated-wood surfaces exposed to artificial weathering using Raman microspectroscopy was recently published. Chemical imaging pointed out different artificial weathering degradation patterns between wood specimens (Japanese cedar) that were clear-coated with impregnating materials and film-forming materials. For the impregnating systems, the surface-distributed coating gradually decreased during irradiation and lignin in the top surface cell walls degraded similarly to uncoated wood [64]. The result is, therefore, that both good penetration and the formation of a top film of the coating are beneficial for improved UV protection.

The preliminary microscopic investigation of our samples highlighted such "composite coating layers" with total thicknesses varying in the range of approximately 25–75 μm, while the top layer film included varied in the range 5–20 μm, with maximum thickness in ash early-wood (vessels). From the waterborne coating materials employed in this research, the polyurethane lacquer (F3) was the most fluid and ensured the best penetration into the wood substrate, approximated to around 50 μm for maple, 30–60 μm for ash (with the high values in early-wood vessels) and 25–60 μm for walnut (large values in vessels). This might explain, at least partially, the best protective effect against light-induced colour changes demonstrated in this research by the F3 finish for all the three wood species tested. However, more research and complementary methods are necessary for a conclusion with this respect.

4. Conclusions

The paper investigated comparatively the light-induced colour changes for three hardwood species, namely, European maple (*Acer pseudoplatanus* L.), European ash (*Fraxinus excelsior* L.) and European walnut (*Juglans regia* L.), as well as the influence of finishing with three types of clear, colourless waterborne lacquers (F1: acrylic-polyurethane, F2: acrylic, F3: polyurethane) on their colour stability.

Colour measurements in the CIELab system in conjunction with an artificial accelerated ageing test under the action of UV-VIS radiation, simulating natural light passing through window glass, and two types of test samples were employed to highlight the influence and contributions of the wood substrate and of the coating films to the global colour modifications.

Coating films applied on 1 mm clear glass slides were employed as a sort of "detachable" finish for this purpose. A different transmittance of the UV and VIS radiation from the artificial source through this glass support and a decrease in light intensity reaching the substrate is a limitation of the experimental method employed, which has to be considered in future research.

Direct exposure to UV-VIS light caused visible colour changes for both uncoated and coated wood surfaces, the values of the calculated colour differences (ΔE) after 72 h exposure being dependant on both the wood species and the coating material, which were found to be statistically significant influencing factors in interaction.

Excepting two situations for walnut, statistically significant higher colour differences were obtained for the uncoated samples: maple (9.36 units), ash (8.39 units), walnut (6.20 units), compared to the coated ones: maple (4.92–5.71 units), ash (2.25–3.94 units), walnut (4.74–7.70 units), depending on the type of coating.

From the three coating materials tested, the waterborne polyurethane lacquer represents a good option for all the wood species. The effect of each coating material tested depended on the wood species, as a more powerful influencing factor. The interaction between the wood substrate and the coating material at the interface level needs further investigation.

The wood substrate underlying the coating film was found to bring the maximum contribution to the overall colour changes in the clear-coated surfaces, the actual ΔE values being influenced by the wood species and the coating film as significant factors. The colour changes ΔE in the wood substrates under the simulated "detachable" coating films varied in the ranges 4.44–6.02 units for maple, 2.97–4.43 units for ash and 4.51–5.93 units for walnut.

The coating films employed in this research demonstrated a fairly good colour stability to UV-VIS light exposure, with colour changes (ΔE) up to only 1.30 units, with no statistically significant differences.

These results will be employed as a basis for further research looking at efficient methods to stabilise the natural colour of the clear coated wood surfaces against the effect of light in indoor applications, considering the specific contributions of the wood substrate and of the coating films, as well as the particularities of each wood species in relation to their chemical composition.

Author Contributions: Conceptualisation: M.-J.T. and M.C.T.; methodology: M.-J.T. and M.C.T.; supervision: M.C.T., preparation of test samples: M.-J.T.; colour measurements: M.-J.T.; data processing: M.-J.T. and M.C.T.; graphs: M.-J.T.; writing and editing draft version: M.-J.T.; revisions: M.C.T. All authors have read and agreed to the published version of the manuscript.

Funding: This research is part of a budget funded Ph.D. research project undertaken by the principal author within the Doctoral School of the Transilvania University of Brașov-Romania.

Institutional Review Board Statement: Not applicable.

Informed Consent Statement: Not applicable.

Data Availability Statement: The data presented in this study are available within the article.

Acknowledgments: The authors would like to express their sincere thanks to the company "Craiul Muntilor SRL", Stupărei village, Valcea county, Romania, for providing the wood material for this research. Furthermore, the authors would like to express their gratitude to the KronColoR company from Brasov, Romania for their professional advice and material support by providing the wood coating materials employed in this research. Special thanks are addressed to our colleague Bogdan Bedelean for advice and assistance in implementing the statistical analysis of data during the paper revision process. We also acknowledge the PRO-DD structural funds project (POS-CCE, O.2.2.1., ID 123, SMIS 2637, no. 11/2009) for providing the infrastructure used in this work at the ICDT Research Institute of the Transilvania University of Brasov-Romania.

Conflicts of Interest: The authors declare that there are no conflicts of interest. The funders had no role in the design of the study; in the collection, analyses, or interpretation of data; in the writing of the manuscript; or in the decision to publish the results.

References

1. Oltean, L.; Teischinger, A.; Hansmann, C. Wood surface discolouration due to simulated indoor sunlight exposure. *Holz Als Roh-Und Werkst.* **2007**, *66*, 51–56. [CrossRef]
2. Kropat, M.; Hubbe, M.A.; Laleicke, F. Natural, Accelerated, and Simulated Weathering of Wodd: A Review. *Bioresources* **2020**, *15*, 9998–10062. [CrossRef]
3. Spear, M.J.; Curling, S.F.; Dimitriou, A.; Ormondroyd, G.A. Review of Functional Treatments for Modified Wood. *Coatings* **2021**, *11*, 327. [CrossRef]
4. Rajkovic, J.V.; Miklecic, J. Enhancing Weathering Resistance of Wood—A Review. *Polymers* **2021**, *13*, 1980. [CrossRef]
5. Blanchet, P.; Pepin, S. Trends in Chemical Wood Surface Improvements and Modifications: A Review of the Last Five Years. *Coatings* **2021**, *11*, 1514. [CrossRef]
6. Yi, T.; Morrell, J.J. Role of α/γ Fe_2O_3 and ZnO nano-particles in reducing photodegradation of wood components. *Wood Sci. Technol.* **2023**, *57*, 427–446. [CrossRef]
7. Lee, H.-W.; Lee, E.-J.; Shin, Y.-J.; Jo, H.-Y.; Song, D.-Y. Surface Discoloration of Ultraviolet (UV)-Irradiated *Phyllostachys bambusoides* Bamboo. *J. Korean Wood Sci. Technol.* **2023**, *51*, 173–182. [CrossRef]
8. Fengel, D.W.G.; de Gruyter, W. *Wood—Chemistry, Ultrastructure, Reactions*; Gruyter, W.D., Ed.; Walter de Gruyter: Berlin, Germany; New York, NY, USA, 1985. [CrossRef]
9. Müller, U.; Rätzsch, M.; Schwanninger, M.; Steiner, M.; Zöbl, H. Yellowing and IR-changes of spruce wood as result of UV-irradiation. *J. Photochem. Photobiol. B Biol.* **2003**, *69*, 97–105. [CrossRef]
10. George, B.; Suttie, E.; Merlin, A.; Deglise, X. Photodegradation and photostabilisation of wood—The state of the art. *Polym. Degrad. Stab.* **2005**, *88*, 268–274. [CrossRef]
11. Williams, R.S. Weathering of Wood. In *Handbook of Wood Chemistry and Wood Composites*; USDA, Forest Service, Forest Products Laboratory: Madison, WI, USA, 2005. Available online: https://www.fs.usda.gov/research/treesearch/22122 (accessed on 8 September 2023).
12. Evans, P.; Chowdhury, M.; Mathews, B.; Schmalzl, K.; Ayer, S.; Kiguchi, M.; Kataoka, Y. Weathering and Surface Protection of Wood. In *Handbook of Environmental Degradation of Materials*; William Andrew Inc.: New York, NY, USA, 2005; pp. 277–297. [CrossRef]
13. Teacă, C.A.; Roşu, D.; Bodîrlău, R.; Roşu, L. Structural Changes in Wood under Artificial UV Light Irradiation Determined by FTIR Spectroscopy and Color Measurements—A Brief Review. *BioResources* **2013**, *8*, 1478–1507. [CrossRef]

14. Kataoka, Y.; Kiguchi, M.; Williams, R.S.; Evans, P.D. Violet light causes photodegradation of wood beyond the zone affected by ultraviolet radiation. *Holzforschung* **2007**, *61*, 23–27. [CrossRef]
15. Živković, V.; Arnold, M.; Radmanović, K.; Richter, K.; Turkulin, H. Spectral sensitivity in the photodegradation of fir wood (Abies alba Mill.) surfaces: Colour changes in natural weathering. *Wood Sci. Technol.* **2014**, *48*, 239–252. [CrossRef]
16. Agresti, G.; Bonifazi, G.; Calienno, L.; Capobianco, G.; Lo Monaco, A.; Pelosi, C.; Picchio, R.; Serranti, S. Surface Investigation of Photo-Degraded Wood by Colour Monitoring, Infrared Spectroscopy, and Hyperspectral Imaging. *J. Spectrosc.* **2013**, *2013*, 380536. [CrossRef]
17. Zahri, S.; Belloncle, C.; Charrier, F.; Pardon, P.; Quideau, S.; Charrier, B. UV light impact on ellagitannins and wood surface colour of European oak (*Quercus petraea* and *Quercus robur*). *Appl. Surf. Sci.* **2007**, *253*, 4985–4989. [CrossRef]
18. Chang, T.-C.; Chang, H.-T.; Wu, C.-L.; Chang, S.-T. Influences of extractives on the photodegradation of wood. *Polym. Degrad. Stab.* **2010**, *95*, 516–521. [CrossRef]
19. Chang, T.-C.; Chang, H.-T.; Wu, C.-L.; Lin, H.-Y.; Chang, S.-T. Stabilizing effect of extractives on the photo-oxidation of Acacia confusa wood. *Polym. Degrad. Stab.* **2010**, *95*, 1518–1522. [CrossRef]
20. Chang, T.-C.; Lin, H.-Y.; Wang, S.-Y.; Chang, S.-T. Study on inhibition mechanisms of light-induced wood radicals by Acacia confusa heartwood extracts. *Polym. Degrad. Stab.* **2014**, *105*, 42–47. [CrossRef]
21. Passauer, L.; Prieto, J.; Müller, M.; Rössler, M.; Schubert, J.; Beyer, M. Novel color stabilization concepts for decorative surfaces of native dark wood and thermally modified timber. *Prog. Org. Coat.* **2015**, *89*, 314–322. [CrossRef]
22. Oltean, L.; Hansmann, C.; Németh, R.; Teischinger, A. Wood surface discolouration of three hungarian hardwood species due to simulated indoor sunlight exposure. *Wood Res.* **2010**, *55*, 49–58. Available online: https://www.researchgate.net/publication/299019578 (accessed on 11 October 2023).
23. Timar, M.C.; Varodi, A.M.; Gurău, L. Comparative study of photodegradation of six wood species after short-time UV exposure. *Wood Sci. Technol.* **2016**, *50*, 135–163. [CrossRef]
24. Liu, X.Y. Contributions to the Study of Ageing Phenomena of Wooden Substrate and Traditional Materials for Transparent Finishes—A Comparative Approach for Europe and China with Applicability in Furniture Conservation/Restoration. Ph.D. Thesis, Transilvania University, Brasov, Romania, 2017.
25. Liu, R.; Zhu, H.; Li, K.; Yang, Z. Comparison on the Aging of Woods Exposed to Natural Sunlight and Artificial Xenon Light. *Polymers* **2019**, *11*, 709. [CrossRef] [PubMed]
26. Cirule, D.; Kuka, E.; Sansonetti, E.; Hivrica, M.; Andersone, I.; Andersons, B. Wood photosensitivity to different artificial light sources. *Eur. J. Wood Wood Prod.* **2022**, *80*, 1507–1519. [CrossRef]
27. Cirule, D.; Kuka, E.; Andersone, I.; Andersons, B. Wood discoloration patterns depending on the light source. *Herit. Sci.* **2022**, *10*, 158. [CrossRef]
28. MacLeod, I.T.; Scully, A.D.; Ghiggino, K.P.; Ritchie, P.J.A.; Paravagna, O.M.; Leary, B. Photodegradation at the wood-clearcoat interface. *Wood Sci. Technol.* **1995**, *29*, 183–189. [CrossRef]
29. Forsthuber, B.; Müller, U.; Teischinger, A.; Grüll, G. Chemical and mechanical changes during photooxidation of an acrylic clear wood coat and its prevention using UV absorber and micronized TiO_2. *Polym. Degrad. Stab.* **2013**, *98*, 1329–1338. [CrossRef]
30. Saha, S.; Kocaefe, D.; Boluk, Y.; Pichette, A. Surface degradation of CeO_2 stabilized acrylic polyurethane coated thermally treated jack pine during accelerated weathering. *Appl. Surf. Sci.* **2013**, *276*, 86–94. [CrossRef]
31. Aloui, F.; Ahajji, A.; Irmouli, Y.; George, B.; Charrier, B.; Merlin, A. Inorganic UV absorbers for the photostabilisation of wood-clearcoating systems: Comparison with organic UV absorbers. *Appl. Surf. Sci.* **2007**, *253*, 3737–3745. [CrossRef]
32. Ncube, E.; Meincken, M. Surface characteristics of coated soft- and hardwoods due to UV-B ageing. *Appl. Surf. Sci.* **2010**, *256*, 7504–7509. [CrossRef]
33. Rao, F.; Chen, Y.; Zhao, X.; Cai, H.; Li, N.; Bao, Y. Enhancement of bamboo surface photostability by application of clear coatings containing a combination of organic/inorganic UV absorbers. *Prog. Org. Coat.* **2018**, *124*, 314–320. [CrossRef]
34. Esposito, C.C.; Frigione, M. Novel UV-Cured Nanocomposite Used for the Protection of Walnut Wood Artworks. *Wood Res.* **2014**, *59*, 229–244. Available online: https://www.researchgate.net/publication/287350116_Novel_UV-cured_nanocomposite_used_for_the_protection_of_walnut_wood_artworks (accessed on 22 October 2023).
35. Chou, P.L.; Chang, H.T.; Yeh, T.F.; Chang, S.T. Characterizing the conservation effect of clear coatings on photodegradation of wood. *Bioresour. Technol.* **2008**, *99*, 1073–1079. [CrossRef]
36. Capobianco, G.; Calienno, L.; Pelosi, C.; Scacchi, M.; Bonifazi, G.; Agresti, G.; Picchio, R.; Santamaria, U.; Serranti, S.; Monaco, A.L. Protective behaviour monitoring on wood photo-degradation by spectroscopic techniques coupled with chemometrics. *Spectrochim. Acta Part A Mol. Biomol. Spectrosc.* **2017**, *172*, 34–42. [CrossRef]
37. Mahltig, B.; Swaboda, C.; Roessler, A.; Böttcher, H. Functionalising wood by nanosol application. *J. Mater. Chem.* **2008**, *18*, 3180–3192. [CrossRef]
38. Wu, X.; Yang, F.; Gan, J.; Kong, Z.; Wu, Y. A Superhydrophobic, Antibacterial, and Durable Surface of Poplar Wood. *Nanomaterials* **2021**, *11*, 1885. [CrossRef]
39. Panek, M.; Oberhofnerova, E.; Hysek, S.; Sedivka, P.; Zeidler, A. Colour Stabilization of Oak, Spruce, Larch and Douglas Fir Heartwood Treated with Mixtures of Nanoparticle Dispersions and UV-Stabilizers after Exposure to UV and VIS-Radiation. *Materials* **2018**, *11*, 1653. [CrossRef]

40. Olsson, S.K.; Johansson, M.; Westin, M.; Östmark, E. Reactive UV-absorber and epoxy functionalized soybean oil for enhanced UV-protection of clear coated wood. *Polym. Degrad. Stab.* **2014**, *110*, 405–414. [CrossRef]
41. Bielerman, J. *Lackadditive*; Wiley-VCH: Weinheim, Germany; New York, NY, USA; Chichester, UK; Brisbane, Australia; Singapore; Toronto, ON, Canada, 1998; p. 395. Available online: https://epdf.tips/lackadditive-german-edition.html (accessed on 19 September 2023).
42. Rosu, D.; Teaca, C.A.; Bodirlau, R.; Rosu, L. FTIR and color change of the modified wood as a result of artificial light irradiation. *J. Photochem. Photobiol. B* **2010**, *99*, 144–149. [CrossRef]
43. Essoua, G.G.E.; Blanchet, P.; Landry, V.; Beauregard, R. Maleic Anhydride Treated Wood: Effects of Drying Timeand Esterification Temperature on Properties. *BioResources* **2015**, *10*, 6830–6860. [CrossRef]
44. Dong, Y.; Yan, Y.; Ma, H.; Zhang, S.; Li, J.; Xia, C.; Shi, S.Q.; Cai, L. In-Situ Chemosynthesis of ZnO Nanoparticles to Endow Wood with Antibacterial and UV-Resistance Properties. *J. Mater. Sci. Technol.* **2017**, *33*, 266–270. [CrossRef]
45. Yuan, B.; Guo, M.; Huang, Z.; Naik, N.; Hu, Q.; Guo, Z. A UV-shielding and hydrophobic graphitic carbon nitride nanosheets/cellulose nanofibril (gCNNS/CNF) transparent coating on wood surface for weathering resistance. *Prog. Org. Coat.* **2021**, *159*, 106440. [CrossRef]
46. Torcătoru, M.-J.; Timar, C.M. An experimental method to evaluate the contribution of the wood substrate and of the coating film in the global light in-duced colour changes of wood surfaces in indoors conditions. *Bull. Transilv. Univ. Bras. Ser. II For.-Wood Ind.-Agric. Wood Eng.* **2023**, *16*, 205–226. [CrossRef]
47. Deka, M.; Petric, M. Photo-degradation of water borne acrylic coated modified and non-modified wood during artificial light exposure. *BioResources* **2008**, *3*, 346–362. [CrossRef]
48. Liu, X.Y.; Timar, M.C.; Varodi, A.; Nedelcu, R.; Torcătoru, M.-J. Colour and Surface Chemistry Changes of Wood Surfaces Coated with Two Types of Waxes after Seven Years Exposure to Natural Light in Indoor Conditions. *Coatings* **2022**, *12*, 1689. [CrossRef]
49. Available online: https://www.renneritalia.com/en/ (accessed on 13 October 2023).
50. Available online: http://www.kroncolor.ro/contact.html (accessed on 18 October 2023).
51. Available online: https://www.avantes.com/ (accessed on 27 October 2023).
52. *ISO-4582-2017*; Plastics—Determination of Changes in Colour and Variations in Properties after Exposure to Glass-Filtered Solar Radiation, Natural Weathering or Laboratory Radiation Sources-Requirements. ISO: Geneva, Switzerland, 2017.
53. Timar, M.C.; Beldean, E.C. Modification of Shellac with Clove (*Eugenia caryophyllata*) and Thyme (*Satureja hortensis*) Essential Oils: Compatibility Issues and Effect on the UV Light Resistance of Wood Coated Surfaces. *Coatings* **2022**, *12*, 1591. [CrossRef]
54. *ISO 16474-3:2013*; Paints and Varnishes—Methods of Exposure to Laboratory Light Sources Part 3: Fluorescent UV Lamps. ISO: Geneva, Switzerland, 2013.
55. Persze, L.; Tolvaj, L. Photodegradation of wood at elevated temperature: Colour change. *J. Photochem. Photobiol. B Biol.* **2012**, *108*, 44–47. [CrossRef]
56. Duarte, I.; Rotter, A.; Malvestiti, A.; Silva, M. The role of glass as a barrier against the transmission of ultraviolet radiation: An experimental study. *Photodermatol. Photoimmunol. Photomed.* **2009**, *25*, 181–184. [CrossRef]
57. Serrano, M.A.; Moreno, J.C. Spectral transmission of solar radiation by plastic and glass materials. *J. Photochem. Photobiol. B* **2020**, *208*, 111894. [CrossRef]
58. Dawson, B.S.W.; Singh, A.P.; Kroese, H.W.; Schwitzer, M.A.; Gallagher, S.; Riddiough, S.J.; Wu, S. Enhancing exterior performance of clear coatings through photostabilization of wood. Part 2: Coating and weathering performance. *J. Coat. Technol. Res.* **2008**, *5*, 207–219. [CrossRef]
59. Kúdela, J.; Sikora, A.; Gondáš, L. Wood Surface Finishing with Transparent Lacquers Intended for Indoor Use, and the Colour Resistance of These Surfaces during Accelerated Aging. *Polymers* **2023**, *15*, 747. [CrossRef]
60. Meijer, D.M. A Review of Interfacial Aspects in Wood Coatings: Wetting, Surface Energy, Substrate Penetration and Adhesion; 2005. In Proceedings of the COST E18 FInal Seminar, Paris, France, 26–27 May 2005. Available online: https://www.researchgate.net/publication/260601859_A_review_of_interfacial_aspects_in_wood_coatings_wetting_surface_energy_substrate_penetration_and_adhesion (accessed on 7 December 2023).
61. Meijer, D.M.; Thurich, K.; Militz, H. Comparative study on penetration characteristics of modern wood coatings. *Wood Sci. Technol.* **1998**, *32*, 347–365. [CrossRef]
62. Meijer, D.M.; Thurich, K.; Militz, H. Quantitative measurements of capillary coating penetration in relation to wood and coating properties. *Holz Als Roh-Und Werkst.* **2001**, *59*, 35–45. [CrossRef]
63. Özgenç, K.Ö.; Hiziroglu, S.; Yildiz, U. Weathering properties of wood species treated with different coating applications. *BioResources* **2012**, *7*, 4875–4888. [CrossRef]
64. Kanbayashi, T.; Matsunaga, M.; Kobayashi, M.; Maeda, K. Elucidation of the degradative behavior of coated-wood surfaces exposed to artificial weathering using Raman microspectroscopy. *Prog. Org. Coat.* **2024**, *187*, 108184. [CrossRef]

applied sciences

MDPI

Article

Designing Multi-Functional Magnetic Storage Cubes for Use in Modern Homes and Schools

Martin Zach [1,*], Jiří Tauber [2], Přemysl Dohnal [3] and Jaroslav Svoboda [4]

[1] Expert Engineering Department, Institute of Lifelong Learning, Mendel University in Brno, 613 00 Brno, Czech Republic
[2] Department of Industrial Design, Faculty of Mechanical Engineering, Brno University of Technology, 616 00 Brno, Czech Republic; jiri.tauber@vut.cz
[3] Department of Theoretical and Experimental Electrical Engineering, Brno University of Technology, 616 00 Brno, Czech Republic; dohnalp@vutbr.cz
[4] Department of Design and Furniture, Faculty of Forestry and Wood Technology, Mendel University in Brno, 613 00 Brno, Czech Republic; jaroslav.svoboda@mendelu.cz
* Correspondence: martin.zach@mendelu.cz; Tel.: +420-739-372-443

Abstract: Using modern methodologies in the sectors and subareas of industrial design, where they currently find only marginal application, brings a potential for interrelating technology, the arts, and fine arts. To illustrate this, we present model procedures and options for designing a versatile storage cube that integrates magnetic structural components facilitating easy and quick assembly. In addition to being an item of real furniture, portable and readily convertible into a table or soft stool, the cube supports children's creative games and helps to develop their overall skills in the present-day household and in pre-primary and primary education. The basic material rests in birch plywood, and the joint edges between the individual walls are covered with smooth plastic guards manufactured via additive 3D printing from corn fodder-based filament. Thus, an interesting structural detail, namely, plastic edges, is generated, reinforcing the entire product. The walls comprise decorative, multicolor, polyurethane foam-based elements that can be removed and reinserted. Regarding the manufacturing technology, CNC machining and laser shaping are widely employed on the main parts, and the plastic edges are 3D-printed. In terms of the original idea, the product responds to customer requirements within a specific design project. The robustness and stability tests have proved that the cubes fully satisfy the relevant standards.

Keywords: storage cube; furniture design; magnetic structural components; play element; 3D printing; structure robustness; functional testing

Citation: Zach, M.; Tauber, J.; Dohnal, P.; Svoboda, J. Designing Multi-Functional Magnetic Storage Cubes for Use in Modern Homes and Schools. *Appl. Sci.* **2024**, *14*, 829. https://doi.org/10.3390/app14020829

Academic Editors: Emilia-Adela Salca, Lidia Gurau and Mihaela Câmpean

Received: 29 November 2023
Revised: 8 January 2024
Accepted: 9 January 2024
Published: 18 January 2024

1. Introduction

In the development of a product, the actual design plays a crucial role. Manufacturers, renowned ones in particular, usually collaborate with designers, developers, and structural engineers to deliver functional products that satisfy mechanical safety and strength requirements. Product and service making, however, also involves non-human factors, which markedly affect the design process from the outset. Such factors, by definition, are natural and/or artificial, comprising physical objects, materials, digital technologies, and algorithms. Importantly, artificial intelligence tools are widely relied upon, especially as they have revolutionized the entire design domain [1,2]. Content-wise, product designing incorporates and combines multiple aspects of a novel item to be marketed. This item is, above all, expected to fulfill the purposes for which it has been developed and manufactured, reflecting the needs of the target user; in this context, the product has to be conceived with respect to the relevant ergonomics, complying with the user's anthropometric data. Further, the set of major product properties prominently includes a visually attractive layout and technical feasibility [3].

Design is frequently employed as a marketing instrument, in addition to its basic roles, which generally subsume innovation, aesthetics, visuals, ergonomics, psychological action, and social integrability; the principal task of the designer then rests in interpreting the client's visions. Within technology, constant innovation is an imperative, and effective designing strategies embody an indispensable precondition for competitive market performance and viable manufacturing economy. The mechanisms are analyzed and theoretically refined through expert design management, a domain broadly represented in economics study programs that teach production control, corporate finance processing, and project procedures in industrial manufacturing [4–7]. Regarding the practical outputs, these are then embedded in design development and implementation as fundamental branches centered on innovative approaches.

A popular scheme is custom manufacturing, where the client may exert substantial influence on the concept and fabrication of the product by proposing preliminary drafts to set the final shape and appearance. Such a method was employed also in the storage cube designing activities described herein, one of the customer-specified requirements being the use of nontraditional technology, namely, 3D printing with corn starch-based filament to reduce the product's ecological impact. The techniques that enabled us to implement the project are discussed in greater detail below, within Section 2. Presently, the manufacture and ecology of a product, when considered jointly, embody a widely discussed subdomain that, to a large extent, determines the success rate of an item to be marketed. Ecological, high sustainability materials have become an essential prerequisite for the economic and social acceptance of industrially manufactured items, and the environmental impacts generated by industrial processes are a focus of multiple debates on the possibilities of circular economy. In the Czech Republic, for example, the ecolabel *environmentally friendly product* [8] has been approved by the Ministry of the Environment to denote products or services posing low risk to the environment and human health throughout their lifecycles [9]; in the Czech market, the national and EU ecolabels are usually granted upon recommendation by the state-owned Cenia agency [10].

The marketing functions and qualities of design, together with its perception by the general public, currently form a major topic in the field [11–14]; interestingly, in some cases the marketing is more impressive than the actual design [15], Figure 1.

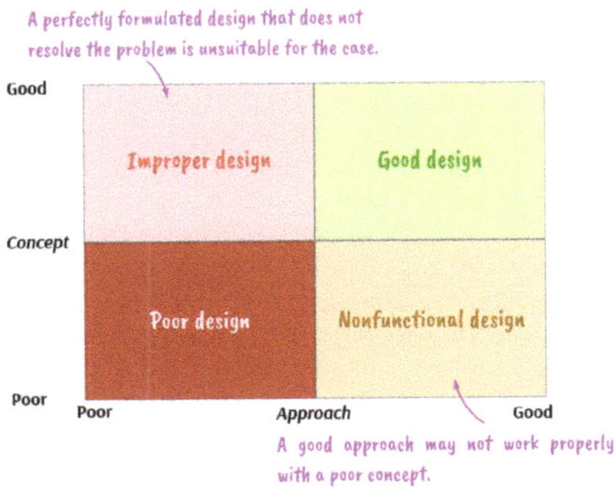

Figure 1. Design interpretations [15].

A poor design often fails to resolve the given problem due to an inadequate formulation, while a nonfunctional one offers a fully applicable approach where the actual concept complicates the implementation and, typically, causes other minor issues. A well-conceived and processed design operates smoothly with the proposed solution, delivering not only a reliable instrument to solve the task advantageously but also real and practical implementability [15].

Another role of design lies within psychology, namely, understanding the current needs of the user. In Norman's view, design embodies an act of communication, and the creative activity is usually meant and performed to completely understand the person intended to be the target recipient of the product [16].

In terms of the diverse facets and attributes associated with design, the visual characteristics and practical properties are exposed at the *user interface* and *user experience* levels, respectively; the latter of these criteria involves formulating various approaches in such a manner that they provide the best achievable user service and comfort [17]. According to Reiss [18], *user experience* is the impression acquired through a person's interactions with devices and other humans, including the handling of events that relate to these two entities.

Bar and Neta [19] propose that a product makes itself attractive to the prospective user through not only its aesthetic and color qualities but also the contours. The balance between the subjective perception and the aesthetics is then obtained via combining sharp outlines with round ones. The project procedures and results set out in this article interconnect applied research, experimental development using an additive technology (3D printing), and nonconventional machining methods, to deliver outputs readily applicable in industrial manufacturing. Additive technologies have been available for several decades, finding use in mechanical engineering, the automotive and aeronautical industries, healthcare sectors, education, the production of consumer goods (such as electrical appliances for the home), and multiple other domains [20–26].

In recent years, additive manufacturing has experienced major progress, with a broad set of newly introduced techniques and materials, inclusive of innovative polymers and photopolymers (fluid resins) [27]. A procedural option utilizing such materials is 3D printing, where the substances are usually connected layer by layer, in agreement with the data of the relevant 3D model [28].

Photosensitive resins can be hardened through laser or sunlight, and their fluid form is employed in, for instance, the healthcare industry to manufacture dentures and other prostheses. In the given context, it is possible to refer to corn-based filament, which has become popular thanks to its non-toxic composition. The most widely applied and cost-effective substance rests in polylactic acid (PLA), derived from plant matter. A frequently encountered variant, the PLA-CC, exhibits improved hardness, toughness, compressive strength, wear resistance, and other favorable parametric changes, as is documented by Fouly et al. [29] on a hip joint implant.

The overall technological and scientific advancement in various domains and subsectors has intensified the interest in nonconvenional machining options (laser/CNC plotter). These approaches facilitate cutting, slicing, or engraving in different materials (cardboard, paper, wood, plywood, textiles, composites, and metals) by means of a laser beam. The actual operations are commonly preceded by traditional machining; in wood, for instance, the refining steps comprise sawing, milling, turning, carving, and grinding. Laser cutting and CNC machining embody comparatively modern procedures, and, together with the traditional routines, they are subsumed under subtractive manufacturing [30,31].

Utilizing such preconditions, the project herein was formulated to deliver technical drawings and a functional procedure that allowed building a specially designed element to customer requirements; this item, namely, an attractive, safe-to-use play cube, was materialized with current manufacturing recipes and options (including additive technologies, such as 3D printing) as well as non-conventional machining. The customer-related details are specified below, in Section 2. To reach the preset goals and to comply with the legal regulations concerning the safety of the technological and structural solutions, we respected, above all, the Czech Republic's Act no. 102/2001 Coll., on General Product Safety [32], and standard ČSN 91 0100 Furniture—Safety requirements [33]; the product testing at an accredited laboratory was then performed accordingly and subjected to a relevant analysis.

The applied research exposed in this article focuses on the properties and use of a new birch plywood-based minor furniture component. To promote recycling and circular economy, the following materials and parts were preferred: joinery plywood, corn-based filament, polyurethane foams, and neodymium magnets. The resulting cube is multifunctional; suitable for the present-day home, school, and kindergarten; has been tested for robustness and stability; and exhibits a potential to influence the designing schemes in its sector. A similar scenario is outlined in a previously published paper by Svoboda et al. [34], which discusses the mechanical capabilities of design products in relation to the matter employed (plastic, wood, metal, and ceramics). The researchers examined 3D-printed samples from polycarbonate/acrylonitrile butadiene styrene (PC-ABS; PC) and alder wood, focusing on their impact toughness. In 3D printing, an essential factor rests in the typology optimization, assuming partial cutouts through a side panel of the component; the aim of such a modification is to reduce the weight of the product while improving its economy and maintaining the physical performance. Regarding the impact toughness, Svoboda et al. [34] propose that 3D-printed structural joints show values similar to those of solid wood (alder), expanding on the conclusions set out by Tauber et al. within their rod connector designing experiment [35]. Interestingly, 3D printing also finds application in anthropometry [36]. Based on the outcomes presented in the referenced studies, we then opted for the 3D-printed corner joint.

2. Materials and Methods

Within the presented concept, we employed novel technologies to produce a multifunctional play cube to custom specifications, as already mentioned above. The actual project, including the technical solution, drawings, and fabrication, had been ordered by the South Moravian Regional Council (Moravia, the Czech Republic), the basic material and structural requirements being as follows: an item fully usable by pre-school children, 3 to 7 years of age; play factors combined with a presentation of the region; colors to specification; a share of recycled plastics (polythene; polypropene; polyamide) and rubber; a sleek plastic corner joint to reinforce the panel edges; state-of-the-art technologies integrated in the manufacturing process; a firm structure and simple assembly steps, even for children; and surface finish. To fulfill the modern technologies stipulation, we employed 3D printing with starch-based, biodegradable polylactic acid.

The fundamental requirements comprised applying such a color scheme that could allow the customer to propose further steps in agreement with the Council-formulated visual style manual. Regarding the selected base colors, these subsumed blue and red, following the graphical layout of the South Moravian Region's logo. The structure and surfaces exploited commonly available materials, inclusive of the hardware and fittings, and the overall aim involved eliminating hazardous waste at the manufacturing, use, and disposal stages.

Considering the target application and the fact that the component is to be used by children, testing embodied a key fabrication phase. We selected three fundamental tests, the assumption being that the cubes would not be regularly sat on. The procedures were as follows:

- stability testing according to the Czech Republic standard ČSN EN 1022:2020 Furniture—Seating—Determination of stability [37], Art. 7.3.1 Forwards overturning;
- static load on the seating surface to the standard ČSN EN 16139:2013 Furniture—Strength, durability and safety—Requirements for non-domestic seating [38], Art. 6, Table 1;
- impact on the seating surface to the standard ČSN EN 16139:2013 Furniture—Strength, durability and safety—Requirements for non-domestic seating [38], Art. 6, Table 1.

Table 1. The dimensions of the cube and accessories.

Component	Width (mm)	Depth (mm)	Frontal Height (mm)
Cube	380	380	380
Cube with gliders	380	380	390
Cushion	400	400	60
Table panel	600	600	15

Source: the authors (2020).

In all of the cases, we relied on the services of the accredited Testing Laboratory for Furniture, a division of the Testing Laboratory for Joinery Products and Furniture.

When selecting convenient technologies to achieve the goals, we opted for 3D printing (an additive technology), CO^2 laser, and CNC machining. The last-named approach embodies one of the most widespread procedures, with the manufacturing tool movements being controlled by a pre-compiled program. In wood and plywood, single and combined CNC units or centers find broad application; the centers, by definition, are machines expanded to perform multiple tasks simultaneously, meaning that they constitute multi-purpose tools to execute complex operations at a time on the same location. The advanced five-axis machining then facilitates fabricating even the most demanding forms and shapes.

Considering the furniture presented in this article, suitable resources comprise solid wood or plywood. In these materials, however, CO^2 laser cutting results in surface soot deposits and dark edges; such spots subsequently require refining (grinding). To process the main plywood panels, we therefore employed a CNC miller, obtaining smooth edges. The PUR foam elements, namely, the M-shaped letters, nevertheless did not show any burns or dark stains. Overall, the CO^2 laser method offers major advantages, including but not limited to clean machining, low noise, high precision, fine detailing, effective engraving, and fast operation (at speeds of around 500 mm/s).

Further, we utilized fused deposition modeling-based 3D printing, the device being an Original Prusa i3 MK3S printer (manufactured by the company Prusa Research a.s., the Czech Republic, Prague, Czech Republic).

The individual concepts proposed by the customer were analyzed and extensively consulted.

The source materials, i.e., furniture grade birch plywood (multiplex), polyurethane foam boards, corn starch-based PLA filament, and neodymium magnets, were then selected upon the analysis of the submitted customer requirements.

2.1. Furniture Grade Plywood

The individual wood-based materials are defined in diverse knowledge databases and on web portals, such as the Czech website titled *The Furniture Maker's Information System* (Nábytkářský informační systém, [39]).

This class of materials includes plywood, and the main subclasses are defined as:

- multi-ply materials;
- agglomerated materials;
- agglomerated matter containing non-wood substances;
- modified wood;
- special wood-based matter.

The classification and terminology of plywood (Multiplex) follow the standard ČSN EN 313-2 (490013) Plywood boards [40]. Multi-ply plywood sheets are characterized as large-sized, wood-based components that comprise several layers of peeled veneer pieced together with an urea-formaldehyde glue. Thanks to their favorable properties (high strength, toughness, and elasticity; low volume weight; and size stability), the sheets have found wide and intensive application in joinery, furniture making, packaging, toy manufacturing, and the designing of musical instruments [41].

In the furniture industry; the most frequently used type of plywood is the joinery variant, whose parametric features are high strength, toughness, elasticity, and size stability, resembling those of the above-mentioned general category. The plywood (Figure 2) was employed in the manufacture of the play cube within our research.

Figure 2. Joinery birch plywood. Source: The authors' archive.

If the material is to be placed in a humid environment, especially in an outdoor location, the most convenient approach lies in opting for the waterproof version (Figure 3). Such plywood exhibits good strength, low volume weight, and sufficient size stability.

Figure 3. A waterproof birch plywood sheet. Source: NIS (2014) [39].

2.2. Polyurethane Foam Boards

The plywood boards are traditionally employed in civil engineering, the packaging industry, toy production, mattress designing, and the manufacturing of orthopedic aids.

A large number of manufacturers, such as the Czech Republic-based company Gumex, also use microporous boards (EVA-PE foam); these contain enclosed microtubes, have an ultra-fine cell structure, and offer superior elasticity [42].

A similar target applicability is characteristic of VYLEN® (Žďár nad Sázavou, Czech Republic) a semi-solid cellular plastic with very good physico-chemical and thermal insulation properties and a long lifecycle, one of tens of years; unlike competing options, VYLEN® is meshed, features a low absorption rate, and delivers high chemical resistance (reliable with oils, water-based liquids, salty solutions, and non-oxidative acids). Another major benefit lies in the absence of adverse health impacts, as proved through a relevant sensor test. The material differs from comparable concepts in that it is finely structured, easily washable, and markedly strong (Figure 4). Further assets include high elasticity, uneven surface damping, formability upon heating, and the ability to be refined via grinding, cutting, and sawing [43].

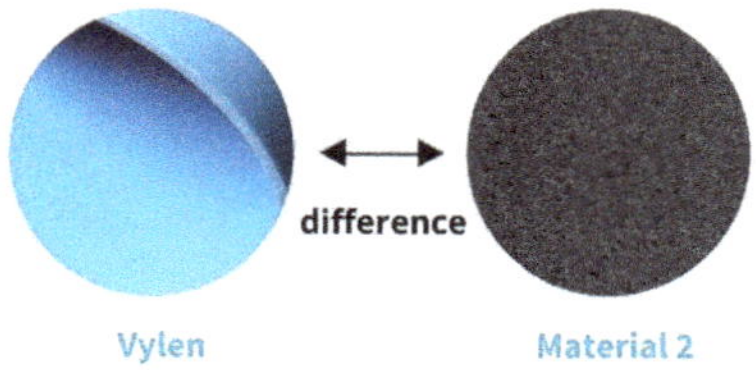

Figure 4. Comparing VYLEN®, with another substance [43].

Being non-toxic, as proved by relevant testing and certification, VYLEN® is employed in foam flooring, carpets, and children's puzzle sets [43]. The board surfaces bear a manufacturing mold imprint and are delivered in 25 colors (see the samples in Figure 5); other details can be obtained from the eponymous Czech manufacturer. The material is discussed herein because it suitably represents the new developments in the sector and could be utilized as a viable alternative in manufacturing the cube panel insertables; neither VYLEN® nor its samples, with the sole exception of the blue and red colors, were nevertheless employed in the product.

Figure 5. The VYLEN samples [44].

2.3. Corn Starch-Based PLA Filament

Filaments are polymeric fibers for FDM 3D printing, usually with a thickness of either 1.75 mm or 2.85 mm. At present, a wide range of filaments are available on the market, featuring multiple types and colors.

PLA (polylactic acid) filaments embody a basic option that, thanks to its major popularity, has prevailed over the alternatives; the source materials being corn or potato starch and/or sugar cane the item can be classified as a biopolymer. The PLA is frequently combined with other substances, including wood, plaster, PET-G, and—in some cases—carbon; the last-named approach produces tough, rugged models [45].

Combining the PLA with fresh wood, however, is not the only method to exploit the material. Tao et al. [31], for instance, propose that leftovers from subtractive manufacturing (laser cutting and CNC machining) might provide inputs for 3D printing: "... leftover veneers could be used to produce wood powder for wood/polylactic acid filaments". Thus, the set of input sources would expand significantly, meaning a principal step towards supporting secular economics.

Further, it is worth noting that many printed products exhibit poor stability in the target environment. This problem is discussed by, among others, Popescu et al. [46], whose related study was conceived to determine whether the actual choice of a material (PLA or ABS), printing device, or slicing software influences the dimensional accuracy of 3D-printed openings.

2.4. Neodymium Magnets

The superior attractive force makes the magnets suitable in any furniture-making application where the parts have to be held together reliably and safely. The magnetic items can be glued or screwed into the component. In terms of the shape, the segments are available as cubes, cylinders, rods, and round or flat tablets. To provide an example of the embedding, let us refer to the cylinders, which are easily inserted in a pre-milled or drilled circular opening that has a specific depth, as needed. Regarding the magnet diameters, these range from 2 mm to 50 mm. Being embedded in the wood, the magnets are shielded to a certain extent and remain visible only partially, observing the aesthetics of the product. In this context, and by general definition, it is worth noting that the high-performance neodymium magnets have been used for a multitude of purposes to date in diverse branches of industry, including but not limited to electrical and electronics engineering, the healthcare sector, toy manufacturing, and jewelry designing [47].

Neodymium magnets are the strongest embodiment of the material, their performance being up to tenfold that of regular ferrites. Some neodymium parts are capable of holding a load exceeding their own weight even a thousand times or more. Overall, the quantity to be considered in magnets is the tearing force, defining the action required to separate the magnet from a surface. When joining the children's furniture elements in our project, we apply magnets having a tearing force of 2 kg (approx. 18.63 N) or less and weighing 3.5 g; the related dimensions correspond to D20 × 1.5 mm in weight. With four magnets, the tearing force will reach slightly below 8 kg (74.52 N). In order for a magnet to be firmly seated in place, gluing or screwing are available, albeit a metal counterpart has to be used concurrently.

2.5. *Multi-Functional Play Cube*

2.5.1. Material, Components, Machining, and Mounting

The product subsumes the actual cube and the accessories, namely, a table panel, a removable cushion, and insertable foam elements (Figure 6a). The entire setup may comprise several variable, foldable cubes, whose schemes are rendered in Figure 6b. Interestingly, the group may serve for exhibition purposes too and is occasionally employed in this manner by the customer.

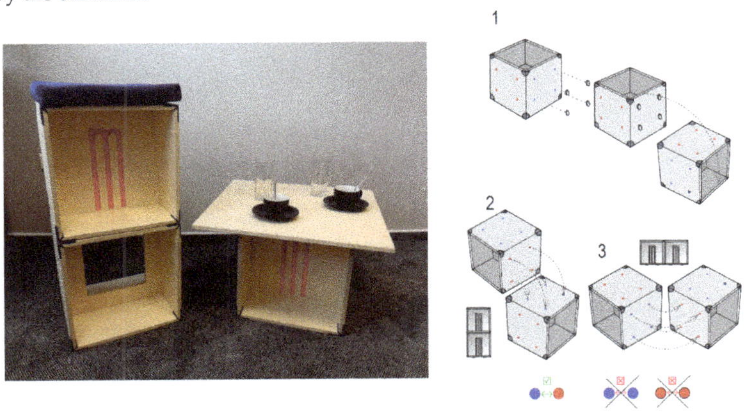

(a) (b)

Figure 6. (**a**) A model play cube setup; (**b**) the folding schemes. In the latter image, segment 1 exposes the use of the neodymium magnets in a manner that combines individual cubes into larger units. To connect two cube walls, the magnets (represented with the red spots) are glued into these walls; the other surfaces then incorporate sheet metal items (located by means of the blue spots). The diagram shows that the walls are invariably fixed to each other over the magnet-to-metal pattern, excluding the magnet-to-magnet and metal-to-metal options. Such a procedure simplifies the actual manufacturing, as the polarities do not need to be observed in the gluing operations. the magnets into the cube. Segments 2 and 3 then illustrate the vertical and horizontal folding, respectively. Source: the authors' design (2020–2022).

In Table 1, the product dimensions are specified in relation to both the whole unit and the individual components.

The multi-functional play cube is intended for pre-school children at 3 to 7 years of age; however, the use by school age children (6 to 12 years) is also possible. In older persons, not all of the product's functions may be convenient, as the children already exhibit well-developed motor skills; thus, the interest in inserting the foam letters will probably have declined by that stage of a child's life. The weight of the baseline version is 6.3 kg.

- Joinery birch plywood (multiplex): thickness of 15 mm; the high precision profiling was performed via CNC machining. The plywood contains 11 layers and complies with the E1 emission class. All of the surfaces and edges have been sealed with a lacquer.
- PUR foam: rendered in blue and red; the high precision profiling was carried out with a laser beam.
- Neodymium magnets: 20 mm in diameter and 1.5 mm in thickness; one cube contains four magnets.
- Corner joint: dimensions of 50 × 50 mm, including the bend; the joint is 3D-printed from corn starch-based PLA.
- Glider: 3D-printed and screwed to the bottom of the cube.

2.5.2. Structure and Joints

In terms of component connectability, the set of key structural segments comprises, above all, the corner joint; the area is mitred, with 30 mm wood slats. The surface has been milled to incorporate a 3D-printed plastic corner joint item; this design element is glued into place with a binder suitable for PUR. Importantly, in addition to reinforcing the edge section, the item embodies a distinctive visual feature. The perforated side walls carry push-in M letters made of PUR foam boards through laser cutting; this processing technology seals the edges of the material, leaving them tough and strong. The profiles in the joinery plywood have been CNC-milled. Regarding the product testing, the unit and its parts have complied with the Czech Republic technical standards ČSN EN 1022 (910234) and ČSN EN 16139 (910650) A [37,38].

In the cushion, the PUR foam is wrapped in glued-on nonwoven fabric and inserted into a zipped cloth cover; sewn into the cover are neodymium magnets to fix the cushion onto the top of the cube. All outside edges have been rounded to deliver Rmin. = 2.5 mm, and the firm structural joints are glued together over dowel pins to ensure good stability. The surface is treated with a water-dilutable, semi-matt lacquer. To characterize the neodymium magnets, we can emphasize their tearing force of 8 kg and the fact that the magnets are high-sheened, similarly to the sheet metal counterparts in the related surfaces. In the testing, the tearing force proved sufficient.

The product provides a value added in that neodymium magnets are incorporated into the upper panel of the cube. The magnets are small and have great contact force, allowing the user to safely attach to the cube a table panel or a seat cushion (Figure 6); thus, the cube is converted into a practical desk or a comfortable stool.

2.5.3. Drawings and Related Documentation

To facilitate manufacturing the sample product via CNC machining and laser cutting, we completed a set of drawings and technical documents; the individual profiles and details are shown in Figures 7 and 8.

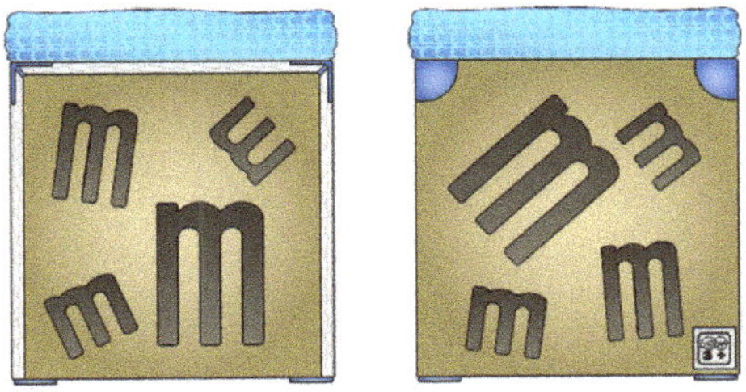

Figure 7. The full drawings of the multi-purpose cube with perforated sides and an upholstered top surface. Source: the authors (2020).

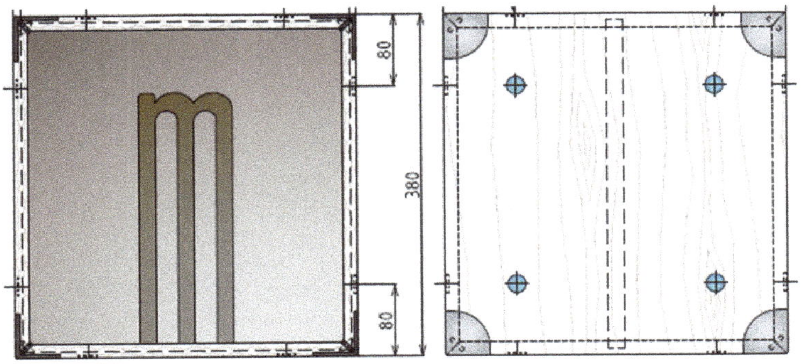

Figure 8. A part of the technical documentation: the basic side and bottom plan. Source: the authors (2020).

2.5.4. Manufacturing Technologies and Approaches

The cube is based on a simple, pure design. The fundamental material is birch plywood (PDP-BR), the boards having 15 mm in thickness. In terms of the technological procedures, we applied CNC and laser machining. Further, we employed neodymium magnets sized 20 mm to mill them into the board surfaces in such a manner that they protrude 1 mm above. The structure comprises 3D-printed protective corner joint components, which also reinforce the actual joints and are embedded in seats countersunk in the boards' corner edges and adjacent surface segments (Figure 9). Simultaneously, in addition to their reinforcing function, these plastic components play an aesthetic role, making the cube's appearance more attractive and being modifiable to the customer's wishes concerning the shape and color.

Figure 9. The 3D-printed component on the corner edge area. Source: the authors (2022).

The versatile design enables the multi-functional cube to be easily converted into a plain or cushioned stool, a small desk, or a play base with large M-shaped color letters made of PUR foam that are insertable into corresponding openings in the side boards to help children develop their motor and combinatorial skills (Figure 10a,b).

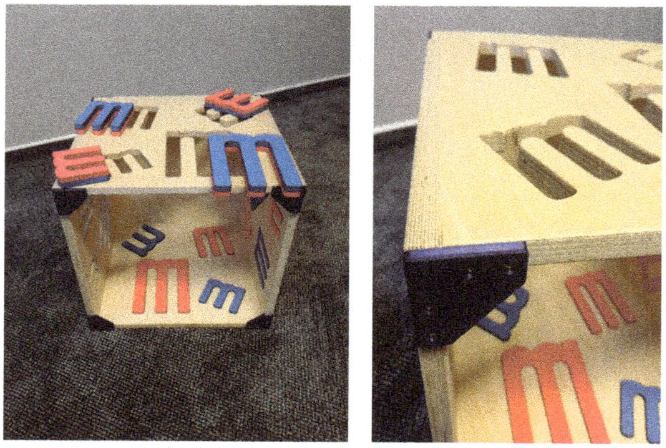

(**a**) (**b**)

Figure 10. The bottom sectors: (**a**) an overall view of the 3D-printed gliders; (**b**) the gliders detailed on a corner edge. Source: the authors (2022).

3. Results

3.1. Verifying the Functionality: Product Testing at an Accredited Furniture Testing Laboratory

The cube with perforated side walls was subjected to testing, due to the product's structure and bottom gliders being different from those of the original baseline model. Compared to this initial version, the tested cube does not include a center part but its inner space is enclosed with walls on all sides; the side boards feature M-shaped openings (as shown on the lower item in Figure 11a); no bottom board is used (Figure 11b, variant A); and the structural joints are identical with those in the prototype, including the plastic corner edge elements. Further, in the tested item the gliders are not held in place by means of magnets (Figure 11b, variant B) but rather screwed into the edges (Figure 11b, variant A).

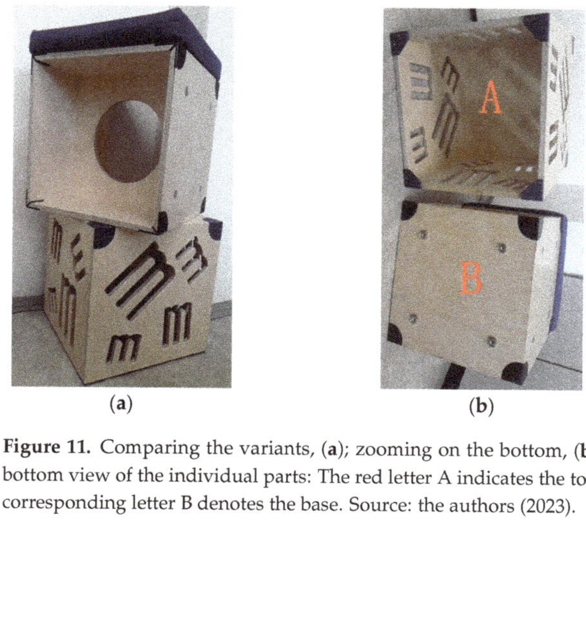

(**a**) (**b**)

Figure 11. Comparing the variants, (**a**); zooming on the bottom, (**b**). The latter image presents a bottom view of the individual parts: The red letter A indicates the top section of the cube, while the corresponding letter B denotes the base. Source: the authors (2023).

Based on the intended modes of use, we performed three basic testing cycles (as specified in Sections 2 and 3.1.1–3.1.3), assuming that the cubes would not be permanently employed to function as seating furniture. The tested product was equipped with a cushion and metal gliders sized D20 × 7 mm; the gliders were fixed to the bottom with magnets.

Alternatively, the cube was subjected to testing together with the baseline version, the assumption being that the products would also be utilized as stools. All of the test cycles were executed at a dedicated laboratory for joinery products and furniture, accredited according to the Czech Republic standard ČSN EN ISO/IEC 17025:2018 [48]; the facility is operated by the Faculty of Forestry and Wood Technology, Mendel University in Brno.

3.1.1. Stability Testing

At the first stage, we carried out stability tests in agreement with the Czech standard ČSN EN 1022:2020 (Furniture—Seating—Determination of stability) [37], art. 7.3.1 (Forwards overturning); the standard covers seating furniture for all locations and types of application. In the procedures, the load on the seating surface at the distance of 60 mm from the edge of the supporting structural board amounted to 600 N, and the applied overturning force was 20 N, both applied as per the standard (Figure 12).

Figure 12. The stability testing configuration. Source: the authors (2023).

Outcomes: The product was not overturned by the relevant force, thus satisfying the stability requirements stipulated by the standard ČSN EN 1022:2020 [37].

By extension, we also tested the strength and durability of the structure, observing the Czech Republic standard ČSN EN 16139:2013 (Furniture—Strength, durability and safety—Requirements for non-domestic seating) [38]. With respect to the modes of use presumed in the product, the selected procedures comprised applying a static load and an impact on the seating surface; both of the test schemes conformed to the provisions of the standard ČSN EN 1728:2020 (Furniture—Seating—Test methods for the determination of strength and durability) [49].

3.1.2. Static Load Testing

Respecting the predefined procedural measures (as set out in the standard ČSN EN 16139:2013, art. 6) [38], we applied on the seating surface, in its relevant axis, the vertical load of 1300 N at the distance of 100 mm from the edge of the supporting structural board; the force acted on the surface through a circular wood pad, invariably for 10 s, and the action was repeated 10 times (Figure 13a,b).

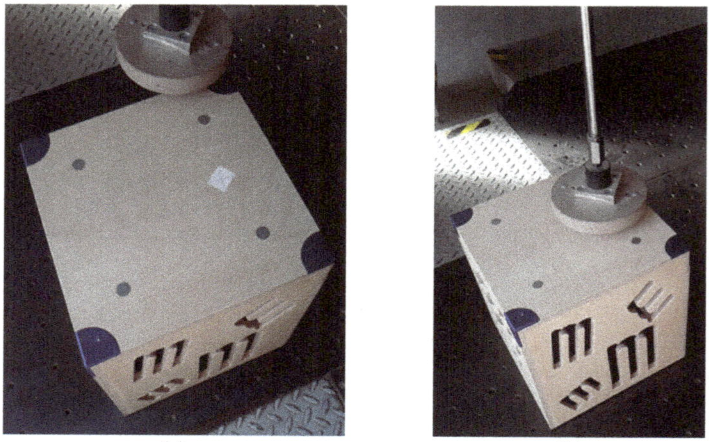

(a) (b)

Figure 13. The tests: (**a**) static loading on a preselected spot; (**b**) overall static loading. Source: the authors (2023).

3.1.3. Impact Testing

Conforming to the provisions set forth in the standards ČSN EN 16139:2013 [38] and ČSN EN 1728 [49], we simulated a fall from the height of 240 mm, repeating the process ten times. A shock generating tool was employed to perform the action, delivering the impact weight of 25 kg (Figure 14a,b).

(a) (b)

Figure 14. The impact testing: (**a**) before the action; (**b**) applying the force. Source: the authors (2023).

Outcomes: The cube did not exhibit any damage after either the static or the impact test, thus satisfying the requirements defined within the standards. In terms of the strength, durability, and stability, the structure can therefore be characterized as fully satisfactory.

4. Discussion

The testing procedures showed that when gliders are mounted, the cube tends to partially yield to the applied force in that it leans on the bottom edge of a structural board (Figure 15a); nevertheless, total overturning does not occur, as already indicated above.

(a) (b)

Figure 15. The stability testing: (**a**) with the gliders; (**b**) without the gliders. Source: the authors (2023).

In this context, it is then somewhat problematic to determine clearly whether the product conforms to the standards. If tested without the gliders, however, the cube exhibits superior stability, satisfying all the requirements (Figure 15b).

Similarly to the images in Figure 13a,b, the graphics in Figure 16a,b below represent the load and impact tests; in the latter figure, however, the testing is performed on a cushioned cube.

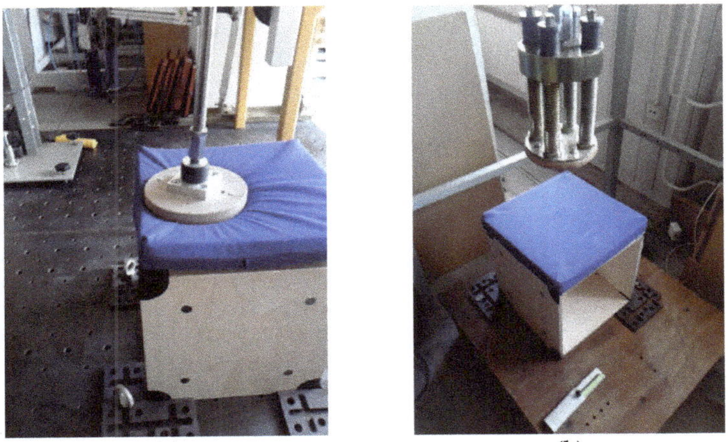

(a) (b)

Figure 16. The load and impact tests: (**a**) static loading; (**b**) impact action. Source: the authors (2023).

The obtained results confirmed the structural strength of the cube and also led to a modification of the glider mounting in response to the ambiguous overturning performance; presently, the gliders are embedded in countersunk seats and protrude beyond the surface line by 2 mm at the maximum. Another alteration lay in the gliders being fixed to their seats with screws in place of the magnets used in the initial versions (Figure 10a,b).

In furniture, a significant parameter rests in the carrying capacity. To verify this property, we first set up a standardized stability test to load the seating surfaces with a weight of 600 N at a distance of 60 mm from the edge of a main structural panel; such a configuration represents a force of 61 kg.

The measurement process and its outcomes confirmed that the product satisfies the requirements on children's furniture in terms of the structure, ruggedness, and stability. The firm fasteners are completed with plastic, 3D-printed edge joints to provide a load capacity of between 150 and 200 kg; thus, the capacity limits for the targeted age group do not need to be estimated. Regarding the physical and safety attributes, the product is well-conceived, combining a glued structure with reinforcing plastic corner elements; in many cases, these edge parts exhibit superior strength values, as also emphasized by Svoboda et al. [34], who examined the parameters that exert an impact on furniture safety. By a more detailed definition, the researchers focused on the impact toughness testing of samples integrating 3D-printed plastics (PC-ABS; PC) with alder tree wood; the printed elements included structural joints in particular [34,35].

From the utility perspective, the cube was assessed in terms of whether its use by pre-school children can be regarded as safe. The set of monitored sectors involved, among others, the openings milled in the panels to accommodate the PUR-based letters; in this context, we reviewed the size of the gaps to avoid children's fingers getting wedged and, subsequently, hurt or broken.

Furniture safety is generally provided for in the Czech Republic's Act no. 102/2001 Coll., on General Product Safety [32], and standard ČSN 91 0100 Furniture—Safety requirements [33]. A product designated as safe must comply with specific national and international legal regulations; where no such regulations are in force, the attribute of *safe* shall denote any product that satisfies the Czech technical standards and/or expresses the state of the art at the time of being introduced to the market [50].

Overall, regarding the concept, functional capabilities, and safety, the furniture set designed within our project may be compared to items such as the multi-purpose Montessori cUbe series [51] and the Whitney Brothers Play House Cube [52]. The Montessori cUbe, however, appears to best resemble the solutions adopted by us: the traditional Montessori's versatility is similar to ours, albeit more sophisticated, and both schemes share a portion of the basic material (birch plywood), with the comprehensive test performance outcomes being very close [53,54]. The visions and purposes formulated within the previous research and requirements on children's furniture design [55,56] reflect in the project discussed herein, even though our team worked, in fact, independently from these: the ideas centered on supporting the play factor to facilitate effective learning and development in children are broadly (and indirectly) shared by diverse designers, manufacturers, and retailers who employ or sell, for instance, the Montessori method. Further, depending on the interpretation, the color scheme in our cube might suitably complement Montessori's patterns, which may seem rather dry in some of the series. Arguably, the perspectives outlined within this article lie in a relatively warm and friendly rendering of the product, whose marketing potential is thus markedly enhanced. In this manner, the cube might then be perceived by other designers in the field as inspiring. By contrast, we assume that some of the mechanical properties of our cube, such as the edge stability, could be improved to match the performance of mass-produced furniture.

With respect to the examined physical quantities and product capabilities, and considering the relevant standards, we compared our procedural testing approaches discussed herein with those applied by other researchers. In seating furniture, interesting static and dynamic options, including progressive loading, are set out in the very recent study by Diler et al. [57]. These authors tested a large set of items under varying conditions and loads to establish that the strength and capacity of a product markedly depend on the loading direction. In contrast to our concept, Diler et al.'s methods are markedly broader, more detailed, and involve factors such as inherent material properties; the techniques have been formulated specifically to evaluate chairs, whose basic function nevertheless largely corresponds to that of the play cube [57]. Further, a comprehensive, optimization-oriented testing strategy had been proposed by Güray et al. [58]; in this case, a high number of wood matter and impact force types were employed, forming a model for the future studies. In the Czech-based laboratories within this article, by comparison, the scope of corresponding tasks was relatively modest, as our project had been conceived significantly less generously, due to the limited amount of materials. The associated development, refining, and (to a limited extent) basic tests of children's or purpose-specific furniture are all exposed in the research by Zhang et al. and Jai-ai et al. [59,60], with the former outlining the role of artificial intelligence at the diverse stages of the furniture designing, manufacturing, and verification procedures; such an approach is a blueprint for the next phases of our efforts, as it provides novel baseline ideas to rethink the testing process.

In addition to the pre-formulated central function-related goal, namely, verifying the structure of the cube with respect to the stability, static load, and impact capacities, other major factors of influence on persons such as the users and manufacturing operatives have to be recognized. Prominently included in this set are chemicals, represented by, for instance, volatile organic compounds. The relevant substances then comprise, among others, formaldehyde, an ingredient of the glue that is used in diverse types of plywood. Formaldehyde, by extension, finds application also in surface finish lacquers, fabrics, carpets, stationery, and electronics. Importantly, materials for the indoor environment are subjected to testing before entering the market.

To characterize formaldehyde in a balanced manner, we should describe the substance's benefits too: a colorless glued splice joint, superior shear strength, favorable cost, and hardening temperatures between 10 and 180 °C. Yet formaldehyde is released into the air, and this drawback reflects in the legal requirements on children's furniture; within the European Union, the applicable regulation consists in the standard EN 71-3 [61], its Czech Republic equivalent being ČSN EN 71-3+A1 (94 3095) [62], that sets forth stipulations concerning the safety of toys and the migration of selected chemical elements [62].

During the 1970s, the emission value reached 100 mg/100 g of the mass. Since 2004, European producers have been supplying chipboard and plywood boards in two emission classes, E1 and E2. In view of the released volumes of formaldehyde, plywoods are categorized into the A group (ČSN EN 1084, canceled in 2014) [63], and multiplex plywoods are classified according to their response to fire with the C designation in the standard ČSN EN 13501-1 [64]. The class E1 stipulates emissions below 0.1 ppm (0.124 mg/m^3, as set out in ČSN 717-1/EN 717-1) [65]; in chip- and plywood boards, formaldehyde contents are specified as up to 8 mg/100 g of mass, according to EN 120 (an already invalid standard that has been substituted with ČSN EN ISO 12460-5 [66]. Materials characterized in the class E2 are suitable for outdoors only because the formaldehyde volume may attain 8 to 30 mg/100 g mass. Another finding lies in that the European Commission has set a new limit for the concentration of formaldehyde in "ambient" air: Regulation 2023/1464, published on 14 July 2023. This provision reduces the previous values to a half by setting the limit of 0.062 mg/m^3 on the emissions, particularly with regard to wood-based products and furniture as possible "sources" of formaldehyde. Currently, there is chipboard/plywood furniture on the European market that has not been treated with formaldehyde or even has lower formaldehyde emissions than wood. The manufacturer or seller of such furniture shall provide an indication of the emission class. Each product and plywood material has to be described with a technical data sheet containing information on the properties and applicability of the product [41,67,68].

The Multiplex birch plywood complies with the E1 emission class; the central Europe-based company Demos Trade may serve in this context as a convenient reference subject, declaring that formaldehyde-free E0 plywood is available upon request (boards made from beech wood being the most affordable ones). An optional solution for the cube would rest in using beech plywood multiplex with an E0 emission class glue. To prevent the release of formaldehyde, all surfaces and edges of relevant products must be sealed with lacquer. The multifunctional play cube is protected with a water-based polyurethane varnish, meaning that even when using an E1 class plywood the release of formaldehyde is avoided [41,69].

In general, the path to offering and supplying a product in which the amount of emissions is clearly demonstrable rests in certification, i.e., labeling according to the volume of VOC emissions. This step has already been taken by France, Germany, and some

Russian-speaking countries, to provide applicable examples. In France, products are tested according to EN ISO 16000, A+ being the best rating [70]. In Germany, all products introduced to the market must be tested. Other states use a recognized label, such as the Finnish M1 certificate, which monitors the TVOC for more than 2000 items [67,71].

Regarding the structure, strength, ruggedness, and stability, the product appears to be excessively massive to be used by children, considering the 15 mm plywood. In further research, we intend to examine the capabilities of 12 mm thick multiplex plywood, reducing the overall weight of the basic element down to approximately 5.0 kg.

A potential direction for the future investigation may lie in other items where the technologies involved (CO^2 laser, 3D printing, CNC milling) at both the development and the application stages facilitate design progression. Using the combination of options specified herein often results in a multitude of convenient materials, new shape possibilities, innovative structural solutions, different designs, and a higher product strength.

Utilizing laser technology (PUR perforation cutting) together with PUR foam and birch plywood multiplex delivers a final product that supports and develops the motor skills of pre-school children. Here, the fact that (as a new direction for the research) a similar solution could support fine motor skills in pre-school children experiencing developmental dyspraxia and/or related disorders should be seriously considered.

5. Conclusions

The article discusses a multi-faceted product design that is based on functional tests at an officially approved furniture testing laboratory. The aim lies in exposing various application options for state-of-the-art designing technologies that facilitate the fabrication of a novel and interesting product with value-added features. In the case of the storage cubes, such properties are embodied especially in the actual use of the neodymium magnets and corner components, the latter being manufactured through the additive technology of 3D printing. The magnets function as fastening elements to hold either further parts of the play cube or a whole other cube. Despite their small size, the magnets exhibit a great contact force, and the tearing force equals 8 kg. The original design relied on the tearing force of 2.5 kg, but this lower value did not prove sufficient to satisfy the relevant testing requirements. Regarding the proportion of plastic materials, the largest amount is found in the product-specific corner components, as these were 3D-printed from the PLA (polylactic acid obtained via enzymatic hydrolysis from corn starch); importantly, the plastic is fully biodegradable. The acid, however, despite being the most favored option for the purpose thanks to its non-toxic character, lack of printing deformities, and easy recycling, is somewhat hampered by its relatively low temperature resistance of up to 65 °C. In comparable materials, relevant deformations occur at higher levels (ABS = 100 °C, PETG/HT-PETG = 110 °C). The reason to substantiate the use of PLA filament rests in the recyclability, ruggedness, and positional stability, the latter two properties being ensured by the edge joint. The recycling is commonly performed in industry-scale composting plants, where the conditions allow bioplastics to decay.

From the child's perspective, the most significant aspect is the play factor, which, by extension, supports the motor skills and helps to develop the overall set of abilities. A further major asset is embedded in the visual and shape characteristics of the cube, being most prominent on the side surfaces through the insertion and extraction of the PUR foam-based, M letter-shaped color segments. The cube is a multi-purpose, structurally attractive, and playful item. In the ruggedness, stability, and safety testing, the sample involved did not show any damage or breakage, thus complying with the standards and confirming the structural design approach selected. As a matter of fact, when evaluated via the results, the product appears to be excessively massive, meaning that the future fabrication efforts might rely on thinner birch plywood (12 mm in thickness, for instance) to reduce the weight and to obtain a less bulky but sufficiently rugged product.

In the development procedures, we have combined suitable applied research and experimental manufacturing options, using nonconventional machining processes (laser and CNC plotter tasks) and additive technologies, namely, product design-oriented 3D printing. Such technologies then bring substantial benefits to both the designer and the market in terms of the economic, social, and environmental impacts.

Author Contributions: Conceptualization, M.Z. and P.D.; methodology, M.Z.; software, J.T.; validation, M.Z., J.S. and P.D.; formal analysis, M.Z.; investigation, J.T.; resources, M.Z.; data curation, M.Z.; writing—original draft preparation, J.S., J.T. and M.Z.; writing—review and editing, M.Z. and P.D.; visualization, J.T.; supervision, M.Z. and J.S.; project administration, J.S.; funding acquisition, J.T. All authors have read and agreed to the published version of the manuscript.

Funding: This research received no external funding.

Institutional Review Board Statement: Not applicable.

Informed Consent Statement: Not applicable.

Data Availability Statement: The data created within this study are fully incorporated in the article and will be fully available from the corresponding author upon request.

Acknowledgments: The presented design tasks and operations were performed to complete a customer order (a play cube development project). The authors thank Mendel University in Brno for all their support and for permission to use the university's official furniture testing laboratory, a part of the industrial Testing Laboratory for Joinery Products and Furniture, in the series of tests to verify the properties of the product.

Conflicts of Interest: The authors declare no conflict of interest.

References

1. Černý, J. *Ne-Humánní Aktéři a Jejich Role v Designu (Non-Human Factors and Their Roles in Design)*; Design Kisk: Brno, Czech Republic, 2023; p. 6. Available online: https://medium.com/design-kisk/ne-hum%C3%A1nn%C3%AD-akt%C3%A9%C5%99i-a-jejich-role-v-designu-c4779236d4be (accessed on 15 September 2023).
2. Huang, M.H.; Rust, R.T. Artificial Intelligence in Service. *J. Serv. Res.* **2018**, *21*, 155–172. [CrossRef]
3. Produktový Designér vs Konstruktér (Product vs Structural Designers). Blog. 2018, Article 41, p. 3. Cotu s.r.o. Available online: https://www.cotu.cz/blog/41/produktovy-designer-vs-konstrukter (accessed on 21 September 2023).
4. Utomo, C.; Astarini, S.D.; Mertha, D.M.S.W.; Rahmawati, Y.; Aqsha; Nurcahyo, C.B.; Nahdiyah, M. Decision Factors of Stakeholder Integration in Connected Construction for Circular Economics. *Sustainability* **2022**, *14*, 16200. [CrossRef]
5. Nam, K.; Kim, G.; a Borja de Mozota, B. Design Management Knowledge: Identifying Learning Objectives of Various Stakeholders for Needs-Driven Learning. In Proceedings of the Academy for Design Innovation Management, Hong Kong, 7–9 June 2017; Volume 4, pp. 1439–1455. Available online: https://www.designresearchsociety.org/articles/design-management-academy-2017-conferee-proceedings-available-online (accessed on 25 August 2023).
6. Bruce, M.; Cooper, R.; Vazquez, D. Effective design management for small businesses. *Des. Stud.* **1999**, *20*, 297–315. [CrossRef]
7. Miller, C.Z.; Palsikar, S.; Spears, J.M. Evolving Praxis in Design Management: The transdisciplinary trajectory. *Des. Manag. J.* **2022**, *17*, 91–102. [CrossRef]
8. Ekoznačení Výrobků (Product Ecolabeling). Zo Čsop Veronica. Veronika. 2020. Available online: https://www.veronica.cz/ekoznaceni-vyrobku (accessed on 16 September 2023).
9. Envic. Co Znamená Ekologicky Šetrný Výrobek? (What Is an Environmentally Friendly Product?). In *Environmentální Informační Centra Plzeňského Kraje*; Envic: Plzeň, Czech Republic, 2013; Available online: http://www.envic.cz/faq/co-znamena-ekologicky-setrny-vyrobek.htm (accessed on 18 September 2023).
10. Czech Environmental Information Agency. Ekoznačení (Ecolabeling). In *Česká Informační Agentura Životního Prostředí*; Cenia: Praha, Czech Republic, 2023; Available online: https://www.cenia.cz/spolecenska-odpovednost/ekoznaceni/ (accessed on 18 September 2023).
11. Kostík, O. Design Jako Marketingový Nástroj, Jeho Role a Vnímání u Veřejnosti (Design As a Marketing Instrument: Roles and Perception by the General Public). Master's Thesis, Vysoká Škola Ekonomická v Praze, Praha, Czech Republic, 2008.
12. Roper, S.; Micheli, P.; Love, J.H.; Vahter, P. The roles and effectiveness of design in new product development: A study of Irish manufacturers. *Res. Policy* **2016**, *45*, 319–329. [CrossRef]
13. Henseler, J.; Guerreiro, M.; de Matos, N. The interplay of marketing and design. *Rev. Manag. Sci.* **2021**, *15*, 1129–1137. [CrossRef]
14. Henseler, J.; Guerreiro, M. Design and marketing: Intersections and challenges. *Creat. Innov. Manag.* **2020**, *29*, 3–10. [CrossRef]
15. Maňák, M. Dobrý a Špatný Design (The Good and the Bad Design). Online. Blog. 2015, p. 8. Available online: https://www.manakmichal.cz/blog/user-experience/dobry-a-spatny-design/. (accessed on 1 October 2023).
16. Norman, D.A. *Design of Everyday Things*; Perseus Books: New York, NY, USA, 2013.

17. Leška, Š. Co je to UX a UI design? (Defining UX and UI design). *BlueGhost* **2020**, *1*, 3. Available online: https://www.blueghost.cz/clanek/co-je-to-ux-a-ui-design/ (accessed on 21 September 2023).
18. Reiss, E. *Usable Usability: Simple Steps for Making Stuff Better*; John Wiley & Sons: Hoboken, NJ, USA, 2012.
19. Bar, M.; Neta, M. Visual elements of subjective preference modulate amygdala activation. *Neuropsychologia* **2007**, *45*, 2191–2200. [CrossRef] [PubMed]
20. Ngo, T.D.; Kashani, A.; Imbalzano, G.; Nguyen, K.T.; Hui, D. Additive manufacturing (3D printing): A review of materials, methods, applications and challenges. *Compos. Part B Eng.* **2018**, *143*, 172–196. [CrossRef]
21. Kim, H.-S.; Lee, D.-Y.; Park, J.-H.; Kim, J.-H.; Hwang, J.-H.; Jung, H.-I. Optimization of electrohydrodynamic writing technique to print collagen. *Exp. Technol.* **2007**, *31*, 15–19. [CrossRef]
22. Chung, H.; Das, S. Processing and properties of glass bead particulate-filled functionally graded Nylon-11 composites produced by selective laser sintering. *Mater. Sci. Eng. A* **2006**, *437*, 226–234. [CrossRef]
23. Chung, H.; Das, S. Functionally graded Nylon-11/silica nanocomposites produced by selective laser sintering. *Mater. Sci. Eng. A* **2008**, *487*, 251–257. [CrossRef]
24. Eshraghi, S.; Das, S. Mechanical and microstructural properties of polycaprolactone scaffolds with one-dimensional, two-dimensional, and three-dimensional orthogonally oriented porous architectures produced by selective laser sintering. *Acta Biomater.* **2010**, *6*, 2467–2476. [CrossRef] [PubMed]
25. Kim, M.-S.; Chu, W.-S.; Kim, Y.-M.; Avila, A.P.G.; Ahn, S.-H. Direct metal printing of 3D electrical circuit using rapid prototyping. *Int. J. Precis. Eng. Manuf.* **2009**, *10*, 147–150. [CrossRef]
26. Im, Y.; Cho, B.; Seo, S.; Son, J.; Chung, S.; Jeong, H. Functional prototype development of multi-layer board (MLB) using rapid prototyping technology. *J. Mater. Process. Technol.* **2007**, *187*, 619–622. [CrossRef]
27. Bourell, D.; Kruth, J.P.; Leu, M.; Levy, G.; Rosen, D.; Beese, A.M.; Clare, A. Materials for additive manufacturing. *CIRP Ann.* **2017**, *66*, 659–681. [CrossRef]
28. *ASTM F2792-12a*; Standard Terminology for Additive Manufacturing Technologies. ASTM International: West Conshohocken, PA, USA, 2012.
29. Fouly, A.; Assaifan, A.K.; Alnaser, I.A.; Hussein, O.A.; Abdo, H.S. Evaluating the Mechanical and Tribological Properties of 3D Printed Polylactic-Acid (PLA) Green-Composite for Artificial Implant: Hip Joint Case Study. *Polymers* **2022**, *14*, 5299. [CrossRef]
30. Wimmer, R.; Steyrer, B.; Woess, J.; Koddenberg, T.; Mundigler, N. 3D printing and wood. *Pro Ligno* **2015**, *11*, 144–149. Available online: https://publik.tuwien.ac.at/files/PubDat_243647.pdf (accessed on 28 September 2023).
31. Tao, Y.; Yin, Q.; Li, P. An Additive Manufacturing Method Using Large-Scale Wood Inspired by Laminated Object Manufacturing and Plywood Technology. *Polymers* **2021**, *13*, 144. [CrossRef]
32. Zákon, č. 102/2001 Sb., o Obecné Bezpečnosti Výrobků a o Změně Některých Zákonů (Zákon o Obecné Bezpečnosti Výrobků) (Act No. 102/2001 Coll., On General Product Safety and on the Amendment of Certain Laws (Act on General Product Safety)). Available online: www.aspi.cz (accessed on 15 October 2023).
33. *ČSN 91 0100 (910100)*; Nábytek—Bezpečnostní Požadavky (Czech Technical Standard ČSN 91 0100 (910100); Furniture—Safety requirements). Český Normalizační Institut: Praha, Czech Republic, 2006. Available online: http://csnonline.agentura-cas.cz/. (accessed on 15 October 2023).
34. Svoboda, J.; Tauber, J.; Zach, M. 3D print application in furniture manufacturing. In Proceedings of the 12th WoodEMA Annual International Scientific Conference—"Digitalisation and Circular Economy: Forestry and Forestry Based Industry Implications", Union of Scientists of Bulgaria and WoodEMA, Varna, Bulgaria, 11–13 September 2019; pp. 131–140, ISBN 978-954397042-1. Available online: https://www.woodema.org/proceedings/WoodEMA_2019_Proceedings.pdf (accessed on 1 December 2023).
35. Tauber, J.; Začal, J.; Šustr, M.; Brabec, M. Design and mechanical performance of 3-D printed T-shaped rod connector manufactured by FDM technology. In Proceedings of the 12th WoodEMA Annual International Scientific Conference—"Digitalisation and Circular Economy: Forestry and Forestry Based Industry Implications", Union of Scientists of Bulgaria and WoodEMA, Varna, Bulgaria, 11–13 September 2019; pp. 155–160, ISBN 978-954397042-1.
36. Paramés-González, A.; Prieto-Lage, I.; Gutiérrez-Santiago, A. 3D printing in fitting furniture to student anthropometry. *ESHPA—Educ. Sport Health Phys. Act.* **2021**, *5*, 76–89. [CrossRef]
37. *ČSN EN 1022 (910234)*; Nábytek Bytový—Sedací Nábytek—Hodnocení Stability (Czech Technical Standard ČSN EN 1022 (910234) Domestic Furniture—Seating-Determination of Stability). Úřad Pro Technickou Normalizaci, Metrologii a Státní Zkušebnictví: Praha, Czech Republic, 2020. Available online: https://csnonline.agentura-cas.cz/Detailnormy.aspx?k=509494 (accessed on 14 October 2023).
38. *ČSN EN 16139 (910650)*; A Nábytek—Pevnost, Trvanlivost a Bezpečnost—Požadavky na Nebytový Sedací Nábytek (Czech Technical Standard ČSN EN 16139 (910650) A Furniture—Strength, Durability, and Safety—Requirements for Non-Residential Seating). Úřad Pro Technickou Normalizaci, Metrologii a Státní Zkušebnictví: Praha, Czech Republic, 2013. Available online: https://csnonline.agentura-cas.cz/Detailnormy.aspx?k=93760 (accessed on 12 October 2023).
39. INT, S.R.O. *Nábytkářský Informační System (Furniture Maker's Information System)*; Brunecký, P., Hála, B., Jičínský, M., Jančová, V., Tauber, J., Svoboda, J., Zach, M., Šimek, M., Dvouletá, K., Káňová, D., et al., Eds.; Software. INT, S.R.O.: Brno, Czech Republic, 2014; Available online: http://www.n-i-s.cz/. (accessed on 12 December 2023).

40. *ČSN EN 313-2 (490013)*; Překližované desky—Klasifikace a terminologie—Část 2: Terminologie (Czech Technical Standard ČSN EN 313-2 (490013); Plywood—Classification and terminology—Part 2: Terminology). Český Normalizační Institut: Praha, Czech Republic, 2000. Available online: https://www.technicke-normy-csn.cz/csn-en-313-2-490013-207631.html# (accessed on 14 December 2023).
41. Překližované Materiály (Plywood Materials). Nábytkářský Informační Systém. (Furniture Maker's Information System) 2014, article 215, p. 33. Licence: Nábytkářský Informační Server (Furniture Maker's Information Server). INT, S.R.O.: Brno, Czech Republic, 2013. Available online: https://www.n-i-s.cz/cz/preklizovane-materialy/page/215/ (accessed on 1 October 2023).
42. GUMEX, spol. *s r.o. Mikroporézní Deska Eva 100—Červená (The EVA 100 Microporous Board)*; GUMEX: Strážnice, Czech Republic, 2023; Available online: https://www.gumex.cz/h/mikroporezni-deska-eva-100-cervena-10501#sekce=variant-table-placeholder (accessed on 30 September 2023).
43. Materiál VYLEN®(The VYLEN®Material). Blog. 2023, article 2, p. 2. Available online: https://www.penovysvet.cz/blog/vylen/ (accessed on 1 October 2023).
44. VYLEN, s.r.o. *Vzorník Barev—Kulatý (The Material Colors: A Ring of Samples)*; Pěnový svět: Žd'ár nad Sázavou, Czech Republic, 2023; Available online: https://www.penovysvet.cz/podlahy/vzornik-barev/ (accessed on 1 October 2023).
45. PLA Filament. Materialpro3d.cz. Available online: https://www.materialpro3d.cz/pla-1-75/pla-filament-1-75-mm-bily-regshare-1-kg/ (accessed on 1 October 2023).
46. Popescu, D.; Gheorghe Amza, C.; Marinescu, R.; Cristiana Iacob, M.; Luminiţa Căruţaşu, N. Investigations on Factors Affecting 3D-Printed Holes Dimensional Accuracy and Repeatability. *Appl. Sci.* **2023**, *13*, 41. [CrossRef]
47. Cui, J.; Ormerod, J.; Parker, D.; Ott, R.; Palasyuk, A.; Mccall, S.; Paranthaman, M.P.; Kesler, M.S.; McGuire, M.A.; Nlebedim, I.C.; et al. Manufacturing Processes for Permanent Magnets: Part I—Sintering and Casting. *JOM* **2022**, *74*, 1279–1295. [CrossRef]
48. *ČSN EN ISO/IEC 17025 (015253)*; Všeobecné Požadavky na Kompetenci Zkušebních a Kalibračních Laboratoří (Czech Technical Standard ČSN EN ISO/IEC 17025 (015253); General Requirements for the Competence of Testing and Calibration Laboratories). Český Institut pro Akreditaci, o. p. s.: Praha, Czech Republic,, 2018. Available online: https://csnonline.agentura-cas.cz/Detailnormy.aspx?k=504810 (accessed on 28 December 2023).
49. *ČSN EN 1728 (910235)*; Nábytek—Sedací nábytek—Metody zkoušení pro stanovení pevnosti a trvanlivosti (Czech Technical Standard ČSN EN 1728 (910235); Furniture—Seating—Test methods for the determination of strength and durability). Textilní zkušební ústav s. p.: Brno, Czech Republic, 2020. Available online: https://www.technicke-normy-csn.cz/csn-en-1728-910235-239251.html (accessed on 28 December 2023).
50. *Bezpečnost (Safety). Nábytkářský Informační Systém. (Furniture Maker's Information System) 2014, article 183, p. 2. Licencse: Nábytkářský Informační Server. (Furniture Maker's Information System). INT, S.R.O.: Brno, Czech Republic, 2013.* Available online: https://www.n-i-s.cz/cz/bezpecnost/page/183/ (accessed on 14 December 2023).
51. Brubi, s.r.o. *Big Set of Montessori Cube Chairs 1+2+3—3 PIECES*; Brubi: Banská Bystrica, Slovakia, 2023; Available online: https://brubi.eu/kids-furniture/big-set-montessori-cube-chairs-123-3-pieces (accessed on 14 December 2023).
52. Today's Classroom. *Play House Cube—Whitney Brothers*; TODAY'S CLASSROOM: Canton, OH, USA, 2023; Available online: https://www.todaysclassroom.com/play-house-cube-whitney-brothers/ (accessed on 14 December 2023).
53. Montessori Furniture—A Guide for Every Room. KATANABANA. © 2023, KATANABANA: Montessori Learning. Blog, 2023. Available online: https://katanabana.com/blogs/montessori-learning/montessori-furniture (accessed on 14 December 2023).
54. Woodandhearts. *Montessori Cube Chair. Wood and Hearts*; Woodandhearts: Wilmington, DE, USA, 2023; Available online: https://woodandhearts.com/products/montessori-cube-chair (accessed on 14 December 2023).
55. Lillard, A.S. Playful Learning and Montessori Education. *Am. J. Play.* **2013**, *5*, 157–186. Available online: https://www.proquest.com/scholarly-journals/playful-learning-montessori-education/docview/1459141902/se-2 (accessed on 14 December 2023).
56. Mohammed, M.; Qi, W. Exploring Conceptual Play Furniture for Children. *Int. J. Sci. Technol. Res.* **2020**, *9*, 255–259. Available online: https://www.academia.edu/64427823/Exploring_Conceptual_Play_Furniture_For_Children (accessed on 14 December 2023).
57. Diler, H.; Kasal, A.; Kuşkun, T.; Erdil, Y.Z.; Güray, E. Strength Classification of Wooden Chairs under Cyclic Loads Based on an Experimental Study. *Materials* **2023**, *16*, 6580. [CrossRef] [PubMed]
58. Güray, E.; Ceylan, E.; Kasal, A. Weight-strength optimization of wooden household chairs based on member section size. *Maderas, Cienc. Tecnol.* **2022**, *24*, 31. [CrossRef]
59. Zhang, W.; Li, S. The Integrated Development of Furniture Design and Children's Characteristics Based on Artificial Intelligence. *Sci. Program. Vol.* **2022**, *2022*, 6036160. [CrossRef]
60. Jai-ai, N.; Seviset, S.; Saribut,U. The Study and Development for the Set of Activity Table to Support the Activity Learning of Grade 4 Students by Department of Arts. In Proceedings of the 2019 International Conference on Modeling, Simulation and Big Data Analysis (MSBDA 2019), Wuhan, China, 23–24 June 2019; pp. 432–439. [CrossRef]
61. *European Standard EN 71-3:2019+A1:2021*; Safety of Toys—Part 3: Migration of Certain Elements. European Committee for Standardization—CEN-CENELEC: Brussels, Belgium, 2021. Available online: https://standards.cencenelec.eu/dyn/www/f?p=CEN:110:0::::FSP_PROJECT,FSP_ORG_ID:73357,6036&cs=1873D18444192856EC27A3175B45C822D (accessed on 14 December 2023).

62. *ČSN EN 71-3+A1 (943095)*; Bezpečnost Hraček—Část 3: Migrace Určitých Prvků (Czech Technical Standard ČSN EN 71-3+A1 (943095) Safety of Toys—Part 3: Migration of Certain Elements). Úřad Pro Technickou Normalizaci, Metrologii a Státní Zkušebnictví: Praha, Czech Republic, 2021. Available online: https://csnonline.agentura-cas.cz/Detailnormy.aspx?k=513409 (accessed on 14 December 2023).
63. *ČSN EN 1084 (492407)*; Překližované Desky—Třídy úniku Formaldehydu Podle Metody Plynové Analýzy (Czech Technical Standard ČSN EN 1084 (492407) Plywood—Formaldehyde Release Classes Determined by the Gas Analysis Method). Český Normalizační Institut: Praha, Czech Republic, 1997. Available online: https://www.technicke-normy-csn.cz/csn-en-1084-492407-208081.html# (accessed on 14 December 2023). Norma Není Platná (The Standard Is Not Valid).
64. *ČSN EN 13501-1 (730860)*; Požární Klasifikace Stavebních Výrobků a Konstrukcí Staveb—Část 1: Klasifikace Podle VýSledků Zkoušek Reakce na Oheň (Czech Technical Standard ČSN EN 13501-1 (730860) Fire Classification of Construction Products and Building Elements – Part 1: Classification Using Test Data from Reaction to Fire Tests). Úřad Pro Technickou Normalizaci, Metrologii a Státní Zkušebnictví: Praha, Czech Republic, 2019. Available online: https://www.technicke-normy-csn.cz/csn-en-13501-1-730860-222113.html# (accessed on 14 December 2023).
65. *ČSN EN 717-1 (490163)*; Desky ze Dřeva—Stanovení úniku Formaldehydu–Část 1: Emise Formaldehydu Komorovou Metodou (Czech Technical Standard ČSN EN 717-1 (490163)/European Standard EN 717/1 Wood-Based Panels—Determination of Formaldehyde Release—Part 1: Formaldehyde Emission by the Chamber Method). Český Normalizační Institu: Praha, Czech Republic, 2005. Available online: https://www.technicke-normy-csn.cz/csn-en-717-1-490163-207748.html# (accessed on 14 December 2023).
66. *ČSN EN ISO 12460-5 (490163)*; Desky na Bázi Dřeva—Stanovení úniku Formaldehydu—Část 5: Extrakční Postup Zvaný "Perforátová Metoda" (Czech Technical Standard ČSN EN ISO 12460-5 (490163) Wood-Based Panels — Determination of Formaldehyde Release—Part 5: Extraction Method (Called the Perforator Method)). Úřad Pro Technickou Normalizaci, Metrologii a Státní Zkušebnictví: Praha, Czech Republic, 2015. Available online: https://www.technicke-normy-csn.cz/csn-en-iso-12460-5-490163-207759.html# (accessed on 14 December 2023).
67. Havel, M. Vnitřní Prostředí Staveb a Chemické Látky. (Indoor Building Environments and Chemicals). Portál TZB-Info. 2014, Article 11134, s. 4. Available online: https://stavba.tzb-info.cz/podlahy-pricky-povrchy/11134-vnitrni-prostredi-staveb-a-chemicke-latky#pozn4 (accessed on 14 December 2023).
68. Čapka, R. Nové Limity Pro Emise Formaldehydu—Poloviční Hodnoty Evropské Klasifikace E1. (New Formaldehyde Emission Limits—European Classification Values Reduced to a Half). Online. 2023, s. 1. Licensee: © 2023 Dřevařský Magazín—drevmag.com. Trendwood—twd, s.r.o. Available online: https://drevmag.com/cs/2023/08/09/nove-limity-pro-emise-formaldehydu-polovicni-hodnoty-evropske-klasifikace-e1/ (accessed on 14 December 2023).
69. Démos Trade, A.S. Překližky a Laťovky: Velký Přehled Dřevin, Kvalit a Jejich Vlastností. (Plywoods and Blockboards: A Major Survey of Woody Plants, Qualities, and Properties). PDF. Ostrava-Kunčičky. Available online: https://www.demos-trade.cz/file/39328/ (accessed on 14 December 2023).
70. *European Standard EN ISO 16000-1-11:2006*; Indoor Air—Part 1: General Aspects of Sampling Strategy. European Committee for Standardization—CEN-CENELEC: Brussels, Belgium, 2006. Available online: https://standards.cencenelec.eu/dyn/www/f?p=CEN:110:0::::FSP_PROJECT,FSP_ORG_ID:25725,6245&cs=1EEBCCC04C6EF14384DC9473A1702C577 (accessed on 14 December 2023).
71. Eurofins. French VOC Regulations. PDF. Galten and Folsom, 2012. © Eurofins Scientific 2023. Available online: https://www.eurofins.com/media/1579/french-voc-regulations-en.pdf.https://www.product-testing.eurofins.com/ (accessed on 15 December 2023).

MDPI AG
Grosspeteranlage 5
4052 Basel
Switzerland
Tel.: +41 61 683 77 34
www.mdpi.com

Applied Sciences Editorial Office
E-mail: applsci@mdpi.com
www.mdpi.com/journal/applsci

www.ingramcontent.com/pod-product-compliance
Lightning Source LLC
Chambersburg PA
CBHW061545220926
51850CB00005B/258
9783725817832